Understanding the Universe

A conversational introduction to cosmology

Online at: https://doi.org/10.1088/2514-3433/ade963

About the program:

AAS-IOP Astronomy ebooks is the official book program of the American Astronomical Society (AAS) and aims to share in depth the most fascinating areas of astronomy, astrophysics, solar physics and planetary science. The program includes publications in the following topics:

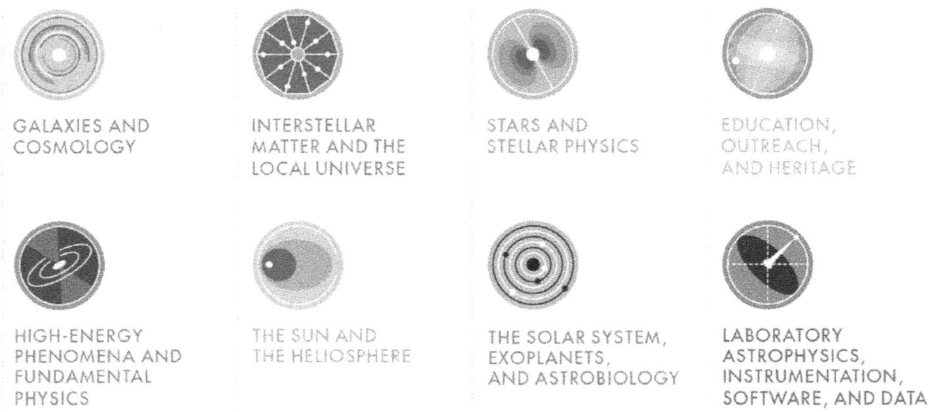

Books in the program range in level from short introductory texts on fast-moving areas, graduate and upper-level undergraduate textbooks, research monographs, and practical handbooks.

For a complete list of published and forthcoming titles, please visit iopscience.org/books/aas.

About the American Astronomical Society

The American Astronomical Society (aas.org), established 1899, is the major organization of professional astronomers in North America. The membership (~7,000) also includes physicists, mathematicians, geologists, engineers, and others whose research interests lie within the broad spectrum of subjects now comprising the contemporary astronomical sciences. The mission of the Society is to enhance and share humanity's scientific understanding of the universe.

Editorial Advisory Board

Understanding the Universe

A conversational introduction to cosmology

Yen Chin Ong

Center for the Cross-disciplinary Research of Space Science and Quantum-technologies (CROSS-Q), College of Physics, Nanjing University of Aeronautics and Astronautics, 29 Jiangjun Road, Nanjing City, Jiangsu Province 211106, China

IOP Publishing, Bristol, UK

ISBN 978-0-7503-6219-1 (ebook)
ISBN 978-0-7503-6217-7 (print)
ISBN 978-0-7503-6220-7 (myPrint)
ISBN 978-0-7503-6218-4 (mobi)

DOI 10.1088/2514-3433/ade963

Version: 20251201

AAS–IOP Astronomy
ISSN 2514-3433 (online)
ISSN 2515-141X (print)

British Library Cataloguing-in-Publication Data: A catalogue record for this book is available from the British Library.

Published by IOP Publishing, wholly owned by The Institute of Physics, London

IOP Publishing, No.2 The Distillery, Glassfields, Avon Street, Bristol, BS2 0GR, UK

US Office: IOP Publishing, Inc., 190 North Independence Mall West, Suite 601, Philadelphia, PA 19106, USA

This book is dedicated to Fat Grey (肥灰) *and* Plump Ginger (胖橘).

Contents

Preface		**xiii**
Foreword		**xviii**
Acknowledgments		**xx**
Author Biography		**xxi**
1	**Science as a Human Endeavor**	**1-1**
1.1	Our Attempts to Understand the Universe	1-1
1.2	Motivations and Questions: How Science Evolves	1-5
1.3	Cautionary Tales from "Failed Science"	1-13
1.4	The Role of Mathematics	1-14
1.5	Mathematical Modeling: A Crucial Skill for Physics	1-19
1.6	Exercises	1-26
	References	1-27
2	**Celestial Spheres and Planetary Motions**	**2-1**
2.1	Modeling the Heaven: Geocentrism vs Heliocentrism	2-3
2.2	Case Study: Kepler's Models	2-12
2.3	Our Changed View of the Cosmos	2-23
	2.3.1 Galileo: Casting Doubts on a Perfect Heaven	2-23
	2.3.2 Descartes: Casting Doubts on Everything	2-29
	2.3.3 Newton: First Successful Theory of Gravity	2-35
2.4	Exercises	2-42
	References	2-43
3	**The Infinite Universe and the Dark Sky**	**3-1**
3.1	Is the Universe Finite or Infinite?	3-3
3.2	Newtonian Cosmology and Its Problems	3-8
3.3	The Dark Sky and Olbers' Paradox	3-13
3.4	Exercises	3-19
	References	3-21
4	**Thermodynamics: The Science of Heat**	**4-1**
4.1	The Laws of Thermodynamics	4-3
4.2	Equipartition Law, Virial Theorem, and Partition Function	4-9

4.3	The Arrow of Time	4-19
4.4	Exercises	4-25
	References	4-26
5	**Our Changing Cosmos**	**5-1**
5.1	General Relativity: Gravity as Spacetime Curvature	5-2
	5.1.1 From Pythagoras Theorem to Spacetime Metric	5-2
	5.1.2 Geodesics and Curvature	5-6
5.2	Equivalence Principle and Its Perils	5-12
5.3	Dynamical Universe: The Friedmann Equation	5-14
	5.3.1 Cosmic Expansion and Acceleration	5-14
	5.3.2 Example: Matter-Dominated FLRW Universe and the Size of the Universe	5-22
5.4	Modern Cosmology at a Glance	5-26
5.5	Exercises	5-43
	References	5-45
6	**The Dark Sector**	**6-1**
6.1	Dark Matter	6-2
6.2	Dark Energy	6-9
6.3	Modified Gravity	6-16
6.4	Some Lessons From History	6-22
6.5	Exercises	6-27
	References	6-28
7	**Light: Particle, Wave, and Both**	**7-1**
7.1	The Nature of Light	7-2
7.2	Light Bending from Newtonian Gravity to Cosmology	7-9
7.3	Electromagnetism, Special Relativity, Field and the Æther	7-18
7.4	From the Dual Nature of Light to Atomic Models	7-29
7.5	Exercises	7-37
	References	7-38
8	**From Quantum Mechanics to Imaginary Time**	**8-1**
8.1	The Schrödinger Equation and Imaginary Time	8-3
8.2	Path Integral and Wick Rotation	8-7

8.3	Quantum Entanglement, Bell Inequality and Its Cosmic Test	8-15
8.4	Quantum Mechanics in Astrophysics and Cosmology	8-20
	8.4.1 How Does the Sun Shine?	8-20
	8.4.2 Degenerate Pressure Inside White Dwarfs and Neutron Stars	8-24
	8.4.3 Neutrino Oscillation, Dirac Equation, and Majorana Neutrino	8-29
	8.4.4 Quantum Cosmology	8-33
8.5	Exercises	8-34
	References	8-36
9	**Staring into the Abyss: Black Holes**	**9-1**
9.1	Black Holes in General Relativity	9-2
9.2	Physics Near Black Holes	9-8
9.3	Black Hole Shadow, Accretion, and Jet	9-17
9.4	Compact Objects and Black Hole Mimickers	9-23
9.5	Exercises	9-27
	References	9-29
10	**Beyond Black Hole Horizons**	**10-1**
10.1	What's Inside a Black Hole?	10-2
10.2	Thermodynamics of Black Holes: More than an Analogy?	10-5
10.3	Hawking Radiation, Singularities, and Paradoxes	10-16
10.4	Different Types of Horizons	10-22
10.5	Exercises	10-30
	References	10-31
11	**Making Sense of Spacetime**	**11-1**
11.1	Properties of Spacetime	11-2
11.2	Gravitational Waves	11-9
11.3	What Is Spacetime?	11-19
11.4	Exploration	11-23
	References	11-23
12	**Modifying Physics: Case Studies**	**12-1**
12.1	From Modified Gravity to Modified Entropies	12-2
12.2	Barrow Entropy and Cosmological Implications	12-4
12.3	How Generalized Entropy Affects Gravity	12-12

12.4 Generalized and Extended Uncertainty Principles 12-18
12.5 Exercises 12-27
References 12-29

13 The Big Questions 13-1

13.1 Why Is There Something Rather Than Nothing? 13-3
13.2 Universe or Multiverse? 13-5
13.3 Was There An Inflation? 13-7
13.4 How Might the Universe End? 13-12
13.5 Are We Alone? Where Is Everybody Else? 13-15
13.6 Are Wormholes, Hyperdrives, and White Holes Possible? 13-20
13.7 Are We Living In a Simulation? 13-22
13.8 Are Physical Constants Really Constant? 13-24
References 13-26

14 Afterthoughts: Humanity and the Cosmos 14-1

References 14-15

Preface

"A 'scientific' interpretation of the world, as you understand it, might therefore be still one of the most stupid of all possible interpretations of the world, meaning that it would be one of the poorest in meaning [...]. Assuming that one estimated the value of a piece of music according to how much of it could be counted, calculated, and expressed in formulas: how absurd would such a 'scientific' estimation of music be! What would one have comprehended, understood, grasped of it? Nothing, really nothing of what is 'music' in it!" — Friedrich Nietzsche, 'The Gay Science'.

This book is somewhat unusual in many ways. Its subtitle may indicate that the focus is to *teach* cosmology. This is, however, not exactly accurate. Cosmology is a branch of science that studies the large-scale structures of the Universe,[1] as well as the beginning of the Universe itself, and even its eventual demise. This book covers some of that, but also a whole lot more, including gravitation, black holes and some quantum gravity phenomenology, and also relatively "mundane" subjects like thermodynamics and quantum mechanics. These are of course not unrelated, and I think it is important that students be exposed to a variety of subjects—a cosmologist should know some black hole physics and vice versa. More importantly, while aimed to be a textbook, this book does not really "teach" students most of the nitty-gritty details about the subject; in fact you cannot hope to master any of the physics topics from this book alone. This book should be treated as a supplemental reading alongside more traditional textbooks. My main objective of writing this book is to provide students with some contexts that I think are equally important but rarely mentioned in traditional textbooks—the human elements of doing science. This is especially important for Masters/PhD students, who are just embarking on the journey to perhaps becoming a physicist or an astronomer. It is easy to bury oneself in professional work, striving for the numbers: one more paper, one more grant. Sure, those things are important, even necessary for our career (and arguably, survival) in science. However, at the end of the day we must ask ourselves why we are doing what we do. If we can see our work as part of that edifice, that continual progress of humanity's attempt to contemplate the Universe, it would endow us some sense of belonging, of purpose, and perhaps of meaning. For what is a life good for if we don't do something we ourselves think is meaningful?[2]

[1] I will use the capital lettered "Universe" to refer to our actual Universe, and "universe" to mean any other possible universes.

[2] Whether there is such a thing as the true meaning in life or not is quite beside the point!

Having said that, this is *not* a book about the history of science or cosmology (which I am not an expert in), but rather *how* certain concepts came to be—what were the motivations at the time? How did old incorrect beliefs get replaced by newer understanding of things? Often we will see that progresses were slow and not as linear as usually perceived by students (or even some practicing scientists). Often when teaching a technical subject like general relativity or cosmology, there is little time available to teach the relevant history. Any historical mention would be brief and simplified and can indeed be misleading. I will do what I can to give the correct course of history, and correct the historical fictions in textbooks that have been passed down for generations, when possible.[3] But even then I am sure there are still mistakes that I am not aware of. Historical contexts are also important as we will see how the prevalent thoughts and philosophy of the era influenced scientific theories. A typical student pursuing a scientific career may be less concerned about humanities and may wonder if taking all these detours is simply time-wasting ("why not just start from what we already know today?"). The answer is a resounding no. If nothing else, we should learn from history, and there are plenty of examples we can learn from by looking at how our predecessors approached the great problems of their respective era. As a species, our ways of thinking and intelligence have not changed that much for the past few hundred years, so we can even see that similar ideas were proposed now that are reminiscent of the past (in different contexts). In addition, many great scientists are also great thinkers. Einstein, for example, read plenty of philosophy (from Ernst Mach to David Hume and many others) and actively conversed with philosophers. In short, science is a human endeavor, and as such is necessarily influenced by other human activities, including philosophy and art. Some scientists and science students have a "superiority complex" and think that their subject is better and above that of philosophy and other fields, but there are things we can learn by being humble. As Einstein once wrote,

> "All religions, arts and sciences are branches of the same tree. All these aspirations are directed toward ennobling man's life, lifting it from the sphere of mere physical existence and leading the individual towards freedom."

In this book, we will see that some great problems of the past, like the nature of the æther[4], were neither "solved" nor "falsified" but went away, replaced by new

[3] Spoilers: Planck did not set out to solve the "ultraviolet catastrophe"; Rutherford did not propose the "solar system model" of the atom; Hubble wasn't the first to propose "Hubble's law"; and the cosmological constant did not all of a sudden come as a "surprise" after the supernovae observations in 1998. I hope you will be enticed to read on and learn more about what really happened!

[4] This refers to the hypothetical substance that was thought to permeate the entire Universe. It is sometimes spelled "ether" or "aether". This is similar to the word esthetics, which is sometimes written as aesthetics. The former is more often used in American English; while the latter is used in British English. In terms of the aesthetics, I prefer to write "æther" instead. (This was used by, among other, Paul Dirac, in his 1951 *Nature* letter entitled "Is there an Æther?")

ways of looking at Nature. Thus we must keep in mind that problems of today may share the same fate, as Corsland wrote in his book *The Science of Matter*, "study of the history of science does not require a moral justification, but if it did, it might be to teach humility". Only time will tell.

Indeed, the main aim of the book is to show what science is in spirit, for which the elements of humanities are indispensable. Science is not just about cold hard facts and data points, it is the story of humanity to understand the world around us. In the age of artificial intelligence, we often see the debate concerning whether AI art really qualifies as art (by the way, no AI was used in writing this book), and how it is different from just a tool like a camera (as no doubt, when photography was first invented, there was also debate regarding whether photography can be an art form). The central issue here concerns what roles *humans* play in creating an art piece. In mathematics there is the "four color theorem", which says that no more than four colors are required to color the regions of any map so that no two adjacent regions have the same color. The proof, however, involves the use of a computer. This ignited a debate in the mathematical community—if no human can ever check the proof and understand its details (other than checking to ensure that the programming is correct), does it count as a proof? Unlike the AI art debate, this happened back in the 1970s. Some mathematicians are alright with calling it a proof as long as it is logically correct; but for some others, mathematics should be more than logic, and they insist that proofs should be able to be appreciated to see how it works intuitively (by a mathematician's standard). For those in the latter camp, a proof that evades human's appreciation just doesn't *feel* the same. Thus we see that even in the sciences, human elements are more important than most people may realize. Say, in theoretical physics, physicists enjoy devising new theories. If in the future AI can do so (not impossible, as they can learn the literature and try to improve on existing models and maybe even suggest entirely new ones[5]) and science is reduced to testing whether the AI's models or theories are correct, would that remove all the excitements that draw people into doing science in the first place? I certainly think so. One attraction of science is that it allows us to investigate and explore the world, solving one mystery after another. We have always been fascinated by stories of great adventures, whether a geographical one or a scientific one. So why deny our students the exciting developments of the ideas of scientific theories? Experiments and calculations, theorems and derivations, they are all crucial elements of science, but without good stories that remind us of our humanity, to paraphrase Nietzsche, what would one have comprehended, understood, grasped of it? Nothing, really nothing of what is *science* in it!

The plan of this book is as follows. We will start with a further discussion on how science is done and some of the human elements involved, then we will explore how

[5] Science fiction author Ted Chiang's short story *The Evolution of Human Science*, written at the dawn of the 21st century, contemplates the role of human scientists in an age when the frontiers of scientific inquiry have moved beyond the comprehensibility of humans.

our view of the cosmos has drastically changed throughout the ages, from the models of the Solar System to the evolving Universe governed by the laws of general relativity, and how various physical concepts like thermodynamics, statistical mechanics, and quantum mechanics led to our current understanding of the Universe. Through these subsequent chapters we will encounter some important episodes in the history of physics and cosmology, each with its own lesson to teach. For example: why did Kepler once propose that the five platonic solids have any relevance to the Solar System? (Despite it not working out, this is a very good example of model building in science!) Why is the night sky dark? (A seemingly trivial but profoundly deep question, which serves as an example that in science, sometimes we have to ask questions of things that everyone sees but none *perceives*.) My aim is to teach students how to *think*, so each chapter (except Chapter 13 and 14) also comes with some exercises or questions to be explored at the end. The style of this book is rather informal and conversational, much like an elaborated chat a professor might have with students over a cup of coffee. There are suggestions and advice for students and young researchers scattered here and there when appropriate. Additional interesting information is sometimes provided in boxes if they are too long to be a footnote. Since this is a textbook and not a research monograph, I try to minimize the number of references, only providing them when either the source is not as well-known, or if I think the readers will benefit from following up on those articles or books. Thus at times, some books or articles may be mentioned but not explicitly referenced.

For a book like this it is hard to state what are the precise prerequisites. I try to keep mathematical details to a minimum. For example, although general relativity requires differential geometry, this is not required for our purpose. As long as one is willing to take certain things for granted, much can be conveyed with just knowledge of calculus and basic differential equations (and the somewhat mysterious quality often referred to as being "mathematically matured", which among other things, essentially means that the student must be willing to put in serious efforts to think and derive some of the equations). By their nature, some chapters are more technical than others. There are also boxes that contain additional information that don't fit quite well within the main narratives. Some sections like the one on path integrals can even be skipped by readers who have no prior experience with the topic as it might be too technical. Nevertheless, the emphasis is always on the big picture, the main ideas involved. The full mathematical treatments are of course indispensable if one wishes to truly learn physics, but there are plenty of excellent textbooks for that already. This book is somewhere between a popular science book and a traditional textbook. Still, if I must specify an audience, I think this book is probably most suitable for a senior undergraduate student or a beginning Masters/PhD student, one who has learned some basic physics and wants to learn more about our Universe and contemplate our existence in it; but even experts should be able to learn a new thing or two. Gravitation, astrophysics, and cosmology are beautiful subjects; I hope that you, my dear reader, will enjoy this book and start contemplating the Universe (Figure 0.1).

Figure 0.1. Possibly an astronomer in his study, looking out of the window, seemingly in deep thought. Lithograph by M. Jackson after J. G. *Credit:* Astronomy: an astronomer [?] in his study, looking out of the window. Lithograph by M. Jackson after J. G. Wellcome Collection. *Source*: https://wellcomecollection.org/works/jk7kjqru/images?id=f43m2xgp

Foreword

When I was a university student, in those far-off days before the internet, there was a dire shortage of intelligible information regarding advanced topics in physics and mathematics: that is, regarding precisely those topics which interested me most but which our courses covered least. This was a serious defect of the educational systems of that time. We were in effect preparing ourselves for a scientific career while knowing extremely little about what being a scientist actually entails. Nobody would embark on any other career in such a state of ignorance: imagine the analogous situation for students of medicine or law.

The student of today, raised in the internet era, is confronted with the opposite, and even worse, predicament. He or she is constantly assailed not just with deliberate scientific untruth, but also, what is perhaps worse, with sincere reports of one scientific "revolution" after another. "Einstein was wrong, say scientists"; and they say it once a week at least, only for the "revolution" to sink into well-deserved obscurity.

This in itself is nothing new. In 1947, Einstein himself was moved to the following declaration, which might have been written yesterday:

> "As an incidental remark I want to stress the following. It seems undesirable to me to present such preliminary attempts to the public in any form. It is even worse when the impression is created that one is dealing with definite discoveries concerning physical reality. Such communiqués given in sensational terms give the lay public misleading ideas about the character of research. The reader gets the impression that every five minutes there is a revolution in science, somewhat like the coup d'état in some of the smaller unstable republics."

All that has changed is that the situation Einstein deplored has become infinitely more deplorable, aided and abetted by electronic means of propagating sensational "discoveries" and by AI-generated "hallucinations". The student of today is a pitiable object, who can be forgiven for descending into a state of permanent uncertainty as to what scientists believe about essentially everything.

There is thus a desperate need for a book about "cutting-edge" scientific research which is both authoritative and yet appealing to eager youth. It must be written by someone who is an active researcher in one or more of the most interesting research topics, and yet not such a specialist that the work is incomprehensible. I do not mean that this work should be so simple as to be completely understandable by any undergraduate; on the contrary, it should be on such a high level as to stimulate the fascination of a young (or young-at-heart) person, without putting him or her off altogether. It does not matter if that person does not understand everything: the gaps will fill in of themselves in the course of time. Above all, the book must sharply distinguish ideas which are reasonably universally accepted from those which, while interesting, are likely to be ephemeral.

That book, in the case of theoretical physics, is the one you hold in your hands. It covers a vast range, from basic ideas of dynamics and thermodynamics (which are nevertheless treated with great care, and not dismissed) right up to clear explanations of the latest in black hole theory and cosmology. At the same time, the author has injected just the right amount of historical and even philosophical sidelights to keep the reader engaged. Students who are primarily interested in physics should resist the temptation to skip these, as the book aims to teach not only physics, but the ways of thinking.

This is exactly the book I wish I had been able to read when I was a student, and it is the book for students of the present day who are befuddled by scientific information overload. It should be a strongly recommended reading for every undergraduate or even graduate student (and by no means only in physics or indeed in science), and in fact not a few of their professors would also benefit from close attention to it. You who are about to embark on reading this unique resource are in for a rare intellectual feast.

Prof. Brett McInnes
Department of Mathematics
National University of Singapore

Acknowledgments

I thank IOP Publishing for inviting me to write a book. Instead of a research monograph, which I am not quite yet ready to tackle until I have accumulated enough insights, I decided to write something that is in between a traditional textbook and a popular science book. I have long felt that there are not enough books that convey to students the ideas and the passions behind all the technical results, and how ideas have changed yet sometimes remain similar through centuries of scientific thoughts. Many popular science books are either too "watered down" with inappropriate analogies, or simplify the history so much that the narrative becomes misleading. When I was a student, I always wanted a book that could teach me not just the technical stuff, but how scientists actually think about problems, past and present. As the popular saying goes, "write the book you want to read". Hence this book. Much of the material is based on my existing personal notes (hence I also include clarifications on things that once confused me as a student), and things learned from discussions with various colleagues and students. Still, writing a book calls for a more comprehensive approach, more detailed elaborations, and through consolidating more information—both historical studies and latest findings—to prepare for the writing, I have learned a great deal.

I also wish to express heartfelt gratitude to my Master's supervisor Prof. Brett McInnes, and my PhD supervisor Prof. Pisin Chen, both of whom have instilled in me the importance of having a strong motivation when choosing a research topic, and the art of asking penetrating questions. I still vividly remember how Prof. Chen would insistently say: "We don't do *'what if'* physics!" Despite his busy schedule, Prof. Chen often makes time for landscape sketching. This strengthened my view that life is more than just about science (or any job for that matter). We also need to learn to perceive beauties. Thus I tried to incorporate some aspects of art history in this book. Incidentally, Section 1.5 on mathematical modeling is partially based on the lecture notes of Prof. McInnes. His relatively informal writing styles no doubt also influenced my own.

Parts of this book were written during my time with the Center for Gravitation and Cosmology (CGC) at Yangzhou University. I thank Prof. Bin Wang, the director of CGC, for all his support and valuable advice. I also thank all past and present colleagues of CGC, especially Prof. Rui-Hong Yue, Prof. Jian-Pin Wu, Prof. Xiao-Mei Kuang, Prof. Cheng-Jun Xia, Assoc. Prof. Hao Xu, Dr. Guoyang Fu, Dr. Dan Zhang, and Mr. Yong-Ming Yan for their assistance in one form or another.

I am deeply indebted to the various inspirations, guidance, and advice received through dreams and synchronicities, especially during difficult times.

Prof. Yen Chin Ong
22 May 2025
Kuala Lumpur, Malaysia

Author Biography

Yen Chin Ong

Yen Chin Ong (王元君) is a professor and a vice director of the Center for the Cross-disciplinary Research of Space Science and Quantum-technologies (CROSS-Q), College of Physics at Nanjing University of Aeronautics and Astronautics, China. He obtained the prestigious Excellent Young Scholars Fund from National Science Foundation of China in 2019. In the same year he was awarded the Jiangsu Specially-Appointed Professorship. He is also affiliated with the Center for Gravitation and Cosmology at Yangzhou University.

Professor Ong is a Malaysian, graduated from the Department of Mathematics, National University of Singapore, with B.Sc. (Hons) and M.Sc. degrees. When in Singapore, he taught at various levels, including as a mathematics teacher at Nanyang Girls' High School, a lecturer at Nanyang Polytechnic, and a tutor at National Institute of Education in Nanyang Technological University. Subsequently, on the support of Taiwan scholarship, he obtained a PhD in Astrophysics from National Taiwan University, where he was a member of the Leung Center for Cosmology and Particle Astrophysics (LeCosPA). During that time he was awarded the university Philosophy Laurel Prize (2nd place for the natural sciences category). In 2014 he won the Dean's Award for Excellent PhD Thesis, which was subsequently published by the Springer Theses Award series, entitled "Evolution of Black Holes in Anti-de Sitter Spacetime and the Firewall Controversy". In the same year he was awarded the Excellent Teaching Assistant award. After graduation, he joined the Nordic Institute for Theoretical Physics (Nordita) in Stockholm, Sweden, as a Nordita fellow. He was also a visiting Riemann Fellow at the Riemann Center for Geometry and Physics, Leibniz Universität Hannover, Germany.

Professor Ong has broad research interests in general relativity and gravitation, especially in black hole physics (classical solutions and thermodynamics, spacetime singularities, as well as quantum properties like Hawking radiation, and aspects of gauge/gravity duality). He is interested in the interplay between gravity and thermodynamics, such as the notion of gravitational entropy, and how modifications of entropy affect theory of gravity and cosmic evolution. His wide ranging yet thought-provoking research has furthered our understanding of the deep connections between black hole thermodynamics, cosmic censorship conjecture, and quantum gravity phenomenology such as generalized uncertainty principle, with astrophysical and cosmological implications. Occasionally he also works on modified gravity theories and quantum information.

Professor Ong enjoys drinking coffee, reading, tending to his plants, cooking, and contemplating the meaning of life, art, and philosophy in general. His personal website is oyc.space and he can be reached at ongyenchin@nuaa.edu.cn.

"*I have come to believe that the whole world is an enigma, a harmless enigma that is made terrible by our own mad attempt to interpret it as though it had an underlying truth.*"

—Umberto Eco, "Foucault's Pendulum".

"*There is a theory which states that if ever anyone discovers exactly what the Universe is for and why it is here, it will instantly disappear and be replaced by something even more bizarre and inexplicable. There is another theory which states that this has already happened.*"

—Douglas Adams, "The Restaurant at the End of the Universe".

"*A man said to the universe:*
'*Sir, I exist!*'
'*However,*' *replied the universe,*
'*The fact has not created in me*
A sense of obligation.'"

—Stephen Crane, "A Man Said to the Universe".

Chapter 1

Science as a Human Endeavor

"The effort to understand the universe is one of the very few things that lifts human life a little above the level of farce, and gives it some of the grace of tragedy." —Steven Weinberg, "The First Three Minutes".

We begin our journey by looking at how the ways we, the human species, employed to comprehend the world evolved from ancient magic and alchemy to modern science. Despite the huge differences in the approaches, we have to keep in mind that biologically speaking, we are the same as people from a few centuries and even a few thousand years ago. Our advantage is that now we have in possession advanced technology that can help us to measure things that our ancestors couldn't even imagine, and that our "baseline" of understanding how things work is higher. However, how we *think* about problems is not so different, and to do science requires us to be able to think critically and more importantly, creatively. For this reason, lessons from the past are still valuable to us today. In this chapter we will also look at some examples of "failed science" and the fallibility of scientists. We are humans after all, and every scientist should keep that in mind. Students should learn early in their career that science does work by replacing outdated knowledge with new knowledge, but this is a convoluted nonlinear process that takes time, sometimes over centuries. We also explain the role of mathematics in doing physics, and introduce the art of mathematical modeling at the end of the chapter.

1.1 Our Attempts to Understand the Universe

Humanity has looked up and marveled at the starry night sky for many millennia, and has tried to decode the meaning of it all ever since. Myths and legends were born, to give us a place in the vastness of existence. As our tools and understanding become more and more sophisticated, modern scientific methods guided us to appreciate the world as it really is, without mystical interpretations. Modern science

doi:10.1088/2514-3433/ade963ch1

only has a few hundred years of history. Yet in this short amount of time, we have achieved quite remarkable stuff—we discovered that the Universe is expanding and the Earth is constantly changing (plate tectonics movement only became widely accepted by the 1960s!); we decoded DNA and use it as tools in forensic investigation and to understand biological evolution; we sent probes that venture far out into outer space; we discovered the laws of quantum mechanics that is vastly different from those that govern the macroscopic world (and as a consequence, built modern technologies like computers and lasers); we figured out the cause of many diseases and proceeded to cure and prevent them; AI and machine learning led to new algorithms that revolutionized our understanding of protein folding; we even detected the minute signals from gravitational waves that vibrate the very "fabric" of space and time. Our modern 21st century civilization would seem like magic to someone from 500 years ago. These achievements, and many more to come, is a testament of how powerful the scientific method is in understanding the world.

Poets and artists sometimes complain that science is too cold and austere, devoid of emotion. They say that the world has lost its sense of mystery and awe, even disenchanted. Viewed from the right perspective, this is certainly not true. Astronomy, for example, tells us that we are made of stardust, hence through us, the Universe contemplates itself (as the science communicator and astronomer Carl Sagan (1934–1996) famously put it). How more poetic do you want? How about heavier elements like gold, mostly produced from colliding neutron stars—stars as massive as the Sun yet are only the size of a city about 20 km across? Also, scientists are people, and are motivated by reasons no less passionate from that of an artist. Yet something is not quite right with the state of our education. It's not only poets and artists who don't understand science, but arguably many scientifically trained professionals do not see science beyond its technicalities and findings. We risk focusing too much on the crafts and forget about the arts, for in science too there are aspects of humanities—history, philosophy, and how these influenced the development of scientific ideas throughout the ages. Somehow this aspect has been gradually neglected ever since science replaced its predecessor—natural *philosophy*.

It is true that science can be taught and practiced without ever knowing the history of science. However, just like in any field, we can—and should—learn from history. In fact, it is often quite illuminating (not to mention inspiring) to read the original papers than the modern textbooks. Of course, we all have limited time, so it is not realistic to read too many original papers. Still, I would encourage students to choose at least some "landmark papers" to read. Great scientists like James Maxwell and Albert Einstein shared the same sentiments:

> "It is of great advantage to the student of any subject to read the original memoirs on that subject, for science is always most completely assimilated when it is in the nascent state." —James Maxwell

> "There is always a certain charm in tracing the evolution of theories in the original papers; often such study offers deeper insights into the subject matter than the systematic presentation of the final results, polished by the words of many contemporaries." —Albert Einstein

More importantly, being familiar with how scientific thinking has changed over time, and having the awareness that even well-accepted theories can be overthrown would make us a better scientist, one who is skeptical of new claims yet always open-minded. Being aware of how scientific theories have philosophical underpinnings would help us spot the assumptions and dogmas, and with luck, could guide us to better theories, even a new paradigm shift. To quote Albert Einstein from his correspondence to Robert Thornton in 1944,

> "I fully agree with you about the significance and educational value of methodology as well as history and philosophy of science. So many people today—and even professional scientists—seem to me like somebody who has seen thousands of trees but has never seen a forest. A knowledge of the historic and philosophical background gives that kind of independence from prejudices of his generation from which most scientists are suffering. This independence created by philosophical insight is—in my opinion—the mark of distinction between a mere artisan or specialist and a real seeker after truth."

Indeed, we should aim at understanding the Universe, not just at publishing yet another paper. At the very least these two objectives should be balanced. In today's "publish or perish" culture, it is important not to lose sight of the main objective of science. To understand the Universe, in my opinion, we should not neglect understanding how fellow humans *think*. Furthermore, for a scientist and especially for a beginning PhD student, the history of science can help to put some meaning in one's own research. That is, we can put our work in contexts and see how it is connected to all the past attempts to understand the Universe, that we are carrying on the mission, passing on the torch of humanity. This way, we can bring back some romance and passion into scientific pursuits.

However, while looking at the history of science, it is important not to judge people of the past through our modern lenses; after all, everything is clearer with hindsight. Astronomers like Tycho Brahe and Johannes Kepler also practiced astrology,[1] and Sir Isaac Newton practiced alchemy. Alchemists also discovered some chemical elements (Figure 1.1). One might be tempted to ask: *How could people,*

[1] The usual objection raised against using astrology to tell someone's personality is that due to precession of the equinoxes, the astrological sign no longer matches the actual sky. That is, when we say that someone is a "Gemini", it means that he was born when the Sun is in the constellation of Gemini, which supposedly happens between May 21 and June 21. This has since shifted to June 21 and July 20. However, this cannot really be used as an argument that astrology is wrong (I am not claiming that it is right, either!). This is because despite its name and philosophy, astrology may have nothing to do with the stars. Instead, it could simply be based on "big data" to find common traits of people born in different months. For example, perhaps seasons (or something else that correlate with seasons) can affect personalities, and the ancient people would associate certain qualities with different figures, and then used their imagination to draw those figures onto the stars that were associated with the Sun *at that time*. In other words, the stars never had anything to do with influencing people's personalities. Such a view was also proposed by the psychoanalyst Carl Jung (1875–1961). Of course, astrology may not work at all even as averages from big data, but my point is we cannot use such an argument to refute it. To be a good scientist, we must be careful of whether our argument is logically sound, and whether there is any loophole, even when criticizing fields that are deemed unscientific.

Figure 1.1. Hennig Brand (1630–1710), the German alchemist who discovered phosphorus by distilling urine, while searching for the "philosopher's stone", a mysterious substance that was believed to transmute base metals into gold. Etching, 19th century. *Credit:* Hennig Brand, the German alchemist, discovering phosphorus. Etching, 19th century. Wellcome Collection. *Source*: https://wellcomecollection.org/works/k4gzk252/images?id=eqnnq6af

even great scientists, believe in nonsenses like these and waste their precious time? But there was no clear distinction between astronomy and astrology, or between chemistry and alchemy, back then. Knowledge evolves; it's only with the benefit of hindsight that we know what kind of problems are time-wasting. We will see more examples like this in future discussion, which were less obviously "time-wasting" at the time. The point is that, with hindsight we can understand what was doomed from the beginning, but we do not have the luxury to know that in advance for contemporary research. (There is another reason for astronomers in the old days to dabble in astrology even if they might prefer not to—they were often employed by the royal or funded by aristocracy who wanted their horoscopes read. In that sense, the activity is only as time-wasting as modern grant writing.) There is no certainty in scientific careers to guarantee that we are not working on something that would be proven wrong later.

Let us consider another example: Aristotle's view that a heavier object falls faster than a light one. Every physics student knows that Galileo eventually found that the rate of fall is independent of the object mass, and indeed astronauts on the Moon had demonstrated that a hammer falls at the same rate as a feather. Many students would then wonder: why were people so stupid? How could they accept Aristotle's view for more than a thousand years? This is not being fair to Aristotle, as he never considered what happened when there is no air resistance. As pointed out in Rovelli (2015), "Aristotelian physics lasted long not because it became dogma, but because it is a very good, empirically grounded theory." On the other hand, misleading thought experiments like Einstein's elevator used to argue for the bending of light in a gravitational field, continues to be taught in many textbooks. Yet a typical physics student, who thinks that people were dumb for believing in Aristotelian physics, more often than not, accepts what they are being taught *currently* without much critical thinking. To think for oneself requires "that kind of independence from prejudices of his generation" that Einstein mentioned, and is incredibly difficult. Nevertheless, one must try. Otherwise, as is often the case, some misunderstandings can be passed down from a generation onto the next unchallenged.

To understand the Universe also requires penetration[2]—the ability to recognize key issues or essences of a problem. Even in pure mathematics, in which rigorous proofs are important, one requires insights. These only come with sufficient deep thinking about the issues at hand, and has less to do with mastery of technical details. This is not to say that the latter is not important, but that it is necessary but not sufficient to make significant progress.

In the subsequent chapters, we will explore major concepts in cosmology and related subjects, putting more emphasis on the crucial ideas behind each topic and pointing out subtleties that are not always taught in traditional courses. Hopefully through these explorations, a keen student can develop crucial critical thinking skills.

1.2 Motivations and Questions: How Science Evolves

People have been doing science, and discoveries were made, well before modern science as we know it was a thing. This much is clear. But often we tend to look at history with modern lenses and "imagine" Newton or Kepler as scientists pretty much like how their modern counterparts are. They were not quite. As already mentioned, they were immersed in their contemporary "culture" that dealt equally with the occult[3] as well as what we today characterized as legit science. Belief in the

[2] Closely related is what the polymath Johann Wolfgang von Goethe's (1749-1832) idea of *aperçu*, which is essentially an insight that allows one to recognize a general phenomenon or the whole from a special case or a particular instance. He wrote: "Everything in science depends on what one calls an aperçu, on becoming aware of what is at the bottom of the phenomena. Such becoming aware is infinitely fertile." Among Goethe's better known works is his play *Faust*, which was about a man who sells his soul to the Devil in exchange of knowledge.

[3] The word, from Latin *occultus*, originally simply means "hidden". Any properties that are not understood and thus "hidden" were described as "occult" in medieval Europe, an example of this being the magnetic force. Similarly, "occultation"—a term that is still in use today in astronomy, means one celestial object hides another from view. That said, today the term "occult" almost exclusively refers to supernatural beliefs and practices.

existence of God and divine influences was the norm back then, and to understand how Nature works was viewed as a means to get to know the Divine. This was the predominant view right through the 18th century and even the 19th century. This was also the view in the Islamic civilization, whose golden age spanned the 8th to the 13th century. Many scientific pursuits before modern science were based explicitly on "philosophical" beliefs, which sometimes border on the mystical. For example, the ancient Greek notion that everything is fundamentally composed of four elements: fire, air, earth, water. This point of view influenced medicine, as illnesses were believed to be caused by the imbalance of the four bodily fluids (the "humors"): yellow bile, blood, black bile, and phlegm, which correspond to the four elements, respectively. (Interestingly, ancient Chinese developed the concept of five elements: metal, wood, water, fire, and earth; and similarly constructed a medical system based on that.) Speaking of medicine, even the existence of pathogens ("germ theory") was only widely accepted pretty recently in the 1890s. In the 19th century, it was common for patients to die from post-surgery infections. Not knowing that microorganisms exist (in fact doctors actively dismissed the idea of germs as "unscientific"), surgeons did not have sterilization practices and worse still, surgical instruments were shared from patient to patient. Science has certainly changed the world for the better!

Returning to our story about the fundamental constituents of matter, Paracelsus (1493–1541), the first physician who introduced chemistry into medicine in the 16th century, proposed that primary substances are instead sulfur, mercury, and salt. This was built on the 9th century idea of the Arabic scholar Abu Mūsā Jābir ibn Hayyān (died c. 806–816), who introduced sulfur and mercury to the traditional four elements as "principles" to explain various properties of metals. He called them the *Tria Prima* or the "three primes". These do not merely refer to the common substances of the same name, but rather more abstractly refer to properties that they represent. Salt represents something solid; sulfur represents combustibility, color, and smell; mercury represents volatility. For example, a burning branch releases fire and smoke and ashes, these are the manifestation of sulfur, mercury, and salt, respectively. All this sounds extremely vague by modern standard, yet it inspired many to examine properties of matter. Modern students of physics and chemistry would be familiar with Robert Boyle (1627–1691). But even he was fascinated by alchemy. Through his experiments he gradually realized that there was no evidence to support either the 4 elements view, nor the tria prima view. For example, why would burning a branch be necessary to reduce it to its fundamental constituents? If we instead steam the branch, what then? He even doubted that there is such a thing as "fundamental constituents". Gradually though, the modern atomic theory would be formulated and science finally understood matter.[4] Though the idea that matter is composed of tiny particles or atoms has been around since the ancient Greeks, to make a quantifiable and

[4] This is only up to a point. The physics of fundamental particles like quarks are not yet fully understood.

predictive theory, instead of just a descriptive one, is a true hallmark of a scientific theory. This was then followed by atomic physics that peered into the structure of an atom. We have since discovered that all known matter is made of only a handful of elementary particles, characterized into quarks and leptons, and fundamental forces are carried by bosons. The story of the standard model of particle physics is remarkable in itself but would carry us too far from our discussion right now. The point is that humanity has been trying to understand the Universe for a long time, even before modern science came into being. Progress was made despite some strange things that people used to believe in.

Though things are quite different today, science as a human activity is still more complicated than what non-scientists might think. The so-called "scientific process"—start with observation, formulate hypothesis, test said hypothesis, and revise until a sound theory is formed—is the public perception of how science operates. However, things are not quite as straightforward. The most important step remains unanswered: how does one actually formulate an idea, and come up with a theory? Unfortunately this is not usually taught in schools or even universities; students are often only taught known established theories as they are, and perform well-designed experiments to confirm the taught theories. There is no room for exploration in the standard curriculum. Therein lies the danger that a student may accept everything taught as correct, and that is not how a scientifically literate person should behave. On the contrary, they should always be critical and ask plenty of questions. By asking questions,[5] I don't just mean ask teachers and professors to clarify doubts, but that when studying books or reading scientific publications, one should always *ask oneself*: is this really so? Is the evidence and argument strong enough to justify what the authors claimed? Is the logic sound? Is there any loophole? Such a learning method takes a lot of time and can unfortunately slow down a student's progress and this may affect their grades (which was what happened to me!), so a balance must be sought. Nevertheless, to learn how to ask the right question is the biggest asset a researcher can have, and a beginning PhD student should master this skill as soon as possible. Often with the right question, one is led to a new research project.

There are examples to illustrate why we should not accept everything we are taught.[6] First, a rather extreme but important example from the history of science is how we figured out human anatomy. In the Middle Ages, the study of anatomy was conducted by "professors" who read from the texts of Galen, the eminent Roman physician from the 2nd century, seated in a chair high above the students who gathered around a human cadaver. While the lecture went on, an assistant—the barber-surgeon—would dissect the corpse to point out the relevant parts dictated by the professor. However, as Galen's findings were not based on actual

[5] The Chinese word for knowledge is 学问, which literally means "learn to ask", but is also short for "learn as well as ask".

[6] We must also keep in mind that sometimes we inherit "knowledge" that we assume to be correct but is not. For example, many people think that birds will abandon their young if it has "human scent". This is not the case as most birds rely on sight and sound instead of smell.

human body dissection but rather that of animals, there were many mistakes. The professors high up supposedly did not notice this, and any failure of the barber-surgeon to find a matching structure would be dismissed. Surely the text of the great Galen couldn't be wrong? Finally, Andreas Vesalius (1514–1564) changed this. As a physician, he decided to "get his own hand dirty" to conduct dissection himself and only then Galen's mistakes were gradually corrected—via actual observations. This is a good example that shows why we should not blindly accept "contemporary knowledge", even if it is rooted in the findings of authoritative figures. The modern attitude that we can do better and continue to think of something better was not the norm back then. The Renaissance period in Europe was sometimes mistaken as the era in which people started to strive for something new, a rational approach to understand the world. But Renaissance means "rebirth"; they were trying to re-establish what they believed to be the golden era—culture of the ancient Greeks, to improve their conditions that had deteriorated since then. In that sense it was a return to the old ways, though in doing so it gradually did lead to modern science. Still, the thinking that the ancients might already have figured things out persisted for a while. This may seem strange to the modern readers, who are used to thinking that we improve our understanding of the world over time, but behind this thinking is a religious one: the further back in time a civilization was, the closer it was to "the Fall" (of the Christian sense), and knowledge was less corrupted. The original pristine knowledge (*prisca sapientia*) covers both the natural and the spiritual aspects, but later most were lost and some garbled through human sin, as well as the accumulated errors introduced via copying through the ages. Even Newton believed so, and treated his findings as re-discovering lost knowledge, perhaps at best, as providing a firmer basis for ancient teachings pre-Socrates, including the inverse square law of gravitation, which he believed was already implied in the work of Pythagoras, hidden in the theory of musical scales (Feke 2024)!

Another example, less dramatic, is one that some readers might have been taught at some point. Given a column a water in a tall cylinder, at which height should you drill a hole, so that the water jet that shoots out from the hole can reach the farthest? You might have been taught the correct answer should be the bottom-most position, since water pressure is the highest there. This seems to make sense, until you actually perform the experiment and found that this is *not necessarily* the case. If you just put the cylindrical container filled to the top with water on the floor, water jet shooting out from the *middle* of the cylinder would travel the farthest. In fact, this can be shown mathematically with a simple modeling (we will introduce mathematical modeling formally in the last section of this chapter). Suppose we drill a hole somewhere at the side of the container. Let h_1 denote the level of water above the hole, and h_2 the water height below the hole. Let d denote the horizontal distance of the water jet, and let v denote the initial velocity of the jet as it exits the hole. This velocity is given by Torricelli's theorem[7] (a consequence of energy conservation):

[7] Named after Evangelista Torricelli (1608–1647), a student of Galileo who also invented the barometer.

$v = \sqrt{2gh_1}$, where g is the gravitational acceleration. It is now useful to consider a general trajectory equation that describes the path taken by a projectile, launched at height y_0 from the ground at an angle of elevation θ (we are going to ignore air resistance and other environmental effects). Elementary Newtonian mechanics tells us that its x and y position would be governed by $x = vt\cos\theta$ and $y = y_0 + vt\sin\theta - gt^2/2$, respectively. Combining these yields the trajectory equation

$$y = y_0 + x\tan\theta - \frac{1}{2}\frac{gx^2}{v^2\cos^2\theta}. \tag{1.1}$$

For our scenario, when the water hits the ground (we are treating the water stream as a series of projectiles!), $y = 0$. This would lead to

$$d = \frac{v\cos\theta}{g}\left[v\sin\theta + \sqrt{(v\sin\theta)^2 + 2gy_0}\right]. \tag{1.2}$$

For our case, the water shoots out with a velocity vector parallel to the ground, and so $\theta = 0$, and we get simply

$$d = \frac{v\cos\theta}{g}\sqrt{2gy_0}. \tag{1.3}$$

In the context of our problem, $y_0 = h_2$, and $v = \sqrt{2gh_1}$, so

$$d = \frac{\sqrt{2gh_1}}{g}\sqrt{2gh_2} = 2\sqrt{h_1h_2}. \tag{1.4}$$

To find the farthest distance the water jet can shoot, we need to maximize d. In this case we can apply the AM–GM inequality, which says that the arithmetic mean of two non-negative real numbers x and y is always larger or equal to its geometric mean, i.e., $(x + y)/2 \geqslant \sqrt{xy}$. So the maximum value of d is attained at equality, which only happens when $h_1 = h_2$, for which $d = h_1 + h_2$. Thus the water jet has the farthest range when the hole is drilled at the middle position, as can be verified by experiment. Yet many physics textbooks still have wrong figures that illustrate the water jet from the bottom hole shooting the farthest when the container is put on the floor (even Leonardo Da Vinci made this mistake! But we have known for a long time, at least since Torricelli's time, the correct physics. Unfortunately sometimes mistakes are deeply rooted and hard to get rid of). Of course, if you put the container on a tall table, you can get the hole drilled at the bottom to shoot the farthest, which is kind of a cheat. You can explore this variation in the Chapter exercise.

Science therefore progressed as doubts grew over the old theories or models, as new data and evidence became available.[8] But given that there are so many questions one can possibly pursue, how can one choose a research topic? What motivates a scientist to pursue a certain research direction? There is no clear answer, of course, since everyone has their favorite subjects to work on. However, broadly speaking, a good research topic has to be well motivated. That may sound like something obvious to say explicitly. However, the literature is overwhelmed with journal papers that are just publications for publication's sake. It is easy in theoretical physics to generalize models, add another variable, and combine various proposals, sometimes without clear motivations. Given the limited amount of time we have in our life, it should not be wasted on cheap generalizations that serve no purposes in advancing our understanding. It is not that generalizations are bad, but rather one should have clear motivation in mind as to why a certain generalization is warranted, not simply a "what if". The mathematician Hermann Weyl (1885–1955) once wrote (Weyl 2009):

> "The mathematics of the last few decades has reveled in generalizations and formalizations. But to think that mathematics pursues generality for the sake of generality is to misunderstand the sound truth that a natural generalization *simplifies* [original emphasis] by reducing the number of assumptions and by thus letting us understand certain aspects of a disarranged whole. Of course, it can happen that different directions of generalization enable us to understand different aspects of a particular concrete issue. [...] In recent years mathematicians have had to focus on the general and on formalization to such an extent that, predictably, there have turned up many instances of cheap and easy generalizing for its own sake. Pólya has called it generalizing by dilution. It does not increase the essential mathematical substance. It is much like stretching a meal by thinning the soup. It is deterioration rather than improvement."

[8] For that reason one has to keep up with the latest development. For example, readers who have interests in wildlife who watched documentaries or read books on Komodo dragons prior to the 2000s may recall that Komodo dragons hunt by biting the prey and letting the bacteria in its saliva infect the wound. As the prey has been injured it cannot go far, and eventually succumbs to the infection, and that's when the dragon starts to feed. However, starting in late 2005, new discoveries have uncovered that a Komodo dragon's bite is venomous, which may help to disrupt blood clotting in the prey's wound. Still, the exact purpose of the venom, and how useful it is for a successful hunt, are still being disputed even today. Such is the progress of science. During the COVID-19 pandemic, the public saw how science worked in the limelight, and yet many were frustrated at how medical scientists kept revising their understanding of the virus. People want certainty, because that's the impressions they got from learning science in classrooms—a bunch of facts. The public trust in science has to start from understanding how science works. How to teach that in our classrooms is a crucial issue that educators should seriously consider because ultimately it would impact government policies and fundings.

Regarding simplification, most readers may be familiar with "Occam's razor[9]", the philosophy that in science, a simpler explanation is more likely to be the correct one. Most students of science take this for granted, and never give it a deeper thought. But why do we prefer a simpler explanation? When two or more competing theories are available that proclaim to explain a phenomenon, and when all else being equal (they explain all data equally well), why is it that we should choose a "simpler" one? What does it mean to be "simpler"? What confidence do we have that Nature should prefer the simpler explanation? Indeed we do not. This is only a philosophical guide. However, it is a good guide, in the sense that usually the more complicated a theory is, the more assumptions there are that do not correspond to how the world really is. The founding father of logical positivism and the Vienna Circle, Moritz Schlick (1882–1936), once wrote to Einstein the following (as quoted in Buchwald 2020):

> "In very many cases, the greater simplicity of a theory rests upon its containing fewer arbitrary elements. [...] But it is clear now that the greater the number of arbitrary elements that a theory contains, the more in it results from my willful choice, the less from what the facts compel. But naturally we must say that a theory represents reality only to the extent that it is determined just by the objective facts. [...] Naturally, however, we want to exclude from our theories, as far as possible, not only the false but also the superfluous accessories, our own addition. We do this by choosing those with a minimum of arbitrary assumptions, that is, the simplest. Then we are certain to stray at least no farther from reality than required by the limits of our knowledge."

Yet one must be careful of the simplicity requirement. Hermann Weyl wrote in *Philosophy of Mathematics and Natural Science* that scientific explanations should be sought such that they are simplest relative to the known experience. He then elaborated (Weyl 2009):

> "It often happens that for some partial domain an explanation A is simpler than B; but while A becomes increasingly complicated as the circle of experience widens, the same does not apply to B, with the result that eventually B emerges as the superior theory. Furthermore the required simplicity is not necessarily the obvious one, but we must let Nature train us to recognize the true inner simplicity."

[9] Also spelled "Ockham's razor", named after William of Ockham (c. 1287–1347). The phrase first appeared in the work of Libert Froidmont, *On Christian Philosophy of the Soul*, in which he mentioned "novacula occami" (1587–1653). Ockham did not invent this principle, though he made frequent use of it in his philosophy and theology. In particular, he wrote *Summa Totius Logicae*, a textbook on Logic in which he mentioned: "For it is pointless to do with more what can be done with less." (*"Quia frustra fit per plura quod potest fieri per pauciora"*).

Indeed, it is important to keep in mind that, as Edward Grant wrote in *Physical Science in the Middle Ages* (Grant 1977):

> "Not even Ockham's own famed "razor", or principle of economy, could be applied to the radically contingent world that he envisioned. For it was not possible to determine by any cognitive means whether the world in which we live in is simple or complex. God could willfully choose to produce events and things in ways that are complicated rather than simple. [...] the cutting edge of the razor was reserved solely for philosophical entities and descriptions used to explain the world and its processes."

In fact, foreshadowing Descartes' famous argument that led him to "I think, therefore I am" (see Section 2.3.2), Ockham argued that God could produce in us the belief that a nonexistent object existed.

Strictly speaking, we can never know the "truth". We understand the world via our theories. That's it. For example, why does an apple fall? Newtonian theory would describe it as being the consequence of gravitational force. A force is only a name we give to a quantity that satisfies certain mathematical properties. It still does not quite explain what it *is*. In Einstein's theory of general relativity, gravity is no longer a force, but a manifestation of spacetime curvature. In the limit of a weak gravitational field, general relativity recovers Newtonian theory, so Newton was not wrong, merely that his theory is only limited to certain situations. In the same way, we cannot say that general relativity is "the truth"; perhaps there is a better theory that would encompass general relativity, in which gravity is described by entirely different mathematical structures. In other words, it is important to distinguish observational fact—apples fall—from the theories we formulated to explain it. We use mathematical structures to formulate theories, but it does not mean that the Universe is really like that, it only means that our theories work well for now—gravity may not be due to spacetime curvature, only that it can be described by such a formulation. To be aware of the distinction between facts and theories is important, so that we are always ready to examine our theories and revise them if necessary, with new facts. By keeping in mind that theories are just our constructs, we are empowered to go beyond them when necessary. This is how science continues to progress and evolve.

It is worth mentioning that it is not easy for a new theory to take over. Scientists have high standards regarding new data and evidence, for good reasons: we must ensure their validity. Only when there are a sufficient number of data that cannot be explained under the existing paradigm would scientists start to seriously consider the possibility that a new theory may be required. Even then, it is not easy to formulate a new theory. In the subsequent chapters, we will encounter a few paradigm shifts (this phrase was coined by the historian and philosopher of science Thomas Kuhn (1922–1996), in his attempt to understand how scientific viewpoints evolve), including Einstein's discovery that time is relative, which was completely unexpected; and the discovery of quantum mechanics, forced upon by "anomalies" that classical physics could not explain. In each such episode in the history of science, we will try to appreciate how our understanding of the cosmos radically changed, and how those changes were initiated. How did people start to think in a different way? Before that, however, we should caution the readers that the history of science is fraught with failed theories, and even observations that were not what it seemed. This is why we must be careful validating new claims.

1.3 Cautionary Tales from "Failed Science"

The discovery of X-rays by Wilhelm Röntgen (1845–1923) in 1895 was one of the most remarkable episodes in the history of science. A ray that can penetrate the human body and make images of the bones! We are so used to this concept nowadays, but it must have been quite a shock when X-rays were first discovered. Gamma radiation was discovered soon after in 1900 by Paul Villard (1860–1934). So it seemed that the electromagnetic spectrum could potentially contain more surprises. In 1903, Reé Prosper Blondlot (1849–1930), a physicist at the University of Nancy in France, who was also a member of the French Academy of Sciences, claimed that he had discovered a new radiation, which he named the N-ray, after his town "Nancy". The N-ray was incredibly difficult to observe, even with the later "improved" technique that utilized screens with a phosphorescent coating. To observe the purported glows of the screen under the effect of N-rays, Blondlot instructed that one had to stay in a dark room and wait until one's eyes acclimated to the darkness. Only then one could observe the faint glow, but even then not directly, only by looking out of the corner of one's eye! It is well known by astronomers that it is indeed easier to spot faint stars this way (technically known as averted vision) since rod cells in our eyes that are more light sensitive are more populated in the peripheral vision. Nevertheless, the difficulty to spot N-rays meant that one should be careful about the claim.

Indeed, many other physicists could not observe N-rays following Blondlot's procedures. Yet Blondlot defended his discovery, claiming instead that those colleagues must have visions not sensitive enough to pick up the faint glow. There were no shortage of scientists who *did* say they confirmed Blondlot's results, and hundreds of papers were published on the strange properties of N-rays. For example, apparently it could penetrate solid material like metals and wood, yet was blocked by water. Possible applications of N-rays were also discussed. Amidst all this, it became all the more urgent to determine whether N-rays did in fact, exist, and if so how to make its detection easier. The physicist and inventor Robert Wood (1868–1955) was sent to investigate Blondlot's experiments in 1904. In the near total darkness of the room, Wood sneakily removed a prism in the set-up, which was supposedly necessary for the N-ray experiment to work. Yet with or without the prism, Blondlot and his team continued to observe the effect of N-rays. It was then clear to Wood that N-rays were nothing but the result of self-delusion. Once Wood's investigation was reported by *Nature* (Wood 1904), the craze for N-rays quickly dwindled and eventually no one seriously believed in it any more.

Blondlot remained adamant that N-rays exist; he simply could not accept that he had managed to trick himself into seeing something that wasn't there. In science, we must first ensure that we are not tricking ourselves. This is often easier said than done.

In the field of astronomy, there are similar examples. The most famous is probably the story about the "Martian canals". The astronomer Percival Lowell (1855–1916) wrote numerous books in which he proposed that the features of lines that criss-crossed the surface of Mars were not natural, but rather canals designed by an intelligent extraterrestrial civilization on Mars to irrigate Martian soils. The word "canals" was actually translated from the Italian *canali*, which was used by Giovanni Schiaparelli (1835–1910) in the 19th century to describe the features he saw on Mars, which to him resembled a system of straight lines. "Canali" was better translated as "channels", which would not sound as "man-made" as canals. Yet this was what Lowell insisted until his death in 1916 despite there being no evidence for it other than mere appearances that "the structures look artificial". The idea of Martian canals was debated for years, but it was not until 1965 that we finally had images of the Martian surface from NASA's Mariner 4 probe. There were no canals, and no Martians. In the end, Lowell and his supporters were seeing what they wanted to see, much like Blondlot and his N-rays.

A more recent example is the claim of Peter van de Kamp (1901–1995) in 1963 that a star known as the Barnard's star has a planet. If true, this would have been the first ever detection of an exoplanet. He inferred the existence of the planet by the wobbling of the star. Since a star and its planet would orbit their common center of mass, a sufficiently massive planet would cause the star to "wobble". This technique is still in use today. However, van de Kamp's results were refuted after his colleague John Hershey discovered, 10 years later, that the wobbling was present also in all the stars in the field of view near Barnard's stars. Surely they cannot all have planets that affect their motion in an identical manner? Worse, Hershey discovered that the wobbling apparently happened after each time the objective lens of the telescope was removed, cleaned, and replaced. Other observatories also did not observe what van de Kamp claimed to have measured. Peter van de Kamp continued to believe in his "discovery" till his death.

Why did so many scientists like Blondlot, Lowell, and van de Kamp continued to defend their findings despite evidence to the contrary? Perhaps one of the reasons is that scientists are just people, and like everyone else, most scientists also find it hard to admit mistakes. Another reason is that research demands perseverance. But perseverance and stubbornness are two sides of the same coin. This explains why it is often difficult to give up on one's vision in science, even at the face of evidence to the contrary. It is precisely because science is a human endeavor, that we must always be cautious to avoid pitfalls of human nature.

1.4 The Role of Mathematics

In 1960, the physicist Eugene Wigner (1902–1995), who was awarded the Nobel Prize in Physics in 1963 "for his contributions to the theory of the atomic nucleus and the elementary particles, particularly through the discovery and application of fundamental symmetry principles", wrote a famous article entitled *The Unreasonable Effectiveness of Mathematics in the Natural Sciences* (Wigner 1960), in which he remarked that

> "The miracle of the appropriateness of the language of mathematics for the formulation of the laws of physics is a wonderful gift which we neither understand nor deserve."

Today, mathematics is widely used in many areas, not just in the hard sciences. The use of mathematics is especially heavy in physics and astronomy, so it would perhaps come as a surprise to many students if I were to say that this wasn't always the case!

As we will see in the next chapter, astronomy was, for a long time, only concerned with explaining the observed data, and to make calculations to predict the motion of the planets etc. This part of the activity was no doubt mathematical. However, it was considered separate from *physics*, the understanding of how the real world works. The many convoluted epicycles of Ptolemy were so complicated that modern students often wonder how would anyone seriously believed that the solar system works like that? Well, the ancients probably didn't. The complicated models may just be tools for calculations, *not* a representation of reality. Admittedly, whether this was how the ancients really thought about their models is still debatable, but it is true that the role of mathematics in physics remained controversial at least till Kepler's time. In particular, Kepler had a heated argument with Robert Fludd, also known as Robertus de Fluctibus (1754–1637), about whether mathematics has a role in guiding our understanding of the Universe. Fludd believed in the "old school" of way that alchemical tradition expressed in qualitative and symbolic manners were more essential than quantitative mathematics. We should not view this debate as science versus superstition, as there was *no* such thing as science as we know it back then. Kepler himself was deeply rooted in Hermetic influences, and dabbled in many activities we today would dismiss as superstition. The debate between Kepler and Fludd was discussed at length in Wolfgang Pauli's article *The Influence of Archetypal Ideas on the Scientific Theories of Kepler* (Pauli 1994), in which Pauli also emphasized that Kepler's confidence in heliocentrism was primarily fueled by his spiritual and philosophical belief, and not a scientific one:

> "[...] because he looks at the Sun and planets with this archetypal image in the background he believes with religious fervour in the heliocentric system—by no means the other way around, as a rationalistic view might cause one erroneously to assume. This heliocentric belief to which Kepler remained faithful from early youth, impels him to search for the true laws of the proportion of planetary motion as the true expression of the beauty of creation."

Kepler also viewed the heliocentric universe as a reflection of the Holy Trinity: the Sun at the center represents God the Father, the outer sphere of fixed stars represents Jesus Christ, while the space filled with æther between them the Holy Spirit[10] (Kozhamthadam 2002). Still, Kepler recognized that mathematics is a great tool for astronomy and physics, and in that sense he was "progressive".

In addition to astronomy, other fields of study like optics and mechanics, which today we will classify as physics, were thought to be "mixed science" or "mixed mathematics" under the old tradition. Physics was thought to be a study of the *natural* world, whereas optics and mechanics often involved man-made devices, and the concerns were focused on the applications and not on the *nature* of the properties observed. Even today, mechanics is still a part of A-level *mathematics.*

Today, we are so used to mathematics it is difficult to imagine doing physics without it. Mathematics allows us to do quantitative science, and to make precise predictions once we have a theory or model, which then allows us to test whether said theory or model is correct. A recent example can be used to illustrate this. A *ring galaxy* is a galaxy that has ring-like structures around it. It is thought that a ring galaxy is formed when a galaxy collided with another one that "punches through" its center, and like throwing a stone into a pond, the "ripple" of gaseous material triggered new star formation that forms a ring or multiple rings. Usually there are only a few rings, no more than 3. Still, theory of ring formation and evolution is available, which allows one to predict the location of the rings. A practical model proposed in 2010 by Curtis Struck (Struck 2010) led to the simple relation

$$\frac{r_i}{r_{i+1}} = \frac{2i+1}{2i-1}, \tag{1.5}$$

where i is the ring number and r_i is the radius of the ith ring. (The first ring is the outermost.) Thus, the ratio of the radii between the first and second-created rings should be 3, and the successive ratios for rings 2 to 10 are about 1.67, 1.40, and so on. The "Bullseye Galaxy", officially designated LEDA 1313424 (Figure 1.2), was recently analyzed by the Hubble Space Telescope and at first appeared to have 8 rings. However, the outermost of the 8 was found to fit best with being "ring 3" of the aforementioned relation. This implies that there should be more, less visible rings unseen in the Hubble image.

[10] Concepts similar to æther are ubiquitous in the ancient world and have been applied in various contexts. One example is the ancient Greek's concept of "quintessence", an essence of the celestial world in addition to the four elements of the terrestrial worlds (the name "quintessence" means the fifth element), which we will discuss later. Heinrich Cornelius Agrippa (1486–1535), a polymath and occultist, argued that a medium inferior to the soul but finer than the body is required as a medium for the soul to join with the body. (After all, the soul is so different from the physical body, how do they interact? This "interaction problem" is a famous one faced by the philosophy of mind-body dualism. One should compare this to the latter concept of the "luminiferous æther", a medium that was thought to be necessary for electromagnetic waves to propagate through. People thought: Otherwise what is waving if there is no medium?) Likewise, Agrippa argued that the *Spiritus Mundi*, the "World Soul", permeates all space and links everything together, and can be distributed by the rays of the stars.

Figure 1.2. LEDA 1313424, the "Bullseye Galaxy", located at some 567 million light-years away from us, is 2.5 times larger than our Milky Way and has 9 rings. This is the first galaxy observed that has more than 3 rings, which allowed researchers to test the theory of ring formation and evolution (that the rings will gradually spread out and disintegrate; the outermost rings are therefore more spread-out and less visible). The rings are caused by star formation in the aftermath of a galaxy collision. The blue dwarf galaxy at center-left is likely the one that collided with the main galaxy. *Credit:* NASA, ESA, Imad Pasha (Yale), Pieter van Dokkum (Yale).

The research team found one of the missing rings with the Dragonfly Telephoto Array (the outermost ring is likely to have dissipated and no longer detectable), which is designed to image objects with extremely faint surface brightness. This allows us to test whether Equation (1.5) is correct, and thus whether our understanding of ring formation theory is acceptable. The discrepancy between the predicted and "observed" ring semimajor axis is about 10% larger, which given other uncertainties, is quite remarkable (Pasha et al. 2025). Such is the power of mathematics. In the next section, we will spend some time introducing the idea of mathematical modeling, which will prepare us for the chapters ahead.

It is possible that some readers are mathematics students who want to learn some physics, and for such readers it is important to appreciate that physics and mathematics are quite different. I myself was a pure mathematics major, and it took me quite some time to get used to the different kind of thinking used in physics during my PhD training. Sure, there is much less rigor in physics, and mathematics is treated as a tool, not as a subject in itself.[11] But that is the point, otherwise we wouldn't need a separate discipline. Physics aims to understand the real world, while mathematicians study mathematical structures, which sometimes are applicable to the real world, but need not be. Galileo once wrote to Belisario Vinta, Secretary of State of the Grand Duchy of Tuscany, about the terms of his future position as a court mathematician, in which he wished that his title be a *philosopher* in addition to being a mathematician (Palmerino 2016). By philosopher he of course meant natural philosophy. He explicitly mentioned that he had "spent more years studying philosophy than months studying pure mathematics", thus distinguishing pure mathematics that investigates mathematical structures from physics. Galileo was not against the use of mathematics in physics, but he realized the difference between the two, as he further wrote that (Palmerino 2016):

> "Not that there is anything wrong with inventing at pleasure some kind of motion and theorizing about its consequent properties, in the way that some men have derived spiral and conchoidal lines from certain motions, though nature makes no use of these paths. [...] But since nature does employ a certain kind of acceleration for descending heavy things, we decided to look into their properties so that we might be sure that the definition of accelerated motion which we are about to adduce agrees with the essence of naturally accelerated motion."

Mathematicians *can* make contributions to physics. For example, general relativity can be studied more rigorously utilizing techniques from differential geometry and partial differential equations, which constitutes an entire sub-field called *mathematical relativity*. This allows *theorems* in a mathematical sense to be proved (for example, the various singularity theorems). But theorems always have some assumptions built in, and whether these assumptions conform to the actual Universe cannot be determined by mathematics alone. David Hilbert (1862–1943), who attempted to axiomatize physics in the early 1900s, in much the same way as geometry was axiomatized, later allegedly commented that "Every boy in the streets of Göttingen understands more about four-dimensional geometry than Einstein. Yet, in spite of that, Einstein did the work and not the mathematicians." While this quote remains unverified, it has a lot of truth in it—physicists who do research can utilize mathematics as a tool without fully understanding the inner workings of the tool. A good physicist would of course need to know about the limitations of the tool and not misuse it or use it in the wrong way.

[11] Even in pure mathematics, proofs are not the most important part, but rather the *ideas* behind the theorems and the proofs.

1.5 Mathematical Modeling: A Crucial Skill for Physics

Most students of physics and astronomy have taken a course or two on "mathematical methods for physicists". These courses are aimed to provide students with an arsenal of tools that they can apply in physics. However, arguably more important is the skill of mathematical modeling, which unfortunately is usually not as widely taught, though no doubt some basic examples are covered in other courses. Indeed, many students are good at solving a given mathematical problem, but not as skilled when presented with a "raw problem" that has no mathematical equations and are asked to formulate a mathematical model for it. (Even in elementary and high schools, many students dread the so-called "word problems".) This skill is, however, the most important one in research in the applied sciences.

What exactly is mathematical modeling? In physics, we need to model the systems of interest quantitatively (it's not enough to know that things fall, but how fast it falls and where it will fall). However, actual physical systems are often extremely complicated with all the nuances and subtleties. For an object that falls under gravitational pull, there are also effects from drag, from wind that blows sideways, even from the rotation of the Earth, the fact that the Earth is not really a sphere (it is a *geoid*), the density of Earth at different depths is not uniform, etc.[12] Another example is the water jet that we modeled with a projectile equation (also called the "ballistic equation") in Section 1.2. The *art* of mathematical modeling is to simplify the problems, extracting only its essence to be analyzed. This is an art because if it is overly simplified, it will fail to describe the system sensibly.[13] If it is not simplified enough, it will be too complicated to solve. In practice, one always starts with the simplest model one can think of (the "spherical cow"), and then gradually adds more elements to it if the results do not quite match the actual systems.

From personal experience, students tend to over-complicate models; they often worry about whether we should consider this and that effect. One has to learn that it is acceptable, even preferable, to simplify things *at first*, to at least get a feeling for the problems at hand. Too much complication and we will be lost before we can even get started. In latter chapters, we will see how mathematical models are used to study various problems, so in preparation for that in this section we will provide some easy examples to demonstrate this art.

The first example is: how should cans be made? Consider a tin can, i.e., a container made of thin metal that stores food. The producer of tin cans would definitely want to minimize the cost of the production to make more money. Let us call the cost a function C. What should C depend on? One can think of many factors that would affect C, including the price of the metal, the cost of welding the

[12] An interesting book that considers various effects of falling objects is *Hesiod's Anvil: Falling and Spinning through Heaven and Earth* by Andrew J. Simoson (Simoson 2007).

[13] A quote often attributed to Einstein is "Everything should be made as simple as possible, but not simpler". This is likely to be a compressed version of what Einstein said in a 1933 lecture: "It can scarcely be denied that the supreme goal of all theory is to make the irreducible basic elements as simple and as few as possible without having to surrender the adequate representation of a single datum of experience." (Robinson 2018).

parts together, the shape of the can, etc. However, we should start with the *simplest* model, which is the one that is sometimes encountered in an elementary calculus course:

Model 1: We only consider the amount of metal that is actually on the can, so $C \propto A$, where A is the area of the metal. Since a can is cylindrical, its area is

$$A = 2\pi r^2 + 2\pi rh, \tag{1.6}$$

where r is the radius and h the height of the can. This can be minimized by setting $r = 0$, which means that there is no can. This actually makes sense because if you don't make any can you don't spend any cost! But this is not what we want, we insist on making cans, and so the question is actually: how to minimize the function given that we want to store a fixed volume of content. Therefore, there is a constraint that the volume $V = \pi r^2 h$ is constant. Thus, we can re-write the area as

$$A = 2\pi r^2 + \frac{2V}{r}. \tag{1.7}$$

Differentiating this function and setting it to zero, we have

$$\frac{dA}{dr} = 0 \Rightarrow 4\pi r - \frac{2V}{r^2} = 0, \tag{1.8}$$

and so

$$\pi r^2 h = V = 2\pi r^3 \Rightarrow h = 2r. \tag{1.9}$$

In other words, to minimize the amount of metal used, tin cans should be made with height equal to their diameter, i.e., as high as they are wide. However, this is *not* what we observe in supermarkets. Typically a tin can is taller than its diameter. This is however, quite true for larger cans, for example, those that store paint. Therefore there *must* be other important factors that our simplified model did not take into account. Let us now make it more complicated (and more realistic).

Model 2: If we consider the process of making the cans, it would become obvious that the circular top and bottom parts need to be cut from rectangular sheets of metal. Since the manufacturer needs to pay for the entire sheet, it is crucial to minimize wastage. This therefore becomes a question on how to maximally pack circles. Circle packing is an interesting area of mathematics by itself, which extends to higher dimensional sphere packing. In the case of two-dimensional plane, the result is well known: the densest packing is given by a hexagonal arrangement. That is to say, we first tile the plane with hexagons, like a bee's honeycomb, then in the center of each hexagon we punch out a circle whose circumference touches the sides of the hexagon (in mathematics, we say that the circle is *inscribed* by the hexagon). Thus, instead of just the circle, we have to pay for the hexagon, whose area can be shown to be $2\sqrt{3}r^2$ by elementary trigonometry. Therefore the relevant area that we need to maximize is

$$A = 4\sqrt{3}r^2 + \frac{2V}{r}. \tag{1.10}$$

Note that the cylinder part is simply a rectangle rolled up, so there is no wastage involved. Again, with a constant $V = \pi r^2 h$, we can differentiate this and find that minimization requires that

$$h = \frac{4\sqrt{3}}{\pi}r. \tag{1.11}$$

Numerically, $4\sqrt{3}/\pi \approx 2.21 > 2$, which means that cans should be about 10% higher than wide. This seems to be a better model for food cans, as we mentioned before. However, this does not explain why the first model describes a can of paint better while the second model describes a can of food or drink better. An even more complicated model is needed that can describe these two situations.

Model 3: What other cost is crucial in making tin cans? Well, the parts all need to be welded together! The cylindrical part is obtained by rolling up a rectangle and welding up along a seam of height h. The top and bottom parts are then welded along the boundary circles of the cylinder, each of circumference $2\pi r$. It makes sense to assume that the cost of welding is proportional to the length with some proportionality constant K, with units \$/cm, where \$ is some currency involved. The total cost is then given by the function:

$$C(r) = J\left(4\sqrt{3}r^2 + \frac{2V}{r}\right) + K\left(4\pi r + \frac{V}{\pi r^2}\right), \tag{1.12}$$

where we have expressed h in terms of r using the fixed volume constraint, and J is the cost of metal per unit area. Differentiating this and setting it to zero, we have

$$\frac{dC(r)}{dr} = J\left(8\sqrt{3}r - \frac{2V}{r^2}\right) + K\left(4\pi - \frac{2V}{\pi r^3}\right) = 0. \tag{1.13}$$

From here we can show that minimization requires that the ratio h/r satisfies

$$\frac{h}{r} = \frac{4\sqrt{3} + \frac{2\pi K}{rJ}}{\pi + \frac{K}{rJ}}. \tag{1.14}$$

We note that K/J has dimension of length, as it should, to make K/rJ dimensionless. We also notice that if $r \gg K/J$, then $h/r \to 4\sqrt{3}/\pi \approx 2.21$, and we recover the result of Model 2. On the other hand, if $r \ll K/J$, then $h/r \to 2\pi \approx 6.28$. The fact that we are comparing r with K/J means that the latter sets the *scale* of the problem. Only by identifying the relevant scale can we say what "large" and "small" mean. Thus, we see that if r is large (with respect to K/J), the tin should be made almost as wide as it is tall. For a small can, r is small and it should be made about 3 times taller than its width. Model 3 is therefore a better model.

The example above demonstrates the process of mathematical modeling. If we are still not satisfied with the results, we can add more and more realistic complications. The point is that we should never start with a complicated model; the simpler a model is, the easier it is to handle, and by understanding simple models we can often identify the essence of the problem.

Another typical example of mathematical modeling is population dynamics. The simplest model is the Malthusian model, named after the economist cleric Thomas Robert Malthus (1766–1834), which describes the growth of a population by the ordinary differential equation

$$\frac{dN}{dt} = kN, \tag{1.15}$$

where $k = B - D$ is the difference between the birth rate per capita B and the death rate per capita D, and $N = N(t)$ is the number of the population. Evidently this has an exponential solution, so either N grows exponentially (if $k > 0$) or decays exponentially (if $k < 0$). Note that exponential growth is *unphysical* because resources are finite, thus once N is sufficiently large, there would be over-crowding and resources begin to run thin (and most likely lead to environment problems and conflicts). Thus, more realistically there would be a *cutoff* such that when as $t \to \infty$ the growth is capped at some constant. This is the "logistic model", which can be obtained by assuming that the death rate D is linearly proportional[14] to N, i.e., $D = sN$ for some constant s, which leads to

$$\frac{dN}{dt} = BN - sN^2. \tag{1.16}$$

Assuming first that $B > sN$, then it can be shown that the solution is[15]

$$N(t) = \frac{B}{s + e^{-Bt}\left(\frac{B}{N(0)} - s\right)}. \tag{1.17}$$

When the death rate is small, this would recover the exponential growth model. In fact, when $t < 0$, when resources are plentiful, the growth is very close to being exponential. However, as $t \to \infty$, we see that $N \to B/s$ asymptotically. The shape of the solution is what is called a "logistic curve".[16] Clearly, the important scale here is B/s, which is known as the "carrying capacity". If $B < sN$, then we have instead

[14] Why linear? Well, this is the *simplest* model!

[15] Students should verify this.

[16] The name was coined by the mathematician Pierre François Verhulst (1804–1849), but the reason for this choice of name is unknown, though it was probably in analogy to "logarithmic" (not in the modern sense of logarithmic function, but rather a geometric growth curve, which in modern language would be called the exponential curve).

$$N(t) = \frac{B}{s - e^{-Bt}\left(\frac{B}{N(0)} - s\right)}. \tag{1.18}$$

In this case the population would start with $N(0)$ already larger than the carrying capacity, and the population will decay asymptotically towards B/s. The case $B = sN$ is the borderline case in which N remains constant, precariously balanced by the available resources.

More complicated population dynamics can be applied to understand interactions between species of animals in an ecosystem, or the spread of a disease in a population. The latter is important to know whether intervention like lockdown is needed, as many of us have experienced during the COVID pandemic. One of the simplest and most well-known models is the "SIR model", where S is the number of people who are susceptible to the disease, I is the infected number, and R is the number removed from the population either because they have recovered and are now immune to the disease, or because they sadly passed away. Of course, these are all functions of time. The SIR model is given by a set of coupled differential equations:

$$\begin{cases} \dfrac{dS}{dt} = -\dfrac{\beta}{N}IS, \\ \dfrac{dI}{dt} = \dfrac{\beta}{N}IS - \gamma I, \\ \dfrac{dR}{dt} = \gamma I, \end{cases} \tag{1.19}$$

where $\beta > 0$ and $\gamma > 0$ are some constants. The typical behavior for I is that it will increase, reach a peak, and finally decrease. The disease spreads in the population until enough people have immunity or passed away, then it will die off. On the other hand, R is an increasing function, while S is a decreasing function. The scale of interest in this problem is the dimensionless number β/γ, which is usually denoted as R_0, called the "reproduction number" of the infection. It is the number of secondary infections caused by a single index infection in an otherwise susceptible population. That is, if $R_0 = 10$, it roughly means that on average one person can infect 10 others. If $R_0 < 1$, the outbreak will die out relatively quick (the function I will not have an increasing phase), whereas if $R_0 > 1$, the outbreak will expand.

Clearly this model can be made more complicated, for example, some people can be infected again after recovery. Note also that the sum of these equations is zero, which means that the total population $N = S + I + R$ is constant. This means that the SIR model does not include "vital dynamics" like birth rate, death rate (deaths not caused by the disease), or migration. Birth rate and other death rates typically change the number of the population at a different timescale than a disease, so to first approximation they can be neglected. (This is also common in physics. For example, one speaks of a "quasi-static" system if the rate of change is small compared to another timescale of interest.[17]) Migration *can* be crucial, however, as many people may choose to fled the regions plagued by a dangerous disease, and inadvertently bring the disease with them to somewhere else. An improved model can also include the incubation period of a disease, the probability that some people of the population are naturally immune due to genetic mutations, the different rate of infection for different age group, etc. As in all mathematically modeling, start simple and modify as needed.

Another famous model in ecological population study is the Lotka–Volterra model,[18] which describes how two species interact. The coupled differential equations for the interaction between a predator and prey species are

$$\begin{cases} \dfrac{dx}{dt} = \alpha x - \beta xy, \\ \dfrac{dy}{dt} = -\gamma y + \delta xy, \end{cases} \tag{1.20}$$

where x and y are the population densities of prey and predator, respectively. The parameters are positive and represent: the growth rate per capita for prey α, the growth rate per capita for predator β, the death rate per capita for the predator γ, and the death rate per capita for the prey δ. The solutions for $x(t)$ and $y(t)$ are periodic, since if prey is plentiful, then the population of the predator will grow, but then there won't be enough prey and the predator's population starts declining, which in turn leads to more prey and so on. This is clearly an oversimplified model because in real life, species *can* go extinct if there are too few individuals left to reproduce.

From these various examples, we have seen that mathematical models, no matter how sophisticated, are only approximations. This is best summed up by the common aphorism "all models are wrong, but some are useful". Yet, models should not aim to be unnecessarily complicated. Often a simple model that can capture the essence of the problems at hand is good enough.

This ends the first chapter. In the subsequent chapters, as we explore numerous topics, the readers should keep in mind the general theme: the progress of science relies heavily on *ideas* and creativity, while at the same time, it has to make concrete and quantitative predictions. Philosophical thoughts and artistic fervor need to balance with mathematical modeling (Figure 1.3).

---

[17] Another example is the water cycle. When we first learned it in school, more often that not, it is the simplified version in which underground water in the form of aquifers are not included although they make up about 30 percent of fresh water content. The reason is water moves very slowly underground depending on their depth and the permeability of the soil. The timescale is longer than the surface water.

[18] Alfred James Lotka (1880–1949) initially proposed the model in the context of autocatalytic chemical reactions, not ecological population. Vito Volterra (1860–1940), on the other hand, was studying fish population problem.

Figure 1.3. *The Sleep of Reason Produces Monsters* by the Spanish artist Francisco Goya (1746–1828), created between 1797 and 1799, during the age of Enlightenment. Humans are good at telling stories and fantasizing all sorts of things. Without the guide of reasons and the scientific methods, we would be stuck in a "demon-haunted world" (in the words of the astronomer and science communicator Carl Sagan) plagued by ignorance and superstitions. In fact, genuine fear of witchcraft and demons were widespread in Europe, neighbors excusing each other for being in league with the Devil, resulting with murders and witch trials up until the early 18th century. (Sadly, witch hunts still occur in some parts of the world today.) On the other hand, Goya's own caption reads: "Imagination abandoned by reason produces impossible monsters; united with her, she is the mother of the arts and source of their wonders". Science should go hand-in-hand with imagination; it *requires* imagination to come up with new ideas and theories. It is not a purely deductive, rigidly rational subject, but a passionate one sustained by human's desire for exploration and understanding the world. *Credit:* Image credit: The Charles Deering Collection. *Source*: https://www.artic.edu/artworks/44743/the-sleep-of-reason-produces-monsters-plate-43-from-los-caprichos

1.6 Exercises

(1) In Section 1.2, we have mathematically derived that in a cylindrical container completely filled with water, the middle hole drilled into the cylinder would shoot a water jet the farthest if the container is on a flat surface. Now, consider instead putting the container at the edge of a table of height h_3 above the floor. Consider three possible hole positions: the middle (x_1) of the cylinder, right near the bottom (x_3), and somewhere in between those (x_2). What condition should h_3 satisfy in order for x_3 to shoot water jet the farthest? Would it be possible to get the water jet shooting out from hole x_2 to go the farthest?

(2) The water jet example is only one of many mistakes that unfortunately are still often taught to students in some countries. Some other misconceptions, once propagated into the public perception, are hard to correct. One example is the myth that we only use 10% of our brain. You might also have heard of the "fact" that our tongue is separated into different regions that are more sensitive to different tastes. For example, the tip of the tongue is more sensitive to sweetness. This "tongue map" is, however, another myth. Look up the history and learn how this misunderstanding came to be.

(3)

(a) Show that the Lotka–Volterra equations admit an invariant H that is time-independent, given explicitly by

$$H(x, y) = \delta \log x - \gamma x + \alpha \log y - \beta y. \tag{1.21}$$

Additional note: This function is denoted as H because it is a *Hamiltonian*, in analogy to the situations in Newtonian mechanics in which a Hamiltonian is conserved (when the Lagrangian has no time dependence). The tools of Newtonian mechanics can thus be used to analyze biological interactions. (Readers who have no prior knowledge in classical mechanics can safely ignore what I said here.)

Hint: Note that the Lotka–Volterra equations imply

$$A\frac{dx}{dt} + B\frac{dy}{dt} = 0. \tag{1.22}$$

What are the values of A and B in terms of the parameters and the variables?

(b) Numerically solve the Lotka–Volterra equations for your choice of parameters and convince yourself that the solutions are periodic. The period can actually be numerically computed without first solving the differential equations. See Waldrogel (1983).

(c) Find more information on how the Lotka–Volterra equations can be further generalized. For example, What if we include a "fear" effect in the prey? In a natural environment, prey species would often take precautionary measures against being eaten. How do we model this effect? What if there are more than two species interacting with one another? What if .... Models can be made as complicated as you like!

References

Buchwald, J. Z. 2020, Einstein Was Right: The Science and History of Gravitational Waves (Princeton, NJ: Princeton Univ. Press)
Feke, J. 2024, SHPSA, 107, 91
Grant, E. 1977, Physical Science in the Middle Ages (Cambridge: Cambridge Univ. Press)
Kozhamthadam, J. 2002, RPF, 58, 887
Palmerino, C. R. ed. G. Gorham 2016, The Language of Nature: Reassessing the Mathematization of Natural Philosophy in the Seventeenth Century (Minneapolis, MN: Univ. Minnesota Press)
Pasha, I., van Dokkum, P. G., Liu, Q., et al. 2025, ApJL, 980, L3
Pauli, W. ed. C. P. Enz, & K. von Meyenn 1994, Writings on Physics and Philosophy (Berlin: Springer)
Robinson, A. 2018, Natur, 557, 30
Rovelli, C. 2015, JAPA, 1, 23
Simoson, A. J. 2007, Hesiod's Anvil: Falling and Spinning through Heaven and Earth, (Washington, DC: Mathematical Association of America)
Struck, C. 2010, MNRAS, 403, 1516
Waldrogel, J. 1983, SJNA, 20, 1264
Weyl, H. 2009, Philosophy of Mathematics and Natural Science (Princeton, NJ: Princeton Univ. Press)
Wigner, E. 1960, CPAM, 13, 1
Wood, R. W. 1904, Natur, 70, 530

AAS | IOP Astronomy

Understanding the Universe
A conversational introduction to cosmology
Yen Chin Ong

Chapter 2

Celestial Spheres and Planetary Motions

> *"I know that I am mortal by nature, and ephemeral; but when I trace at my pleasure the windings to and fro of the heavenly bodies I no longer touch the earth with my feet: I stand in the presence of Zeus himself and take my fill of ambrosia."*[1] —Ptolemy, 'Almagest'.

In this chapter, we review the change of our world view from geocentrism to Copernican heliocentrism, with a special focus on how Kepler attempted to model the solar system via geometry—first plane geometry based on inscribed circle and then solid geometry of platonic solids. Despite its eventual failure, this episode of the history of science is a valuable lesson on how science is done. We also explain the contributions of Galileo, Descartes, and Newton, all of whom played important roles in our understanding of the cosmos. Some of the nuances and subtleties in their stories will probably be new and surprising to many readers who were only familiar with the "usual textbook re-telling".

Let us set the stage for our journey. The main story—the gradual shift from geocentrism to heliocentrism and eventually the Newtonian paradigm—is perhaps well-known to most students of astronomy. However, students were also taught that the reason geocentrism held for so long was because Aristotle was such an important figure that no one ever questioned what he said. On top of that we were told that the Catholic Church had always been on the opposite side of science, hindering its progress. In particular, scholastic philosophy attempted to reconcile Aristotelian physics and classical philosophy with Catholic Christianity. One of the leading figures in scholasticism was St. Thomas Aquinas (c. 1225–1274). Since not all of Aristotle's philosophical thoughts and worldview are compatible with Christianity,

[1] Ambrosia refers to the food of the gods.

doi:10.1088/2514-3433/ade963ch2

the scholastic philosophy approach had to reject some Aristotelian ideas, such as the world being eternal (no beginning). So Aristotelian physics *were* challenged in one way or another.

How then was there a sudden shift by the time of Copernicus? The history is of course rather complicated. Numerous people have objected to some aspects of the Aristotelian worldview across history, and while the Church did present some hindrances, religion did play positive roles at times, albeit perhaps not intentionally. One important event that should be more well-known among science students is the so-called "Condemnations of 1277", a series of actions taken by the Bishop of Paris, Étienne Tempier (1268–1279), to prohibit the teachings of 219 philosophical and theological ideas at the University of Paris. On one hand this seems to modern readers like a censorship (which it was by modern standard), but the fact that some beliefs in the Aristotelian worldview (such as the world is eternal; and that a void or vacuum couldn't exist; and that there could only be one Universe with Earth being at its center) were challenged by the religious authority probably made it easier for Copernicus and other astronomers to challenge geocentrism later on. In addition, though how God really created the Universe was considered beyond our comprehension given our finite capability, Tempier endorsed God's absolute power to do whatever He wills. Paradoxically, this opened up the freedom to discuss alternative ideas of what the cosmos could be like, which the Church tolerated as long as they were not claimed to describe the real world.

For the most part these exercises in the hypotheticals did not lead anywhere, but at least now there was freedom to discuss "non-standard" ideas. As Edward Grant pointed out in his *Physical Science in the Middle Ages* (Grant 1977):

> "Until 1277, the possibility of other worlds had not been seriously entertained by Christian authors. [...] After 1277, the intellectual context was dramatically altered and the question about other worlds, as well as a host of other unusual and striking questions, [...] became commonplace. Article 34, which denied that God could make more than one world, was one of the articles condemned in 1277 [...]"

Grant also mentioned that William of Ockham,

> "By emphasizing logical rigor and making no claim about existential implication, Ockham may have encouraged a tendency at Oxford and Paris to imagine all manner possibilities—and even seeming absurdities—without regard for physical reality or application. The characteristic sign of this approach was the phrase *secundum imaginationem*—'according to the imagination'."

Post 1277, we also see more serious criticisms on Aristotelian physics. For example, the philosopher Jean Buridan (1301–1358) wrote *Questions on the Eight Books of the Physics of Aristotle*. Pierre Duhem (1861–1916), a physicist and historian of science, would go as far as to claim that the Condemnation of 1277 led to the birth of modern science, though this view is debatable.

However, the *secundum imaginationem* approach of not caring about the real world—believing it to be beyond humanity's capability—remained strong during the time of Copernicus. Andreas Osiander (1498–1552), the theologian who oversaw the production of Copernicus' book, added a preface (quite possibly without Copernicus' knowledge) that the idea in the book was just an alternative idea that can be applied for calculations but not necessarily representing reality. Of course, he might have done so to prevent the wrath of the Church.

With this historical background in mind, let us begin our story.

2.1 Modeling the Heaven: Geocentrism vs Heliocentrism

Civilizations and cultures all over the world have their own myths and stories of the night sky. Observing the stars soon revealed the existence of celestial bodies that despite being "star-like" to the naked eyes, exhibit peculiar motions. While the entire night sky moves as a whole, with stars rising from the east and setting in the west as the night progresses, these objects move individually on specific paths with respect to the background stars. These are the planets, or *wanderers* in Greek. If we trace their motion across many nights, it soon becomes clear that their paths are far from simple. In particular, they periodically switch direction, looping back and forth (the technical term is "retrograde motion"). Making sense of these observations turned out to be extremely difficult. Nowadays, most of us are familiar with the idea of geocentric model and heliocentric model, i.e., whether the Earth or the Sun is the "center" of the solar system. However, to change the commonly accepted view from geocentrism to heliocentrism was not so straightforward. There were even hybrid models in between as the science progressed. In this chapter, we can only discuss a simplified history. Our main objective is to understand *model-building*, and to appreciate human preconceptions, prejudices and ideals that went into that process. In the later chapters, when we deal with contemporary theoretical physics, we will see that much the same process is still ongoing, although the types of questions being asked are different. Yet, we can still learn a great deal from these historical examples.

Let us begin with the geocentric view of the cosmos. Despite the term "cosmos" brings to our modern mind a vast universe filled with countless number of galaxies, this view is relatively new. The cosmos was a rather small place in comparison in the concentric celestial spheres model, for example. One starts with Earth at the center, which are surrounded by ever larger spheres on which the celestial bodies revolve on—the Moon, Mercury, Venus, the Sun, Mars, Jupiter, and Saturn. Only five planets (excluding Earth itself) were known in the ancient time, and so beyond Saturn is the last sphere, the sphere of fixed stars. The stars are not just fixed on the sphere, but also in time, i.e., no new stars are ever born. These spheres are often thought to be physical, crystalline shells instead of abstract mathematical spheres. This order was probably first proposed in the work of Greek philosopher Eudoxas of Cnidus (390–340 BCE). Note that the modern heliocentric order of the planets can be obtained easily by simply interchanging the Earth-Moon system with the Sun. However, it would be incorrect to say that the geocentric and heliocentric views are equivalent, being just a change of perspective. The details in the various models revealed the various underlying assumptions and the different understanding of the solar system. We first note that the idea of the concentric celestial spheres is not really a "scientific model" in the sense that we cannot make any calculation to describe, much less predict, the motions of the celestial objects at all. In fact, people thought that the celestial spheres were moved by divine forces or angels,[2] as described in Dante Alighieri's (1265–1321) *Convivio* ("The Banquet"[3])

> "And so we must first know that its movers are substances separate from matter, namely Intelligences, which the common people call Angels."

The idea is therefore more of a philosophical thought of how the universe *should* be. On the other hand, precise calculations were needed for many reasons, including season tracking and astrology. A good calendar requires precise mathematical models of the heavens.

One remarkable achievement that allowed detailed calculations, and hence predictions, was the Ptolemy model, named after Claudius Ptolemy (c. 100–c. 170 AD) of Alexandria. The model was built for utility, not elegance. In order to explain the apparent retrograde motions of the planets, the Ptolemiac system utilizes two different spheres. Each planet has a "deferent" sphere centered on the Earth. However, the planets themselves do not directly orbit the Earth on this sphere. Instead, a planet orbits around a point that is the center of a small sphere, which in turn orbits the deferent. This is the "epicycle". The radius of the epicycle can be set to match observations. Since the epicycle rotates within the deferent, this causes the planet to periodically move farther and closer from Earth, accounting for the looping behavior observed. However, in order to fit observational data, the center of the epicycles must not move at a constant speed around the deferent. Instead, its speed should be greater when the planet is closer to Earth and vice versa. This is not ideal from a philosophical point of view. Since at least the time of Plato, the idea of uniform circular motion had always been preferred, its high symmetry is aesthetically pleasing and arguably divine. The way out of this conundrum —the need to both fit actual data and at the same time not deviate too far from the commonly accepted philosophy of the time—was to introduce the concept of the "equant". Although we categorized the Ptolemy model as "geocentric", the Earth is actually not at the center of the cosmos. Rather, the Earth is slightly off from the true geometric center. At the opposite side of the geometric center, at the same distance of the earth, one finds a purely mathematical point called an "equant" (Note already at this stage a hint of Kepler's *ellipse*, which has two foci, though the shape was not there yet!). Then, the center of the epicycles moves at a constant distance from the geometric center, while at a constant speed from the equant. In this sense the model is still "circular and uniform", just that the properties are with respect to a different point in space! Note that

[2] This concept gradually gave way to accepting "natural motion" without the need of a constant push of an outside agency. One of the earliest attempts to explain motion without a constant force acting on it can be found in Francis of Marchia's (c. 1290–after 1344) commentaries on *Sentences*, a compendium of Christian theology written by Peter Lombard (1100–1160) around 1150. Francis of Marchia's work dated to around 1320, proposed the idea of a "stored motion", i.e., the continual motion of an object is caused by a force "left behind" by the object that acted upon it. His idea is often referred to as *virtus derelicta*, "the force left behind".

[3] Accessible online at https://digitaldante.columbia.edu/text/library/the-convivio/book-02/.

the Earth is located at neither of these special points. One could argue whether this really counts as a "way out" or just a contrived argument. However, to do so would be to miss an important point, which is that the ancient astronomers may not necessarily view their geometric models as how the Universe actually works, but only as a computational tool to predict heavenly motions. Astronomy was therefore more mathematics than physics, aimed only to describe the *appearances* of the night sky, rather than what are the "true motions", which are believed to be beyond the grasp of mere mortals anyway. Such an approach has been referred to as "saving the appearances" or "saving the phenomena".

It should be emphasized that even the picture of concentric celestial spheres that most of us are familiar with today has been greatly simplified (and no wonder it cannot be used to perform useful calculations). In the original Aristotle's system, there were 55 concentric celestial spheres. Edward Grant wrote in *Physical Science in the Middle Ages* that (Grant 1977)

> "[...] the planet Saturn, for example, was assigned a total of four spheres to account for its motion: one for the daily motion, one for the proper motion along the zodiac or ecliptic, and two for its observed retrograde motions along the zodiac. [...] In order to prevent Saturn's proper zodiacal and retrograde motions from being transmitted to Jupiter, the next lowest sphere, he introduced three unrolling spheres whose function it was to counteract the motion of the three spheres which controlled all but the daily motion. Since the daily motion was common to all planets, each was assigned a special sphere for that purpose. In all, seven spheres, instead of four, were now required for Saturn. Similarly, Aristotle thought it necessary to add counteracting spheres for all the planets except the Moon."

Thus we see that the celestial spheres were indeed used for model-building, before they were replaced by Ptolemy's epicycles. If these details make you confused, do realize that this is a testament to how sophisticated the ancients were in model-building! The *simplified* celestial spheres then became a philosophical, idealistic *representation* of the physical cosmos. If one requires both epicycles and celestial spheres to be consistent, one has to worry about whether the epicycles can cross another sphere and so on. Since the spheres were thought to be solid, this calls for very complicated models that involve "truncated spheres". Grant wrote:

> "Averroes[4] readily conceded that the use of eccentrics and epicycles was appropriate for the computation and prediction of planetary positions,

[4] Averroes is the Latinized name of Ibn Rushd (1126–1198), one of the great figures of philosophy of the Islamic world. He had wide interests and made contributions to philosophy, linguistic, Islamic jurisprudence, astronomy, mathematics, physics, medicine, and geography. We mentioned at the start of this chapter that one of the articles of the Condemnation of 1277 was regarding the view that the world is eternal, which is against the Abrahamic religious view of creation. Ibn Rushd, despite being a Muslim, believed that *both* views are compatible in the sense that the "creation" was not to be interpreted as an event *in time*, but that God "actualized" the Universe in some way.

> that is, for saving the astronomical phenomena, but vigorously opposed to the opinion, held by a few, that epicycles and eccentrics actually had physical existence within the cosmos.[5] [...] endorsed by almost all scholastics who argued for the probable physical truth of the simplified eight sphere version derived from Aristotle, while granting that Ptolemiac astronomy was essential for saving the astronomical phenomena."

In the 13th century, Robert Anglicus wrote a commentary on the *Sphere of Sacrobosco* written by Johannes de Sacrobosco (1195–1256), in which he mentioned the celestial spheres (he referred to them as "orbs"):

> "As to these orbs, there is disagreement between the naturalists and mathematicians, because naturalists suppose that all orbs are concentric, but the mathematicians do not."

Apparently in the text, the mathematicians did not question whether solid orbs exist or not, but did engage in a disquisition on the meaning of "solid"! (Westman 1980) (Modern students of mathematics would appreciate this, given how obsessed mathematicians are about defining things correctly.)

In any case, even back then in ancient Greece, Aristachus of Samos (c. 310–c. 230 BCE) already proposed the possibility that the Sun is located at the center of the Universe. This heliocentric view, however, was not popular, and it was not until the time of Nicolaus Copernicus (1473–1543), Figure 2.1, that the debates between geocentrism and heliocentrism reignited at a more serious level. Modern readers may find it difficult to appreciate why the heliocentric view was often so readily rejected. However, it was not mere matter of taste or preconception, but rather *observational evidence* or rather the lack of it that played a major role. If the Earth were to revolve around the Sun, argued the critics, surely we would have seen effects of this motion? The argument that we should have *felt* the motion was based on incorrect understanding of physics, but more crucially no *stellar parallax* had ever been observed until the 19th century. The apparent shift of position (parallax) of any nearby star against the background of distant stars, caused by the Earth's changing position around its orbit, are now routinely used to measure our distance to these relatively close stars via simple trigonometry. However, the effect is too tiny and therefore not measurable in the old days (the stars are way more distant than expected). It was the lack of this parallax that led to the idea that all stars are fixed on a celestial sphere, and also as a reason to doubt the heliocentric picture. An obvious lesson here is that it may take some time for the relevant observations to turn the tide for which scientific theory is preferred. So how did Copernicus come to believe in the heliocentric picture?

[5] Yet somehow this was how it is usually presented to modern students in a course like "Introduction to astronomy".

Figure 2.1. *Astronomer Copernicus, or Conversations with God*, completed in 1873 by the Polish artist Jan Matejko. In this painting Copernicus appeared to be observing the night sky. Note a diagram that depicts a heliocentric solar system (with *Sol* written at the center, while the Moon is orbiting the Earth, or *Terra*). Yet a bright light is bathed upon him, almost like he is experiencing a divine epiphany. The instrument to the right is called a triquetrum ("equatorial triangle"). It was used to determine the parallax of the Moon and calculate its distance from the Earth, here positioned in a way that almost reminds one of a crucifix. The painter has exercised his artistic license as the painting is not historically accurate (for example, spot the telescope on the floor hidden in shadow, which wasn't available during Copernicus' time). What it shows, however, is the passionate pursuit of an astronomer to understand the cosmos. For a religious person like Matejko, this means understanding the works of God, a divine pursuit. *Credit:* Jagiellonian University Museum, Collegium Maius. *Source*: https://muzea.malopolska.pl/en/objects-list/2829

It was *not* true that the Copernican heliocentric model was more accurate than the geocentric model[6]. In most popular science books, and even in astronomy classes, it was presented in a misleading way, as if Copernicus just put the Sun at the geometric center of the cosmos, and representing the orbits of all the planets as perfect circles at uniform speed around the Sun, and that somehow "improved" things. This is an overly simplified picture. In fact, if he were to do that, the model's prediction would have been worse than the Ptolemy's epicycles. To achieve the same level of accuracy, Copernicus still needed the epicycles, despite the fact that the Sun is now at the

[6] In 1902, a shipwreck off the cost of the Greek island Antikyther was found to contain a mysterious device, now called the "Antikythera mechanism". Built in the late second century BCE or the early first century BCE, it is often described as the oldest "analog computer". Based on the geocentric worldview, it can calculate and predict astronomical events and cycles (which were known from Babylonian astronomy), such as the phase of the Moon, the synodic phases of the planets, and eclipses (Freeth et al. 2021).

center instead of the Earth. The geocentric model *is* simpler in the sense that though small epicycles are still required to fine-tune the orbits to match observations, it no longer relies on epicycles to explain the retrograde motion of the planets. This simply follows from the geometry of the orbits—as Earth overtakes a planet as they both revolve around the Sun, the planet seems to stop and change direction with respect to an earthbound observer. The main reason that motivated Copernicus wasn't that observational accuracies of the data somehow preferred heliocentric model, but rather philosophically he found the geocentric model unpleasant and he sought a simpler, more elegant model. But *why* insist on simplicity? Copernicus, like most people at that time, was very religious. In his opinion, an all-powerful God with limitless intelligence would choose the simplest way to create the cosmos. In addition, it has been claimed that Copernicus was also heavily influenced by Hermeticism, which to put it simply, is a part of a larger culture under the umbrella of "Western esotericism". Copernicus wrote in his *De revolutionibus orbium coelestium*:

> "At rest, however, in the middle of everything is the Sun. For, in this most beautiful temple, who would place this lamp in another or better position than that from which it can light up the whole thing at the same time? For, the Sun is not inappropriately called by some people the lantern of the universe, its mind by others, and its ruler by still others. The Thrice Greatest labels it a visible god, and Sophocles' Electra, the all-seeing. Thus indeed, as though seated on a royal throne, the Sun governs the family of planets revolving around it. Moreover, the earth is not deprived of the moon's attendance."

The "Thrice Greatest" refers to Hermes Trimegistus, a legendary figure of Hermeticism who is a combination of the Greek god Hermes and the Egyptian god Thoth. In a collection of writings alluded to him, the *Corpus Hermeticum*, one can find suggestions of heliocentricism such as: "For the Sun is situated in the center of the cosmos, wearing it like a crown." However, it is quite clear that the paragraph quoted above was meant to motivate heliocentrism; this does not necessarily mean that Copernicus *believed* in Hermetic thoughts or that he was secretly a Hermetist.

As we have repeatedly emphasized, astronomical history presented in textbooks are often too simplified. One has to also remember that other cultures too have made contributions to astronomy, and they influenced each other. Science is not an isolated field now, and it certainly wasn't back then. Copernicus was well aware of scientific thoughts and achievements of the astronomers of the Islamicate civilization. His *De Revolutionibus* explicitly cited their works. In particular he cited Al-Battānī (858–929) 23 times (Guessoum 2008). Unlike in Europe, there were active discussions in the Islamicate civilization about different models of the cosmos, beyond that of Ptolemy. There is an argument that Copernicus probably was influenced, at least in part, by the "Maragha revolution", an astronomical tradition that originated in the Maragha observatory established in the middle of the 13th century under the patronage of

Hulegu Khan (the grandson of Genghis Khan) and directed by the polymath Nasīr al-Dīn al-Tūsī (1201–1274). They managed to devise an alternative model to Ptolemaic one, which got rid of the equant and eccentrics, and were also more accurate in computations. Still, the Maragha school did not make the paradigm shift to heliocentrism. For more discussions on the possible connections between Copernicus and his Islamic predecessors, see Ragep (2007).

As mentioned, one of the complaints against the model of Ptolemy was that it is not a *bona fide* uniform circular motion; the motion is only uniform when we take the equant as the center. Yet the Earth is supposed to be the center of the Universe. So there is a "multi-center" problem. Note that Copernicus, in some sense, made this problem *worse*. Now that the Sun is the center of the Universe, why do things fall towards the Earth, and not the Sun? Also, why does the Moon alone orbit the Earth, while everything else orbits the Sun? One could say that the latter question was settled after Galileo observed Jupiter's moons, so that Earth is no longer the "odd one out". But note that this does not resolve the fundamental question regarding the multi-center problem. To resolve this, one needs to recognize that the problem is actually based on the premise that there is a distinguished center that acts as the "natural position" that things should fall towards. A resolution was then achieved by a reasoning that got rid of the need of such a location.

Such an argument was provided by the philosopher Nicole Oresme[7] (1325–1382). To understand his argument, we must first describe the accepted picture of the Earth back then based on the four-element philosophy (see Section 1.2). The Earth is surrounded by layers of atmosphere: the layer of water[8] is closest to Earth, and as the latitude increases, one encounters the layer of air, followed by the layer of fire. The Moon is believed to be located at the boundary of the fire layer, at the edge of outer space. (For an illustration, see Figure 3.1.) Earth occupies the center, the location for which all things tend towards. In *Livre du ciel et du monde* (1377), Oresme imagined a long tube that connects the center of the Earth to the very edge of the last layer of the atmosphere. In this thought experiment, Oreseme filled the tube with fire except for a small amount of air at the very top. Since a heavier body would descend below the lighter body, the air would drop all the way to the center of the Earth. This suggested to Oresme an alternative possibility: air rises or falls not in

[7] He also proved that the harmonic series $1 + 1/2 + 1/3 + 1/4 + \cdots$ diverges. Remarkably, he also knew that the set of odd natural number has the same cardinality (size) as the set $\mathbb{N}$ by establishing a one-to-one correspondence between the sets. The fact that there exist infinities of different cardinalities was later established by Georg Cantor (1845–1918), who suffered from depression, likely made worse by criticism of his work. During his depressed periods, Cantor would avoid mathematics and turn towards contemplating philosophy.

[8] The *firmament* separates the heavenly body of water from the Earth below. Such a worldview probably influenced how people interpreted unknown phenomena seen in the sky. In various Irish annals in the 8th century, there were records of sightings of ships with their crews in the sky. If there's an ocean overhead, it's not too far-fetched to imagine that there could be people traveling in ships in that ocean. The story evolved through the Middle Ages and in one version, an occupant of the ship "swam down" to retrieve an anchor and drown in our air, which is like the bottom of the ocean to them, once captured by the crowd. (In some versions of the legend the sailor was saved in time and swam back up to his ship.)

accordance with the location it is occupying, but rather in accordance with its environment (whether it is surrounded by fire or air). Therefore, he argued, there was no need for any "natural position" that things fall towards to. Each planet can therefore have its own atmosphere, lighter matter surrounding heavier bodies. As there is no light body such as fire and air in between planetary bodies, "up" and "down" are not distinguishable, so other planets do not fall towards us. In other words, Oresme provided an argument for the idea that the cosmos can have multiple "centers" without getting into contradictions.

Pierre Gassendi[9] (1592–1655) also argued that the source of gravity is the Earth itself, and not the distinguished *location* that Earth is occupying.[10] He asked in his *Syntagma philosophicum* (published posthumously in 1658) what would happen if God were to put Earth where the Moon is, and argued that things drop in the vicinity of Earth will fall towards Earth still, and not Earth's originally occupied position:

> "Therefore, when this stone moves, it moves toward the Earth primarily, and toward some 'center' only accidentally."

We have previously mentioned that in addition to the geocentric and heliocentric pictures, there were also hybrid models. For example, the Danish astronomer Tycho Brahe[11] (1546–1601) proposed a model in the early 1580s in which every planet revolves around the Sun *except* the Earth. To be more precise, the Sun, the Moon, and the sphere of fixed stars circle around the Earth, while the five planets revolve around the Sun. Tycho probably saw the advantages of Copernicus model, but was reluctant to give up on geocentrism completely, hence the hybrid model that incorporated features from both. Sometimes it is simply very hard to give up on

[9] Pierre Gassendi envisioned the corpuscular theory of light. In fact, he thought that interactions of different kinds of corpuscles give rise to various phenomena, including gravity. Specifically, the Earth emits corpuscles which hook onto and pull on orbiting bodies. Largely forgotten, Pierre Gassendi made significant contributions to the concept of space and time, and the principle of inertia, which he understood better than Galileo (Pancheri 1978). We will come back to this later.

[10] It should be emphasized to the modern readers that the mainstream viewpoint was that everything massive contains a property called *gravitas*, and would seek its natural position. Earth being such a body, naturally sits with that distinguished location at its center. So things were not attracted to the Earth, but simply sought their natural positions. There is therefore, traditionally, no such concept as a gravitational *force*. Similarly the word "mass" understood as a modern concept of gravitational mass did not appear until much later.

[11] Tycho had an astronomy observatory called Uraniborg, built on the island Hven, which was part of Denmark at that time (now belongs to Sweden) under the rule of Frederick II of Denmark. The island was granted by the king to Tycho in 1576, "to have, enjoy, use and hold quit and free, without any rent, all the days of his life, and as long as he lives and likes to continue and follow his *studia mathematices* ...." However, Tycho Brahe fell out of favor of Frederick II's successor Christian IV. Losing support, he decided to leave the island in 1597 and subsequently took up a position in Prague, under the rule of Holy Roman Emperor Rudolf II, who we will mention again later. It is interesting to mention that Tycho was once offered a rectorship of a university in 1577 but he turned it down. He seemed to think that the atmosphere in a university was not suitable for conducting research of the sort he was pursuing. He even characterized professors as "shadow chasers" and "those engaged in idle formalities and empty processions of words." (1576 December 8, Tycho Brahe, *Opera Omnia*; cited in Westman (1980), pp. 122–123.)

old idea that had taken roots over thousands of years.[12] Tycho made very precise observations of the night sky, and the data would eventually help Johannes Kepler (1571–1630) to formulate his laws of planetary motion. We will discuss these later on. For now, let us mention how astronomical events like comets and supernovae can change our models of the cosmos.

In all the models described thus far, the universe is finite, with a sphere of fixed stars that marks its boundary (this is not always the belief shared across various cultures, of course, we are being Euro-centric at the moment). At the time of Tycho Brahe, a new phenomenon appeared in the night sky of the year 1572: a new, bright star that no one had ever seen. Today we know that this was the explosion of a dead star, a supernova. Now referred to as SN 1572, the event presented a puzzle to Tycho. After all, if the stars were fixed, how can there be any new changes?[13] Tycho made a detailed study of the supernova in his *De nova et nullius aevi memoria prius visa stella* ("Concerning the Star, new and never before seen in the life or memory of anyone"), as did many other astronomers of the time[14]. They concluded that the "star" was truly a celestial event located at least further than the Moon, instead of say, an atmospheric phenomenon. Indeed, meteors were thought to be an atmospheric phenomenon much like lightning, and reports of meteorites fallen from the sky were often dismissed even centuries later, by eminent scientists like Antoine Lavoisier (1743–1794), since "clearly there is no stone in the sky". Thus we see that preconceived notion based on the prevalent world view can affect a scientist's judgment on what is possible or not, and being "open-minded" is not so easy. (On the other hand, recent study has indicated that the ancient Egyptians already understood that meteorites came from outer space (Almansa-Villatoro 2020), and they even made good use of the iron-rich meteorites. In *Shiji* (《史记》), or "Records of the Grand Historia" written by Sima Qian (c. 145–c. 86 BCE) of China, meteorites were explained as a "stone resulted from a fallen star"—"星坠至地，则石也".) After SN 1572, in a short few years, there came another astronomical event: the Great Comet of 1577. Tycho again meticulously observed and studied the comet as it passed through the night sky. He realized that to account for the motion, the comet had to *literally* pass through the celestial spheres of some planets, as though they weren't there. In other words, the celestial spheres could not have been solid. It was then just a matter of time before the notion of celestial spheres as a

[12] Even in more recent history, radical notions like wavefunction in quantum mechanics is hard to fathom: what does it mean that a particle only exists probabilistically? There are many different interpretations. An example being the "pilot wave" (Bohmian mechanics), with some kind of wave guiding a classical, objectively real particle, which is also a hybrid view that incorporates both the wave picture and the particle picture. This does not necessarily mean that it is "wrong", merely that some people would find it philosophically unpleasant.

[13] New stars appearing may not be that unacceptable to the Church, after all there was supposed to be "the Star of Bethlehem" that guided the magi to the birthplace of Jesus. In Aristotelian worldview, however, the heaven is fixed and unchanged.

[14] Prior to SN 1572, there was another supernova event observed in 1054 in the constellation Taurus. SN 1054, also known as the Crab Nebula Supernova, was also recorded by Chinese, Japanese, and Arabic astronomers. The event seemed to have escaped written records in Europe although no doubt some would have noticed it (Stephenson & Green 2003).

device for heavenly bodies to attach to became obsolete. In fact, this explains why Tycho didn't bother with intersecting orbits in his hybrid model.[15]

This led to another question. Now that the celestial spheres are no longer needed, what happened to the fixed stars? Without a physical sphere to serve as a boundary, maybe not all stars are situated at the same distance to Earth, and that means there could be stars that are simply too far away for us to see. Maybe there are stars all the way out into an *infinite* world! Suddenly the cosmos became potentially much bigger. Whether the Universe is finite or infinite is one of the great debates in the history of science, but even today we do not know the answer. We will discuss this debate further in Chapter 3. Here it is worth mentioning that Thomas Digges (1546–1595) was one of the first astronomers to contemplate the idea of an infinite universe, extending the stars far beyond into space in the Copernican picture. After the death of his father, Digges was taken care of by John Dee (1527–1608/1609), who was not only an astronomer but an occultist and alchemist, who at some point made attempts to communicate with the angels. Together with his friend the alchemist Edward Kelley (1555–1597), they developed a new system of Renaissance magic in a mysterious script ("Enochian") that was said to have an angelic origin. Versions of Enochian magic are still being practiced by some occultists today. John Dee was also an advisor to Queen Elizabeth I of England, being her court "conjurer" and astrologer (there was a suspicion that he also served as her spy in foreign nations). Dee also helped to provide navigation techniques with his mathematical skill, enabling the various voyages that England took on in the 16th century. The history of science is intertwined with that of magic (and politics).

2.2 Case Study: Kepler's Models

In this section we will take a look at Kepler's attempts to understand planetary orbits. Most students of physics and astronomy will have heard of Kepler's three laws of planetary motion. However, this is not our main focus, precisely because it is already well-known. What we want to focus on is the paths that Kepler took but then *abandoned* prior to his eventual success. Often, we only teach the final successful models and theories to students, which carries with it the risk that students might end up with the wrong impression that scientific discoveries are only meant for people who have genius insights. This is far from the truth. By telling the stories of failure, we can teach students that struggles are real and commonplace in science, even for "great scientists". More important than that, we can see and appreciate the hallmark of a good scientist—the willingness to *give up* on bad models when they are not supported by real data from observations or experiments.

Kepler first noted that there is a pattern to Jupiter–Saturn conjunction, i.e., when the two planets appear closest together in the sky. This event, also called the "Great Conjunction", happens every 19.6 years or so. Since most planets are confined to the plane of the solar system, as viewed from the Earth, they move on a

[15] The Ptolemiac model also has intersecting orbits due to the various epicycles, but they didn't worry about it for a different reason. As we mentioned, it was viewed only as a mathematical device to allow computations of the locations of the celestial bodies, not as a realistic model of what the cosmos is like.

particular curve called the *ecliptic*, which is also the path the Sun traces out. The constellations it crosses are called the Zodiacs. The ecliptic is essentially a great circle in the sky. Suppose we mark the position the Jupiter–Saturn conjunction that occurred in a certain year, and do so again in 19.6 years. These two points would be roughly 117 degrees apart in the Zodiac constellations. If they were exactly 120 degrees apart, they would evenly spread along the ecliptic and form an equilateral triangle. However, because they are not exactly 120 degrees, the fourth time the conjunction happens, the position would be close to the first event, but not exactly at the same position. Similarly the fifth event would be close to the second event but not exactly at the same place. If we join the lines from the first event to the second event and so on, we would obtain a diagram that looks like a drifting triangle (not really a triangle of course, since the diagram does not close). See Figure 2.2.

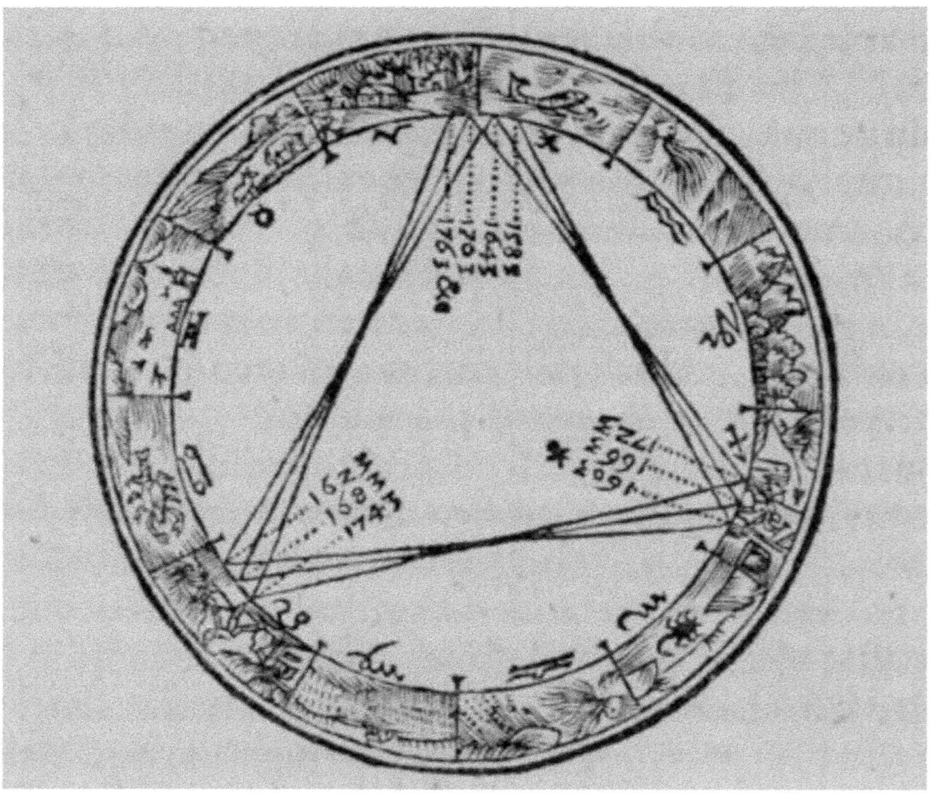

Figure 2.2. A diagram of successive Great Conjunctions, which occur almost at 120 degrees apart in the Zodiac. Because the triangle formed by three consecutive conjunctions does not close, it drifts slowly across the Zodiac. Kepler discussed at length in his *De stella nova* the possible *astrological* interpretations of the transition of the conjunction from the "Watery Trigon" at the top (which contains the constellations of Cancer, Scorpio, and Pisces) to the "Fiery Trigon" (which contains Aries, Leo, Sagittarius) in 1603. *Credit:* Reproduced from (Kepler, 1606) CC BY 4.0. Courtesy of The Linda Hall Library of Science, Engineering & Technology.

If you are a scientist back in the day,[16] you would no doubt ask *why*, if God created the world, he did not make this conjunction perfect so that their positions on the sky match that of an equilateral triangle? Perhaps there is a hidden message? Perhaps thinking along this line, Kepler noticed that once we joined enough lines following the conjunction points that drift along the ecliptic, this construction will inscribe a circle whose radius is half of the original (the ecliptic); see Figure 2.3. The "crazy" idea of Kepler was that these correspond to the orbits of Saturn and Jupiter, since Jupiter's orbit is about half the size of Saturn.

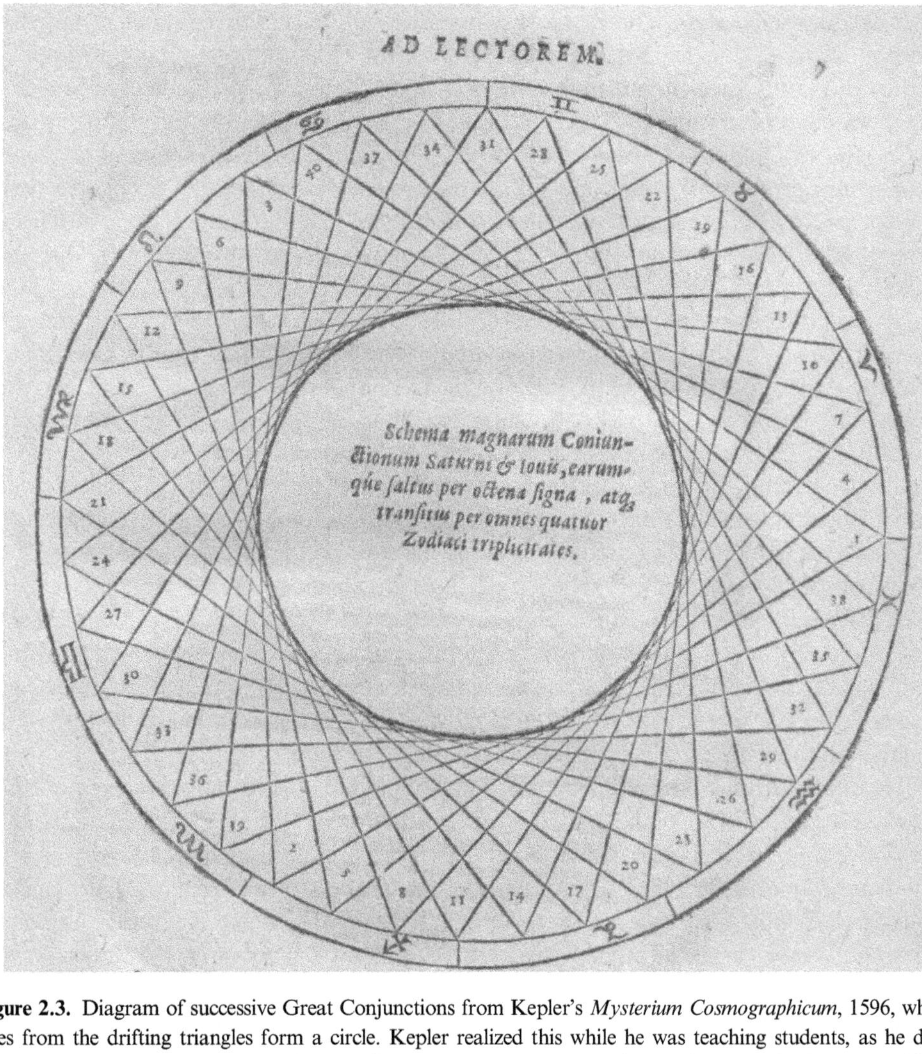

Figure 2.3. Diagram of successive Great Conjunctions from Kepler's *Mysterium Cosmographicum*, 1596, whose lines from the drifting triangles form a circle. Kepler realized this while he was teaching students, as he drew numerous inscribed triangles in a circle. Sometimes research ideas came during teaching! *Credit:* Reproduced from (Kepler, 1596) CC BY 4.0. Courtesy of The Linda Hall Library of Science, Engineering & Technology.

[16] The concept of a scientist did not exist back then, but let us stop being pedantic about the term from now on.

Unfortunately this does not really fit the actual orbits well enough. A good scientist typically does not give up on one idea easily, so the next step is usually to try to save the model. Kepler did this by considering three dimensions instead of two. It is true that orbital motion is effectively a two-dimensional one, but space in the real world is three-dimensional, so maybe there is a constraint that comes from that? To consider three-dimensions means instead of two concentric circles, we need to consider two concentric spheres. In the case of circles, the outer circle circumscribes a triangle, which in turn inscribes the inner circle. For the spheres, Kepler replaced the triangle with a regular tetrahedron. Once tetrahedron is involved, it is quite natural to think of the other platonic solids. There are only five after all: tetrahedron, cube, octahedron, dodecahedron, and icosahedron. Their faces consist of identical regular polygons that meet at the same three-dimensional angles. Kepler then noticed that there are also six planets in the solar system (Mercury, Venus, Earth, Mars, Jupiter, and Saturn; planets beyond Saturn were not known at the time). Thus, if we consider layers of concentric spheres supported by the platonic solids (Figure 2.4), perhaps the six spheres would correspond to the planetary orbits! In this model, Jupiter and Saturn no longer correspond to the spheres supported by the tetrahedron, but rather by the cube (namely, the inner sphere inside the cube corresponds to Jupiter's orbit, while the sphere outside it corresponds to the orbit of Saturn). There is one major advantage of this proposal over the initial two-dimensional one: it could potentially explain why there are exactly six planets. In addition, it is philosophically rather neat since the ancient Greeks have proposed to identify the platonic solids with the "fundamental elements" of the universe: earth (cube), air (octahedron), water (icosahedron), fire (tetrahedron), and the ether (dodecahedron). The first four are the classical elements known since the time of Plato, while the last one was added by Aristotle. The ether in this context is what the gods used to "arrange the stars" and not to be confused with the —now obsolete—concept of æther used later in physics. (Hence my choice for a different word "ether".) Alas, the planetary orbits do not completely match observational data either even in this three-dimensional construction. Kepler seemed to already know this before he published his *Mysterium Cosmographicum*, but he thought the idea was beautiful and thus worth publishing anyway. In fact he continued to be fond of this model even after his discovery of planetary orbits being elliptical.

This rather curious piece of work, despite its failure, is historically quite important.[17] Prior to Kepler's efforts, planetary orbits are pretty much "given" by observations; there was nothing to be explained. What Kepler pioneered is the idea that maybe *there are underlying physical laws* that determine how the planets move, that maybe planetary motions are not so "free". There are also valuable lessons from *Mysterium Cosmographicum*, especially concerning the number of planets in the solar system. Nowadays we know that there are other planetary bodies beyond Saturn. Not only that, we also know that there is *no* physical law that dictates the number of planets. On the

[17] In this work, Kepler also noted that the Sun is not exactly the center of the Universe in heliocentrism. Instead that center is rather nearby the Sun. In his words, "circa Solem esse centrum mundi". This is partially correct; the barycenter of the solar system constantly changes depending on where the planets are. It can sometimes be close to the center of the Sun, but sometimes slightly *outside* the Sun's surface.

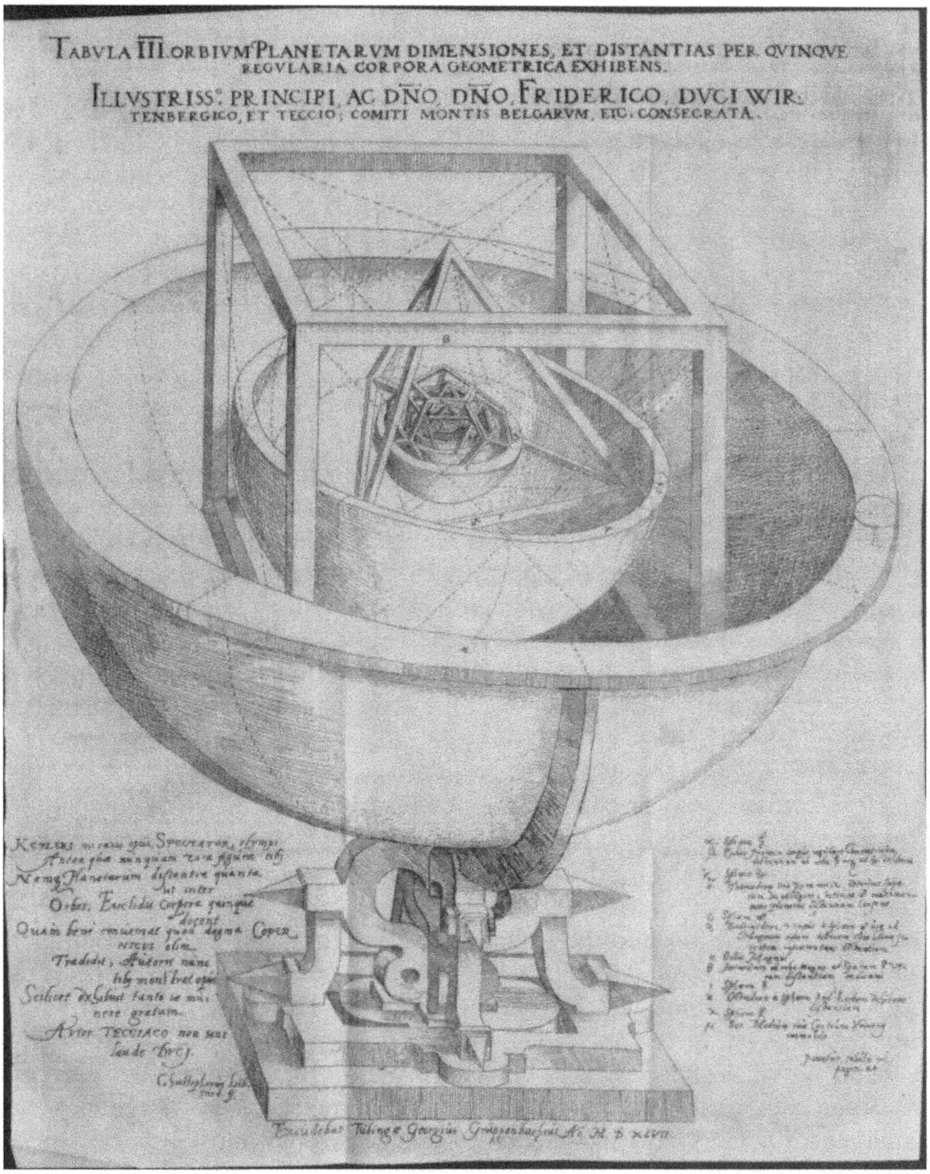

Figure 2.4. Diagram of the nested Platonic polyhedra model of the solar system from Kepler's *Mysterium Cosmographicum*, 1596. *Credit:* Reproduced from (Kepler, 1596) CC BY 4.0. Courtesy of The Linda Hall Library of Science, Engineering & Technology.

contrary, how many planets a star system can have is determined by the initial conditions, such as how the gas clouds around a star is distributed, the size and mass of the planets formed, etc. This does not mean that the number of planets was not a sound scientific question to pursue. We can only know from hindsight if a given problem is a "non-issue". Take for example the cosmological constant, which we will discuss in a later chapter on cosmology. There have been many attempts to explain why its value is so small compared to what one expects from quantum field theory, if it is indeed the energy of "empty space". Yet, it is also possible that its value is just a boundary condition or initial condition of the Universe, and there is nothing much to explain. But because we do not know for sure, research in this direction is still important. It is a haunting thought that one may spend an entire career pursuing a question that eventually turns out to be a non-issue, but that is how science progresses (and why should we worry that we are "wasting our life"? Is it any different from, say, failing a business?). Also related is the issue: when does one give up on a research project? When observational and experimental data are available, it is relatively easy to rule out certain models, as Kepler was able to. But what if you have a theory that makes certain predictions that are never observed in Nature? Maybe the theory is only slightly wrong, and modifying it a bit might lead to success. Yet maybe the whole idea is completely off and no amount of modification can save the theory. When, then, should a good scientist abandon the theory? A good example of this in modern theoretical physics is the idea of "supersymmetry" in particle physics, which predicts a whole suite of "super-partners" to the usual particles. The simplest models were already ruled out since no such particle has ever been observed in our particle colliders. But does this mean that the idea of supersymmetry is wrong? It is difficult to say. In such a situation, whether to continue working in supersymmetry becomes a question of personal taste for the scientists as an individual.

We also need to address the seemingly strange notion of trying to fit planetary orbits using platonic solids. However, this is actually not that unusual. As S. Sternberg once stated,

> "It is customary to sneer at Kepler for this. It is instructive to compare this with the current attempts to 'explain' the zoology of elementary particles in terms of irreducible representations of Lie groups."

While we will not get into the technicalities of particle physics or group theory, the main point is that modern theoretical physics (not just particle physics) often appeals to various symmetries. These symmetries are often much more abstract than that of Kepler's, but the underlying philosophy of applying mathematical structures to explain physical phenomena is the same. We must be careful, however, since Nature does not necessarily conform to our preferred choice of symmetry. Since antiquity astronomers had insisted that planetary orbits must be circular. This obsession is due to the belief that a circle exhibits perfect symmetry (which in turn reflects divine creation), yet this turned out not to be the case.

For completeness, we should mention the eventual success of Kepler by listing his laws of planetary motion, which he discovered by analyzing the detailed data collected by Tycho Brahe (the interactions between the two men were complicated, and together with the stories of other astronomers and philosophers of the time in the court of the Holy Roman Emperor Rudolf II, there are probably enough material to make a TV series,[18] but we will not go into those here). This would not have happened without

[18] Especially fierce was the rivalry between Tycho and Nicholas Reymers Baer (1551–1600), also known as Ursus. Tycho complained that the latter had plagiarized his works, including the Tychonic model for the solar system.

Mysterium Cosmographicum, as Tycho no doubt was impressed by Kepler's ability via this prior work, and thereby enlisting his assistance.

Rudolf II (1552–1612) was himself a very interesting character. He was more interested in art, culture, and the occult, than being an emperor. He would often retreat into his "Kunstkammer" (art chamber), which housed a collection of treasures and rare items collected from around the world (i.e., his "cabinets of curiosity"), instead of taking care of governance. Despite being the Holy Roman Emperor of the Catholic faith, he was tolerant of Protestantism and other religions such as Judaism, though he found it hard to keep the tension at bay. Being a curious Renaissance man, Rudolf II even sought to read books that were deemed heretical and banned by the Roman Catholic Church. He provided strong support for the learned men in arts,[19] astrology, alchemy, and natural philosophy. John Dee, whom we already introduced, was one of the famous visitors to his courts. The environment in Prague under Rudolf II was conducive to the development of science and arts. Among the many alchemists in Rudolf II's court was Michael Sendivogius (1566–1636), who discovered that air is not a single substance but a mixture, with one of the substances being necessary for life (oxygen; Severinghaus 2016). Nevertheless, Rudolf II's "hands-off" governance[20] and his interests in the occult were frowned upon by many, including the Catholic Church, who eventually supported his brother Matthias to usurp the throne in 1611. Rudolf II would pass away just a year later in 1612. The tension between the various faiths finally erupted into the Thirty Years' War under Ferdinand II (who succeeded Matthias after the death of the latter in 1619), who unlike Rudolf II, tried to force Catholicism across his domains.

Rudolf II's contribution to the development of astronomy cannot be overstated. He didn't just provide an open atmosphere conducive for scientific research and free discussions, but his position as the Holy Roman Emperor meant that he could shelter his astronomers from the criticism of the Catholic Church, and they could pursue their research without being afraid of being accused of heresy. Kepler found the first two of his laws when he was in Prague in 1609. Kepler had to leave Prague after Rudolf II's death since he was a Lutheran by faith. Shortly after Kepler had to defend his mother Katrina, who in 1615 was accused of being a witch in Leonberg.[21]

[19] The Italian painter Giuseppe Arcimboldo would paint the famous *Vertumnus* in 1591, which consists of multiple fruits, vegetables and flowers that come together to create a portrait of Holy Roman Emperor Rudolf II. See Figure 2.5.

[20] Readers who are familiar with Taoism may be reminded of *wuwei* (无为), the philosophical idea that the best governance is achieved by doing nothing. Perhaps instead of escaping his duty and being preoccupied with the occult, he was simply delicately balancing the complicated politics to maintain harmony as much as he could—sometimes it may be better not to do too much.

[21] Though not the main reason for the accusation, some historians have speculated that one of the contributing factors may have been a novel entitled *Somnium* ("Dream") that Kepler had written in 1608 (but not formally published until much later after his mother's ordeal). The story is what we today would call science fiction; it was about a trip to the Moon, and Kepler imagined what astronomy would be like if practiced on the Moon. The fictional narrator in the story (who worked with Tycho!) has a mother who practices magic and communicates with *daemons* (not to be confused with demons) to learn the secrets of the Universe.

Figure 2.5. Emperor Rudolf II as Vertumnus, the Roman god of the seasons, growth, plants, and fruit. The painting is by Giuseppe Arcimboldo (1527–1593), who served as the court painter for not only Rudolf II but also Ferdinand I and Maximilian II. It might be surprising that such a "portrait" of an emperor was from the 16th century, given that its style was similar to that of the modern surrealism. On the other hand, at that time the style of mannerism (itself a rather vague and hard-to-define concept, but in general less concerned about balanced perfection) was popular in Europe, and artists were more open to challenging the traditional styles of the Renaissance. As with the arts, natural philosophy was changing, with Kepler in the midst of the shift. Given Rudolf II's fascination with the occult and mysticism, it was not totally unexpected that he would appreciate such a portraiture of himself. Another interesting work of Arcimboldo is "The Librarian", in which the librarian was composed of books. Arcimboldo also had a self-portrait in which his features were rendered as stacks of strips of papers. *Credit:* Arcimboldo, Archimboldo, Giuseppe, Skokloster Castle/SHM (PDM)

It took 6 years for this accusation to be dropped. Kepler's third law was only published in 1619. The laws are:

(1) *First law:* Planets move in elliptical orbits with the Sun as one of the foci of the ellipse.
(2) *Second law:* As a planet moves along its orbit, the area that the line segment joining the planet and the Sun traces out is the same in a fixed duration of time.
(3) *Third law:* The square of the orbital period of a planet is proportional to the cube of the semimajor axis of the elliptical orbit. (In modern form: $T^2 = (4\pi^2/GM)a^3$, where M is the mass of the Sun.)

Planet Mars played a major role in helping Kepler to pin down the correct shape of planetary orbits. His method is ingenious as well. *If* Mars is fixed in space instead of orbiting the Sun, we could draw a line that connects Mars to Earth, as well as another line that connects the Sun and the Earth. As the Earth orbits the Sun, the intersection of these two lines would trace out the orbit of the Earth. Mars is not fixed in space, of course. But we could choose one specific night and record the position of Mars in the night sky. We draw the two lines as explained above. Then we wait until Mars returns to the same position again and repeat. As viewed by observers on Earth, Mars returns to the same position on the sky every 687 days or so. After a few times this would give us some discrete points of Earth orbit. By continuity, we know that if Mars appears in location X on the night we did our measurement, then it would have moved only a little bit along its orbit the next day, and so we can complete the orbit of the Earth this way. Once the shape of the Earth's orbit has been fixed, we can figure out the shape of the orbit of Mars. Again, choose a few nights that are 687 days apart, so that Mars appears to be at the same location as viewed from Earth, although Earth itself is not on the same point along its own orbit. This means we can "triangulate" the position of Mars by the intersection of the lines drawn between Mars and the various Earth's location. As Mars moves along the night sky, this point tracks its locus of motion and we can finally draw its orbit. It should be clear that these require a lot of astronomical data for the planet positions. Kepler was lucky that he had access to Tycho Brahe's detailed data after the death of the latter,[22] as well as ancient data that dated back to the Babylonian. Mars was important because it deviates the most from circular orbit, just after Mercury (which can only be observed near the Sun and the horizon, which means no good data available), and Tycho's accurate data were enough to make this deviation clear. As Kepler himself wrote in his *Astronomia nova* (as cited in Evans 1998):

> "I therefore once again think it to have happened by divine arrangement, that I arrived at the same time in which he[23] was intent upon Mars,

[22] Tycho was over-protective of his data and Kepler was only able to have full access after his death in 1601. Even then, there was a long dispute with Tycho's relatives on the rights to the data.

[23] "He" refers to Tycho Brahe's chief assistant, Christian Longomontanus (1562–1647), who was working on the orbit of Mars at that time. This problem was relegated to Kepler upon his arrival.

> whose motions provide the only possible access to the hidden secrets of astronomy, without which we would remain forever ignorant of those secrets."

Note that these laws were, at the time, purely phenomenological (or empirical). Kepler did not know how to explain why these laws are valid, only that there are laws. In science, we can always keep on asking whether there is a deeper, more fundamental explanation, which at some point we have to stop due to our limit of understanding. It should be emphasized that for Kepler, discovering these laws was already a remarkable achievement. Nowadays we have mathematical tools that allow an undergraduate student to analyze an elliptical orbit and to do some simple computation exercises. Every mathematically trained student should know that an ellipse is described by the equation $x^2/a^2 + y^2/b^2 = 1$, right? How hard can it be to fit observational data on the curve? Yet Kepler did not know; he didn't even know it was an ellipse and tried to fit an oval first. In fact, the idea that we can use algebraic equations to analyze geometrical figures—the analytical geometry of Descartes—was still a generation away; at the time of Kepler, even the equal sign "=" taken for granted today had only been invented one generation ago (first used by Robert Recorde (c. 1510–1558) in 1557)! Kepler did not even have logarithms (invented later in 1614 by John Napier (1550–1617)), which would have made computations easier. Things always seem easy once accomplished, but every generation of scientists have their own challenges to face and overcome.

Kepler can be considered to be the first *astrophysicist*. His laws were published in the book *Astronomia nova*, the full title of which is translated as "New Astronomy, Based upon Causes, or Celestial Physics, Treated by Means of Commentaries on the Motions of the Star Mars, from the Observations of Tycho Brahe, Gent." Note the terms "celestial physics". The work completely transformed the field of astronomy. Following the ancient Greeks, before Kepler astronomy was viewed as more related to mathematics than physics, and the role of an astronomer was merely to calculate positions of planetary bodies for various applications (calendar, astrology etc.). After Kepler, astronomers increasingly concerned themselves with explaining the actual physics of the planetary motions. It is this fundamental shift in involving physics in astronomy that would lead to subsequent revolutions in our understanding of the cosmos. Today we take for granted that astronomy is a branch of physics, how could it have been otherwise? Yet it *was* otherwise. Kepler had to argue for the role of the astronomer to be more than calculating the positions of the planets. He wrote (Westman 1980):

> "[…] the astronomer ought not to be excluded from the community of philosophers who inquire into the nature of things. He who predicts as accurately as possible the movements and positions of the stars performs well the duty of the astronomer. But he who employs true opinions about the form of the universe does it better, and is held worthy of greater praise. The former, to be sure, draws conclusions that are true as far as what is observed is concerned; the latter not only does justice in his

conclusions to what is seen, but also, as I explained above, in order to draw his conclusions embraces the innermost form of nature."

This change of the role of an astronomer is reminiscent of how the painters, sculptors, and architects, once considered mere craftsmen of menial trades, gradually transitioned into one of the "liberal arts" during the Renaissance. Artists in the modern sense, who were sought out by patrons based on their individual styles and reputations, only became commonplace by the 17th century, the same era that Kepler was pushing for the role of the astronomers to involve more that mere computations.

Thus we see that how we characterize branches of knowledge can in fact, in turn, influence how we see the world, and might even impede our understanding of how the world actually works. This is why even nowadays we have to pay attention to other fields, especially those that are closely related to our own field of expertise.

In 1610, Kepler wrote to Galileo about his vision for the future of the human species (Rosen 1965):

"But as soon as somebody demonstrates the art of flying, settlers from our species of man will not be lacking. Who would once have thought

Figure 2.6. Image of Jupiter and Io, one its four Galilean moons, photographed by NASA's Cassini spacecraft on Dec. 1st, 2000. The "Great Red Spot", first observed in 1831, is a high-pressure region in the atmosphere of Jupiter, i.e., a long-lasting storm. It produced wind that can blow up to 432 km/h. Galileo would be amazed at how far we have come in terms of technology, to be able to take high resolution pictures of the planet using multiple wavelengths of light, and even study the meteorological phenomena in the planet's atmosphere, not to mention sending space probes there. *Credit:* NASA/JPL/University of Arizona.

> that the crossing of the wide ocean was calmer and safer than of the narrow Adriatic Sea, Baltic Sea, or English Channel? Given ships or sails adapted to the breezes of heaven, there will be those who will not shrink from even that vast expanse. Therefore, for the sake of those who, as it were, will presently be on hand to attempt this voyage, let us establish the astronomy, Galileo, you of Jupiter, and me of the Moon."

The "art of flying" was demonstrated by the Wright brothers, Orville and Wilbur, in 1903, three centuries after Kepler's letter. Landing on the Moon became reality in 1969 with Apollo 11, and Pioneer 10 became the first space probe to fly by Jupiter in 1973, while the Galileo spacecraft was the first to orbit Jupiter (from 1995–2003). ESA's Jupiter Icy Moons Explorer (JUICE) is on its way to Jupiter at the time of writing, scheduled to arrive in 2031 to study Jupiter and its Galilean moons: Ganymede, Callisto, and Europa are known to harbor underground oceans, while Io is volcanically active. Jupiter and Io are shown in Figure 2.6. Some 400 years after Kepler and Galileo, we are finally exploring the solar system.

2.3 Our Changed View of the Cosmos

2.3.1 Galileo: Casting Doubts on a Perfect Heaven

Kepler's achievement is remarkable. However, there is only so far we can go relying on naked-eye observations. The next revolution in the history of astronomy is the use of telescopes to study heavenly bodies. The views that a few simple glass lenses provided revealed so much about the cosmos that it changed our world view. Galileo Galilei (1564–1642) made various discoveries with his telescope, published later in his famous 1610 work "Sidereal Messenger" (*Sidereus Nuncius*). For example, he realized that there are many more stars in the sky, they were simply too faint to be seen with the naked eyes. He observed that the planet Venus has phases just like the Moon, although to an observer on Earth its position is always near the Sun.[24] Geometrically this is best explained by a heliocentric model, rather than a Ptolemaic one. Thus Galileo's observation helped to establish the heliocentric view, with a certain caveat we will soon mention.

Galileo also observed that there are a few "stars" aligned in a line at the side of Jupiter. These "stars" were invisible to the naked eyes, and their numbers and positions changed as Galileo observed them night after night. If you have never observed this, find a telescope and do so. It is a remarkable sight! It soon occurred to him that these objects are *orbiting* Jupiter. Galileo had discovered the first natural satellites of other planets. Today we call these the "Galilean moons" of Jupiter; they

[24] This means that it appears either at sunrise or sunset. In many cultures, two different names were therefore given to Venus, even if they were aware that it was the same planet. For example, in ancient China the planet was referred to as "Qi-ming" (启明) at dawn, and "Chang-geng" (长庚) at dusk, meaning the beginning of day and (the coming of) a long night, respectively. Similarly, in old English, Venus was known as *morgensteorra* (morning star) and *æfensteorra* (evening star).

are the four largest moons of the planet: Io,[25] Europa,[26] Ganymede, and Callisto. As these moons revolve around Jupiter, some of them are sometimes behind the planet as observed from Earth, thus explaining why their number is changing. This was a confirmation that not everything revolves around the Earth. However, it should be emphasized that this, by itself, did not necessitate the validity of heliocentrism, *only* that the simplest geocentric model was wrong. For example, this did not rule out hybrid models like Tycho's model, which at the time of Galileo, was the one widely accepted by the Church (not the old geocentric model). In fact, the phases of Venus can also be accountable for in Tycho's model.

Another great discovery concerns our own Moon. Telescope revealed to Galileo that the Moon's surface is not smooth; it has "blemishes" consisting of mountains and craters like that once imagined by the philosopher Plutarch (c. AD 46–c. AD 119), whose shadows changed according to the position of the Sun (this also allowed him to estimate the height of the mountains with trigonometry). This may not seem like a big deal to many modern readers, and indeed Venus phases and Jupiter moons seem so much more profound. However, this is because in the modern age most people who have some astronomy education were only exposed to "geocentrism versus heliocentrism", and in such a clear dichotomy it would indeed be the case that both the phases of Venus and Galilean moons should have shifted people's world view to a heliocentric one. However, this was *not* our history. It wasn't just "simple geocentrism" versus heliocentrism, but also Tycho's model. Since the Church has accepted Tycho's model then, these observations couldn't have changed anyone's mind to suddenly abandon Tycho's model. Instead, the impact of Galileo's discoveries was more subtle but arguably more important—he had to change the mainstream *philosophy*, and this was achieved by what he found out about our seemingly humble Moon[27].

Galileo began his lunar observation in 1609; by and large the common belief at that time was that the physical laws on Earth are different from that in the Heaven. More specifically, the sublunar realm (space below that of the Moon) is governed by the physical laws we are used to in everyday life. Things can change and even be chaotic. Beyond that, in the planetary realm beyond the Moon ("supralunar"), the cosmos is pristine and perfect. There is no reason why the same physical laws should hold in the celestial realm.[28] Galileo's discovery showed that the Moon's feature is

[25] It is worth mentioning that Io is the most geographically active body of the solar system, with erupting volcanoes raging on its surface. This is due to the tidal force of Jupiter acting on its crust and upper mantle.

[26] Current planetary science suggests that Europa has a large body of ocean under its icy surface, with water content perhaps twice than that of Earth's. This is impressive given that the diameter of Europa is only about a quarter of Earth's. The existence of the ocean is inferred by the presence of a magnetic field, induced by the interaction with Jupiter's magnetic field as the moon orbits it.

[27] Sometimes I also wonder if we subconsciously feel that discoveries concerning far away objects must be more important than that of a closer one, as if anyone squinting hard enough could have seen the rough lunar features!

[28] The polymath Giambattista della Porta (1535–1615) wrote a book which today we would call "popular science", *Magia Naturalis* in 1558, which covered various topics. He believed that thunder in the sky shared the same mechanism with earthquakes. However, regardless of whether the claim is wrong, note that the *sky* is still sublunar. Thus this view still conformed to the traditional dichotomy that the celestial realms are different from the sublunar one. There *were* some thinkers who considered otherwise, but only Galileo's observational evidence finally established this universality of physical laws.

not so different than the Earth, and therefore started to shake this notion. If the Moon looks so similar to Earth, would the same physical laws apply there? If so, can it apply elsewhere, even beyond the Moon? *This* was the big change in the view of the cosmos—the realization that physical laws could be universal, and that we do not occupy a special place in the Universe. One often reads from popular science books that people used to believe we occupied the central stage in the cosmos, as if our world is somehow privileged and special in a good way. This was not completely true. It was considered a special place, but not in a good way! The common view was instead that our world was in a chaotic mess of inexorable change, unlike the clean orderly heavenly realm of permanence. Galileo's findings therefore brought heavens down to our level—the whole cosmos is a gutter![29]

We should note that even this discovery was rather quickly accepted by the Church authority with no issue. What they had issue with was Galileo's insistence that heliocentrism is correct. As we have mentioned, the Church was against the idea that mere humans can claim to know "the truth"; they could accept treating alternative theories and ideas as "mere conjectures". Modern students of science often have the impression that the Church was stubborn and did not even look through Galileo's telescope. As usual, history is complicated. This is not the entire story. Indeed, as we have seen, based on these observations alone it was very difficult to make the case for heliocentrism. Heliocentrism was *not* competing against the straw man version of the simplest geocentrism, but the more sophisticated Tycho's model. Therefore, in the absence of precise observational data (like the one analyzed by Kepler), the debate couldn't be settled that easily. It was not that the Church did not bother with examining the physical evidence, but that there simply wasn't strong enough evidence to look at. In fact, in his book "Dialogue Concerning the Two Chief World Systems" (*Dialogo sopra i due massimi sistemi del mondo*) published in 1632, Galileo really did not have solid arguments for heliocentrism. Some arguments were in fact incorrect. For example, he argued that tides were produced entirely by the earth's motion (he also presented this idea in an essay letter "Discourse on the Tides" or *Discorso Sul Flusso E Il Reflusso Del Mare*) in 1616 to the cardinal Alessandro Orsini. Galileo presented this argument precisely because he was good enough a scientist to realize that all astronomical discoveries he made can be equally explained by Tycho's model, and he needed something extra to distinguish between Tycho's model and heliocentrism; or in modern technical jargon, something "to break the degeneracy" of the data.

Concerning telescopes, we should keep in mind that Galileo's telescope was rather crude; nowadays we have binoculars that can see sharper and clearer images than his device. In other words, the images that he saw were not the kind of images we are used to seeing nowadays (and even today, students who first use a small telescope to look at the planets may be disappointed at how small the image is as they expect something like NASA images). Thus, it is natural for his contemporaries—some who *did* look through the device—to doubt how much they could *trust* the

[29] "We are all in the gutter, but some of us are looking at the stars."—Oscar Wilde

contraption. Different observers often didn't agree with what they were seeing, especially at night. Also note that these contemporaries were not just "church people", but also fellow mathematicians and philosophers. Among them was Cesare Cremonini (1550–1631), whom was said to have refused even to look through the telescope and insisted that Aristotle had definitely proved the Moon could only be a perfect sphere. The truth was more complicated, and he *did* look through the telescope.

In addition to deformed images created by possible defective lenses (not to mention spherical and chromatic aberrations), could the lenses themselves have *created* some features that were not there, perhaps influenced by the thick atmosphere or other means, even if the images were faithful when used to observe terrestrial objects like a faraway ship? Unfamiliar with the physics of optics, people were naturally skeptical when any observation was *not direct*, but *through* lenses. Consider the *camera obscura*, which is a darkened room or box with a small hole at one side through which an image can be projected onto the opposite wall. This phenomenon has been well-known since ancient times (it was described by Greek philosopher Aristotle (384–322 BCE) and Chinese philosopher Mozi (墨子) in 400 BCE; more on him in Chapter 8). The image produced is upside-down. If light rays through a hole can produce such a drastic effect, what is to say that lenses would not create even stranger effects? It is interesting to note that philosopher Margaret Cavendish (1623–1673) also raised similar skeptism regarding microscopes. Margaret and her husband acquired some microscopes themselves to experiment with, so she was not criticizing without ever looking through the device; the same was true for many who doubted Galileo's conclusions.

Not everyone was a doubter, of course. Kepler, a contemporary of Galileo, supported him. Yet their relationship was also a complicated one. They did not completely agree with each other on many issues: Kepler did not believe that Earth's motion causes the ocean tides as Galileo claimed, rather he believed that a force between the Moon and ocean could explain the tides. Nevertheless, the concept of forces acting like this is quite bewildering and unacceptable to many, and we will return to this issue shortly when we discussed Newton. Galileo, on the other hand, rejected such a "force", as he viewed tides as a major supporting evidence for heliocentrism—he wanted tides to be due to the motion of the Earth, not gravity of the Moon or the Sun. The lesson here is that a scientific position can itself be correct even though what was claimed to be its supporting evidence turned out not to be. Gravitational force, acting at a distance, likely also appeared like a dubious idea to Galileo, especially because of his knowledge of mechanics, that forces need to be in contact with something to be transmitted (not withstanding the magnetic force was well-known). This was the old, so-called "mechanistic view", which we will have more to say later. He thought that the planets move due to inertia, not realizing that a circular orbit involves acceleration.

Galileo never accepted Kepler's discovery that planetary orbits are ellipses, instead he clung to the old Aristotelian philosophy that requires planetary orbits to be a perfect circle; this despite his own discovery that put some doubts on the perfectness of the Heaven. Galileo also maintained that comets were a sublunar atmospheric phenomenon, not celestial. Thus we see how certain ideas, when deeply rooted into the mainstream, is extremely difficult to be rid of, even by those who are a "doubter" or critical thinker. On the other hand, one must ask whether Galileo's doubt about elliptical orbit was simply due to him being unable to give up old tradition. Perhaps he thought that Kepler only obtained the shape of the orbit by data fitting, and thus it is not philosophical enough. (In today's words, maybe he considered Kepler's work to be purely phenomenological and was not convinced unless some underlying theories can explain the phenomenology.)

Incidentally, Kepler tried to obtain a copy of Galileo's telescope but for some reason Galileo never sent one to him. Kepler did manage to get one indirectly. Kepler, when he learned about Jupiter's moons that Galileo observed, realized that this didn't only provide an evidence for heliocentrisim (an indirect evidence; for showing that there are bodies that do *not* orbit the Earth), but more importantly, it proved that Earth is not the only planet with a moon—and therefore Earth is not so special. Based on this he conjectured that other planets too have moons, and that Mercury's moon is too small or too closed to the Sun to be observed. He was right that many planets do have moons, but he was wrong about Mercury—it has none.

Galileo eventually got into trouble with the Church when he insisted that heliocentrism is "the truth", as opposed to being just a hypothesis; the latter view would have been tolerated for discussion. The Roman Inquisition tried him in 1633, and found him "vehemently suspect of heresy". Galileo was sentenced to house arrest from 1634 onwards until he passed away in 1642.[30] His *Dialogue* book was also banned. In addition, Galileo was prohibited from teaching or discussing heliocentrism. During the house arrest period, Galileo therefore turned his attention to another research subject: mechanics, an old interest which he somewhat sidelined during his work with telescopes. It is mechanics that would provide the next piece to cosmic puzzles: how do the planets move?

Among other mechanics questions, Galileo investigated falling bodies since 1589, performing various experiments (but dropping balls from the leaning tower of Pisa most probably did *not* happen). This led to the discovery that objects fall at the same rate regardless of its mass. Galileo wrote "Discourses and Mathematical Demonstrations Relating to Two New Sciences" (*Discorsi e dimostrazioni matematiche intorno a due nuove scienze*) which was published in 1638. In this book, through the character Salviati (who represents the views of Galileo himself), he doubted whether Aristotle himself had ever experimentally tested the claim that heavier bodies will fall at a faster rate than lighter objects. More importantly, through another character Sagredo (named after Galileo's friend Giovanni Francesco Sagredo (1571–1620)), stated that:

> "But I, Simplicio, who have made the test can assure you that a cannon ball weighing one or two hundred pounds, or even more, will not reach

[30] In 1992, the Church finally acknowledged it had persecuted Galileo unfairly.

the ground by as much as a span ahead of a musket ball weighing only half a pound, provided both are dropped from a height of 200 cubits."

The importance of this remarkable discovery cannot be overstated.[31] For more than a thousand years the Aristotelian view that a heavier object falls at a faster rate persisted. This is not without reason, of course. It took careful measurements of brilliantly designed experiments to show Aristotle was wrong, while everyday experience would have seemingly supported Aristotle. This is not to say that no one before Galileo had made similar claims regarding falling bodies. For example, in the 14th century, Thomas Bradwardine (1290–1349), among others, had concluded that two *homogeneous* bodies of different size and thus also weight, would nevertheless fall in a void with equal speed. The argument was based on the assumption that each unit of matter in every homogeneous object is identical. From this they argued that every unit of matter has the same ratio of heavy-to-light elements (following the Aristotelian worldview that heaviness and lightness are innate properties; Grant 1977).

On the other hand, perhaps Galileo's belief that circular motion is natural, together with his findings that objects in motion tend to stay in motion (i.e., this is the natural tendency of motion), contributed to his wrong conclusion that planets can move in circles just by inertia, which gave another reason for his refusal to believe in an "occult gravitational force".

Despite Galileo's experiments with the inclined planes that helped him to understand motion of objects under gravitational pull, and his realization that objects fall at the same rate regardless of their weights, Galileo never completely abandoned Aristotle's concept of heaviness/lightness as being an intrinsic property of matter. Gassendi, on the other hand, realized that this view is untenable, based on an analogy with magnetic fields (he studied William Gilbert's (1544–1603) work on magnetism, *De Magnete* published in 1600). Gassendi wrote (Pancheri 1978):

> "We may understand how there can be in the Earth a magnetic force by means of which all bodies, which are said to be heavy, are drawn downward, that is, perpendicularly to the Earth, by means of the action of a magnet upon iron. Hold in the hand a small portion of iron of some few ounces. Put under this hand a powerful magnet. The hand will then experience a weight (heaviness) no longer of ounces, but of several pounds. At this point we are forced to admit that this heaviness is not so much innate to the iron as it is impressed upon it by the 'attrative,' or 'impellant,' force of the magnet. Since as Gilbert avers, we may view the earth as a very large magnet with respect to stones, etc., we must

[31] In a later chapter we will see there is a deeper physics in this discovery, now often called the "equivalence principle", which led Einstein to the general theory of relativity. Galileo also provided a thought experiment that purported to argue for the equivalence principle purely on logical ground. As we will explore in an exercise at the end of this chapter, his argument is unfortunately not quite correct.

conclude that the gravity, or heaviness, of any object is not *in* it, but is *to* it, from the attractive force of the earth." [Emphasis is his original.]

2.3.2 Descartes: Casting Doubts on Everything

We have mentioned that some of Galileo's contemporaries doubted whether astronomical images in the telescope can be trusted. But one man would go to the extreme, doubting whether we can ever be sure that *anything* is real. That man was René Descartes (1596–1650). After all, we learned about the world through our senses, and Descartes knew that for example, our eyes are just a biological lens that received lights from the external world. If we doubt the images of telescope and microscope because we are seeing through lenses, then the same objection would also hold for our own eyes. In addition, vision and other senses are not direct either; the brain has to process the signals received and then interprets it as our experience of the world. This is a highly complicated process. Indeed, we know not everyone experiences the world the same way. For example, people who are colorblind cannot perceive some colors. One may object that this is due to "defective eyes". However, there are also people with *more* cone cells in the eyes than the normal three colors. People with four types of cone cells (such a condition is called tetrachromacy) would see the world differently than most, not to mention other creatures like the mantis shrimp that have as many as 16 types of cone cells! Another example would be the sense of taste. Some people quite enjoy coriander, while some others are repulsed by its "soapy taste". While the reason is complex, variation in the OR6A2 gene has been identified as one likely reason. In that sense, being "normal" is relative. There is only one physical world but potentially everyone perceives it differently. After all, we cannot compare subjective experiences (or what the philosophers of mind call *qualia*).

In summary, there are properties that do not truly exist in the world, only in one's mind. Now, imagine an evil scientist in an arbitrary advanced civilization, who put a human brain in a liquid-filled container and hooked it up to electrical wires that stimulate the brain in such a way that it thinks it has a full body, watching a sunset on a beach somewhere. This is conceivable. But then how are you so sure that *you* are not such a brain, and that your memories and perceptions are just what the machine provided? This brain-in-a-vat scenario has been discussed for a long time in philosophy literature, and variations of it appear again and again in various contexts. Descartes contemplated a similar scenario, leading him to doubt everything.[32] But if one cannot trust anything is real, how does one function in this world? Eventually he convinced himself that to be able to doubt, at least he himself must exist. This is his famous line: "I think, therefore I am". Descartes also went on to claim that since he was able to conceive of God, who is infinitely more powerful than him, surely this cannot be conceived by himself, and therefore the concept of God is

[32] Cartesian skepticism was also explored by the Danish philosopher Søren Kierkegaard (1813–1855) in his work *De omnibus dubitandum est*, which translates to "everything must be doubted", published posthumously.

real and external. Then, because a kind God would not deceive him, the external world must exist. This argument always seemed quite dubious to me, but we shall not discuss it here as I am not an expert on the philosophy of religion.

What we should emphasize is this: we don't have to go to such an extreme in doubting everything. But like Descartes, we should be bothered with the fact that humanity believed in Aristotelian physics for millennia before Galileo showed that it is not correct. If we can believe in something fundamentally wrong for so long, what else that we believe in could be wrong? To doubt as far as possible is important to make sure that we do not accept anything for granted. As a result of his doubts and questioning, Descartes made one of the first attempts to give modern science a solid foundation through the separation of the mind and the physical world, otherwise known as "mind-body dualism". From a modern point of view, many scientists prefer to think that only the physical world is real, and that the mind arises from the brain, that consciousness is an emergent phenomenon that can ultimately be explained once we understand the brain well enough. In other words, there is no need to talk about the soul or spirit. While this was not the view of Descartes, the fact that he separated the mind from the physical world has a huge impact on how science is done ever since—we study the world as if we do not affect it, the world as it is. We want to know how the world *really* is even if we are not there to observe it.[33]

Curiously, despite all this rationality, Descartes was initially motivated to reform knowledge because of his *dreams*. In a series of three dreams, he believed he was given a divine message or guidance. Especially in the last dream, he saw an anthology of poetry on a table, which upon opening at a random page, he read the verse of Ausonius "Quod vitae sectabor iter" (What path shall I take in life?). A stranger then replied to him with "Est et non" (Yes and no). Convinced that this was a divine message, Descartes even went on a pilgrimage. He subsequently interpreted this dream as showing him that the body of knowledge was in need of a reform and unification, and that his mission was to achieve this via *reasoning*. Descartes wanted to set down "foundations of a marvellous science" (*mirabilis scientiae fundamenta*).

Descartes' contribution as the father of modern analytical geometry is well-known and need not be repeated here. We should, however, discuss his cosmological view. He believed that the Universe is infinitely large and filled with a substance that swirls around like vortices. Near stellar objects like the stars the density of this substance is higher. One way to think about Descartes' idea is to picture the Sun and other stars being at the middle of a swirling whirlpool, whose motion carries planetary bodies like the planets around, much like leaves on the surface of a river.

Why propose such an invisible matter? Because it was very hard to explain motion of planets in an otherwise empty space. The common view back then was that motions have to be mediated by contacts. The philosophy of "mechanism" (that of mechanics) argued that in order to move an object, it has to be acted upon by another object. For example, a ball moves if someone kicks it. The notion of action at a distance is frowned upon. Indeed, the Universe was often seen as a gigantic

[33] Things became a lot subtler and complicated after we discovered quantum mechanics.

clockwork-like mechanism. God, the master clockmaker, could design the Universe as a perfect clock that would run on its own for eternity without His constant direct control. Even animals, as well as human bodies, are just biological versions of mechanistic automata. Since clocks and automata move based on gears in contact, it is natural to generalize that in a clockwork universe, motion requires contact as well, and thus Descartes' vortices.

As we have mentioned in Chapter 1, mechanics was traditionally not considered a part of physics, but of mixed mathematics (today we may call it "applied mathematics"). It is therefore somewhat of a change that mechanics is now used to explain physical laws! As Robert Boyle claimed, mechanical principles are "so universal and therefore applicable to so many things" and therefore any hypothesis grounded in Nature will be compatible with mechanical principles or deducible from them. One glaring exception is *gravity*; Boyle saw no need to explain gravity because this was never an issue in the discipline of mechanics (Berryman 2009). As commented in Kochiras (2022):

> "Thus, he is not revoking the optimism he expressed elsewhere about the universality of mechanical principles or giving reason to think that gravitational effects are unlikely to be mechanically explicable. Nor is he invoking the experimentalist's right to demarcate limited research questions, while leaving other questions for later. Rather, he is simply denying the need to determine whether gravitational effects are mechanically explicable and justifying that denial by pointing to disciplinary conventions. Boyle thus leaves the traditional division of labor intact; in his view, giving a mechanical philosophy of nature does not require that mechanics and natural philosophy be integrated. It is this view of mechanics and of its proper relation to natural philosophy that Newton will repudiate."

Thus, we again see that what is considered a question that requires scientific investigations can also be influenced by whatever world view and philosophy one subscribes to. Descartes, on the other hand, while also a proponent of the mechanistic universe, saw the need to mechanize gravity via his vortex theory (Figure 2.7). He criticized the traditional view of Aristotle that mechanics isn't part of physics, and wrote in a letter to Libert Froidmont[34] (1587–1653) that mechanics "took itself off to the mathematicians, since it found no place with the supporters of the common philosophy" (Kochiras 2022).

It was also the belief in a clockwork universe that inspired Descartes to propose *conservation laws* in physics. A wound-up clock cannot run indefinitely—eventually the potential energy in the spring would be exhausted, or rather converted into other forms of energy like heat generated by friction in the gears. So if we envision a divine power setting the Universe into motion, and with infinite wisdom has designed the

[34] Froidmont once complained that the works of Descartes were "excessively gross and mechanical"!

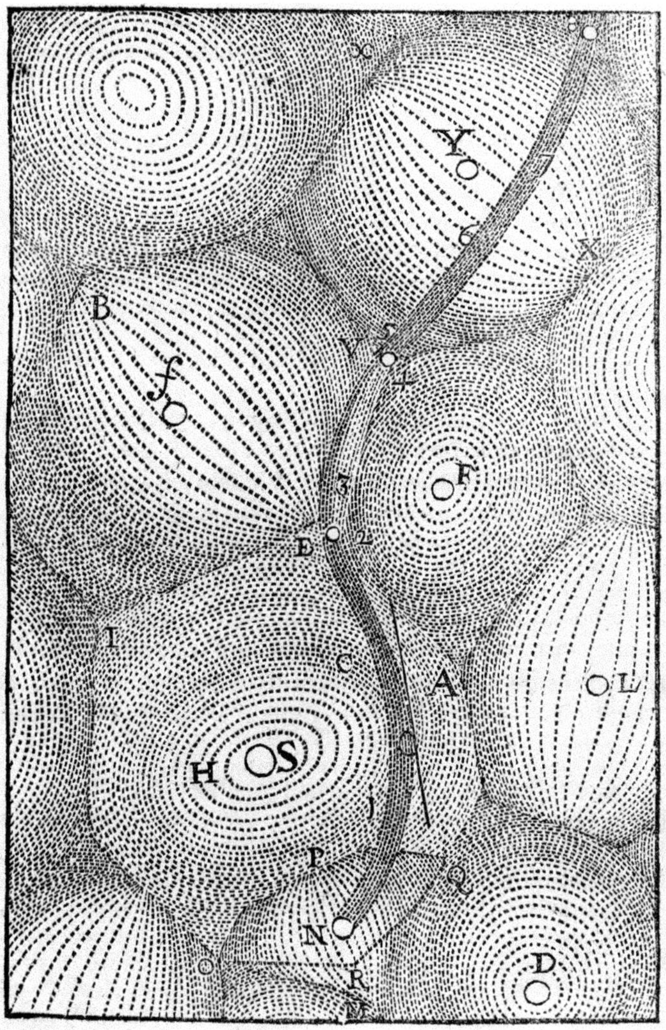

Figure 2.7. Descartes' system of vortices from his *Principia Philosophiae*. The point denoted S is the Sun. The vortex swirling around the Sun carries planets along in their orbits. Every planetary body has a "centrifugal tendency" (the radially outward tendency to escape), which if greater than that of the vortex elements, would cause it to ascend to a "higher level" of the vortex, otherwise it will sink (or rather be pushed down by the vortex elements) further inward. Planetary orbits that are stable are in equilibrium with respect to the vortex elements. In the view of Descartes, this is the reason why there is "gravity". There are other vortices throughout the Universe, and their boundaries press on each other to form cell-like structures like soap bubbles, as shown here. The curve in the figure shows a comet being dragged by the vortices from one system to another. Newton raised doubts concerning this: the comets need to penetrate the layers of any vortex strong enough, yet seemingly have no effect on the vortices (otherwise the planetary orbits would be perturbed as well). In modern language, the question is about the coupling and backreaction between comets and the vortices. Newton would eventually show that Kepler's first and second law could not both hold under Descartes' theory. Credit: M0006150: Descarte's system of vortices. *Source*: https://wellcomecollection.org/works/rxrrnkkv/items

cosmos to run forever without needing further intervention, then some physical quantities need to be conserved, at least in an ideal closed system. In Descartes' words,

> "It is obvious that when God first created the world, He not only moved its parts in various ways, but also simultaneously caused some of the parts to push others and to transfer their motion to these others. So in now maintaining the world by the same action and with the same laws with which He created it, He conserves motion; not always contained in the same parts of matter, but transferred from some parts to others depending on the ways in which they come in contact."

Descartes tried to formulate what we would today call the law of the conservation of momentum, though he failed because he didn't realize that momentum is a signed (vector) quantity. In other words, mass times velocity is conserved but mass times speed is not. Despite this failure, the idea of conserved quantities would become a very important one in physics. Yet it is perhaps surprising to modern readers that the notion can be traced back to reasoning that we would not call scientific by today's standard. We should also note that conservation law is posited by Descartes as a matter of principle, inspired from his personal philosophy. It would be harder to *deduce* a conservation law from mechanical experiments alone as (1) if we don't expect a conservation law in the first place, it takes a keen eye to sieve it out from the data of various experiments that were meant to test for other hypotheses; and (2) experiments are carried out in real life, and so the system loses energy via friction and air resistance, etc. We are often taught as a science student to be diligent and make careful observations of Nature. This is good advice. Yet in this case, a meticulous observation that includes all the complications of the system would instead make it harder to discover something like conservation of momentum or energy.[35] Just like mathematical modeling, it is often useful to start with systems that are as idealistic as possible, even in experiments.

Another ingenuity of Descartes is that the vortex theory allowed him to have planets like the Earth orbiting the Sun, while at the same time claiming that the

[35] Indeed, the concept of kinetic energy originated from Gottfried Wilhelm Leibniz (1646–1716), who called mv^2 the "living force" (*vis viva*) of a system. Vis viva became a rival of momentum, as arguments ensued debating over which one is the correct conserved quantity. This was yet another Newton–Leinbiz argument (in addition to the priority dispute of the invention of calculus); Newton being in the Descartes' side that mv is the correct quantity. John Playfair (1748–1819) also argued against vis viva since he observed that mv^2 is not always conserved (say, during inelastic collisions). Only after much later did people realize that this is because the heat produced during collision was due to some of the kinetic energy being converted into motion of the molecules. It is interesting to note that during the debate, many *engineers* including John Smeaton (1724–1792), the "father of civil engineering" (Morris 2017) found that conservation of momentum was not enough for practical calculation, and resorted to using vis viva concurrently. The engineer community also played an important role in the early development of other ideas, such as "work" (Cardwell 2009). Along with the observation of Willem 's Gravesande (1688–1742), Émilie du Châtelet (1706–1749) played a crucial role in demonstrating via experiments that vis viva indeed is proportional to v^2. The modern prefactor of 1/2 in the definition of kinetic energy was introduced later.

Earth is not moving (as this was the Church's main objection to heliocentrism), in the sense that the Earth does not move with respect to the vortex elements in the band of vortex that Earth is contained within. Descartes wrote:

> "The Earth, properly speaking, is not moved, nor are any of the Planets; although they are carried along by the heaven".

Though perhaps nothing more than a smart move to placate the Church, this is remarkably similar to the modern explanation for the expansion of the Universe—galaxy clusters do not *move* away from each other, they are "carried away" by the expansion of space itself. In modern language, we would have said that in Descartes' world view, the Earth is *co-moving* with the vortex elements.

Note that the idea of Descartes' vortices is similar to the later concept of the æther, in which light—as a wave—is supposed to propagate. Since at that time all the known waves required a medium to propagate in, it was only natural to presume that light too, as an electromagnetic wave, would require one. Therefore, we see that most of the time, scientific ideas are based on the dominant views and philosphical thoughts of the respective era. Speaking of which, we note that viewing the Universe as a clockwork device is itself a rather peculiar thing to do in hindsight; why a clock and not something else? Well, actually there is something else that had been compared to the Universe. Across history, once we understood or invented something, we tended to think that maybe the Universe is also something like that. For example, Galileo described the Universe as a "grand book" written in the language of geometry (more generally, mathematics). Though he might have meant it only as a metaphor, it is easy to imagine God writing the Universe as a divine book, after we ourselves have invented books and printing technology. Likewise, if we with our limited intelligence is able to design clocks and automata, it is not hard to imagine God designed everything as a vastly superior version of the same thing.[36] Nowadays we may not bring divine powers into the picture, but still we often hear sayings along the line "the Universe is like a computer", and then subsequently even a *quantum* computer. These may all be true in some ways, but isn't it interesting that we like to project our latest understanding to the entire cosmos?

Descartes also formulated laws of motion, which would inspire Newton to improve upon. Descartes' laws are as follows (Slowik & Stanford 2021):

(1) *Descartes' First law:* "that each thing, as far as is in its power, always remains in the same state; and that consequently, when it is once moved, it always continues to move."

---

[36] Despite his fascination with clocks and machines, Descartes underestimated how advanced machines can become. He wrote in *Discourse on Method* (1637) that machines cannot possibly hold an intelligent conversation (essentially an earlier notion of a "Turing test"): "It is indeed conceivable that a machine could be so made that it would utter words, and even words appropriate to the presence of physical acts or objects which cause some change in its organs; as, for example, it was touched in some so spot that it would ask what you wanted to say to it; in another, that it would cry that it was hurt, and so on for similar things. But it could never modify its phrases to reply to the sense of whatever was said in its presence, as even the most stupid men can do." These days we often wonder whether we are responding to a comment by a bot on certain online platforms! Descartes would have been astounded with what artificial intelligence can do today.

(2) *Descartes' Second law:* "that all movement is, of itself, along straight lines."
(3) *Descartes' Third law:* "that a body, upon coming in contact with a stronger one, loses none of its motion; but that, upon coming in contact with a weaker one, it loses as much as it transfers to that weaker body."

Note that the first law is essentially the law of inertia. The second law would be replaced by Newton's second law, and in hindsight can be interpreted as the statement that when not acted upon by an external force, an object would either not move (if it was initially stationary) or that its motion is in a straight line. In the third law Descartes tried to formulate his conservation law though it is not quite correct. Like Newton, the first two laws of Descartes concern individual objects, while the third law tells us what happens when two objects interact.

Descartes went to Sweden in 1649, when he was invited by Queen Christina to join her court and teach her philosophy. The lessons, starting early in the morning and thus having to wake up at 5 am, was perhaps too much for Descartes, who had always preferred to stay in bed till midday. Descartes caught a cold that unfortunately led to a pneumonia that resulted in his demise in 1650, barely three months after his arrival in Sweden. Descartes had complained that Sweden was "a land of bears and rocks and ice" where "men's thoughts freeze in winter just as the water does" (Langer 1937).

2.3.3 Newton: First Successful Theory of Gravity

Sir Isaac Newton (1642–1726/27) needs no introduction. Although he didn't "doubt everything" like Descartes, he too wondered how much we can trust our senses. When researching colors in 1665, he would stick a bodkin (a needle-like object, albeit with a blunt point) into his own eye socket through the tear duct, in between the eyeball and the back of the eye to study how pressing his eyeball can create colored circles and bright spots in his vision (these are called "phosphene"). He wrote:

> "I could sometimes perceive vivid colors of blue and red, made by said pressure, and perhaps a critical eye might have discerned this order of colors, from the center: green, blue, purple, dark purple, blue, green, yellow, red like flame, yellow, green, blue, broad purple, dark."

He wanted to know how our eyes observe the world (and so learn about vision and color itself).[37] More importantly, though it is unclear how much Newton

[37] Newton had a debate with Goethe, whom we met in Chapter 1, about the theory of color. Goethe is best known for his literature, but he was also very interested in the sciences. Perhaps partly inspired by his mystical belief (Barsan et al. 2016), Goethe suggested that color is not a property of light per se, but rather deeply influenced by the observer's perception. He objected to Newton's discoveries that white light can be decomposed into the various colors of the rainbow via a prism. Today we know that Newton and Goethe both got something correct: white light *can* be decomposed into various components of different wavelengths, but the *perception* of color is entirely subjective. Note that Goethe also conducted his own experiments, not just arguing based on his beliefs. The Goethe-Newton debate was *not* a debate between an artist's romantic view of the world and the logical and objective view of nature, as it is sometimes presented (Sepper 2009).

understood this, only by understanding how our own sensory organs work can we know what is—hopefully—objectively "real" external to us. In particular, since phosphenes can appear just from stimulation to the retina, this supports the view that the colors we see are not inherent in the objects themselves, but rather what the brain perceives from the light the eyes receive.

Newton studied a great variety of topics, from optics to alchemy and mathematics, inventing calculus independently of Leibniz. Like Descartes, Newton saw the need to investigate gravity. In his view, a study of man-made machines (as in the field of traditional applied mechanics) is incomplete without also investigating gravity and other forces since the whole world is essentially a grand machine created by God. In Newton's worldview, the clockwork universe requires intervention of its creator from time to time, for example, to ensure the stability of the solar system. The Universe, like a clock powered by a spring, will eventually stop and needs to be manually turned to rewind the spring and start ticking again.

Newton's laws of motion are taught to even high school students nowadays:

(1) *Newton's First law:* "An object at rest remains at rest, and an object in motion remains in motion at constant speed and in a straight line unless acted on by an unbalanced force."
(2) *Newton's Second law:* "Force equals mass time acceleration: $F = ma$."
(3) *Newton's Third law:* "For every action, there is an equal and opposite reaction."

The first law is again the law of inertia. It tells us what happens when there is no force acting on an object. This is quite contrary to the old Aristotelian view in which motion is only possible if a force is acting on an object. The Aristotelian view is of course what appears to be true in everyday life; after all a ball doesn't continue on rolling indefinitely after you kick it. But the law of inertia says this would be true in an idealistic situation with no friction whatsoever. Note that Newton's three laws are not mathematical theorems; you *cannot* prove them mathematically, they are more like axioms, statements that we accept for granted and see whether they lead to good physical predictions, which they did. These axioms did not arise out of thin air of course, they were motivated by either thought experiments or actual ones, in one form or another. For example, the law of inertia was actually deduced by Galileo from his experiments with balls rolling down inclined planes. More specifically, consider rolling a ball on a frictionless inverted bowl surface, i.e., a shape that looks like a symmetric valley. The ball would roll down to the bottom and then continues to climb back up the other side, until it reaches the same height as its starting position (because of energy conservation). This is true even if the surface is not quite symmetric: imagine bending the "uphill" side and flattening it out. The ball would have to travel further away in the horizontal direction in order to reach the same vertical height, since the side is not as steep as it was. In the limit, when the gradient reduces to zero, one would conclude that the ball will have to move indefinitely without ever coming to a stop. This last step is of course only achievable in a thought experiment. Note how thought experiments play very important roles in the history of physics.

What *is* a force then? Note that this was never explained. We tend to say that a force is what causes a body to deviate away from its rectilinear motion, its natural state. This is not quite satisfying: this statement seems to only tell us what a force *does*, not what it *is*. In fact, the very equal sign in $F = ma$ does not have any causation built in, it does not say that a force "causes" acceleration, but rather that a force *is* essentially acceleration (multiplied by a dimensionful constant). The difference seems pedantic, but it is not. Note that the same can be said in general relativity: we often say the presence of matter *causes* curvature in spacetime, but again the Einstein field equations only tell us that matter is related to curvature. It seems best that we should not read the equations too much beyond what they actually say, which otherwise can lead to some misunderstanding. For example, in the so-called "gauge/gravity correspondence", a system with gravity is in one-to-one correspondence with a quantum field theory without gravity but in a spacetime with one less spatial dimension. The correspondence is an equivalence. This does not necessarily mean that—as some would claim—gravity is not fundamental but just a property that is emergent. The claim might well be true, but it does not follow from the correspondence.

The first two laws are extremely important in the context of astrophysics as they explained how planets move along their orbit, when Newton subsequently formulated the concept of gravitational force. Once the notion of celestial crystal spheres were abandoned, it was natural to ask the obvious questions: What holds the planet in place, keeping them in orbit? What makes them continuing their motion and why don't they stop? Why don't they all come crashing down to Earth? These are profound and difficult questions. An old Chinese idiom told of a story of a man in the Qi State, who was in constant fear that the "sky" might fall (杞人忧天). More specifically, he asked: if celestial objects like the Sun, the Moon and the stars are just balls of gas (as accordance to one of the cosmological views back then,[38] which was surprisingly accurate in that regard), what if they fall down? "Learned man" told him that there's nothing to worry about since even if glowing balls of gas were to fall down, there would be no physical harm. The idiom has since taken to mean that one should not have imaginary or baseless fears. Yet the advice given was completely wrong of course—if a gas planet like Jupiter falls towards Earth, the chaos it would cause would be unimaginable. The question raised here is far from baseless; Newton asked the same question after he purportedly saw an apple falling down a tree: why didn't the Moon? He came up with an ingenious solution: the Moon *is* constantly falling, but at the rate just right to maintain its orbit. He imagined firing a cannon horizontally in a thought experiment. The cannonball would drop a certain distance away depending on how much firepower the cannon holds. Imagine that we keep on increasing the power, the cannonball will shoot further and further, until it can go all around Earth and arrives at where it started without falling—it has entered an Earth-grazing orbit. Newton formulated his theory of gravitation, which for a

[38] "日月众，星自然浮生虚空之中，其行其止皆须气焉。–《晋书· 天文志》". This view is referred to as 宣夜说, which is interesting in its view that planetary bodies can float in infinite space, as opposed to being attached to the celestial spheres, like that of the earlier proposal (浑天说).

spherical body like the Earth with mass M orbited by another body m at distance r, states that the gravitational force is given by the equation

$$F = -\frac{GMm}{r^2}, \tag{2.1}$$

where G is now called Newton's constant or the gravitational constant. The minus sign is just to emphasize that the force is attractive. The law is more correctly stated using vectorial notation as

$$\overrightarrow{F} = -\frac{GMm}{r^2}\hat{r}, \tag{2.2}$$

where $\hat{r}$ is a unit radial vector. This is often called "Newton's law of universal gravitation"; universal because it applies to anything with a mass. Both the falling apple and the unfalling Moon, though seeming completely opposite, are governed by the same physics. Newton therefore completed the unification of the celestial and the terrestrial, hinted by the findings of Galileo and Kepler. No longer is the sublunar world separate from the heavenly realm. Everything in the universe can be studied by the same laws! This is the basis for modern astrophysics and cosmology.

In 1680, a new comet (known simply as the Great Comet of 1680) became a sensation. It was first noticed by the astronomer Gottfried Kirch (1639–1710), but would soon become so prominent on the sky that everyone looked on in awe (and fear—comets were still widely regarded as a bad omen). Eusebio Kino (1645–1711), the Spanish Jesuit priest, was also among the first to chart the comet's course through the night sky. Based on the observational data of John Flamsteed (1646–1719), Newton applied his theory to calculate the orbit of the comet (Figure 2.8). He found that the orbit matched the observations neatly.[39] The hallmark of a good scientific theory, however, is that it should not only explain observed phenomena, but also make successful new predictions. This would come from the follow-up research of Edmund Halley (1656–1742), who would continue to use Newton's theory to calculate the paths of other comets. By studying old records, he became the first person to suggest that some comets are in fact periodic. One of these is the comet that now bears his name, whose return in late 1758 he successfully predicted (though he didn't live to see it). The orbit of Halley's comet brings it as far as 35 AU from the Sun. (AU is short for "astronomical unit", which is the mean distance between the Sun and the Earth.) The fact that it returned as predicted shows that Newton's law of universal gravitation remains valid at least that far out. Halley's Comet will return in 2061, and makes its closest approach to the Sun on July 28th that year.

---

[39] Newton also speculated in his *Principia* that matter from the tail of the comets can fall down to the Earth and then "condensed" into all types of substances (Dobbs 1982). This reflects his alchemical beliefs. While today we know this is not the case, similar to "stuff falling from comets" is the related, more modern but also speculative idea called "cometary panspermia" that comets can carry viruses or other primitive lifeforms, thus seeding suitable planets with life. Comet bombardments in the early history of Earth could also be the origin of the oceans.

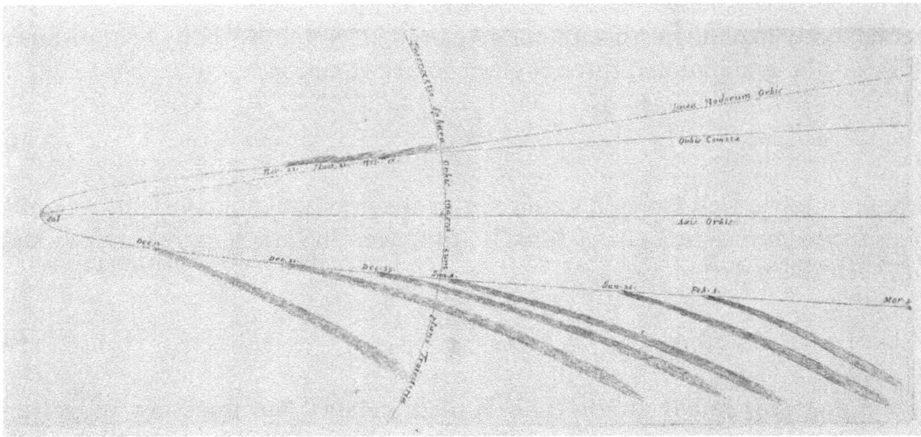

Figure 2.8. The orbit of the comet of 1680, fit to an elliptical orbit (with eccentricity quite close to being a parabola), as shown in Isaac Newton's *Philosophiae Naturalis Principia Mathematica*. The shape of the orbit being elliptical was conjectured by the astronomer bishop Seth Ward (1617–1689), during the comet of 1664–1665 (Chapin 1952). Giovanni Borelli (1608–1679), on the other hand, concluded that the comet was moving in a path "very similar to a parabola" based on his own measurements. He then suggested that comets are subject to the same physical laws that govern the motions of planets around the Sun. His findings were published under the pseudonym Pier Maria Mutoli, possibly to avoid getting into problems with the Church (Boschiero 2009). Comets were regarded as a bad omen that marks various disasters (including war, famine, and plague; or the death of a royal family member) almost universally throughout human civilizations. Newton's theory of gravity was the first successful attempt at understanding the motion of comets, and with that, comets were "tamed" as a natural phenomenon and eventually lost their supernatural menace. Some three centuries later, on 12 November 2014, the lander *Philae* of the space probe *Rosetta*, landed on the Comet 67/Churyumov–Gerasimenko. *Credit:* Philosophiae Naturalis Principia Mathematica, Isaac Newton. Wellcome Collection. *Source*: https://wellcomecollection.org/works/kmzdzr2q/images?id=jxf4tngw

The idea of a gravitational force was, at the time, a rather strange one. Against the mechanistic doctrine, Newton rejected Descartes' vortices in favor of forces that act at a distance. Newton wasn't completely at ease with this idea, he wrote in his letter to Richard Bentley (1662–1742), a classical scholar and theologian:

> "It is inconceivable that inanimate brute matter should, without the mediation of something else which is not material, operate upon and affect other matter without mutual contact, as it must be, if gravitation in the sense of Epicurus, be essential and inherent in it. And this is one reason why I desired you would not ascribe innate gravity to me. That gravity should be innate, inherent, and essential to matter, so that one body may act upon another at a distance through a vacuum, without the mediation of anything else, by and through which their action and force may be conveyed from one to another, is to me so great an absurdity that I believe no man who has in philosophical matters a competent faculty of thinking can ever fall into it. Gravity must be caused by an agent acting constantly according to certain laws; but whether this agent be material or immaterial, I have left open to the consideration of my readers. "

Newton did not want to speculate too much on the nature of gravity, his aim was "merely" to understand how gravity works. Christiaan Huygens (1629–1695) criticized Newton for his "non-explanation". Unlike the vortices of Descartes, Newton's gravitational force is just a description of the phenomenon, and no mechanical or other underlying explanation was given. Some would view this as a distinction between modern science and philosophy: in science, it is often good enough to have a theory that can give a good description and allows us to make testable predictions, and there is no need for a deeper explanation of the "why". Such understanding may come later, but it is not a necessary ingredient for a scientific theory. Christiaan Huygens was essentially complaining that Newton's gravitational force was too "occult" and lacked in natural *philosophy*.

It isn't completely true to say that Newton didn't try to explain the underlying cause of gravity. He had at some point considered something akin to a "gravitational æther", which he discarded much like Descartes' vortices. The problem is that the magnitude of the gravitational force between two bodies of masses M and m is, for a fixed M, proportional to m. Thus, suppose the Moon is orbiting the Earth and it is immersed in the gravitational æther. Since m is a measure of the quantity of matter, the "gravitating cause" needs to operate just as much on the interior particles as on the exterior ones. In addition, like Descartes' vortices, the gravitational æther may introduce unwanted drag and change the particle orbits[40], which would then deviate from the inverse-square law prediction. Newton therefore had to consider that even solid bodies have "pores" or a net-like structure large enough to allow the gravitational æther to seep through. However, such a porous body would cease to be solid and this would also affect its predicted motion. The way out is to replace a physical æther by a spiritual one, i.e., the action of God, or the "penetrating force of spirit" (*vim penetrantem spiritus*) (Dobbs 1988).

This could be inspired by the Hermetic idea that the alchemical active entity "penetrates every solid thing". As Newton wrote in the "General Scholium", an essay appended to his *Principia*, gravity

> "[…] must proceed from a cause that penetrates to the very centers of the Sun and planets, without suffering the least diminution of its force; that operates not according to the quantity of the surfaces of the particles upon which it acts (as mechanical causes used to do), but according to the quantity of the solid matter which they contain."[41]

[40] This should be contrasted with the view of Aristotle that an external medium such as air *must* exist, not only to guarantee movement, but also so that motion is not instantaneous. Without the inhibiting drag of the substance to slow down motion, he seemed to think that objects would travel at an infinite speed. Many had objected to this view, including John Philoponus (490–570), a Greek grammarian. Aristotle wasn't even consistent; he believed that the celestial æther does not give rise to any resistance, yet planets move at a finite speed. This inconsistency was pointed out by the Andalusi polymath Ibn Bājja (c.1085–1138), known in the West as Avempace. Another inconsistency of Aristotle that had been criticized was that he believed that rest is a nobler state than motion. If that is the case, why does the heavenly celestial realm move, while the Earth supposedly stays fixed?

[41] What would Newton think of the modern claim that gravity satisfies the "holographic principle" (the degrees of freedom on the surface encodes all of the interior)?

Later in his *Opticks*, Newton would refer to the cause of gravity as being a certain "active principle", and mentioned that "cohesion of bodies" (namely, "inter-molecular forces") and even "fermentation"[42] are caused by similar active principles. This is again, very similar to alchemical principles and the idea of the Stoic school of thought that the *pneuma* that permeates space not only governs its natural laws but also life.

Prior to Newton, as we have mentioned, Kepler already considered some kind of "influence" emanating out from the Sun that affects the motion of the planets. We may recall from Chapter 1 that Kepler's science was deeply influenced by his religious beliefs. Kepler wrote about "gravity" in his *Mysterium Cosmographicum* (Hecht 2019):

> "[…] either the souls which move the planets are less active the farther the planet is removed from the Sun, or there exists only one motive soul [*motricem animam*] in the center of all the orbits, that is, in the Sun, which drives the planet the more vigorously the closer the planet is, but whose power [*virtutis*] is attenuated when acting on the outer planets because of the long distance."

Kepler had in mind a motive soul that acts on the planets akin to the action of the Holy Ghost. This is similar to Newton's conviction. Kepler would subsequently change his "motive soul" into "motive force" or "motive power". Thus, an idea that originated from theological faith gradually transformed into something more "physical". In the second edition of *Mysterium Cosmographicum*, Kepler clarified (Hecht 2019):

> "If we substitute for the word 'soul' [*Anima*] the word 'force' [*Vim*] then we get just the principle which underlies my celestial physics [Physica coelestis] […] For once I firmly believed that the motive force of a planet was a soul, as I was of course imbued with the doctrines of J.C. Scaliger[43] on moving intelligence. Yet as I reflected that this cause of motion diminishes in proportion to distance, […] I came to the conclusion that this force must be something corporeal, that is, an emanation which a body emits, but an immaterial one."

People were familiar with magnetic force back then, so it wasn't completely unacceptable to imagine that gravity behaves in a similar way. One might ask how did people reconcile magnetic force, which does seem to act without direct contact, with the mechanistic view? Well, they thought, since action at a distance is impossible, whatever appears as such must be explained at the microscopic level by tiny particles that do collide with one another. Alignment of a magnetic compass would then be explained as corkscrew-shaped particles emanating from one

[42] The word fermentation is used in a much wider context here than its modern meaning.

[43] Julius Caesar Scaliger (1484–1558) was an Italian scholar and physician. Kepler was familiar with his work *Exotericae Exercitationes* (1557). The work had motivated Kepler's interests in natural sciences since he was 18 (Sakamoto 2016).

magnetic pole to another, dragging the compass needle along its flow.[44] Thus Kepler imagined that the Sun emits something radially to hold the planets in orbit, though he did not have a Newtonian concept of a force. Galileo criticized this point of view as subscribing to "occult properties"[45]. Kepler's work in gravitation is often overlooked and barely mentioned in modern textbooks on astronomy and physics, but his contribution was an important one. As concluded by Eugene Hecht in his article *Kepler and the origins of the theory of gravity* (Hecht 2019),

> "Before Kepler, 'gravity' simply meant weight or heaviness; after Kepler, 'gravity' became an immaterial something, an emanation that spread through æthereal space, manifested as a mutual attraction of kindred material objects."

2.4 Exercises

(1) We mentioned that in *Mysterium Cosmographicum*, Kepler realized that every Platonic solid can be inscribed and circumscribed by spheres. He then considered nested Platonic solids in the order of (from inside to outside): octahedron, icosahedron, dodecahedron, tetrahedron, and cube. The hope was that the nested spheres would now correspond to the planetary spheres. Using this construction, check how much the relative size of the spheres deviate from the actual planetary orbit (you may take their mean distance from the Sun as a measure of the radius of their orbit).

(2) Assume that the Earth is a non-rotating sphere with constant density. Suppose you drill a hole through the Earth right through the other side (the anti-podal point). Compute the time it would take for someone who drops into the hole to emerge from the other side of the planet, assuming no other complications like air drag. In fact, show that if the person does not grab on to something as he reached the other side, he will drop back down and be subject to a simple harmonic motion back and forth. Show that the period of an "Earth-grazing" satellite (i.e., its orbit is the same as the radius of the Earth) is twice the duration you just found (again, we greatly simplified the situation by ignoring other factors like the atmospheric drag). That is to say, the period of the harmonic motion of the person through Earth's center is the same as the period of the Earth-grazing orbit.

Additional note: If we observe a planetary orbit edge on, we only see the planet move along a line. The circular motion in two dimensions is

[44] The idea that forces are caused by some sort of force-carrying particles is still with us today in particle physics.

[45] How is it that most students today, when learned about Newton's universal gravity in high school or even freshmen physics in university, just accept that it can influence things at a distance without any medium that can transmit the force, while they probably were taught Newtonian mechanics before this, and in kinematics of particles forces always act between bodies with physical contact: collisions, friction, etc.? I wonder what the teachers would have said if some students were to ask this question!

projected onto a one-dimensional simple harmonic motion, which can then be modeled like a spring using Hooke's law.

(3) Galileo had a thought experiment (in addition to his actual performed experiments), published in his *De motu antiquiora*, that purportedly explains why two objects of two different masses dropped from the same height would hit the ground at the same time. The thought experiment goes like this: imagine that we drop a mass M along with another with mass $M + m$. In Aristotelian physics, the latter is supposed to hit the ground first. However, if we imagine that the mass $M + m$ actually consists of two bodies: M and $m < M$, tied together with a rope of negligible mass, then as they fall, the body m should fall at a slower rate than M, thus produced a drag effect, slowing the whole system down. Hence we conclude that the composite mass hits the ground at a later time. This is a contradiction. Therefore, it appears that we can deduce *on pure logical ground* that bodies should fall at a rate independent of their mass! Though widely repeated in many books and websites, this thought experiment is not quite correct. See if you can spot the mistake.

Hint: Note that the argument is far too general.

Additional note: The fact that all objects fall at a rate independent of their mass is because gravitational mass and inertial mass are the same, which is sometimes referred to as the "equivalence principle". Gravitational mass is what appears in Newton's law of gravity, whereas inertial mass has nothing to do with gravity *a priori*, but rather is defined by Newton's second law $F = ma$. Confusions often arise since both are called "mass". It is best to think of gravitational mass as being "gravitational charge", just like the electric charge in Maxwell's theory. An electron has an electric charge, and also has an inertial mass. The fact that the gravitational charge is the same as the inertial mass is nontrivial. (Pedantically, they might be proportional but not equal, but we will never know this since the proportional constant would just be absorbed into the gravitational constant G.)

References

Almansa-Villatoro, M. V. 2020, JEgAr, 105, 73

Barsan, V., Merticariu, A., & Fosl, P. S. 2016, CogAH, 3, 1145569

Berryman, S. 2009, The Mechanical Hypothesis in Ancient Greek Natural Philosophy (Cambridge: Cambridge Univ. Press)

Boschiero, L. 2009, JHA, 40, 11

Cardwell, D. S. L. 2009, Some Factors in the Early Development of the Concepts of Power, Work and Energy (Cambridge: Cambridge Univ. Press)

Chapin, S. L. 1952, ASPL, 6, 221

Dobbs, B. J. T. 1982, Isis, 73, 511

Dobbs, B. J. T. ed. P. B. Scheurer, & G. Debrock 1998, Archives Internationales D'Histoire des Idées/International Archives of the History of Ideas Vol. 123 (Dordrecht: Springer) 55

Evans, J. 1998, The History and Practice of Ancient Astronomy (Oxford: Oxford Univ. Press) 431

Freeth, T., Higgon, D., Dacanalis, A., et al. 2021, NatSR 11, 5821

Grant, E. 1977, Physical Science in the Middle Ages (Cambridge: Cambridge Univ. Press)
Guessoum, N. 2008, Obs, 128, 231
Hecht, E. 2019, AmJPh, 87, 176
Kochiras, H. 2022, PhilSci, 80, 557
Langer, R. E. 1937, AMM, 44, 495
Morris, A. M. A. 2017, HisSC, 56, 196
Pancheri, L. U. 1978, AmJPh, 46, 455
Ragep, F. J. 2007, HisSC, 45, 65
Rosen, E. 1965, Kepler's Conversation With Galileo's Sidereal Messenger, 1st complete translation, with an introduction and notes (New York: Johnson Reprint Corp)
Sakamoto, K. 2016, Julius Caesar Scaliger, Renaissance Reformer of Aristotelianism (Leiden: Brill)
Sepper, D. L. 2009, RIP, 249, 261
Severinghaus, J. W. 2016, AdPE, 40, 370
Slowik, E. 2021, Descartes' Physics (Stanford Encyclopedia of Philosophy) https://plato.stanford.edu/entries/descartes-physics/#LawsMotiCartConsPrin
Stephenson, F. R., & Green, D. A. 2003, JAHH, 6, 46
Westman, R. S. 1980, HisSC, 18, 105

Chapter 3

The Infinite Universe and the Dark Sky

> *"If the stars should appear one night in a thousand years, how would men believe and adore; and preserve for many generations the remembrance of the city of God which had been shown! But every night come out these envoys of beauty, and light the universe with their admonishing smile."* — Ralph Waldo Emerson, "Nature and Selected Essays".

The size of the Universe can be either finite or infinite. In this chapter we will go through some history of this old debate. Then we shall look at why Newtonian gravity, when applied to an infinite static and eternal Universe, gives rise to problems. Lastly, we explained the Olbers' paradox: why is the night sky dark? This problem is usually only mentioned in passing in cosmology textbooks without much elaboration. We will use it as an example to demonstrate how mathematical modeling works.

The infinite has always fascinated us. Even just as an abstract concept in mathematics, it was only until the work of George Cantor in the 19th century that we truly understood the concept of cardinality, the "size" of an infinite set. For example, as we briefly mentioned in the previous chapter, the set of odd natural number has the same cardinality as the set of all natural number. This is because for any $n \in \mathbb{N}$, we have the one-to-one correspondence (injective map) $2n + 1 \leftrightarrow n$. Similarly the set of all even numbers also has the same cardinality because $2n \leftrightarrow n$ is an injective map. The "Hilbert's Hotel" demonstrates how strange infinity can be. Imagine that the hotel has infinitely many rooms, which are all occupied. One day a new guest appeared. The guest can still be accommodated by asking the people in room No. 1 to move to room No. 2, and those in room No. 2 to move to No. 3 and so on *ad infinitum*. The newly arrived guest can then rent room No. 1. In Hilbert Hotel, all rooms can be occupied, while also making new rooms available. This is

doi:10.1088/2514-3433/ade963ch3

very counter-intuitive indeed. This also illustrates how an infinite space can still expand,[1] as it does in modern cosmology.

Some infinities are larger than others. For example, the set of real numbers $\mathbb{R}$ has cardinality larger than that of $\mathbb{N}$, as demonstrated by Cantor's famous "diagonal argument" in 1891. Specifically, the cardinality of $\mathbb{N}$ is denoted $\aleph_0$, and the cardinality of $\mathbb{R}$ is $2^{\aleph_0}$, usually denoted as $\mathfrak{c}$. If we arrange the sets of infinities from the smallest to the largest, $\aleph_0$ is the smallest, followed by $\aleph_1$ and so on. For some time mathematicians wondered whether $\mathfrak{c} = \aleph_1$, or whether its cardinality is between $\aleph_0$ and $\aleph_1$. The *continuum hypothesis*, proposed by Cantor in 1878, asserted that $\mathfrak{c} = \aleph_1$ and that there is no set whose cardinality lies strictly between $\aleph_0$ and $\aleph_1$. It turned out that, as later understood from the work of Kurt Gödel[2] (1906–1978) and Paul Cohen (1934–2007), we *cannot* prove the hypothesis true or false under the framework of the assumed Zermelo–Fraenkel set theory together with the so-called *Axiom of Choice* (ZFC). The latter is necessary to allow us to choose an element from each infinite set to set up a one-to-one correspondence. In other words, the continuum hypothesis is independent of ZFC; we can either add it to the set of axioms of ZFC, or add its negation, both of which would be consistent.

Given that infinity as an abstract entity is already such a hard concept to contemplate in mathematics[3], physicists usually do not think that true infinity can arise in the actual world. For example, the singularity inside a black hole has infinitely large curvature, and most physicists think that this is unphysical; it probably just indicates that our theory failed close to the singularity. Nature abhors infinity. Or does it? What about the Universe itself? Many cosmologists seem to find the concept of an infinitely large cosmos acceptable, although of course the *observable universe* is finite. The issue cannot be decided from observations alone. Nevertheless, the debate concerning the size of the Universe is an old one. An infinite universe does lead to serious philosophical consequences. For example, if the arrangements of matter is finite, we should expect that in an infinite universe, there should be infinitely many copies of worlds that look very close, if not exact copies, of ours. There would be copies of *us* like in a parallel universe, but in this case they are within the same cosmos, just out of reach beyond the observable universe. This is a "Level 1 multiverse" in the classification of Max Tegmark.[4]

In this chapter we will see how the finite view of the Universe gradually gave way to the infinite cosmos. We will also discuss the shortcomings of Newtonian

[1] Incidentally, just like Hilbert's hotel, the Universe is expanding *in* itself. There is no need to have a bigger ambient space that the Universe is expanding *into*.

[2] The famous Gödel's incompleteness theorem.

[3] I certainly struggled a lot when I took a course on axiomatic set theory as an undergraduate student in mathematics!

[4] Level 2 consists of different post-inflationary universes, each of these could be infinite to contain their own level 1's. For more discussions on inflation, see Section 13.3. Level 3 are quantum branches as in the "many-world" interpretations of quantum mechanics. Level 4 consists of alternative mathematical structures that would lead to different physical laws; including different kinds of quantum mechanics. For example, in our Universe quantum mechanics utilizes Hilbert spaces, which are vector spaces over $\mathbb{C}$. We could contemplate the possibility of building a theory on vector spaces over a Galois field $\mathrm{GF}(q)$, named after the tragic mathematician Évariste Galois (1811–1832). Such a theory has been studied in the literature.

cosmology in an infinite static universe and discuss the various proposals to resolve these problems. This is followed by a detailed discussion of the Olbers' paradox—why is the night sky so dark? A seemingly innocent childish question that led to our better understanding of a dynamical cosmos.

3.1 Is the Universe Finite or Infinite?

We have seen from the previous chapter that in various depictions of celestial spheres, the cosmos is depicted as being finite, until the English mathematician Thomas Digges (1546–1595) argued otherwise and his drawing of the celestial spheres no longer have the stars fixed on the outermost sphere, but rather filling the space beyond the planetary spheres; see Figure 3.1. He believed the Universe is indeed infinitely large and that it is filled with an infinite number of stars.

Thomas Digges received his first mathematical training from his father, a fine mathematician by the name of Leonard Digges, who also studied optics and purportedly invented early forms of telescopes (the descriptions are quite vague to be sure, though this raised the possibility that Digges too had pointed telescopes towards the heaven, but there is no evidence for this ever happening). Unfortunately his father passed away when he was 14, and the astronomer-occultist John Dee (see Section 2.2) took him under his care, and provided him more training in mathematics. Digges was a skilled astronomer, as he also conducted study on the "new star" (the supernova SN 1572) that appeared in the constellation of Cassiopeia in 1572, and showed that the star was *supralunar*, i.e., above the Moon, and thus the celestial realm was not immutable. In his meticulous analysis, he even took into account the eccentricity of the eye (the distance of a point on the retina from the center of the eye's fovea). To measure the small parallaxes involved in his astronomical studies, Digges often had to derive new useful trigonometric identities.[5] A believer in the heliocentric model, Digges recognized the opportunity to use SN 1572 to test the theory. The luminosity of SN 1572 fluctuated, that is to say, it was brighter on some nights. Digges and Dee thus devised an idea that the brightening and dimming cycle was due to the *motion* of the star that periodically moved closer and further away from the Earth. Of course, they did not realize how fast and how far a star needs to move in order to account for the changes in the luminosity. Furthermore, their idea could not explain why the star appeared suddenly out of nowhere already being so bright, instead of slowly increasing its brightness as it should if it came from afar. Digges had also experimentally tested the argument against Earth's motion that if the Earth is moving, an object dropped from the top of a tower would not hit the tower's base but at a certain distance to the west, having been "left behind" by the Earth's motion. Digges' illustrations include that of a plummet dropped from the top of a ship mast when the ship is in motion.

The astronomical works of Digges were known, but unfortunately seldom been read fully by others (Johnson & Larkey 1934). In fact, even in modern days, many

---

[5] We learned trigonometry in high school but students are rarely told that the subject was primarily invented to aid in astronomical calculations! *Spherical trigonometry*, a branch once even more useful for astronomy and navigation alike, was once a part of mathematics education but has since disappeared from most curriculum.

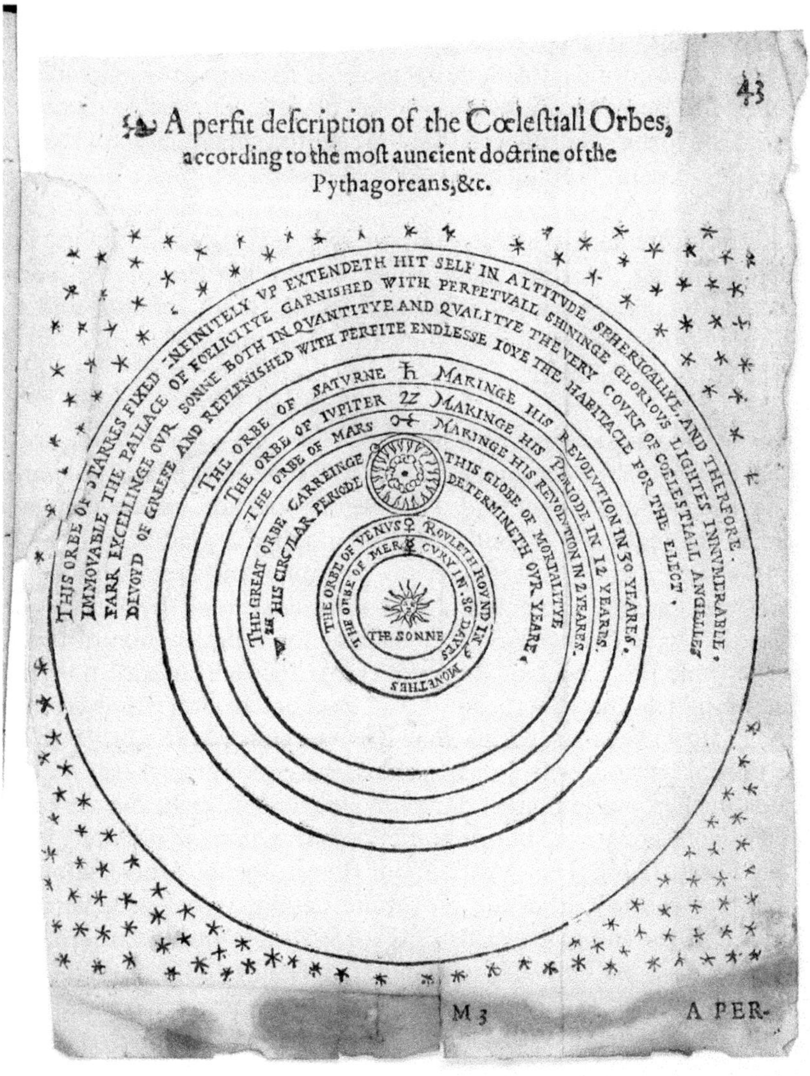

Figure 3.1. Thomas Digges' illustration of an infinite universe with a heliocentric worldview, originally published in 1576. The stars are no longer confined to a "sphere of fixed stars", but rather scattered across the cosmos. The diagram is a curious mix of radical new view and older traditions. For example, the Earth is surrounded by the sphere of the Moon, below which is the sublunar realm of the traditional Ptolemaic view, with the sphere of earth, water, air, and fire. Even the title reads (in modern spellings): "A perfect description of the Celestial Orbs, according to the most ancient doctrine of the Pythagoreans, etc.". In the outermost sphere it is written: "This orb of stars fixed infinitely up extends itself in altitude spherically and therefore immovable; the palace of felicity garnished with perpetual shining glorious lights innumerable, far excelling our Sun both in quantity and quality; the very court of celestial angels, devoid of grief and replenished with perfect endless joy; the habitation of the elect." *Credit:* A prognostication everlasting of right good effect, fruitfully augmented by the author, containing plaine, briefe, pleasant, chosen rules to iudge the weather by the sunne, moone, starres, comets, rainbow, thunder, clowdes, with other extraordinary tokens, not omitting the aspects of planets, with a briefe iudgement for ever, of plentie, lacke, sicknes, dearth, warres, &c. opening also many naturall causes worthie to be knowne ... / corrected and augmented by Thomas Digges his sonne. *Source*: https://wellcomecollection.org/works/pwneanbw/images?id=pcdsggmx

textbooks credited instead the Italian philosopher Giordano Bruno (1548–1600) for the idea of an infinite cosmos. Bruno proposed that the stars are not unlike our own Sun, and they are surrounded by their own planets, on which there could be extraterrestrial life forms. He believed that the celestial spheres are not solid (though not based on any observation) as he wrote:

> "Now behold, the man who has surmounted the air, penetrated the sky, wandered among the stars, passed beyond the borders of the world, effaced the imaginary walls of the first, eighth, tenth spheres, and the many more you could add according to the tattlings of empty mathematicians and the blind vision of vulgar philosophers."

Bruno was burned at the stake in Rome in 1600 by the Church. Some astronomy textbooks have perpetuated the myth that he was burned primarily because of his belief in the infinitude of stars and possible life outside our planet, and thus misrepresented the case as a conflict between religion and scientific thoughts (Bruno's ideas were really just based on metaphysical speculations). This is, however, not the case. The Church was mainly concerned by his spiritual and esoteric ideas that were deemed heretical, such as the reincarnation of the soul.

Around the same time, Nicolai Reymers (1551–1600), also known as Reimarus Ursus,[6] an imperial mathematician to Emperor Rudolf II, wrote in his *Fundamentum Astronomicum* of 1588 that it is not possible to say if the stars are at equal or unequal distances from us, and had suggested that there could be stars that are much further, even infinitely far away, from us. Like Bruno, he entertained the possibility that there are an infinite number of stars filling in infinitely large space (Johnson & Larkey 1934). The debate about the size of the Universe nevertheless predated the 16th century; after all, it is quite natural to ask whether the world extends forever. Nicolaus Copernicus (1473–1543), a century earlier, seemed to harbor the belief that the Universe is infinite (McColley 1936), though he did not show this in his heliocentric celestial sphere diagram. However, at times he seemed to be agnostic about the issue (Johnson & Larkey 1934), and preferred to leave the issue to be discussed by the "philosophers". Another curious thinker who is little known by students in physics and astronomy is the influential figure during the early Renaissance, Nicholas of Cusa[7] or Nicolaus Cusanus (1401–1464), a German Catholic bishop, who was also skilled in philosophy, mathematics, law, and astronomy. Nicholas of Cusa thought that although God is infinite and therefore

[6] A bitter rival and competitor of Tycho Brahe, see Section 2.2.

[7] Nicholas of Cusa was way ahead of his time. He also disowned the long-held belief that celestial objects are perfectly spherical and that planetary orbits are circular. He allowed for deviations in both. Even Kepler, two centuries later, would find it hard initially to accept non-circular orbits until he could no longer deny the observational data; Galileo never believed that orbits can be anything but circular, despite Kepler's findings. Nicholas of Cusa also believed that celestial objects like other planets, the Moon, and the Sun, are not dissimilar to Earth. This is also quite against the commonly held belief back in the day that supralunar stuff is completely different from sublunar material. However, he went too far in the similarity argument and even thought that there are people living on the Moon and the Sun!

cannot be comprehended by mortals like us, we can nevertheless *approach* the unlimited understanding. He provided a geometrical analogy,[8] which essentially amounts to imagining a tangent line to a circle. If the circle is becoming infinitely large, then the tangent line would tend to the circumference of the circle. Thus, even if we cannot ever figure out "the truth", this does not mean that we should give up on trying to understand it as best as we can. Nicholas of Cusa believed that the Universe can be infinitely large, and therefore there is no true *center*, only a perspectival one. Thus a heliocentric model has us at the center only because we viewed the Universe from our location. This is the same view shared by modern cosmology—the Universe is homogeneous, i.e., the same everywhere. Often popular science books would clarify to the readers that the Big Bang was not at the center of the Universe, because there is no center, or rather, every point is as good as any other to be called a "center". Modern textbooks tend to gloss over the history so much that readers are left with the impression that people in the past only argued over the validity of geocentrism versus heliocentrism—both of these have the word "center" in their name! In actual history, a world with no true center has been envisioned for a long time.

Nicholas of Cusa wrote in *De docta ignorantia* ("Of Learned Ignorance") (Small 1983):

> "Therein we have the reason why every man, whether he be on Earth, in the Sun or on another planet, always has the impression that all other things are in movement while he himself is a sort of immobile center [...]. In consequence, there will be a *machina mundi* whose center, so to speak, is everywhere, whose circumference is nowhere, for God is the circumference and center and He is everywhere and nowhere."

Machina mundi is Latin for "world machine", the idea that the Universe is a delicate and magnificent machine created by God (c.f. Descartes' "clockwork Universe" mentioned in the previous chapter[9]). This phrase appeared as early as the 13th century, in the work of the Oxford-trained scholar Johannes de Sacrobosco (1195–1256), also known as John of Holywood. The idea of the cosmos being a sphere with the center everywhere and circumference nowhere is a rather old one (Small 1983). For example, Vincent of Beauvais (c. 1184/1194–c. 1264), a Dominican friar in France, mentioned in his work *Speculum historiale* the same description for a cosmic sphere, which in Latin reads *sphaera cuius centrum ubique, circumferentia nusquam.*

---

[8] Nicholas of Cusa believed that mathematics, with its abstractness, is suited as a tool to understand spiritual matters.

[9] Incidentally, Descartes did not like the word "infinite", which he reserved for God, but he did consider the possibility that the Universe can be as large as needed. The latter concept he called "indefinite". Indefinite things are those in "which, from some point of view, we are unable to discover a limit." Newton would criticize these distinctions, and even remarked that Descartes "should be corrected by the grammarians. For the word "indefinite" ought never to be applied to that which actually is, but always looks to a future possibility", i.e., that the Universe is either finite or infinite. See more discussions in Verelst (2014).

Further back in time, Isidore of Seville (560–636), an archbishop in Seville of Spain, contemplated the idea that—though not quite an infinite cosmos—the sphere of fixed stars has a certain thickness, and that some of the stars could be larger but further away (though he also mistakenly believed that stars do not emit their own lights but reflect light from the Sun).

What did Aristotle think about infinite space back in the days of ancient Greece? He rejected the possibility that the cosmos is infinite in extent. This is because in Aristotle's view, motion—for example, the fact that things fall towards the Earth—is because they tend towards their natural state. In the case of objects falling, there is a distinguished location, namely the center of the Earth, which was argued to be the center of the Universe. But we have just mentioned that there is no "center" in an infinite world. Thus, to keep his philosophy of motion, Aristotle had to reject the notion of an infinite cosmos. If every point is equivalent to any other as per the requirement of homogeneity, then why, argued Aristotle in *Physics*, that an object falls "down", "why [not] rather up or down or anywhere whatever" within the infinite space?

Moving ahead in time instead, one finds the debate still alive during the time of Newton, namely in the 17th century. Isaac Newton (1643–1727), along with Joseph Raphson[10] (1648–1715) as well as the philosophers Henry More (1614–1687), Samuel Clarke (1675–1729), Baruch Spinoza (1632–1677), and the cleric Thomas Bradwardine (1290–1349) would argue for the infinite extent of the Universe. Some of the arguments were based on theology. For example, to limit the size of the Universe is to limit the power of God, since an almighty being would surely be able to create an infinitely large cosmos. Some thinkers like More and Raphson would go further and *divinize* space, which means that the infinite space is now an attribute of God. An empty space without matter (a modern reader can easily imagine this as Euclidean space $\mathbb{R}^3$) is not material, and therefore More viewed it as spiritual, hence divine. Since the immensity of God now constituted the infinite space, God is omnipresent and is contained by everything.

These discussions would appear to a modern scientifically trained reader to be rather bizarre, but it's just a typical way of thought back then in Europe. Note, however, the progress made between the time of Nicholas of Cusa and Newton. The old debate concerned whether there is anything outside the sphere of fixed stars; and if there is even such a thing as a sphere of fixed stars. If one claims that there is *nothing*, not even space or time,[11] outside the sphere of fixed star, or any fixed boundary, what would happen to someone who extend his arm or body beyond that into the ultimate void?

After the geometry of Descartes and physics of Newton became available to the natural philosophers, the debate switched to focus on whether the Universe is truly infinite, and with the Christian God as part of the debate. Sure, the idea that the void despite being devoid of all things can still be filled by spiritual substances can be

[10] Physics and mathematics students would recognize his name from the Newton–Raphson method.

[11] In the words of Aristotle in *De caelo*, Book I, Chapter 9, there can be "neither place, nor void, nor time, outside the heaven". In other words, there would be *nothing*, not just *no-thing*.

traced back to pre-Christian era, e.g., in the Hermetic dialogue *Asclepius* (or the "Perfect Discourse of Hermes Trismegistus") (Grant 1969) written some time between the 2nd or 3rd century. Nevertheless, there was a subtle shift in the debate of the nature of space. As Edward Grant wrote (Grant 1977):

> "Whereas in the Middle Ages God's nondimensionality determined the properties of an imaginary and dimensionless, extramundane, infinite, void space, in the seventeenth and early eighteenth centuries it was the three-dimensional infinite void of the new physics that, unwittingly perhaps, conferred tridimensionality upon the deity."

We will continue our discussion on the nature of space in Section 11.1.

Despite Newton's belief that the Universe is infinitely large, his theory of gravity encounters various problems when it is applied to such a cosmos. In the next section, let us try to understand Newtonian cosmology better (some of its results are analogous to the general relativistic one that we will discuss in Chapter 5), and see what people who have tried to solve the problems encountered.

3.2 Newtonian Cosmology and Its Problems

The Newtonian gravitational potential Φ can be related to matter density ρ via the Poisson equation

$$\nabla^2\Phi = 4\pi G\rho, \tag{3.1}$$

and the gravitational acceleration is given by $\overrightarrow{a}(t) = -\nabla\Phi$. In mathematical literature, the Laplacian ∇^2 is often denoted by Δ. In the special case where $\rho \equiv 0$, we also refer to the equation as the Laplace equation. The Poisson equation is a consequence of the so-called Gauss law (in differential form):

$$\nabla \cdot \overrightarrow{g} = -4\pi G\rho, \tag{3.2}$$

as can be readily verified by substituting $\overrightarrow{a}(t) = \overrightarrow{g}(t) = -\nabla\Phi$ into the Gauss law. Note that this Gauss law is the same as the one for an electric field in electromagnetism, except for an overall minus sign:

$$\nabla \cdot \overrightarrow{E} = \frac{\rho}{\varepsilon_0} \Rightarrow \nabla^2\Phi_E = -\frac{\rho}{\varepsilon_0}, \tag{3.3}$$

where ε_0 is the permittivity of free space and ρ is now the charge density. Similar to the gravitational field, the electric field $\overrightarrow{E}$ is related to its potential by $\overrightarrow{E} = \nabla\Phi_E$. Why the difference in the overall minus sign? This is because in the case of electric charge, like charges repel. For gravity, "gravitational charge" is just the mass, which only comes in one sign, and they always attract.

Now, let us consider an infinitely large universe filled uniformly with matter (i.e., it is homogeneous). Consider a point mass m and set up a spherical coordinate centered there, and a spherical shell of matter surrounding it at distance r. We can arbitrarily split the shell into two halves (say the left and right hemispheres), and ask what is the total force exerted by the hemisphere at r on the point mass. This can be calculated by first computing the total mass in a ring with angular position θ and angular thickness of $d\theta$, and thickness Δr, which yields $\rho \cdot 2\pi r \sin\theta \cdot r d\theta \cdot \Delta r$. Essentially, we are using the fact that "mass = density × volume", where the volume of the "ring" is "radius × height × thickness". (Draw a diagram to see this!) The total force exerted by the hemisphere at r along the x-axis is then given by

$$\overrightarrow{F} = -\frac{GMm}{r^2}\hat{r}, \tag{3.4}$$

where $\hat{r}$ is the unit radial vector, and

$$M = 2\pi G\rho r^2 \Delta r \int_0^{\pi/2} \sin\theta \cos\theta d\theta. \tag{3.5}$$

Note that the amount of matter grows like r^2, but this cancels with the r^{-2} in the gravitational force. Thus $\overrightarrow{F}$ is independent of r. One finally obtains[12]

$$\overrightarrow{F} = -2\pi G\rho\Delta r \cdot \frac{1}{2}\sin^2\theta \,|_0^{\pi/2}\hat{r} = -\pi G\rho\Delta r\hat{r}. \tag{3.6}$$

Clearly, by symmetry the same holds for the left hemisphere but with an opposite sign. This is just the second Newton's shell theorem—if the body is a spherically symmetric shell, no net gravitational force is exerted by the shell on any object inside (which is true even if the body is not situated at the exact geometrical center). Now imagine there are infinitely many discrete shells of matter, because the result obtained above is independent of r, then the total force is just

$$\overrightarrow{F}_{\text{Total}} = \pi G\rho\Delta r(1 - 1 + 1 - 1 + 1 - 1 + \cdots)\hat{r}. \tag{3.7}$$

If we consider the sum of infinitely thin shells we should replace this with an integral, but already we see the problem—the sum is not convergent. $1 - 1 + 1 - 1 + 1 - 1 + \cdots$ is known as the Grandi's series in mathematics, named after the Italian mathematician and priest Guido Grandi (1671–1742). The series appeared in his 1703 book *Quadratura circula et hyperbolae per infinitas hyperbolas geometrice exhibita*. It can be made to "converge" to *any number* we want, and thus is *not convergent*. While we can group it as $(1 - 1) + (1 - 1) + \cdots$ and claim that the answer is zero, we can also group it as $1 + (1 - 1) + (1 - 1) + \cdots$ and argue that it is 1. It is often useful for some purposes to *define* a new kind of summation (e.g., Cesàro summation) in which the series has a value of $1/2$ (this value was also what Grandi prescribes to the

[12] Now we see why we don't want to write dr but instead use Δr instead, otherwise it would be strange to have differential on the right-hand side, but not on the left-hand side! Sorry, but this is a sloppy physicist's way of doing calculus.

series). However, for *our* purpose, the fact that this sum is not well-defined is a problem.

In fact, as pointed out by John Norton (Norton 1999), various useful quantities that involve the integrals

$$\int_{r_1}^{r_2} \rho r \, dr, \qquad \int_{r_1}^{r_2} \rho \, dr, \qquad \int_{r_1}^{r_2} \frac{\rho}{r} \, dr, \tag{3.8}$$

are all non-convergent in the limit $r_2 \to \infty$ if ρ is constant. This renders physics in the homogeneous infinite cosmos ill-defined. (There is of course also the stability issue—perturbation would clump matter together and it is hard to maintain strict homogeneity, but this is not the main problem since we need not claim the Universe is strictly homogeneous, and in fact it is not. There are empty spaces between galaxies, for example. Newton and contemporaries didn't know about galaxy distribution, but they did know that planets have different densities than spaces in between them. The divergent issue arises even if ρ is only constant on average.) To be clear, the calculation above is not strictly speaking, correct.

To understand this, although the previous calculation is in terms of the force, we can also consider the issue from the perspective of the Poisson equation. Mathematically, the Poisson equation can have a non-unique solution, depending on the boundary condition. The problem thus lies in what is the boundary condition involved in this problem. This is often not stated explicitly in physics textbooks, but is crucial for the study of differential equations. Evidently the integral is not convergent because the integrand does not decay sufficiently fast. Indeed, ρ being constant implies that it does not decay at all, unlike the typical physical situations in which we assume $\Phi(r) \to 0$ as $r \to \infty$. In the homogeneous infinite universe, the shell theorem actually fails. In any case, the issue cannot be resolved by making the argument more rigorous. Today we understand that Newtonian cosmology only really makes sense when we regard the Universe as a *finite* ball of galaxies that looks isotropic when we are deep inside it, too far from the boundary to realize otherwise. Force being a vector, hence with a direction, cannot describe a truly isotropic Universe except at the origin of the sphere.

But what can one do about this? How did thinkers in the past deal with this problem? In theoretical physics, whenever we reached a problematic conclusion or an apparent paradox, we should examine the assumptions or hypotheses and see if the conclusion can be avoided by modifying the underlying assumptions. Here, the assumption is that we have: (1) an infinitely large universe that is, (2) uniformly filled with matter, which is (3) governed by Newtonian law of gravitation. Any of these three assumptions can be challenged.

Quite naturally, one of the approaches is to consider matter configurations that are not homogeneous in some clever arrangements. One such example was given by the "hierarchical cosmology" that was proposed by quite a few physicists and astronomers, including Carl Charlier (1862–1934) and Franz

Selety (1893–c. 1933), the details of which can be found in Norton's article (Norton 1999). An earlier approach was by Edmund Fournier d'Albe (1868–1933). In hierarchical cosmology, stars are clustered in many "levels" in a self-similar manner, such that the density is different from the ith level to the $(i+1)$-level. This has the flair of what would later be called fractal structures, although the word "fractal" was only coined in 1975 by the mathematician Benoît Mandelbrot (1924–2010). Another option is to reject the notion that the Universe is infinitely large. Josef Lense (1890–1985) gave an example of a closed spherical universe. In such a world, the gravitational force would not diminish as $1/r^2$, but rather as $[R^2 \sin^2(r/R)]^{-1}$, where R is the radius of curvature of the spherical space. He showed that the various integrals do converge in this case. Lense is better known for the Lense–Thirring effect (or frame dragging) in general relativity, which is a general relativistic correction to the precession of a gyroscope near a large rotating mass. The idea that space can have different topologies, and that it can be finite yet without boundary like a sphere, is a relatively new idea. In the debate regarding whether the Universe is finite or infinite during the Renaissance and the Middle Ages, one can find arguments that amount to asking "if the Universe is finite, then what is beyond its boundary?" They had in mind a finite world bounded by a "wall" of sorts (see Figure 3.2), but this line of thinking did not consider the possibility that the world has a different "shape". If the topology is spherical, for example, we can in principle travel in a straight line in space and end up where we started, i.e., we can circumnavigate the Universe much like we can do so on the surface of the Earth.[13] (In reality, even if this is the case, circumnavigation is not possible now that we know the Universe is expanding, and we cannot travel fast enough to "overtake" the expansion.)

Other more exotic proposals include that of August Föppl (1854–1924), who proposed (Treder 1991) the possibility of the existence of negative gravitational masses, so that the total masses may not be as large as initially thought or even canceled perfectly, and the integral may have a chance of being convergent. In Föppl's idea, there would then be two kinds of "gravitational charges" (masses), not unlike that of electromagnetism. However, in this case, like charges attract, while opposite charges repel. The fact that we do not find any negative mass can be explained by the virtue of this repulsion—the negative masses[14] must have been

[13] Mathematically, we say that the spatial geometry of the Universe could be S^3, instead of S^2 like the Earth's surface.

[14] Between the 17th and the 18th century, the phlogiston theory was used to explain how stuff can burn and oxidize. Phlogiston is a hypothetical element contained inside any combustible substance, which upon being set on fire, is released. However, some reactions produced an end product that is *heavier* than the initial substance. Most chemists tried to explain this by considering absorption of something from the air into the substance, which out-weighed the release of the phlogiston. Some others, however, proposed that phlogiston may come with a variety that has a negative mass! Proposing new type of matter that has unusual properties to explain anomalies is not so unusual throughout the history of science.

Figure 3.2. This is an image whose artist remains unknown, but seems to be first published in the work of the French astronomer and writer Camille Flammarion (1842–1925), *L'atmosphére: météorologie populaire* (1888). It is often referred to as the "Flammarion engraving"' (the original is a wood engraving). It shows a man peering through an opening in the heavenly firmament, and looking outside the sphere of fixed stars, which seems to contain glowing spheres, fires, swirling clouds, and a cosmic machine that looks like a wheel-within-wheel, which is reminiscent of the visions of Ezekiel in the Biblical story. Flammarion wrote: "A naïve missionary of the Middle Ages even tells us that, in one of his voyages in search of the terrestrial paradise, he reached the horizon where the earth and the heavens met, and that he discovered a certain point where they were not joined together, and where, by stooping his shoulders, he passed under the roof of the heavens." *Credit:* This [Flammarion] has been obtained by the author(s) from the Wikimedia website, where it is stated to have been released into the public domain. It is included within this book on that basis.

repelled away far from our positive mass galaxy. Putting aside other problems with negative masses, it is a tricky business to try to "regularize" infinity with minus infinity to yield a finite number in a unique manner, so this proposal wasn't taken as seriously as a solution of the Newtonian gravity problem.

William Thomson (1824–1907), also known as Lord Kelvin, once realized that if the Universe is filled uniformly with æther, which he believed to be the case, then gravitational attraction would cause it to collapse unless it has infinite resistance. His solution was simple—he just *postulated* that æther does not gravitate, or in his words æther[15] is "a substance outside the law of universal gravitation".[16] This certainly does not resolve the problem of divergence for ordinarily matter, unless there are some strange interactions between the æther and the matter. Kelvin simply proposed that the density for ordinary matter is not uniform and diminishes further away, independently of the æther.

Finally, one can contemplate the possibility that Newtonian gravity is incorrect. We defer this discussion to Chapter 6. We shall now move on to another problem that seems quite different at first glance, but is quite similar to the Newtonian cosmology problem. For more discussions on problems in Newtonian gravity, see Wallace (2017).

3.3 The Dark Sky and Olbers' Paradox

The question is deceptively simple: why is the night sky dark? This sounds like a bizarre question at first. One immediate answer is because the Sun is not there to illuminate the sky. The real conundrum comes if we assume a homogeneous universe that is infinitely large, unchanged and eternal (infinite in time also; i.e., has no beginning). In this case every line of sight will end on the surface of a star. The entire night sky should then be bright. In other words, the darkness of the night sky conflicts with the assumption of an infinite and eternal static universe. This is the "Olbers' paradox", named after Heinrich Olbers (1758–1840) who described the problem in 1823. This is, however, a very old problem. Johannes Kepler (1571–1630), for example, had used the fact that the night sky is dark to argue that the Universe must be finite[17] (in the sense that there is a boundary beyond there is nothing). Thomas Digges was also aware of the problem, and so did many others, in one form or another. The first modern form of the "paradox" is attributed to Jean-Philippe Loys de Cheseaux (1718–1751), which predated Olbers' work by almost 80 years.

To appreciate why under the aforementioned assumption the night sky should be bright, we can try to set up some mathematical models. This is a good opportunity

[15] It is interesting to note that throughout 11th-16th century China, philosophers have discussed the nature of the "great emptiness" (太虚), which originally just referred to empty space. Matter is believed to be formed from *qi* (气). The debate concerned whether *qi* is produced in space, or whether *qi* itself *is* space, in the sense that space cannot exists without it. This is similar to the Leibniz–Clarke debate in the 14th century concerning whether absolute space exists. See Section 11.1 for more details.

[16] This approach is similar to the Bohr's model of atoms (see Chapter 7), in which Neils Bohr postulated that electrons in some special orbits simply do not radiate. Sometimes to make progress, one has to make assumptions and postulates even if the *why* still eludes us.

[17] Kepler was against the idea that the Universe is infinitely large for another reason. In his *Epitome astronomiae*, he illustrated his argument that was first developed in his 1606 *De nova stella*, which shows the Sun surrounded by the orbit of the Earth and the sphere of fixed stars, beyond which are many more stars scattered all over the infinite Universe. The caption for the figure reads *Quod habes hujus rei Argumentum?*, i.e., "What argument do you have for this matter?". Kepler thought that this is rather strange—why would the infinite Universe outside be full of stars but there is so much empty space in the solar system? In other words, he questioned the homogeneity assumption of the infinite Universe. This mistake was due to Kepler not being aware of the true distances between stars, and indeed there are empty spaces almost everywhere. Kepler also did not believe that other stars are orbited by planets; in fact he sees the Sun as quite a different celestial object than the stars (Granada 2008).

for us to see how mathematical modeling, discussed in Section 1.5, works. Some of the discussions here are based on an article by Edward Harrison (Harrison 1990). It is also good to mention some details as most cosmology textbooks may mention the Olbers' paradox but quickly glossed over it, giving the incorrect impression to students that there wasn't any serious attempt at trying to solve the problem before modern cosmology came about.

Consider a thin spherical shell of thickness Δr centered at Earth, whose inner radius is $r \gg \Delta r$. We can estimate the number of stars in this shell by $N \sim 4\pi r^2 \Delta r \cdot n$, where n is the number density (number of stars per unit volume). Let l be the luminosity of light emitted by a single star. Assuming all stars are equally luminous for simplicity, how much light from any given star in the shell can reach us? Well, since lights are emitted outward from a star in a spherical light-front, the light along any single ray has luminosity about $l/4\pi r^2$. Thus, the total amount of light we receive from all the stars *in this shell* must be

$$N \times \frac{l}{4\pi r^2} = 4\pi r^2 n \Delta r \times \frac{l}{4\pi r^2} = nl\Delta r. \tag{3.9}$$

Note that this is independent of r. In fact, avid readers would now realize that the calculation is more or less the same as that in the previous section when we discussed matter shells. Thus for the same reason, given that there are infinitely many shells in an infinitely large cosmos, the luminosity must diverge. The night sky should thus be not only bright, but infinitely so! Clearly something has gone wrong here. This divergence, unlike the gravity case, is due to the oversimplification of the model in treating stars as point sources. (This over-simplified model was never seriously considered in the history of the paradox; it is mentioned here, as in some other textbooks, only as a teaching tool.) In a more realistic case, we should not be able to see stars out to infinity since their light would be blocked by the foreground stars (unlike gravity, light can be blocked!). We should consider the fact that (1) stars have finite size, and (2) lights can be absorbed or blocked. Let us do exactly that and see if we can "regularize" the infinity.

Following Harrison (1990), let λ be the distance between the stars, or more accurately the mean free path of a ray of starlight between its emission and absorption by other stars. We can write $\lambda = V/\sigma$, where V is the average volume occupied by a single star[18] and $\sigma = \pi R^2$ is the cross-sectional area of the star. We can assume that R is the same for all stars. Now each star at distance r away would cover $\sigma/4\pi r^2$ of the sky—the further away a star is, the smaller the fraction, so we would see it as a point source. Denote the fraction of the sky covered by all the stars up to a radius $\mathfrak{R}$ away by Ξ. It is given by the integral (now we use dr instead of Δr because we want to integrate)

[18] The mean free path is usually defined as $(n\sigma)^{-1}$, where $n = N/V_{\text{total}}$. Here $V = V_{\text{total}}/N$. In radiative transfer theory, which every student of astrophysics will learn at some point, $1/\lambda$ is often called the "extinction coefficient".

$$\Xi = \int_0^{\mathfrak{R}} 4\pi r^2 n \, dr \times \frac{\sigma}{4\pi r^2} = \int_0^{\mathfrak{R}} \sigma n \, dr = \sigma n \mathfrak{R}. \tag{3.10}$$

The entire sky would be covered if $\Xi = 1$. That is, if $\mathfrak{R} = (n\sigma)^{-1}$. We note that

$$\lambda = \frac{V}{\sigma} = \frac{V_{\text{total}}}{N\sigma} = \frac{1}{n\sigma}. \tag{3.11}$$

Therefore, we have found that there is a *cutoff scale* given by $\mathfrak{R} = \lambda$ such that all of the stars beyond that is obscured. In this model, since the number of stars are no longer infinite, there is no divergence, but the night sky is still not dark. Can we improve on the model further?

A closer look at the model above reveals that it does not take into account that stars within the sphere with radius $r = \lambda$ can also be occulted—there is always some probability that foreground stars can block background stars. In fact, Edmund Fournier d'Albe wrote in his *Two New Worlds* the extreme unlikely possibility that "an infinite line of stars might exist behind every star in the heavens". One simple way to model this is to treat absorption of starlight by foreground star as *attenuation* of light as it travels through some medium. By the homogeneity assumption, the probability of a light beam being absorbed per unit distance must be a constant. This means that we must have an exponential attenuation.[19] In the model above, the fraction integral

$$\Xi = \int_0^{\mathfrak{R}} \sigma n \, dr = \int_0^{\mathfrak{R}} \frac{dr}{\lambda}, \tag{3.12}$$

is then replaced by

$$\Xi = \int_0^{\mathfrak{R}} e^{-\frac{r}{\lambda}} \frac{dr}{\lambda} = 1 - e^{-\frac{\mathfrak{R}}{\lambda}}. \tag{3.13}$$

For $\mathfrak{R} \gg \lambda$, we have $\Xi \to 1$ as expected, whereas for $\mathfrak{R} \ll \lambda$, we recover $\Xi = \mathfrak{R}/\lambda$. This is expected—as well as a consistency check—since for small r, there is less occultation (less foreground stars), so the result should match that of the previous model. However, even for this improved model, the Olbers' paradox is still there. In fact, this model is probably the one that matches the intuitive explanation of the paradox, namely that every line of sight should hit a star, and so the night sky should be bright. We do not "overcount" the contribution of stars that are blocked.

So what can be done to resolve the paradox? Just like the Newtonian cosmology problem, some astronomers had tried to argue that the distribution of stars are not uniform but hierarchical in a way that does not lead to the paradox. Edmund Fournier d'Albe had another proposal: that most stars in the Universe is actually non-luminous. This idea of 1907 is akin to the modern idea of "dark matter" (whose

[19] Students may want to think about why this is so.

gravitational effect we can detect yet cannot be seen), but was used in a different way. If there are enough non-luminous stars, that would block enough starlight to make the night sky sufficiently dark. For this to work, Fournier estimated that to every luminous star like our Sun, there must exist some 10^{22} non-luminous stars! The most sensible, though ultimately incorrect, attempt was provided by de Cheseaux and Olbers. They argued that light beams can be attenuated also by dust, or more generically, what we today would call "interstellar medium" (ISM). Adding this effect into the model amounts to adding another attenuation term, thus changing the fraction integral into

$$\Xi = \int_0^{\mathfrak{R}} e^{-r\left(\frac{1}{\lambda}+\frac{1}{\mu}\right)} \frac{dr}{\lambda} = \frac{\mu}{\lambda + \mu}\left[1 - e^{-\mathfrak{R}\left(\frac{1}{\lambda}+\frac{1}{\mu}\right)}\right]. \tag{3.14}$$

Here μ is the absorption distance in the ISM. If the Universe is infinitely large, then $\mathfrak{R} \to \infty$, and the fraction of the night sky covered is now

$$\Xi \to \frac{\mu}{\lambda + \mu}. \tag{3.15}$$

If there are a lot of stuff in the Universe that can absorb light, we can model this with setting μ to be small (the smaller the absorption distance, the more effective absorption is), i.e.,[20] $\mu \ll \lambda$. In this case, $\Xi \ll 1$ is also small. This would then mean that the night sky is dark. For consistency check, we note that if there is no ISM, then $\mu \to \infty$ and we arrive at $\Xi \to 1$, in agreement with the previous result.

The astronomer John Herschel (1792–1871), who discovered the planet Uranus, realized in 1848 that this proposal cannot work because light also carries energy, and the amount of starlight involved would eventually[21] heat up the ISM so much that they would also radiate brightly. He himself preferred the non-uniform distribution of stars as an explanation of Olbers' paradox. His belief was no doubt based on his own observation that stars do form clusters. Utilizing his 40-foot telescope (Figure 3.3), Herschel even made the first quantitative chart of the Milky Way in 1785. Still, clustering of stars does not spoil *average* homogeneity enough to avoid Olbers' paradox, just like it cannot avoid the Newtonian gravity divergence issues discussed in the previous section.

While the non-strictly uniform distribution of stars does have some contribution to the resolution of the paradox, the main effect comes from the *finite* age of the stars, and in fact of the Universe. Surprisingly this idea was first proposed by the imaginative writer and poet Edgar Allen Poe (1809–1849), who is still famous for his horror and macabre stories. In his work *Eureka—A Prose Poem*, written in 1848, just a year before he passed away, Poe described his idea of how the cosmos came about. *Eureka* is rather ambitious, with the third paragraph stating:

[20] Recall from Section 1.5 that when we say something is small or large, there must be a reference scale of what it is smaller/larger than.

[21] Remember that with the assumption of an eternal cosmos, there is an infinite amount of time to achieve this; so it must already been heated up.

Figure 3.3. William Herschel's 40-foot telescope, constructed between 1785 and 1789 at Observatory House in Slough, England. The 40 foot (12 m) refers to the focal length of its 120 cm diameter primary mirror. For half a century it was the largest telescope in the world. The telescope was constructed by William Herschel and his sister Caroline Herschel. They were both exceptional observational astronomers. Caroline was the first woman to discover a comet: the periodic comet 35P/Herschel—Rigollet. She would go on and discover 7 more comets during her lifetime. She was the first female astronomer to receive salary from King George III. Today the largest single-aperture optical telescope on the ground is the Gran Telescopio Canarias located at the Roque de los Muchachos Observatory on La Palma in the Canary Islands of Spain, with mirror that has a diameter of 10.4 m. *Credit:* Astronomy: a 40-foot telescope constructed by William Herschel, in use outdoors. Coloured etching, 18–. Wellcome Collection. *Source*: https://wellcomecollection.org/works/srz5zy9y/images?id=hvwkmy46

> "I design to speak of the Physical, Metaphysical and Mathematical—*of the Material and Spiritual Universe:—of its Essence, its Origin, its Creation, its Present Condition and its Destiny.* I shall be so rash, moreover, as to challenge the conclusions, and thus, in effect, to question the sagacity, of many of the greatest and most justly reverenced of men."

Poe described the Olbers' paradox and gave its resolution:

> "The only mode, therefore, in which, under such a state of affairs, we could comprehend the voids which our telescopes find in innumerable directions,

would be by supposing the distance of the invisible background so immense that no ray from it has yet been able to reach us at all."

The main point here is that light from some stars simply has not enough time to reach us yet. This is impossible if we assume an eternal static infinite cosmos. However, if the Universe had a beginning, and stars only started to form somewhat "recently", then given that the speed of light is finite (this was known already in the 17th century, see Box 1 below) and the size of the Universe immense (finite or otherwise), there simply isn't time for all the starlight to reach us.[22] Herein lies an idea remarkably similar to the modern Big Bang cosmology. However, note that *the expansion of the Universe is not necessary for the resolution of Olbers' paradox*. All that is required is that stars are not eternal, that they formed at some time in the past and the light from faraway stars hasn't got enough time to reach us[23]. This was the possibility that Fournier deemed unlikely. In *Two New Worlds*, published in 1907, he wrote:

> "But to any one familiar with the line of reasoning derived from atomic conditions, the eternity of stars must appear extremely probable, whatever may be said of their luminosity of their individual fate."

Box 1: Measuring the Speed of Light
The first measurement of the speed of light was made by the Danish astronomer Ole Rømer (1644–1710) in 1676. The idea is simple in hindsight, but is still nothing short of remarkable. Since the speed of light is so fast, it was incredibly difficult to measure by any experiment on Earth at that time. Rømer noticed that the apparent orbit of Io, one of the Galilean moons of Jupiter (see Figure 2.6), is somewhat irregular. On average it takes only about 42.5 hours to orbit around Jupiter. One can observe quite frequently the moment it enters Jupiter's shadow (and so not visible in the telescope) and the moment it emerges from the shadow. Rømer noticed that the time *between* the successive "eclipses" was not constant. Instead, it decreases when Earth moved towards Jupiter (as measured by their relative position in orbits), and increases when Earth moved away from Jupiter. In particular, he noted a 20-minute delay when Earth was furthest away in the orbit, compared to when Earth was closest to Io. He reasoned that this is because light has to travel an additional distance (the diameter of the Earth's orbit) to reach us. Given the estimated Sun-Earth distance back then, he estimated that the speed of light is 226,663 km s^{-1}. Though about 24% lower than the true value of 299,792 km s^{-1}, it is the same order of magnitude. This is remarkable since back then it wasn't even obvious that the speed of light is *finite*, and some even argued that it is infinite.

[22] In fact, in some cosmological spacetime like the de Sitter space, there is a cosmic event horizon, and there are stars that we cannot see, even if we wait forever.
[23] In fact, it would be possible to have Olbers' paradox in an expanding Universe if star distribution remains uniform and unchanging in time. See Harrison's article (Harrison 1990) for more details.

Modern understanding of atomic and stellar physics would correct this view, and we now accept that stars do have a finite age. Of course, if the Universe itself is eternal, it would be harder to explain why there are only stars “recently” given the infinite amount of time available, so a finite age of the Universe itself would appear more natural in this regard. Johann Mádler (1794-1874) also proposed the finite-age idea in 1861. A more detailed scientific analysis of the finite age resolution was carried out by Lord Kelvin in 1901. That said, the expansion of the Universe does contribute another effect—the redshift of light. As the Universe expands, the source of light that is far away from us would appear redder and dimmer, so stars from sufficiently far away would simply be invisible. The effect of expansion and redshift, however, turns out to be relatively minor and cannot resolve Olbers' paradox on its own.

Another crucial factor that resolves the paradox is that there simply isn't enough stars to make the night sky glow. Harrison estimated that the intensity of starlight is about 10^{-13} of the thermodynamic equilibrium intensity required by the Olbers' paradox. By thermodynamic equilibrium, we mean the following: if the sky glows as it would in the Olbers' paradox, we would expect that interstellar space also has temperatures equal to that of the surface of stars. (In reality, space has a temperature of the cosmic microwave background of about 2.7 Kelvin, and indeed the sky glows in the microwave spectrum, a discovery made in 1964 that earned Arno Penzias (1933–2024) and Robert Wilson (1936–present) their Nobel Prize in Physics in 1978; see Section 5.4.)

Ultimately, the Olbers' paradox is historically interesting in that it is the kind of “obvious question” that nevertheless has a nontrivial resolution. Everyone knows that the night sky is dark, so most people would take it for granted. To realize that this is a question in need of answering is itself quite extraordinary. It is satisfying that the answer to such a seemingly mundane question leads all the way to “genesis”, the ultimate birth of our Universe and formation of stars. The Universe (and the stars within it) is not only not-static, but there was a day before which there were no stars, and even a day without yesterday. I wonder how Poe would react to modern cosmology validating—and surpassing—his imagination.

3.4 Exercises

(1) Show that if the Newtonian gravitational potential is $\Phi = -GM/r$, then $\nabla^2\Phi = 0$ provided that $r \neq 0$. Since the Poisson equation is

$$\nabla^2\Phi = 4\pi G\rho, \tag{3.16}$$

this means that all of the mass is concentrated at the center. Thus, in order to obtain the right-hand side, $4\pi G\rho$, we essentially have to ask, ignoring the constant factor $-GM$ for the moment, what is the volume integral of $\nabla^2(1/r)$ over a small volume V that contains $r = 0$. By converting the volume integral to an area integral, show that

$$\nabla^2\left(\frac{1}{r}\right) = -4\pi\delta^3(r), \tag{3.17}$$

and hence

$$\nabla^2\left(-\frac{GM}{r}\right) = 4\pi GM\delta^3(r). \tag{3.18}$$

Additional note: Here $\delta^3(r)$ is the three-dimensional Dirac delta "function" (mathematically it is rigorously defined as a *distribution*). Note that since

$$\int_{\mathbb{R}^3} \delta^3(r)\, dV = 1, \tag{3.19}$$

the physical dimension of $\delta^3(r)$ is inverse volume. Hence $M\delta^3(r)$ is indeed the matter density ρ.

(2) Replace the Newtonian potential above with a slightly modified version that is not singular at the origin as follows:

$$\Phi = -\frac{GM}{\sqrt{r^2 + b^2}}, \qquad b > 0. \tag{3.20}$$

This is known as the *Plummer Potential*, named after Henry Plummer (1875–1946), and is often used to model globular clusters and galaxies. What does the potential look like? How does it behave at large and small r? Use the Poisson equation to find the corresponding matter contribution. How does it behave at large and small r?

(3) Sometimes one may see an argument that tries to explain why Newtonian gravity is instantaneous (namely, if the Sun disappears now, the Earth's orbit would immediately be affected) by appealing to the fact that the Poisson equation contains no time derivative. That is, as soon as the matter distribution ρ changes, the potential also changes concurrently. However, as we mentioned in this chapter, the Poisson equation also appears in electromagnetism. Yet we know that an electromagnetic propagation has a finite speed. Explain why the argument does not work for electromagnetism.

(4) Related to the debate concerning whether the Universe is finite or infinite, is the so-called "Great Debate" in astronomy. Also known as the Shapley–Curtis Debate, the event was held on 26 April 1920 at the Smithsonian Museum of Natural History. Two astronomers Harlow Shapley (1885–1972) and Heber Curtis (1872–1942) presented two opposite viewpoints. Shapley argued that the so-called "spiral nebulae" then observed by telescopes were actually objects inside our own galaxy, albeit at the outskirts. Curtis argued that they were independent galaxies as big as ours. After Curtis' viewpoint was validated by more observations, our view of the Universe changed again—the cosmos is much larger than anyone had ever imagined. Find out more about the arguments of Shapley and Curtis: why did they argue for the opposite positions they held?

(5) For students who are familiar with differential geometry or general relativity, consider learning more about the geometrization of Newtonian gravity: the Newton–Cartan theory. See, e.g., Hartong et al. (2023).

References

Granada, M. A. 2008, JHA, 39, 469

Grant, E. 1969, Isis, 60, 39

Grant, E. 1977, Physical Science in the Middle Ages (Cambridge: Cambridge Univ. Press)

Harrison, E. R. 1990, The Galactic and Extragalactic Background Radiation, Proc. 139th Symp. IAU, ed. S. Bowyer, & C. Leinert (Dordrecht: Kluwer) 3

Hartong, J., Obers, N. A., & Oling, G. 2023, FrP, 11, 1116888

Johnson, F. R., & Larkey, S. V. 1934, HLB, 5, 69

McColley, G. 1936, PA, 44, 525

Norton, J. 1999, The Expanding Worlds of General Relativity: Einstein Studies Vol. 7 (Boston, MA: Birkháuser)

Small, R. 1983, JHI, 44, 89

Treder, H.-J. 1991, AN, 312, 229

Verelst, K. 2014, Sy, 191, 2907

Wallace, D. 2017, SHPS, 57, 35

AAS | IOP Astronomy

Chapter 4

Thermodynamics: The Science of Heat

"Some say the world will end in fire,
Some say in ice.
From what I've tasted of desire
I hold with those who favor fire.
But if it had to perish twice,
I think I know enough of hate
To say that for destruction ice
Is also great
And would suffice." —Robert Frost, "Fire and Ice".[1]

Thermodynamics and statistical mechanics play very important roles in modern physics. They are not only useful in traditional problems of heat engines and pistons, but also in modeling astrophysical systems and cosmology. This chapter aims to provide a quick review of the basic laws of thermodynamics, and then crucial concepts that will be useful in the latter chapters, such as the ideal gas law, the equipartition law, the virial theorem, and the partition function. This chapter ends with some discussions about the second law of thermodynamics and the puzzle of the "arrow of time"—why time "flows" from the past towards the future. This is ultimately a question of cosmology.

We *feel* heat when we huddle around a bonfire outdoors, without which it's freezing cold in the deep winter. But what *is* heat? Science can only make progress if we can make what we want to study more concrete and tangible, which means we must be able to *measure* it. This was achieved by the invention of the thermometer

[1] The poem was apparently inspired by the poet's conversion with the astronomer Harlow Shapley (Hansen 2000), whom we introduced in the Exercise section of Chapter 3.

doi:10.1088/2514-3433/ade963ch4

(and its predecessor, the thermoscope, which could only measure temperature *differences*.) The first modern mercury thermometer with a standardized scale was invented by Daniel Gabriel Fahrenheit (1686–1736) in 1714. The thinking that there are tiny particles (atoms) whose motion gave rise to what we perceive as heat is not so modern; it can be traced back to the ancient Greeks. However, how exactly this "heat" came about from the motion of the atoms were not understood for a long time. For example, Galileo wrote in his book *The Assayer* that (Drake 1957)

> "Those materials which produce heat in us and make us feel warmth, which are known by the general name of 'fire,' would then be a multitude of minute particles having certain shapes and moving with certain velocities. Meeting with our bodies, they penetrate by means of their extreme subtlety, and their touch as felt by us when they pass through our substance is the sensation we call 'heat.' [...] The operation of fire by means of its particles is merely that in moving it penetrates all bodies, causing their speedy or slow dissolution in proportion to the number and velocity of the fire-corpuscles and the density or tenuity of the bodies. [...] I do not believe that in addition to shape, number, motion, penetration, and touch there is any other quality in fire corresponding to 'heat'; this belongs so intimately to us that when the live body is taken away, heat becomes no more than a simple name [...]"

Though still rudimentary, Galileo had the correct idea that *heat* is our subjective perception due to the motion of the atoms in our environment, which is measured by the thermometer. Temperature is physically real, but whether it is hot or cold is not something that is objectively real. But even today we often use "hot" to colloquially refer to a state of sufficiently high temperature and "cold" otherwise. The word "heat" is also still used in the objective sense, in fact, in two senses: (1) the portion of energy that changes with a change in temperature, and (2) the transfer of thermal energy from a hotter to a colder system or body (Buyse 2022).

The *heat equation* in three-dimensional space

$$\frac{\partial u}{\partial t} = \alpha\left(\frac{\partial^2 u}{\partial x^2} + \frac{\partial^2 u}{\partial y^2} + \frac{\partial^2 u}{\partial z^2}\right), \tag{4.1}$$

where $\alpha > 0$ is the "thermal diffusivity" of the medium, describes how heat spreads by diffusion in an environment. Joseph Fourier developed the heat equation in 1807. From this investigation of heat came the concept of what we today know as the Fourier series, the representation of certain functions as sum of cosines and sines. This in turn, caused some stirs in the *mathematical community*. As explained in the book *A Radical Approach to Real Analysis* by David Bressoud (Bressoud 1994),

> "The crisis struck four days before Christmas 1807. The edifice of calculus was shaken to its foundations. In retrospect, the difficulties

had been building for decades. Yet while most scientists realized that something had happened, it would take fifty years before the full impact of the event was understood. The nineteenth century would see ever expanding investigations into the assumptions of calculus, an inspection and refitting of the structure from the footings to the pinnacle, so thorough a reconstruction that calculus was given a new name: Analysis. Few of those who witnessed the incident of 1807 would have recognized mathematics as it stood one hundred years later.[2] […] Here was the heart of the crisis. Infinite sums of trigonometric functions had appeared before. Daniel Bernoulli (1700–1782) proposed such sums in 1753 as solutions to the problem of modeling the vibrating string. They had been dismissed by the greatest mathematician of the time, Leonhard Euler (1707–1783). Perhaps Euler scented the danger they presented to his understanding of calculus. […] Well into the 1820s, Fourier series would remain suspect because they contradicted the established wisdom about the nature of functions. Fourier did more than suggest that the solution to the heat equation lay in his trigonometric series. He gave a simple and practical means of finding those coefficients, the a_i, for any function. In so doing, he produced a vast array of verifiable solutions to specific problems."

This is an example that shows that progress in one field in science (say, physics), can lead to progress in another (say, mathematics).

In this chapter, let us review the laws of thermodynamics and some important results from statistical mechanics, many of which we will apply in subsequent chapters. We also discuss the arrow of time, the still unsolved mystery of why time "flows" in one direction. Ultimately, it makes sense that entropy should increase over time. The hard part is to explain why it was ever so low in the beginning.

4.1 The Laws of Thermodynamics

Thermodynamics is the branch of physics that, roughly speaking, studies heat flow and the related concepts such as entropy[3] and energy. It is a macroscopic theory in the sense that, for example, for a box of gas, thermodynamics studies the relationship between variables such as temperature, volume, and pressure. Subsequently,

[2] Before it was made rigorous, calculus was also criticized by the philosopher George Berkeley (1685–1753), for treating infinitesimals sometimes as zero but sometimes as nonzero. He complained in *The Analyst* that: "They are neither finite Quantities nor Quantities infinitely small, nor yet nothing. May we not call them the ghosts of departed quantities?" What he meant was if we write a derivative $f'(x)$ as the limit $f'(x) := \lim_{h\to 0} \frac{f(x+h)-f(x)}{h}$, then in order for the ratio to make sense, $h \neq 0$. However after some manipulations, h "is set to" zero at the end. For example, with $f(x) = x^2$, we would have $f'(x) = \lim_{h\to 0}(2x + h) = 2x$. Berkeley's concern can only be properly addressed by the "ε-δ" definition of limit in analysis.

[3] The notion of entropy was introduced by Rudolf Clausius (1822–1888) in 1865, having been inspired by the book *Reflections on the Motive Power of Fire* of a French military engineer Nicolas Léonard Sadi Carnot (1796–1832), who unfortunately died young in relative obscurity due to cholera.

mainly due to Ludwig Boltzmann, the underlying theory of the microscopic description involving atoms[4] and molecules was developed more completely, though as we have discussed the idea that heat is due to the motion of tiny particles was already proposed in the early days by other scholars like Galileo. Another example is Francis Bacon (1561–1626), who wrote in his *Novum Organum* in 1620 that "heat itself, its essence and quiddity, is motion and nothing else" (Buyse 2022). This became known as *statistical mechanics*, which forms one of the major pillars in modern physics. Before going into the discussion of statistical mechanics, however, let us review some important concepts in thermodynamics, and also point out some interesting features and properties that are perhaps lesser known.

We start from the four laws of thermodynamics.

(1) *0th Law:* Temperature is uniform in a thermodynamic system in thermal equilibrium. Consequently, if two systems are both in thermal equilibrium with a third system, then they are in thermal equilibrium with each other.

(2) *1st Law:* If the energy of the system (E) is changed, then dE can be related to the temperature T and entropy S by

$$dE = TdS + \text{work terms}. \tag{4.2}$$

An example of a work term is $-PdV$ where P is the pressure and V the volume. Note that the first law is essentially the conservation of energy: energy cannot be destroyed or created but it can change its form.

(3) *2nd Law:* The entropy of an isolated system is non-decreasing over time. This law governs the irreversibility of natural processes, for example we cannot "unfry" an egg, despite the reverse process not violating any fundamental laws of physics. (As will be discussed in Section 4.3, the second law is just a probabilistic statement and thus does not contain any deep fundamental physics—the fact that we see entropy increasing is because entropy was low in the past, and explaining why that is the case is the issue.)

(4) *3rd Law:* The entropy of a system approaches a constant value as its temperature approaches absolute zero. However, $T = 0$ cannot be reached in a finite number of operations (The "unattainability principle" of Walther Nernst (1864–1941)).

[4] The physicist and philosopher Ernst Mach (1838–1916) was very against the reality of atoms. As a philosopher of science, he believed that science should only be based on what could be observed or experimentally probed. To him atoms were merely a mental construct since they cannot be measured or detected at the time, and hence he also doubted the atomic hypothesis itself. This led to a fierce debate between him and Boltzmann. Boltzmann even left his position at the University of Vienna to avoid Ernst Mach (he took up an offer from the University of Leipzig, on the invitation of the chemist and philosopher Wilhelm Ostwald (1853–1932), although the latter also opposed the existence of atoms), and would only return to Vienna after the retirement of Mach.

In my experience, thermodynamics is often loathed by students (compared to other subjects in physics). One reason could be that the traditional way to teach the subject is perceived to be rather archaic, with examples drawn from heat engines that dated back to the industrial revolution,[5] and too disconnected from a typical student's life to engage their interest. The story of industrial revolution is actually interwoven with physics and art, and if told properly, may instil more interests in students. See Figure 4.1.

Perhaps the most important problem is that a lot of the concepts are too vague and too abstract. Take for example, the various differential operators involved, and the many types of energy (internal energy U, the enthalpy $H := U + PV$, the Helmholtz free energy $F := U - TS$, the Gibbs free energy $G := U + PV - TS$ just to name a few). No wonder we can find even quotes from eminent physicists and mathematicians who complained about how confusing the subject is, such as this one by Arnold Sommerfeld (1868–1951):

> "Thermodynamics is a funny subject. The first time you go through it, you don't understand it at all. The second time you go through it, you think you understand it, except for one or two points. The third time you go through it, you know you don't understand it, but by that time you are so used to the subject, it doesn't bother you anymore."

Here is another one by mathematician Vladimir Arnold (1937–2010):

> "Every mathematician knows it is impossible to understand an elementary course in thermodynamics. The reason is that thermodynamics is based—as Gibbs has explicitly proclaimed—on a rather complicated mathematical theory, on the contact geometry. Contact geometry is one of the few 'simple geometries' of the so-called Cartan's list, but it is still mostly unknown to the physicist—unlike the Riemannian geometry and the symplectic or Poisson geometries, whose fundamental role in physics is today generally accepted."

We are not going to introduce, much less explain, contact geometry in this book. However, perhaps it would be useful to tidy up some of the basic concepts one encountered in thermodynamic courses. The various "energies" mentioned above are often introduced together with the so-called *Maxwell's relations*. The internal energy U is what appears in the first law:

$$dU = TdS - PdV. \tag{4.3}$$

[5] The concept itself is of course useful. Even today a lot of machinery makes use of thermodynamics. For example, most commercial airplanes fly at the height of about 10 km because the flight engine compressor has efficiency $\eta = 1 - T_{\text{cold}}/T_{\text{hot}}$, where T_{cold} is the temperature of the cold end of the cycle, or in this case the surrounding air (this formula is known as *Carnot efficiency*). The air temperature is colder higher up, so the engine operates with better efficiency and therefore cut down the fuel cost. Of course one has to optimize by taking into account the fuel used during the ascent, which gives the 10 km height and not higher.

Figure 4.1. Claude Monet (1840–1926), founder of impressionism, painted this work *Arrival of the Normandy Train, Gare Saint-Lazare* in 1877. The Impressionists frequently painted the modern aspects of Paris, including trains, which were one of the innovations back then. Decades before, the Romantic painter Joseph Turner (1775–1851) traveled widely in search of his artistic ideas. During his time, the industrial revolution led to the invention of the steam-powered train. Train travel would revolutionize the world with its hitherto unmatched speed, depicted in many paintings of the era. Machines were starting to give a vibe of sublime, on par with natural grandiose scenarios. In 1844, Turner painted *Rain, Steam and Speed—The Great Western Railway*. The work can be viewed on the National Gallery website: https://www.nationalgalleryimages.co.uk/asset/1574/. If you look hard enough, you should be able to see a hare on the track, itself a symbol of (natural) speed, running away from the (artificial) train. On one hand, the science of thermodynamics grappled with abstract notions like heat (intuitively we all know what heat is, but how to quantify it? How does it relate to useful work?) and eventually, that of entropy; while on the other hand, it served to improve engine efficiency and designs that further pushed the industrial revolution. All these eventually led to social changes and further revolutions in science. Notably, train travel led to the invention of time zones, as regional time was no longer enough when people started to travel between cities at that kind of speed. Accurate time tables were called for, and with it better designs for clocks, and most crucially, time-*synchronizing* systems. In the 1900s there was already an electrical clock network in the city of Bern in Germany (which also housed an astronomical clock tower called Zytglogge), and time synchronization became an engineering challenge. A patent clerk by the name of Albert Einstein was in the midst of these revolutions, and the rest is history (Galison 2000). *Credit:* Mr. and Mrs. Martin A. Ryerson Collection, Art Institute Chicago; CC0 Public Domain. *Source*: https://www.artic.edu/artworks/16571/arrival-of-the-normandy-train-gare-saint-lazare

This is a differential form, from which we recognize that the coefficients of the dS and dV terms must satisfy the equations

$$T = \frac{\partial U}{\partial S}\bigg|_V, \qquad P = -\frac{\partial U}{\partial V}\bigg|_S. \tag{4.4}$$

We also see that

$$\frac{\partial T}{\partial V}\bigg|_S = \frac{\partial}{\partial V}\bigg|_S \frac{\partial}{\partial S}\bigg|_V U = \frac{\partial}{\partial S}\bigg|_V \frac{\partial}{\partial V}\bigg|_S U = -\frac{\partial P}{\partial S}\bigg|_V. \tag{4.5}$$

Likewise, we can calculate also the similar results for the other "energies". These form the Maxwell's relations:

(1)

$$dU = TdS - PdV \Rightarrow \left(\frac{\partial T}{\partial V}\right)_S = -\left(\frac{\partial P}{\partial S}\right)_V. \tag{4.6}$$

(2)

$$dH = TdS + VdP \Rightarrow \left(\frac{\partial T}{\partial P}\right)_S = \left(\frac{\partial V}{\partial S}\right)_P. \tag{4.7}$$

(3)

$$dF = -SdT - PdV \Rightarrow \left(\frac{\partial S}{\partial V}\right)_T = \left(\frac{\partial P}{\partial T}\right)_V. \tag{4.8}$$

(4)

$$dG = -SdT + VdP \Rightarrow -\left(\frac{\partial S}{\partial P}\right)_T = \left(\frac{\partial V}{\partial T}\right)_P. \tag{4.9}$$

Often some tricks are used to keep track of all the various minus signs. (Readers who are familiar with theoretical classical mechanics may recognize that these "energies" $\{U, H, F, G\}$ are just examples of the various "generating functions" in Hamiltonian mechanics, often referred to as $\{F_1, F_2, F_3, F_4\}$, respectively.[6] These generating functions are often introduced in quite an abstract manner via "generalized coordinates", perhaps it would be better to motivate them in classical mechanics courses using Maxwell's relations?) However, there is simplicity behind this mess, as recognized by David Ritchie in 1958 (Ritchie 1968). First, we start with the first law Equation (4.3), and integrate it around a closed loop γ (physically, a cyclic process so that the internal energy returns to its initial value after a cycle):

$$\oint_\gamma (TdS - PdV) = \oint_\gamma dU = 0, \tag{4.10}$$

[6] Some students may never have realized this. Unfortunately in my experience a lot of students only look at each course separately and fail to integrate what they learn into a big picture.

or

$$\oint_\gamma TdS = \oint_\gamma PdV. \tag{4.11}$$

This is just the statement that the heat being added into a system equals the work it does. However, by Green's theorem[7] we can convert the loop integral into an area integral, where Ω is the area bounded by γ. Thus we have

$$\oint_\gamma TdS = \int_\Omega dTdS, \tag{4.12}$$

and similarly

$$\oint_\gamma PdV = \int_\Omega dPdV. \tag{4.13}$$

Thus

$$\int_\Omega dTdS = \int_\Omega dPdV. \tag{4.14}$$

Since this has to hold for any region Ω, one must have $dTdS = dPdV$. Technically this is an oriented area element, so that interchanging the product introduces a minus sign ($dT \wedge dS = -dS \wedge dT$; think of a vector cross product if you are not familiar with the wedge product of differential forms). The claim is that $dTdS = dPdV$ contains everything you need to know about Maxwell's relations!

Without attempting to be rigorous, we now introduce two dummy variables X and Y and divide $dTdS = dPdV$ by $dXdY$ throughout:

$$\frac{dTdS}{dXdY} = \frac{dPdV}{dXdY}. \tag{4.15}$$

Now X and Y can be any of the variables V, S, T, P. For example, let $X = V$ and $Y = S$. Then

$$\frac{dTdS}{dVdS} = \frac{dPdV}{dVdS} \Rightarrow \frac{dTdS}{dVdS} = -\frac{dVdP}{dVdS} \Rightarrow \left.\frac{\partial T}{\partial V}\right|_S = -\left.\frac{\partial P}{\partial S}\right|_V. \tag{4.16}$$

Note that we can only "cancel terms" when they are in the right order (to have the correct minus sign). Similarly, by taking $X = T$ and $Y = V$, we can obtain

$$\frac{dTdS}{dTdV} = \frac{dPdV}{dTdV} \Rightarrow \left.\frac{\partial S}{\partial V}\right|_T = \left.\frac{\partial P}{\partial T}\right|_V. \tag{4.17}$$

The other Maxwell's relations can be similarly obtained.

---

[7] Named after George Green, who was almost entirely self-taught, having only received one year of formal education when he was 8–9 years old. Before pursuing his mathematical studies, he spent his youth as a miller. He was admitted as an undergraduate at Cambridge when he was already 40 years old.

In mathematical literature, $(S, T, V, P) \in \mathbb{R}^4$ can be viewed as coordinates in the *phase space*. The quantity $\omega = dTdS - dPdV$ is called a *symplectic structure*. This can be generalized to more coordinates in $\mathbb{R}^{2n}$. The pair $(\mathbb{R}^{2n}, \omega)$ then forms a *symplectic space*, or more generally a symplectic manifold. The n-dimensional submanifold $\Sigma \in \mathbb{R}^{2n}$ on which any closed region Ω gives $\int_\Omega \omega = 0$ is referred to as a *Lagrangian submanifold.*

Box 2: Maxwell's Demon

In 1867, Maxwell came up with a thought experiment that initially seemed to challenge the second law of thermodynamics. Imagine that there is a box of gas with a partition in the middle that has a door. A "demon" (this name was due to Lord Kelvin) can control whether the door opens or closes. The demon can also see individual molecules and he decides whether a molecule can be allowed through the door based on their speed. For example, if a molecule approaches the door from the left side and has speed above some some arbitrary scale v_c), he opens the door to let it through to the right side. If such a molecule approaches from the right he keeps the door closed. On the other hand, he allowed slow-moving molecules to pass from right to left but not the other way round. Eventually, all the fast-moving molecules are collected on the right, while the slow-moving ones on the left. This means that the left side of the box becomes cold and the right side hot. The action of the demon thus seems to *lower* the entropy of the entire box without applying any work, which seems to violate the second law.

There is, however, no real violation. the demon still needs to expend energy to observe the molecules to determine their speed, so its interaction with the box is not so simple energy and entropy-wise. Most crucially, we now understand that *information* also has entropy. The demon must record down the information regarding the speed of the particle (even if in his brain). If we replace "demon" by "computer" we will appreciate that periodically the storage needs to be erased to make room for data. Then, according to Landauer's erasure principle, named after Rolf Landauer (1927–1999) who later asserted that "information is physical" (Landauer 1991), this will lead to an increase in the system entropy. This rise will compensate for the decrease in the gas entropy so that overall the second law is upheld (Rex 2017).

4.2 Equipartition Law, Virial Theorem, and Partition Function

Most students' first exposure to thermodynamics of ideal gas is probably via Boyle's law and Charles' law.[8] Boyle's law concerns an *isothermal* situation, where the temperature is fixed. In this case the pressure of a box of gas is inversely proportional to the volume. Charles' law assumes an *isobaric* situation, where the pressure is held constant, in which case the volume is linearly proportional to the temperature. In an *isochoric* situation in which the volume is fixed, the pressure is likewise linearly

[8] Jacques Charles (1746–1823) was not only a scientist but also a balloonist.

proportional to the temperature. Later on, we learned that these are all special cases of the ideal gas law

$$PV = Nk_BT, \tag{4.18}$$

where N is the number of particles (usually atoms or molecules) of the gas, and k_B the Boltzmann constant. In chemistry it is usually written as

$$PV = nRT, \tag{4.19}$$

where n is the number of moles, i.e., $n = N/N_A$ with N_A being the Avogadro constant, while $R = k_B N_A$. We can apply the microscopic viewpoint of Boltzmann, that of statistical mechanics, by treating $\{P, V, T\}$ as manifestations of the underlying motion of atoms or molecules. This view makes understanding the ideal gas law quite intuitive. For example, pressure is caused by momentum transfer from the particles to the wall of the container. Thus for a fixed volume, if we increase the temperature, the particles receive more energy to move faster, and so they collide more frequently with greater momentum with the wall. This causes the pressure to increase. Likewise, fixing the temperature but squeezing the box so that its volume decreases amounts to more particles confined per unit volume, thus increasing the frequency of collision with the wall, giving rise to a larger pressure, and hence the Boyle's law. One can also plot in three dimensions the surface defined by the ideal gas law. It is sometimes called the "PVT surface". This can be done also for other (actual) substances like water, for which the surface would look more complicated.

Recall that in mathematical modeling (see Section 1.5) we should start simple, and then improve the model when necessary. Other generalizations of the ideal gas law are common in the literature. For example, one can consider higher-order corrections parametrized as a power series:

$$P = \rho RT[1 + B(T)\rho + C(T)\rho^2 + \cdots], \tag{4.20}$$

where $\rho = n/V$ and B, C, etc. are some functions of the temperature. This is similar to Hooke's law where $F = -kx$ only holds in the elastic linear regime of a spring, and one can model the nonlinear regime by a power series in the displacement x. The coefficients of the terms of course depend on what the spring is made of; different materials behave differently under stretching. For gases, the coefficients also depend on the type of gas.

Real gases are made from molecules that interact with each other, and the molecules have finite sizes. To model these effects, the ideal gas law is generalized to the *van der Waals equation*, usually written in the form

$$\left(P + \frac{an^2}{V^2}\right)(V - nb) = nRT, \tag{4.21}$$

where a and b are constants that should be determined from experiments. The term an^2/V^2 is the correction for molecular attraction, whereas nb is the correction for the volume of molecules due to their size. An improvement to the van der Waals model is the *Redlich–Kwong equation*, due to Otto Redlich (1896–1978) and Joseph Neng Shun Kwong (1916–1998) in 1949, which reads:

$$P = \frac{RT}{V_m - b} - \frac{a}{\sqrt{T}\, V_m(V_m + b)}, \tag{4.22}$$

where $V_m = V/n$ is the "molar volume". In this generalization, we see that a in the van der Waals equation has been replaced by $a/\sqrt{T}$ to take into account that molecular attraction should change when their kinetic energy (governed by the temperature) changes. Also, V_m^2 has been replaced by $V_m(V_m + b)$, which works better at higher density by reducing the strength of the attractive term. Again, we see that mathematical modeling of physics does not need to start from a more fundamental theory and *derive* the equations, but rather posit directly to fit observations. Students in mathematics usually find such an approach too "brutal", but physics is not mathematics, and it can be checked against experiments or observations, and hence does not need to rely on mathematical proofs. Neither Newton's second law $F = ma$ nor the Schrödinger equation was "derived". What derivations can help is *after* a mathematical model has been constructed, we can explore its properties and possibly make new predictions that can be tested via mathematical derivations. Unless otherwise specified, for simplicity we will only deal with ideal gas.

Let us consider N particles in a cubic box of side length L. Denote the velocity of a particle by $\vec{v} = (v_x, v_y, v_z)$, and the associated momentum $\vec{p} = (p_x, p_y, p_z)$. The rate at which a particle hits one of the sides perpendicular to the z-axis is v_z/L (check that this has the physical dimension of $1/\text{time}$), which imparts a momentum of p_z. The rate of momentum transfer is thus $p_z v_z/L$. The pressure on either the top or bottom side of the box is given by

$$P = \frac{N}{L^3}\langle p_z v_z \rangle. \tag{4.23}$$

(Check that this has the correct physical dimension!) The $\langle \cdot \rangle$ denotes an average of all the $p_z v_z$ values of all the particles. By isotropy (motion of the particles are equally likely in all directions), we can write

$$\langle p_x v_x \rangle = \langle p_y v_y \rangle = \langle p_z v_z \rangle = \frac{1}{3}\langle \vec{p} \cdot \vec{v} \rangle. \tag{4.24}$$

Therefore the pressure exerted on each side of the box is

$$P = \frac{n}{3}\langle \vec{p} \cdot \vec{v} \rangle, \tag{4.25}$$

where we have defined the *number density* $n := N/L^3$, which is just the number of particles per volume.

If the motion of the particles is non-relativistic, we can write $\vec{p} \cdot \vec{v} = mv^2$, so that

$$P = \frac{n}{3}\langle \vec{p} \cdot \vec{v} \rangle = \frac{2}{3} n \left\langle \frac{1}{2} m v^2 \right\rangle. \tag{4.26}$$

That is to say, the pressure is $2/3$ times the kinetic energy density. For relativistic particles, however, P is only $1/3$ times the kinetic energy density.

In the discussion above, taking some kind of statistical average is why we refer to the subject as *statistical mechanics.* In classical mechanics, given the initial conditions, we can always calculate with certainty what will happen to a system in the future. This is true even when given a system with a large number of particles interacting in a complicated manner with each other. The computation may not be feasible in practice, but it is possible in principle. *Laplace's demon* is the name given to such a super-human entity, who has access to the precise location and momentum of all particles in the Universe, and hence can calculate with absolute certainty their past and future. As Pierre Simon Laplace wrote in *A Philosophical Essay on Probabilities*, see p.4. of Laplace (1951):

> "We may regard the present state of the universe as the effect of its past and the cause of its future. An intellect which at a certain moment would know all forces that set nature in motion, and all positions of all items of which nature is composed, if this intellect were also vast enough to submit these data to analysis, it would embrace in a single formula the movements of the greatest bodies of the universe and those of the tiniest atom; for such an intellect nothing would be uncertain and the future just like the past could be present before its eyes."

Philosophically this gives rise to the issue of whether we even have free will (which in turn leads to problems in theology and law—if one has no free will how can one be held guilty/sinful for committing a crime?). This issue has continued to haunt many thinkers, even with our current realization that the world is not governed by classical physics but quantum mechanics[9]. Since we are not Laplace's demon, we have to apply statistical tools to deal with the large number of particles involved in our thermodynamical systems.

In the presence of gravity, an important concept called *Virial theorem*[10] is extremely useful. We first consider a gas cloud at *hydrostatic equilibrium*, i.e., the internal pressure of the cloud resists further gravitational collapse. Assume a spherical cloud for simplicity. The inward acceleration (if present) due to a mass element at distance r from the center satisfies

$$\rho(r)\frac{d^2r}{dt^2} = -\frac{Gm(r)\rho(r)}{r^2} - \frac{dP(r)}{dr}. \tag{4.27}$$

[9] In 2006, mathematicians John H. Conway and Simon B. Kochen proved the so-called "free will theorem" (Conway & Kochen 2006), which essentially says that if certain axioms are satisfied, then we have as much free will as elementary particles!

[10] The name "virial" came from "vis" in Latin, which stands for "force" or "energy"; before the concept of kinetic energy was introduced, mv^2 is known by the name of *vis viva*, the "living force" of a system. See Footnote 35 in Section 2.3 for more historical remarks.

This is just Newton's "$F = ma$"; check that $(1/\rho)dP/dr$ has the same physical dimension as acceleration. Since the cloud is at equilibrium, there is no acceleration. Hence

$$\frac{dP(r)}{dr} = -Gm(r)\rho(r)/r^2. \tag{4.28}$$

Assuming a constant density, the density $\rho(r)$ is related to the mass $m(r)$ by

$$\rho(r) = \frac{m(r)}{\frac{4}{3}\pi r^3} = \text{const}. \tag{4.29}$$

Thus, a straightforward integration yields

$$\int_0^R 4\pi r^3 \frac{dP(r)}{dr}\, dr = -\int_0^R 4\pi G m(r)\rho(r) r\, dr = -\int_{m=0}^{m=M} \frac{Gm(r)}{r}\, dm. \tag{4.30}$$

The right-hand side is just the total amount of gravitational potential energy E_{grav}. Integrating by parts, the left-hand side gives

$$4\pi r^3 P(r)|_0^R - \int_0^R 4\pi \cdot 3r^2 P(r)\, dr. \tag{4.31}$$

The first term is zero because $P(r) = 0$ at the surface ($r = R$) of the cloud. The second term can be written as

$$-3\int_0^R 4\pi r^2 P(r)\, dr \equiv -3\langle P\rangle V, \tag{4.32}$$

where $\langle P\rangle$ is called the "volume-averaged pressure", and V the geometric volume of the cloud. Therefore we have arrived at

$$\langle P\rangle = -\frac{1}{3}\frac{E_{\text{grav}}}{V}. \tag{4.33}$$

By our previous discussion surrounding Equation (4.26), a non-relativistic ideal gas in a volume V satisfies

$$\langle P\rangle = \frac{2}{3}\frac{E_{\text{KE}}}{V}, \tag{4.34}$$

where E_{KE} denotes the (averaged) kinetic energy. Combining these two results give the famous Virial theorem:

$$E_{\text{grav}} + 2E_{\text{KE}} = 0. \tag{4.35}$$

A more general virial theorem holds even outside the context of a gravitationally bound system. Indeed, if $\overrightarrow{F_k}$ denotes the conservative force (work done is independent of path) on the kth particle located at $\overrightarrow{r_k}$, then

$$\sum_{k=1}^N \left\langle \overrightarrow{F_k} \cdot \overrightarrow{r_k} \right\rangle + 2E_{\text{KE}} = 0. \tag{4.36}$$

Although we have derived this equation assuming hydrostatic equilibrium, this equation actually has huge implications in the physics of star formation. As a gas cloud contracts to form a star, it is called a "protostar" (see Box (6) in Section 8.4). In this process, half of the gravitational potential energy is radiated away, while the other half is converted into kinetic energy of the falling material. This causes the protostar to heat up. A star can only form if the temperature is hot enough to trigger and sustain thermonuclear reaction (about 10^6 K to burn deuterium). The maximum temperature that can be reached depends on the size of the protostar, so if a protostar is less than 0.08 solar masses, they form a "failed star" known as a brown dwarf. The heat from the gravitational contraction keeps it hot enough to radiate in the infrared and therefore can be detected.

Note also that since $E_{\mathrm{KE}} + E_{\mathrm{grav}} = E_{\mathrm{total}}$, the total conserved energy, the Virial theorem also implies

$$E_{\mathrm{total}} = \frac{1}{2}E_{\mathrm{grav}} = -E_{\mathrm{KE}}. \tag{4.37}$$

A system that is gravitationally bound would need to satisfy $E_{\mathrm{grav}} < 0$, i.e., $E_{\mathrm{total}} < 0$ or $E_{\mathrm{KE}} > 0$. In the usual situation of a box of gas, E_{KE} is proportional to heat, i.e., the larger E_{KE}, the hotter the gas. Thus, by injecting more energy into the system to increase E_{total}, we expect that the system should become hotter. This is not the case in a gravitating system—injecting more energy into it makes the system *colder*, whereas extracting energy from it causes it to become *hotter*. Gravitational systems therefore have a *negative heat capacity*. This causes some interesting phenomena.

For example, a *globular cluster* is a large collection (consisting of tens of thousands to many millions) of stars that is gravitationally bound together. Figure 4.2 is one such example. The stellar density in a typical globular cluster is roughly about 100 stars per cubic light-year. Why are they so tightly bound together? The Virial theorem provides an explanation. Start with a uniformly distributed spherical cluster of stars. We can roughly subdivide this system into two parts: the interior part $r < R$ and the exterior part $r > R$, where R is an arbitrary cutoff that essentially allows us to neglect gravity when $r > R$. The Virial theorem for a gravitating system applies in the region $r < R$. Suppose the system is perturbed so that the interior part has a higher average kinetic energy then the outer part. Then, just like heat flows from a higher temperature region to a lower temperature one, this will cause the average kinetic energy of the exterior part to increase. But since E_{KE} decreases in the interior region, E_{grav} has to go up, causing the stars in the center to become even more closely packed (the potential well becomes deeper). Once "heat" (not actual heat of the stars) flows from the interior region to the exterior region, it gets even "hotter" due to the negative specific heat property, thus this process continues with the potential well becoming ever more deeper. This leads to *gravothermal catastrophe*. The final effect is that the interior region becomes very compact, while the outer region (with positive specific heat; since gravity is weak there) *also* becomes "hotter", meaning its individual stars have large velocities, thus probably become unbounded and escape the system.

Figure 4.2. A globular cluster called NGC 6397 (also known as Caldwell 86) in the constellation Ara, as imaged by the European Space Agency (ESA) space telescope Euclid. Located about 7800 light-years from Earth, NGC 6397 is the second-closest globular cluster to us, first discovered by French astronomer Nicolas-Louis de Lacaille (1713–1762) in 1752. It contains some 400,000 stars. The data in this image were taken in just about one hour of observation. Euclid's main objective is to map the geometry of the Universe and better understand the mysterious dark matter and dark energy. *Credit:* ESA/Euclid/Euclid Consortium/NASA, image processing by J.-C. Cuillandre (CEA Paris-Saclay), G. Anselmi. CC BY-SA 3.0 IGO

This is not something that is counter-intuitive, the same thing happen for systems with only two gravitationally bound objects, say a binary star. If we add energy into the system, the orbit will expand and become wider, and so the stars would orbit slower—thus they have less kinetic energy ("colder"). It's just that with such a relatively simple system we don't usually use the heat flow analogy. A globular cluster is just a system with many stars, thus we can apply the tools of statistical mechanics to it.

The statement that the averaged kinetic energy is proportional to the temperature can be formulated more rigorously. For a monoatomic ideal gas, it is given by[11]

$$E_{\mathrm{KE}} = \frac{3}{2}k_B T. \tag{4.38}$$

The factor $3/2$ comes from the fact that the kinetic energy can be decomposed into the standard Cartesian components

$$\frac{1}{2}m|\vec{v}|^2 = \frac{1}{2}m(v_x^2 + v_y^2 + v_z^2), \tag{4.39}$$

and each of these component contribute $k_B T/2$ to the average kinetic energy in thermal equilibrium. This is known as the *equipartition law*, which more generally states that in thermal equilibrium, the total energy of the molecule is divided equally among all degrees of freedom of motion. In the case of a diatomic molecule, for example, in addition to the translational degrees of freedom above, we also need to take into account both the rotational and vibrational motions (for certain molecules the vibrational motion can be neglected). The rotational degrees of freedom give a total energy of

$$E_{\mathrm{rot}} = \frac{1}{2}I_1\omega_1^2 + \frac{1}{2}I_2\omega_2^2 = 2 \times \frac{1}{2}k_B T = k_B T, \tag{4.40}$$

where I_i is the moment of inertia and ω_i the angular speed. The vibrational energy E_{vib} has both kinetic part and potential part, each also contributes $k_B T/2$. Therefore for a diatomic molecule, the total energy is

$$E_{\mathrm{total}} = E_{\mathrm{KE}} + E_{\mathrm{rot}} + E_{\mathrm{vib}} = \frac{3}{2}k_B T + k_B T + k_B T = \frac{7}{2}k_B T. \tag{4.41}$$

More generally, the equipartition theorem says that if the Hamiltonian of a thermodynamical system in contact with a heat reservoir at temperature T depends on some degree of freedom x, then we have the ensemble average

$$\left\langle x\frac{\partial H}{\partial x}\right\rangle = k_B T. \tag{4.42}$$

We can check that this recovers the monoatomic result since in that case

$$H = \sum_{k=1}^{3}\frac{p_k^2}{2m}, \tag{4.43}$$

and x would be p_k. More generally, for a physical system with Hamiltonian energy function H and degrees of freedom x_m, we have

$$\left\langle x_m\frac{\partial H}{\partial x_n}\right\rangle = \delta_{mn}k_B T, \tag{4.44}$$

[11] The heat capacity is defined by $C := \partial E_{\mathrm{total}}/\partial T$. Thus we see that for a gravitating system, since $E_{\mathrm{total}} = -E_{\mathrm{KE}} = -\frac{3}{2}k_B T$, we have $C < 0$.

where δ_{mn} is the Kronecker delta, which equals unity when $m = n$ and zero otherwise.

Lastly, let us introduce the *partition function*. In a canonical ensemble in which a system is in contact with a heat bath at constant temperature, the partition function is defined as

$$Z = \sum_i \exp(-\beta E_i), \tag{4.45}$$

where β is the inverse temperature $\beta := 1/k_B T$, and E_i the total energy of the system in the ith-microstate. The exponential term $\exp(-\beta E_i)$ is called the *Boltzmann factor*. Z stands for "Zustandsumme", a German word that literally means "sum over states". The quantity Z is such that the probability distribution,[12] the *Boltzmann distribution*

$$\rho_i = \frac{e^{-\beta E_i}}{Z}, \tag{4.46}$$

maximizes the Boltzmann–Gibbs entropy[13]

$$S = -k_B \sum_i \rho_i \ln \rho_i. \tag{4.47}$$

Given that the system is in thermal equilibrium, the average energy is constant and we denote it as

$$U := \langle E \rangle = \sum_i \rho_i E_i. \tag{4.48}$$

It can be shown that

$$S = k_B \ln Z + \frac{U}{T}. \tag{4.49}$$

The free energy can be expressed as $F = U - TS$.

For a continuous system, one would need to replace the various sums with integrals. For example, the classical partition function of a gas of N identical classical particles in three dimensions is[14]

$$Z = \frac{1}{N! h^{3N}} \int d^3q_1 d^3q_2 \cdots d^3q_N \int d^3p_1 d^3p_2 \cdots d^3p_N \left\{ \exp\left[-\beta \sum_{i=1}^{N} H(q_i, p_i) \right] \right\}, \tag{4.50}$$

[12] Thus $\sum_i \rho_i = 1$.

[13] If a system is in a global thermodynamic equilibrium, we have for all N microstates, which are now equally likely, the uniform probability $p_i = 1/N$. The Boltzmann–Gibbs entropy then reduces to $S = k_B \ln(N)$, which is called the *Boltzmann entropy*. The Boltzmann–Gibbs entropy is quite general and does not require an equilibrium. It is not an accident that the Boltzmann–Gibbs entropy has the same form as "information entropy" first defined by Claude Shannon (1916–2001) (Jaynes 1957), given our discussion about Maxwell's Demon in Box 2.

[14] The notation here may be uncomfortable for a mathematics student but is a common one in physics, especially when the integrand is long and complicated. Essentially we write, for example, $\int\int\int f dxdydz$ in $\mathbb{R}^3$ as $\int d^3x f$. This is just convention. The notation d^3x may also look strange, but it is only a convention. In advanced analysis, one sometimes even suppresses the various dx's and simply write $\int f$ if there is no risk of confusion as to what the function is being integrated with respect to.

where H is the Hamiltonian, which is in turn a function of the canonical position q_i and the canonical momentum p_i. The factor h^3 is there to make Z dimensionless. Interestingly h has the same dimension as the Planck's constant in quantum mechanics, and thus *is* taken to be such. (It is related to the reduced Planck's constant $\hbar$ by $h = 2\pi\hbar$.) It may seem strange that h appears in *classical physics*.[15]

On the other hand, the *quantum* partition function is $Z = \mathrm{tr}(e^{-\beta\hat{H}})$, which has no explicit h. Passing to a continuous integral for the classical form, however, one can obtain

$$Z = \int \frac{dxdp}{h} \langle x, p| e^{-\beta\hat{H}} |x, p\rangle. \tag{4.52}$$

The shadow of quantum mechanics in Equation (4.50) is in the h—it is due to the minimum cell size in the phase space. To understand this, consider the three-dimensional case (meaning $N = 1$ in Equation (4.50)). If we consider the quantum state in a finite cubic box with a periodic boundary condition,[16] and with side length L, then the volume is L^3. In quantum mechanics, the momentum p is related to the wavenumber k by the De Broglie's relation $p = \hbar k$. For the wave to fit inside the box, the spacing of the wavenumber $\Delta k = 2\pi/L$ translates to $\Delta p = \frac{2\pi\hbar}{L} = h/L$. Thus, in the continuous version, and in the $L \to \infty$ limit, the volume measure has to be normalized by the density of state, which leads to

$$d^3p\frac{L^3}{h^3} = d^3p\frac{V}{h^3} \longrightarrow \frac{d^3p\, d^3x}{h^3}. \tag{4.53}$$

It should be noted that although the concept of phase space was already used by Boltzmann and Gibbs[17], the idea that we can treat phase space as a bona fide higher-dimensional space with its own "volume" and to study the trajectories in phase space to track the evolution of the system did not become comfortable until later on. Indeed, even though the geometric interpretation was obvious, Boltzmann and Gibbs never described them as higher-dimensional volume integrals. This could be due to the fact that the notion of higher-dimensional space, even in mathematics, was historically not taken seriously, though things started to change right about then (see Section 11.1 for more discussions). For more historical remarks concerning phase space, see Nolte (2010).

If instead of the probability of a system being in a specific energy state, we want to describe the distribution of particle speeds, or equivalently their kinetic energy at a

[15] For another example, consider the "classical" massive Klein–Gordon equation

$$\left(\Box + \frac{m^2c^2}{\hbar^2}\right)\phi(x) = 0, \tag{4.51}$$

which also contains $\hbar$.

[16] Often this is also a useful trick in numerical computation as computers cannot handle infinite size. For an infinitely large space, $L \to \infty$ and the period is not physically relevant.

[17] Gibbs wrote in his *Elementary Principles in Statistical Mechanics* in 1902: "If we regard a phase as represented by a point in space of $2n$ dimensions, the changes which take place in the course of time in our ensemble of systems will be represented by a current in such space."

given temperature T, then we need to use the *Maxwell-Boltzmann distribution*, $f(v)$, which in three spatial dimensions, is given by

$$f(v) = \left(\frac{m}{2\pi k_B T}\right)^{\frac{3}{2}} 4\pi v^2 \exp\left(-\frac{mv^2}{2k_B T}\right), \tag{4.54}$$

where m is the mass of the individual particle that makes up the substance. From here, by taking the derivative $f'(v)$ and setting it to zero, one can find the most probable speed of the particle:

$$v_{\mathrm{p}} := \sqrt{\frac{2k_B T}{m}}. \tag{4.55}$$

The mean speed $\langle v \rangle$, on the other hand, is given by

$$\bar{v} := \langle v \rangle = \int_0^\infty v f(v)\, dv = \sqrt{\frac{8k_B T}{\pi m}}. \tag{4.56}$$

Similarly, the root mean square speed $\langle v^2 \rangle$ is given by

$$v_{\mathrm{rms}} := \langle v^2 \rangle = \int_0^\infty v^2 f(v)\, dv = \sqrt{\frac{3k_B T}{m}}. \tag{4.57}$$

Note that the normalizing factor $(m/2\pi k_B T)^{\frac{3}{2}}$ in the distribution also makes physical sense: if the temperature is higher, we would expect particles to move further at greater speed, thus spreading out the shape of the distribution (i.e., makes the distribution broader), and similarly, if m is larger, the heavier particles would move slower, the occurrence of $m^{3/2}$ in the numerator conforms to that intuition, as the shape of the graph of larger m is "narrower". (Students should learn to check whether an equation "makes sense"!)

4.3 The Arrow of Time

Part of this section have been reprinted with permission from (Gregoris et al., 2020), Copyright (2020) by the American Physical Society. Time has always been an intriguing element in science and philosophy, no doubt partly because *we* are subject to the flow of time—we all eventually age and die.[18] Time, unlike space, has a *direction*, which was called the *arrow of time*[19] by British astronomer Arthur Eddington in 1927. From his 1928 book *The Nature of the Physical World*, which helped to popularize the term, Eddington states:

> "Let us draw an arrow arbitrarily. If as we follow the arrow we find more and more of the random element in the state of the world, then the arrow

[18] There is an entire genre of art called *vanitas* whose purpose is to depict the transcience of life. Vanitas paintings usually show clocks, hourglass, candles, withering flowers, skulls etc., reminders of time passing and the eventual death and decay.

[19] In ancient Chinese idiom, time is also said to be (moving fast) like an arrow (光阴似箭).

> is pointing towards the future; if the random element decreases the arrow points towards the past. That is the only distinction known to physics. This follows at once if our fundamental contention is admitted that the introduction of randomness is the only thing which cannot be undone. I shall use the phrase *time's arrow* to express this one-way property of time which has no analogue in space."

If someone were to videotape a physical process and then play the tape backward, we might be able to deduce that the tape is being played backward by noticing the occurrences of "crazy" things that common sense tells us are ridiculous. For example, a broken egg that "unbreaks" itself, or a cup of water that spontaneously cools down and produces ice cubes at the top. In terms of the second law of thermodynamics, the arrow of time is reflected in the fact that statistically speaking, entropy tends to increase rather than decrease. This is not surprising because as Boltzmann taught us, entropy counts the number of microstates via the formula $S = k_B \log W$ (which was engraved on Boltzmann's tombstone). In the phase space then, a system naturally evolves from a region of smaller phase space volume to one with a larger volume. There is hardly any physics at this stage—what we are dealing with are merely *combinatorics*. For example, there are just a lot more ways to have broken eggs than unbroken one. Therefore it is not at all surprising that one expects entropy to increase merely because there are more ways for the configurations of a closed system to be in high entropy states than in lower entropy ones.[20] The surprising thing is that this argument is time symmetric and so by appealing to combinatorics alone we should also expect entropy to be increasing towards the past! Why then do we have unbroken eggs in the first place? Since entropy is increasing, it must have been lower in the past, all the way back to the Big Bang. It is because of the very low entropy of the very early Universe that we (and unbroken eggs) exist at all; if everything has been in equilibrium at the very beginning, nothing would have happened.

The physics thus comes in by demanding that the *initial condition* of the Universe must be such that it is at very low entropy (this is called the *Past Hypothesis*) (Ainsworth 2008; Callender 1998), and then the combinatorics nature of the second law takes over and naturally evolves it towards a higher entropy future. The question is not why entropy increases, as that was already settled by Boltzmann. The question is: *why* is the very early Universe in such a low entropy state?

In other words, the problem of the arrow of time *is* the problem of the initial condition of the Universe. In *Feynman Lectures in Physics*, Richard Feynman made a comment in his chapter on entropy that

[20] Another example would be a *wired* headphones. If you have carried one in your pocket, no doubt you have experienced an increase in entropy—it always ended up tangled in one way or another, simply because there are a lot of ways to be tangled, only one way to be not tangled. This issue is, thankfully, solved with the invention of wireless headphones.

> "So far as we know, all the fundamental laws of physics, such as Newton's equations, are reversible. Then where does irreversibility come from? It comes from order going to disorder, but we do not understand this until we know the origin of the order."

It is certainly possible that there is no great wisdom in trying to explain the Past Hypothesis—it is possible that the Universe simply started like what it did by chance, however unlikely that is. While it is definitely possible that we got lucky and things just happened the way it did *a lá* anthropic principle, it would be preferable to find a more satisfying answer to explain the Past Hypothesis. This is similar to the fact that we are also stuck with the laws of physics that we have, yet we still search for deep principles underlying their nature. It therefore seems sensible to treat our initial conditions in the same way, and try to understand why we have these conditions rather than some others. This suggests at least two strategies:

(1) Find an explanation for the low entropy past: That is, we should explain the as yet unknown physics that imposed the low entropy initial condition.
(2) Find a dynamical explanation: It is possible that we can start from a generic state and somehow evolve the system dynamically to produce what looks like a low-entropy past.

Note that it is important for any purported solution of the arrow of time problem to pass the "double-standard test" proposed by the philosopher Huw Price (Price 1998). Price noted that if a physical mechanism is supposed to explain past low entropy (which gives rise to what we experience as the passing of time) without itself sneaks in time asymmetry, then that mechanism should also be applicable to the "end state of time", i.e., future conditions. In other words the scenario must make sense if we reverse the arrow of time, *unless* there is some *a priori* "natural" reason that breaks the symmetry and makes the past objectively distinct from the future (if so, we must *prove* this).

We make a remark that there are models that tried to construct a time symmetric universe, which will satisfy the double-standard test explicitly. Thomas Gold, for example, proposed a contracting model that will end with a Big Crunch but in which the arrow of time *reverses* at the onset of the contracting phase (Gold 1962). This is perhaps familiar to science fiction fans. For example, philosopher Craig Callender mentioned (Callender 1998):

> "In Philip K. Dick's *Counter-Clock World* the direction of time flips in 1986, putting the Earth into what its inhabitants call the 'Hogarth Phase'. Named, after the scientist who predicted that 'time's arrow' would change direction, the Hogarth Phase is a period in which entropy decreases instead of increases. During this time the dead call from their graves to be excavated, people clean their lungs by 'smoking' stubs that grow into mature cigarettes, coffee separates from cream, and so on."

If the Universe indeed behaves like this, it must be noted that the inhabitants in the Hogarth Phase would not notice anything different, after all the Gold Universe is time symmetric and it is by convention that one decides in which phase of the Universe we are currently living. If we live in the Hogarth Phase, we would define the arrow in the *other direction.*[21] In other words, the direction of time is always defined as the direction in which entropy increases. The Nobel laureate Murray Gell-Mann (1929–2019), together with James Hartle (1939–2023), have also investigated the possibility of imposing quantum mechanical boundary conditions in both the past and future, and even proposed that the possibility of a time-symmetric Gold Universe may be ruled out by observation from the absence of observed backward-moving radiation emitted by stars in the future (Gell-Mann & Hartle 1994). However, as pointed out by Price in Price (1996), such an observation is not as straightforward as one may expect.

Boltzmann, after introducing his "H-theorem", came under many attacks from critics (mainly Johann Josef Loschmidt (1821–1895) and Ernst Zermelo (1871–1953)[22]) who rightly pointed out that there is no way to deduce a time-asymmetric result from the time-symmetric assumption and that some time asymmetry must have "sneaked in" somewhere in his derivation (the double standard!), in particular, his "molecular chaos" assumption (that the velocities of colliding particles are uncorrelated, and independent of position). This has come to be known as the "Loschmidt paradox". Boltzmann eventually realized that although he could show that entropy should statistically increase, he could not explain the low entropy in the past without also imposing initial conditions. However, he also considered other ways to resolve the mystery of low past entropy, and in one of his attempts he considered time to be eternal.

Boltzmann then proposed that the Universe (or perhaps more accurately the "multiverse"—see also Section 13.2), given an eternity to wait, is already in thermal equilibrium, and thus have maximum entropy. Statistical fluctuations however do happen and cause some regions to naturally fluctuate into a lower entropy state, which then slowly settles back to the maximum entropy state. In Boltzmann's own words (Zermelo 1897),

> "There must then be in the Universe, which is in thermal equilibrium as a whole and therefore dead, here and there relatively small regions of the size of our galaxy (which we call worlds), which during the relatively short time of eons deviate significantly from thermal equilibrium. Among these worlds the state probability increases as often as it decreases. For the Universe as a whole the two directions of time are indistinguishable, just as in space there is no up or down. However, just as at a certain place on the earth's surface we can call "down" the direction toward the center of the earth, so a living being that finds itself in such a world at a certain period of time can define the time direction as going from less probable to

[21] Much like our convention of "clockwise" and "counter-clockwise" was likely due to the fact that in the northern hemisphere, the shadow of the gnomon on a sundial rotates "clockwise".

[22] This is the Zermelo well-known in mathematical logic for Zermelo set theory and Zermelo–Fraenkel set theory; see also the introduction of Chapter 3.

> more probable states (the former will be the "past" and the latter the "future") and by virtue of this definition he will find that this small region, isolated from the rest of the Universe, is "initially" always in an improbable state."

Of course back in the day, the Milky Way galaxy was thought to be all there is in the Universe. Replacing the galaxy with our observable Universe, Boltzmann's idea is that our observable Universe just happens to be in a low entropy region of a much larger region that has attained maximum entropy. In modern words, one might say "life is only possible in the regions of multiverse (or string landscape) that supports life", this is the so-called "anthropic principle" —which we will encounter a few more times in this book—appearing in the 19th Century!

In fact similar arguments can be traced further back: Latin philosopher Lucretius in first century BCE, an atomist in the tradition of Democritus and Epicurus, already proposed an explanation for how the orderly Universe around us could arise from the motion of individual atoms in his *De rerum nature* ("On the Nature of Things"):

> "For surely the atoms did not hold council, assigning order to each, flexing their keen minds with questions of place and motion and who goes where. But shuffled and jumbled in many ways, in the course of endless time they are buffeted, driven along, chancing upon all motions, combinations. At last they fall into such an arrangement as would create this Universe."

Eventually no matter how unlikely, an event is bound to occur. Sean Carroll referred to this view as the Boltzmann–Lucretius Scenario (Carroll 2010).

The problem with this scenario is that as we know, fluctuations are rare, and large fluctuations are even rarer. It is far more likely to produce a galaxy than an entire observable Universe via low entropy fluctuation. Sir Arthur Eddington, in 1931, made the following observation regarding a statistically fluctuated universe (Eddington 1931):

> "A universe containing mathematical physicists will at any assigned date be in the state of maximum disorganization which is not inconsistent with the existence of such creatures."

That is, given the constraints of the anthropic principle, we should find ourselves in a minimum possible deviation from maximum entropy. For Eddington, that meant a single mathematical physicist, suddenly assembled as a random fluctuation out of the surrounding thermal radiation. This sounds ridiculous, but it is far more likely than fluctuating out the entire low-entropy observable Universe! One might mistakenly think that if the early Universe is very small in size before expanding to its current size, then perhaps the probability to fluctuate this tiny Universe out is not that big. However, the early small Universe and the current one are *two configurations of the same system*, and so the smallness of the initial region alone does not explain why its entropy was so low (Carroll 2005).

In fact, one can imagine spontaneous fluctuations that create just a single brain that is conscious and thinks something along the line “I think, therefore I am” before dissolving into the thermal equilibrium environment. One can even imagine that such a brain comes with a false memory of the world around it and thinks that it is someone like us living in a majestic universe full with galaxies, and then it simply dissolves in matter of microseconds. Clearly we ordinary observers are nothing like these so-called Boltzmann’s brains,[23] but there should be more Boltzmann brains than ordinary observers! Furthermore, regardless of ordinary observers or Boltzmann’s brains, in a universe created from statistical fluctuations of thermal equilibrium, it is far more likely that everything else in said universe is in thermal equilibrium. That is, even disregarding Boltzmann’s brains, it is easier to fluctuate out a universe which is anthropically compatible but does not look anything like our Universe that has a very low entropy in the past (for example, just fluctuate out a single galaxy is more than enough to sustain us). This means that our current observed Universe is non-generic in the Boltzmann–Lucretius scenario, and this defeats the purpose of trying to provide a more natural explanation to the Past Hypothesis.

Observer issues aside, the main objection to Boltzmann–Lucretius scenario is the following (McInnes 2007): If we resort to low entropy fluctuation to explain our observed low entropy past, why stop there? Why not just argue that the entire Universe just fluctuated into existence as it is last Thursday completely with false evidences (and our false memories) that fool us into believing that it was very old (This scenario actually goes by the name “Last-Thrusdayism”)? Waiting for an eternity is also not very satisfactory. Indeed one may note that (McInnes 2007)

> “Normally, when we discover some ‘highly improbable’ pattern in Nature, our reaction is that the seeming improbability is really a sign of the existence of some hitherto unknown *law of Nature*. When we discover, for example, that planetary orbits have the approximate shape of an ellipse, nobody would think it an adequate response to point out that such paths are bound to be traced out by randomly moving particles in the fullness of time. Instead we search for a law, or a collection of laws (in this case, Newton’s laws of motion and of gravitation). When these

[23] Unlike the “brain in a vat” scenario, in which one’s experience may just be a product of a brain being stimulated by an evil scientist in a lab, a Boltzmann brain is expected to dissipate back into thermal equilibrium very quickly. Therefore if you have been reading this book until now, you are not a Boltzmann brain. In addition, we infer Boltzmann brains from our knowledge of thermodynamics. If there is no external world and we are just Boltzmann brains with built-in fake knowledge of the world, how can we trust whatever conclusion we drew from any “knowledge”? Indeed, in his book *On Certainty*, the philosopher Ludwig Wittgenstein (1889–1951) argued that skeptical (against the reality of the world) claims like these, including Cartesian skeptisism that led Descartes to his famous motto “I think, therefore I am”, are *unsinn*, i.e., non-sensical, because they imply doubting knowledge that is required in order to make sense of the claims themselves.

laws are understood, the improbable is transformed into a certainty; elliptical orbits are an unavoidable consequence of the laws. Similarly here: surely the normal scientific response to Penrose's observation[24] is that the initial conditions of our Universe seem improbable only because we do not know the underlying law of Nature."

Readers who are interested in the arrow of time can also read the introductory article by Price (2013) to learn more.

4.4 Exercises

(1) Although Virial theorem is most useful when we have a large number of particles, show that for the case of only one planet orbiting the Sun (a "two-body problem"), Kepler's third law—which says that

$$T^2 = \frac{4\pi^2}{GM}a^3, \tag{4.58}$$

where M is the mass of the Sun, T is the period of the orbit, and a the radius of the orbit (assuming it is a circle for simplicity)—is consistent with the Virial theorem. (That is to say, show that Kepler's third law satisfies $2E_{\rm KE} = -E_{\rm grav}$.)

(2) We have shown that a self-gravitating ball of gas in hydrostatic equilibrium satisfies

$$\langle P \rangle = -\frac{1}{3}\frac{E_{\rm grav}}{V}. \tag{4.59}$$

Consider an adiabatic process (no heat transfer), which satisfies PV^γ=const. for some *adiabatic index* γ.

(a) Show that $d(PV) = -(\gamma - 1)PdV$.

(b) Denote the internal energy (kinetic energy along with other possible internal degrees of freedom) by $E_{\rm In}$. By the first law of thermodynamics, we have $dE_{\rm In} = -PdV$. Show, using (a), that $E_{\rm In} = PV/(\gamma - 1)$.

(c) Hence, show that

$$3(\gamma - 1)E_{\rm In} + E_{\rm grav} = 0. \tag{4.60}$$

(d) Compute the total energy, which is the sum of $E_{\rm In}$ and $E_{\rm grav}$, and deduce that the gas cloud becomes gravitationally unbound if $\gamma < 4/3$.

Additional note: Thus, we note that in the case $E_{\rm In} = E_{\rm KE}$ (monoatomic gases), we recover the Virial theorem if $\gamma = 5/3$. This only holds for non-relativistic scenario. As the monatomic particle motion becomes

[24] Sir Roger Penrose argued that the probability that our low entropic Universe is born simply by chance is about 1 in $10^{10^{123}}$. The latter absurdly large number is called the "Penrose number".

faster and faster, $\gamma \to 4/3$ and the gas cloud becomes unstable under perturbation.

(3) We can view galaxies as a system of particles in the six-dimensional phase space, in which each point is given by $(u^i) = (\vec{x}, \vec{v})$, where i runs from 1 to 6. Denote its *distribution function* by $f(\vec{x}, \vec{v})$. In fluid mechanics, the time evolution of the fluid is given by the so-called *continuity equation*. This can be generalized to a phase space version, which is given by

$$\frac{\partial f}{\partial t} + \sum_{i=1}^{6} \frac{\partial}{\partial u^i}(f\dot{u}^i) = 0. \tag{4.61}$$

Find out how this leads to the *collisionless Boltzmann equation*:

$$\frac{\partial f}{\partial t} + \vec{v} \cdot \nabla f - \nabla \Phi \cdot \frac{\partial f}{\partial \vec{v}} = 0. \tag{4.62}$$

This equation is useful in the study of the dynamics of large collections of particles like stars in a galaxy, in which individual collisions are negligible.

(4) In statistics, the central limit theorem explains why the sum of a large number of random variables is approximately normal (under some conditions). This is why the normal distribution is often encountered in statistical mechanics. However, sometimes it is not appropriate to apply the normal distribution. For example, consider the stars in a globular cluster, which orbits the Milky Way. We may be tempted to guess that the star velocities are normally distributed. However, if a star moves too fast and escapes the gravitational influence of the cluster, it will be lost to the cluster. This is why globular clusters have a somewhat sharp boundary. To model the velocity distribution, we need to set a cut-off to the velocity. This is the *King model*, introduced in Ivan King's (1927–2021) landmark 1966 paper *The Structure of Star Clusters. III. Some Simple Dynamical Models* (King 1966). Essentially, the cut-off requirement yields

$$\tilde{f} = \begin{cases} C\left(e^{-\frac{E}{\sigma^2}} - 1\right), & \text{if} \quad E \leqslant 0, \\ 0, & \text{if} \quad E > 0, \end{cases} \tag{4.63}$$

where C and σ are some constants. The radius at which the density drops to zero is called the *tidal radius*. Find out more about how this model was constructed.

References

Ainsworth, P. M. 2008, Sy, 162, 157

Bressoud, D. 1994, A Radical Approach to Real Analysis (Washington, DC: Mathematical Association of America)

Buyse, F. A. A. 2022, Encyclopedia of Renaissance Philosophy, ed. M. Sgarbi (Cham: Springer)

Callender, C. 1998, BJPS, 49, 135

Carroll, S. 2010, From Eternity to Here: The Quest for the Ultimate Theory of Time (Dutton: Boston, MA)
Carroll, S. 2005, GReGr, 37, 1671
Conway, J., & Kochen, S. 2006, FoPh, 36, 1441
Drake, S. 1957, Discoveries and Opinions of Galileo, Translated with an Introduction and Notes (New York: Anchor Books)
Eddington, A. S. 1931, Natur, 127, 447
Galison, P. 2000, CI, 26, 355
Gell-Mann, M., & Hartle, J. B. 1994, Physical Origins of Time Asymmetry, ed. J. Halliwell, J. Perez-Mercader, & W. Zurek (Cambridge: Cambridge Univ. Press)
Gold, T. 1962, AmJPh, 30, 403
Gregoris, D., Ong, Y. C., & Wang, B. 2020, PhRvD, 102, 023539
Hansen, T. 2000, Ex, 59, 27
Jaynes, E. T. 1957, PhRv, 106, 620
King, I. R. 1966, AJ, 71, 64
Landauer, R. 1991, PhT, 44, 23
Laplace, P. S. 1951, Philosophical Essay on Probabilities, translated into English from the original French, ed. F. W. Truscott, & F. L. Emory (New York: Dover)
McInnes, B. 2007, [arXiv:0711.1656 [hep-th]] 2007arXiv0711.1656M
Nolte, D. D. 2010, PhT, 63, 33
Price, H. 1995, Time's Arrows Today: Recent Physical and Philosophical Work on the Direction of Time, ed. S. F. Savitt (Cambridge: Cambridge Univ. Press) 66
Price, H. 1996, Timeas Arrow and Archimedes Point: New Directions for the Physics of Time (Oxford: Oxford Univ. Press)
Price, H. 2013, Time: Poincaré Seminar 2010, Progress in Mathematical Physics, ed. B. Duplantier (Basel: Birkháuser) 63
Rex, A. 2017, Entrp, 19, 240
Ritchie, D. J. 1968, AmJPh, 36, 760
Zermelo, E. 1897, AnP, 60, 392

Chapter 5

Our Changing Cosmos

> *"I am always surprised when a young man tells me he wants to work at cosmology; I think of cosmology as something that happens to one, not something that one can choose."*[1]—William Hunter McCrea, presidential address to the Royal Astronomical Society, February 1963.

In this relatively long chapter, we aim to achieve two objectives. First, we shall introduce the general theory of relativity using the minimal amount of mathematics. After that, we shall apply the theory to understand how the Universe evolves. No longer static, the cosmos governed by the Friedmann equation in general relativity can expand (an accelerated expansion no less!), as verified by observations. The Universe now had a beginning—the Big Bang. We also give a review of what we currently know about the Universe in modern cosmology (the concept of the distance ladder, the cosmic microwave background, the primordial abundance of light elements, etc.). Some unsolved mysteries such as the Hubble tension are introduced towards the end.

This chapter deals with two major topics: general relativity and cosmology. The theory of general relativity is required to understand modern cosmology, as it tells us how our Universe can evolve. In the Newtonian worldview, space and time set up the stage in which physical systems reside. Matter and energy are the actors on this fixed stage. In general relativity, the stage is like a modern one that changes in different scenes; matter and energy interact with space and time.

Spacetime as an entity itself can be described by curved geometry. Mathematically it is treated as a *manifold*. Even in special relativity, once

[1] Yuri Manin (1937–2023) also said that "we don't choose mathematics as our profession, but rather it chooses us" (Manin 2009)

doi:10.1088/2514-3433/ade963ch5

Hermann Minkowski (1864–1909) geometrized the theory and introduced what we now call "Minkowski spacetime", he wrote:

> "Henceforth space by itself, and time by itself, are doomed to fade away into mere shadows, and only a kind of union of the two will preserve an independent reality."

For this reason I prefer *spacetime* written as a joined inseparable single word "spacetime", though sometimes it is written as "space-time" especially in the older texts. Beginning students of physics often find general relativity challenging because of the need to master a certain amount of differential geometry first. This is unavoidable, after all, popular science "explanations" using analogies can only go so far. The challenge is worth it; the fruit of that struggle is sweet, as it opens up an entire new way to appreciate how the very fabric of reality—woven by space and time—works.

We will start by introducing the bare minimum of mathematics that allows a student to appreciate (but not to master) general relativity, from the geometric route of understanding the metric tensor as a fancy generalization of the Pythagoras theorem, and the concept of geodesics and curvature. In the second part of the chapter, we explain the geometry of the Universe and how its evolution is governed by the Friedmann equation, which is arguably the most important ODE to understand in cosmology. Finally, we explain what we currently know about the Universe and what other mysteries await solutions. Clearly it is impossible to discuss these topics without some serious mathematics, however "minimal" we strive for. Hence some degrees of "mathematical maturity" is especially important in this chapter.

5.1 General Relativity: Gravity as Spacetime Curvature

5.1.1 From Pythagoras Theorem to Spacetime Metric

General relativity is an extremely successful theory that describes gravity, and in order to understand cosmology properly, we have to first learn a thing or two about general relativity (it is fair to say that modern cosmology as a science only starts after the birth of general relativity). Of course, general relativity requires an entire course to be properly taught, as such I cannot give it a proper rigorous treatment in just a single section of a chapter. What I will do instead is teach the bare minimum of the *ideas* behind general relativity.

It is often said that in general relativity, gravity is a manifestation of spacetime curvature. In order to understand this, we will start from the very basic: the Pythagoras theorem in $\mathbb{R}^3$, which should be familiar to everyone from high school plane geometry. Suppose we have a right-angled triangle whose two sides that are perpendicular to each other is placed on the standard Cartesian x and y axes, respectively. Denote the side lengths by Δx and Δy. The length of the hypotenuse is then

$$\Delta s^2 = \Delta x^2 + \Delta y^2. \tag{5.1}$$

For an infinitesimal triangle, we will write

$$ds^2 = dx^2 + dy^2. \tag{5.2}$$

This is known as the *metric* (more technically, a metric tensor), while ds is sometimes called the *line element*. The notion of an infinitesimal triangle will be rather uncomfortable to a mathematics student (as well as to most professional mathematicians). Unfortunately we have to deal with such ambiguous notions unless we plan to be rigorous with all the differential geometry foundations. If it makes you more comfortable, you may think of this equation as a shorthand notation for the equation for the speed of a test particle traveling on the hypotenuse, measured with some time t, namely:

$$\frac{ds}{dt} = \sqrt{\left(\frac{dx}{dt}\right)^2 + \left(\frac{dy}{dt}\right)^2}. \tag{5.3}$$

For reasons that will become transparent later, let us rewrite Equation (5.2) as

$$ds^2 = 1 \cdot dx^2 + 0 \cdot dxdy + 0 \cdot dydx + 1 \cdot dy^2. \tag{5.4}$$

All that we have done is introducing two cross terms, which are multiplied by zero so they make no contribution. But now we can write the metric in a more abstract form:

$$ds^2 = g_{xx}dx^2 + g_{xy}dxdy + g_{yx}dydx + g_{yy}dy^2, \tag{5.5}$$

where $g_{xx} = 1 = g_{yy}$ and $g_{xy} = 0 = g_{yx}$. Here we use the notation g_{ij} to denote the coefficient of the $dx^i dx^j$ terms (with $\{x^i\}_{i=1,2}$ denoting the coordinates x, y. Namely $x^1 = x$, $x^2 = y$). These are called the *metric coefficients*. Note also that $g_{ij} = g_{ji}$, i.e., the metric coefficient is symmetric. (This is true in Riemannian geometry, but not necessarily in more general geometries.) Semi-rigorously speaking, the metric tensor $g = g_{ij}dx^i dx^j$ is a linear map that assigns two tangent vectors at a spacetime point p to a real number. In terms of coordinates, the tangent space at p is spanned by the vectors[2]$\{\partial/\partial x^i\}$, sometimes denoted as ∂_i, where i ranges over the coordinate indices. The coefficient g_{ij} is then

$$g_{ij} = g\left(\frac{\partial}{\partial x^i}, \frac{\partial}{\partial x^j}\right) \in \mathbb{R}. \tag{5.6}$$

[2] Beginners often find it strange that partial derivative operators are treated as a vector. We often imagine a tangent space as a plane on top of p (in the case of a two-dimensional surface), yet in differential manifolds we should not think of an ambient space that contains the manifold. The manifold itself is all there is. So instead of "arrows", the idea here is to consider taking *directional derivatives* (since derivatives are tangents at a point; imagine an infinitely small arrow instead). In a tangent space that has d-dimensions, there are d linearly independent directions we can take the derivative in. We then use derivatives *as* vectors. Conversely therefore, a vector in differential geometry becomes a derivative operator.

Readers should refer to textbooks in differential geometry or general relativity for a proper definition of the metric tensor, as well as tensors in general.

Let us further rewrite the metric in a more compact manner by introducing the "Kronecker delta" δ_{ij}, whose value is 1 when $i = j$ and 0 otherwise. Then,

$$ds^2 = \sum_{i,j} \delta_{ij} dx^i dx^j. \tag{5.7}$$

Since this kind of summation occurs so often, it is customary to just not write the summation symbol, so

$$ds^2 = \delta_{ij} dx^i dx^j. \tag{5.8}$$

The summation is implied by the repeated indices: one upper index and one lower index are summed over when they are the same. This is known as the "Einstein summation convention". The notation g_{ij} immediately suggests that one can treat it as an entry of a matrix (g_{ij}). In our example, it is easily checked that

$$(g_{ij}) = (\delta_{ij}) = \begin{pmatrix} 1 & 0 \\ 0 & 1 \end{pmatrix}. \tag{5.9}$$

Why do we write the Pythagoras theorem in such a complicated notation? Well, because this allows us to generalize to more complicated metric tensors. In general, g_{ij} can be functions of the coordinates. Thus, when we want to describe a curved spacetime, the metric is just a direct generalization of the Pythagoras theorem in that spacetime. Note that in general it is very difficult to tell from the metric alone whether the underlying geometry of space or spacetime is curved or not, because the metric is clearly coordinate dependent. For example, if we switch to the polar coordinates (r, ϕ), where $x = r \cos \phi$ and $y = r \sin \phi$, then the metric of a flat plane would instead satisfy

$$ds^2 = dr^2 + r^2 d\phi^2. \tag{5.10}$$

Thus, in this coordinates (just to use a different notation), $\tilde{x}^1 = r$, $\tilde{x}^2 = \phi$, and

$$(\tilde{g}_{ij}) = \begin{pmatrix} 1 & 0 \\ 0 & r^2 \end{pmatrix}. \tag{5.11}$$

Thus, given a complicated-looking metric, it might still be a flat geometry in disguise. How can we know whether or not there exists a coordinate transformation that can turn it into the standard Pythagoras theorem form (namely δ_{ij})? What is needed is the notion of a quantity that measures curvature and does not depend on coordinate transformation. But before getting to that, we should first introduce the notion of a *spacetime* more properly. You might have heard that spacetime is four-dimensional. Often we write it as (3 + 1)-dimensional to emphasize that time is not quite like the other three spatial dimensions (we can only go forward in time). With the inclusion of time, we write the metric as

$$ds^2 = \eta_{ab} dx^a dx^b, \tag{5.12}$$

where now $\{x^a\}$ is the set of four-dimensional coordinates, whose elements consist of $x^0 = ct$, $x^1 = x$, $x^2 = y$, $x^3 = z$, and (η_{ij}) is a diagonal matrix with $\eta_{ij} = \delta_{ij}$ and $\eta_{00} = -1$. In other words,

$$ds^2 = -c^2dt^2 + dx^2 + dy^2 + dz^2. \tag{5.13}$$

Here c is the speed of light in vacuum. Why is there a minus sign for the dt^2 term? This is not so mysterious. By construction, dx^0 is a distance that light travels in an infinitesimal time interval dt. Thus, we must have, for a beam of light

$$c^2dt^2 = dx^2 + dy^2 + dz^2, \tag{5.14}$$

or equivalently,

$$0 = -c^2dt^2 + dx^2 + dy^2 + dz^2. \tag{5.15}$$

It is then natural to take the right-hand side as a definition for our metric. Because of the 0 on the left-hand side, trajectories of massless particles like light in spacetime is called a *null* "worldline" or simply a null curve. For massive particles, their speed cannot exceed the speed of light, so their trajectory in spacetime necessarily satisfies[3] $ds^2 < 0$. This is equivalent to say that the magnitude of c^2dt^2 is larger than the magnitude of $dx^2 + dy^2 + dz^2$, so they are "closer to being like time", hence these are called *timelike* trajectories. To be more precise, if an observer O moves at constant velocity $v < c$, we know from special relativity that O can treat their frame of reference as being stationary, and only the proper time elapses. The proper time is then a measure of the "length" of the worldline of O. Whenever the metric can be measured by a clock, it is "timelike". Written out in equations, in O's frame with coordinates $\{\tau, x', y', z'\}$,

$$ds^2 = -c^2dt^2 + dx^2 + dy^2 + dz^2 = -c^2d\tau^2 < 0. \tag{5.16}$$

A hypothetical particle that can travel faster than light (a "tachyon") would travel on a *spacelike* curve, with $ds^2 > 0$. Such curves are called "spacelike" for the obvious reason that their lengths are measured as ordinary distances (with a ruler, say). The magnitude of c^2dt^2 is smaller than the magnitude of $dx^2 + dy^2 + dz^2$ for a spacelike curve.

The *light cone* can be understood as the trajectories of light rays in spacetime. Consider turning on a light bulb at initial time $t = 0$, which is represented by a point in spacetime. At some time $t = t_1$ later, the lightfront has traveled some distance ct_1. Geometrically this is a sphere in three-dimensional space. Because we cannot draw four-dimensional diagrams, we usually suppress at least one of the spatial dimensions to draw a spacetime diagram. Thus, with one spatial dimension suppressed, the sphere at time t_1 is represented as a circle whose radius is ct_1. At some later time t_2, we

[3] Here we are using the sign convention of $(-, +, +, +)$, with spatial dimensions represented by $+$ signs. Geometrically this is more natural as it reduces to the usual Pythagoras theorem when t is constant. However, the convention $(+, -, -, -)$ is also used in the literature, especially by particle physicists. In this case, timelike curves would correspond instead to $ds^2 > 0$.

will have a larger circle with radius ct_2. Stacking these circles vertically (a spacetime diagram is customarily drawn with the vertical axis being the temporal one) and we end up with an inverted cone; this is the light cone. For massive particles that also start out at the same location at $t = 0$, their trajectories must lie within the light cone.

So far we have only discussed flat spacetime of special relativity. In general relativity, we have to deal with more complicated metric $ds^2 = g_{ab}dx^a dx^b$, but the idea is quite the same. Note that a four-dimensional spacetime entails a 4×4 matrix (g_{ab}). But due to the symmetry $g_{ab} = g_{ba}$, there are "only" 10 independent metric components. Often in specific physical models with more symmetries, we require even less nonzero components. Still, the calculations are often quite tedious if done by hand. For the sake of clarity, let us return to two-dimensional geometries (without considering time as a dimension) to introduce some important ideas.

5.1.2 Geodesics and Curvature

First, let us discuss a quantity called *Lagrangian* from classical mechanics. If you have no previous knowledge of Lagrangian, it suffices to know that it is a quantity from which we can obtain the equation of motion via some form of calculus known as the "calculus of variation". Specifically, given a Lagrangian $L(q(t), \dot{q}(t))$, where q and $\dot{q}$ denotes a particle's position and velocity respectively, we want to know which path it would take from an initial point a to a final point b. A priori, if we just simply draw a curve from a to b, there are infinitely many possible paths. Thus we must "vary" these paths to find the physical one, which happens to be an extremum in a technical sense. We define the *action* by

$$S = \int_a^b L(q(t), \dot{q}(t))\, dt. \tag{5.17}$$

The variation of S is:[4]

$$\delta S = 0 \Rightarrow \int_a^b \left(\frac{\partial L}{\partial q} \delta q(t) + \frac{\partial L}{\partial \dot{q}} \delta \dot{q}(t) \right) dt = 0. \tag{5.18}$$

We can use integration by parts to rewrite this as

$$\delta S = 0 \Rightarrow \int_a^b \left[\frac{\partial L}{\partial q} \delta q(t) - \frac{d}{dt}\left(\frac{\partial L}{\partial \dot{q}} \right) \delta q(t) \right] dt + \int_a^b \left[\frac{d}{dt}\left(\frac{\partial L}{\partial \dot{q}} \delta q(t) \right) \right] dt = 0. \tag{5.19}$$

By the fundamental theorem of calculus, the second term is just the quantity $(\partial L / \partial \dot{q}) \delta q$ evaluated at the end points. The end points a and b are fixed, and therefore the variation is zero. Therefore we are left with

[4] If you are not familiar with variation, just treat δ as a kind of differentiation whose order can be interchanged with the usual derivative; so $\delta \dot{f} = \frac{d}{dt}(\delta f)$.

$$\delta S = 0 \Rightarrow \int_a^b \left[\frac{\partial L}{\partial q} \delta q(t) - \frac{d}{dt}\left(\frac{\partial L}{\partial \dot{q}} \right) \delta q(t) \right] dt = 0, \tag{5.20}$$

which holds for all possible variation δq. Therefore it must be that the integrand vanishes, i.e.,

$$\frac{d}{dt}\left(\frac{\partial L}{\partial \dot{q}} \right) - \frac{\partial L}{\partial q} = 0. \tag{5.21}$$

This is known as the *Euler–Lagrange equation.* It is common to be confused with the fact that in the Lagrangian we are treating q and $\dot{q}$ as independent variables; how does that even make sense? Well, in elementary physics, given the position function q, we can take a derivative with respect to time and get the velocity $\dot{q}$. That is mathematically very true. However, implicitly we are assuming already that we have a physical law that governs the motion (Newton's laws of motion). Given a law of motion, we can know how the particle moves from one instance to the next. The variation of the Lagrangian, however, is meant to derive the equation of motion. So only *after* that q and $\dot{q}$ have the physical meaning of position and velocity; they are related via the physical law. At the beginning, they are just two separate quantities.

Now, let us return to geometry. Given two points on a two-dimensional surface, how do we find the shortest path that connects point A and point B? Recalling the discussion surrounding Equation (5.2), we note that the metric

$$ds^2 = g_{ab} dx^a dx^b \tag{5.22}$$

implies

$$\left(\frac{ds}{d\tau} \right)^2 = g_{ab} \frac{dx^a}{d\tau} \frac{dx^b}{d\tau} = g_{ab} \dot{x}^a \dot{x}^b, \tag{5.23}$$

where τ is a time parameter (in general relativity this would be the proper time). Now this looks very similar to velocity squared, i.e., the kinetic energy. Namely, we notice that

$$\frac{1}{2} m v^2 \sim \frac{1}{2} m g_{ab} \dot{x}^a \dot{x}^b. \tag{5.24}$$

In the simplest classical mechanic systems, the Lagrangian usually takes the form $L = T - V$, where T is the kinetic energy and V the potential energy. So we can define an effective Lagrangian from the metric: $L = g_{ab} \dot{x}^a \dot{x}^b$. (The constant prefactor $m/2$ is irrelevant for deriving the equation of motion.)

Let us consider a simple but nontrivial example: the 2-sphere S^2, whose metric is given by

$$ds^2 = d\theta^2 + \sin^2 \theta d\phi^2. \tag{5.25}$$

The effective Lagrangian is simply

$$L = \dot{\theta}^2 + \sin^2 \theta \dot{\phi}^2. \tag{5.26}$$

Instead of performing the variation all over again, we can simply apply the Euler–Lagrange equation. First, we do this for the θ coordinate:

$$\frac{d}{d\tau}\left(\frac{\partial L}{\partial \dot{\theta}}\right) - \frac{\partial L}{\partial \theta} = \frac{d}{d\tau}(2\dot{\theta}) - 2\sin\theta\cos\theta\dot{\phi}^2 = 0. \tag{5.27}$$

That is,

$$\ddot{\theta} - \sin\theta\cos\theta\dot{\phi}^2 = 0. \tag{5.28}$$

Similarly, for the ϕ coordinate, we can obtain

$$\ddot{\phi} + 2\cot\theta\dot{\theta}\dot{\phi} = 0. \tag{5.29}$$

Equations (5.28) and (5.29) are called the *geodesic equations* for S^2. One can check that the great circles ϕ=const. in fact do satisfy these equations; they are the "shortest" possible paths between two points, at least locally (because one can go "the other way around" the great circle, the geodesic is the same, but the distance traversed would not be the shortest in that case).

In general, geodesic equations take the form[5]

$$\ddot{x}^a + \Gamma^a_{\ bc}\dot{x}^b\dot{x}^c = 0, \tag{5.30}$$

where $\Gamma^a_{\ bc}$ is called the *Christoffel symbols* or the *connection coefficients*. From Equation (5.28) we can read off the Christoffel symbols directly:

$$\Gamma^\theta_{\ \phi\phi} = -\sin\theta\cos\theta. \tag{5.31}$$

Similarly, from Equation (5.29) we can read off

$$\Gamma^\phi_{\ \theta\phi} = \Gamma^\phi_{\ \phi\theta} = \cot\theta. \tag{5.32}$$

Note that the Christoffel symbols are also symmetric in the lower indices: $\Gamma^a_{\ bc} = \Gamma^a_{\ cb}$.

For flat geometry, if we use the standard Cartesian coordinates, the Christoffel symbols vanish. However, they do not vanish if we use, say, the polar coordinates. So having nonzero Christoffel symbols need not necessarily imply that the underlying space (or spacetime) is curved. For that, we need to compute the curvature. Before that, let us take a closer look at the geodesic Equation (5.30). Suppose that the Christoffel symbols are all zero. Then we have just $\ddot{x}^a = 0$. This amounts to the statement that the acceleration is zero, and so the particle moves in a straight line as expected. Now we can appreciate the effect of the Christoffel symbols. In curved geometries, it describes the correction to the notion of "straightness". In fact,

---

[5] In this section I don't usually distinguish spatial and spacetime indices since they will be clear from the contexts (and also that I am using two-dimensional spatial geometries to provide concrete examples, but the formulae are true also for four-dimensional spacetimes). But in later chapters, I will usually use $a, b, c, \ldots$ to denote spacetime indices, and $i, j, k, \ldots$ to denote spatial indices.

$$\ddot{x}^a = -\Gamma^a_{bc}\dot{x}^b\dot{x}^c \tag{5.33}$$

is analogous to the equation $F = ma$, where the right-hand side plays the role of a "force". Thus in general relativity, a gravitational "force" is such an effect from the geometry. If the geometry is flat, but we are using a non-Cartesian coordinates, then the right-hand side is analogous to a fictitious force. However, in general relativity, even gravity is not a true force because it is simply the result of the geometry. In fact, this is why the effective Lagrangian associated with the metric only has a kinetic term—there is no potential term associated with any force. In other words, a test particle in spacetime in general relativity is a free particle.

Given two nearby geodesics, we can quantify the curvature of the underlying geometry. For example, two particles that start out parallel at the equator of S^2, their paths along great circles would converge at the North Pole, which means that the geometry cannot be flat—in Euclidean geometry, parallel lines cannot possibly meet. At this point, it should be noted that a common analogy in popular science is quite misleading. The analogy states that "gravity is geometry" can be appreciated as being analogous to the two observers being attracted by a gravitational force, thereby decreasing their distance as they move towards the North Pole. This is not the case. The situation we have is that of two test particles, whose own mass and gravitational field is negligible. Therefore both of them only "feel" the overall gravitational field of the background, not their mutual gravitational attraction.

In any case, to quantify curvature, we therefore start with the geodesic equation for a particle whose worldline is described by the coordinates $\{x^a\}$. To make the equation clearer, we restore the derivative operator instead of using an over-dot:

$$\frac{d^2}{d\tau^2}x^a + \Gamma^a_{bc}(x)\frac{d}{d\tau}(x^b)\frac{d}{d\tau}(x^c) = 0. \tag{5.34}$$

Consider also the geodesic equation for a nearby particle whose worldline is given by $x^a + \delta x^a$. Here δ is not a variation but denotes a tiny displacement. That is:

$$\frac{d^2}{d\tau^2}(x^a + \delta x^a) + \Gamma^a_{bc}(x + \delta x)\frac{d}{d\tau}(x^b + \delta x^b)\frac{d}{d\tau}(x^c + \delta x^c) = 0. \tag{5.35}$$

We then subtract these equations to obtain an equation for the displacement (ignoring all higher-order terms in δx). The derivation is somewhat tedious, but the end result is that we ultimately arrived at the so-called *geodesic deviation equation*

$$\nabla^2_\tau\ \delta x^a = R^{\ a}_{b\ cd}\frac{d}{d\tau}(x^b)\frac{d}{d\tau}(x^c)\delta x^d, \tag{5.36}$$

where $\nabla_\tau\ \delta x^a := \frac{d}{d\tau}\delta x^a + \Gamma^a_{bc}(x)\frac{dx^b}{d\tau}\delta x^c$ is called the *covariant derivative*, and

$$R^{\ a}_{b\ cd} = \partial_c\Gamma^a_{bd} - \partial_d\Gamma^a_{bc} + \Gamma^e_{bd}\Gamma^a_{ec} - \Gamma^e_{bc}\Gamma^a_{ed}, \tag{5.37}$$

is known as the (components of) *Riemman curvature tensor*. We can also define R_{abcd} by contracting with the metric: $g_{ae}R^{\ e}_{b\ cd} = R_{abcd}$.

Historically, Riemann came up with this expression by considering the conditions under which a metric can be transformed into the flat space form $ds^2 = \delta_{ij}dx^i dx^j$ (for spacetime just replaces δ_{ij} with η_{ab}). Riemann actually had the foresight that the geometry of the real world is a matter of physics, and that one needs to check whether it is non-Euclidean (actually, his advisor, Gauss, also considered such a possibility). Riemann considered the possibility that the geometry of space may be given by some metric $g_{ij}dx^i dx^j$, but he didn't include the dimension of time.[6] That idea would have to wait for Minkowski, after Einstein published his theory of special relativity.

For S^2 the unit 2-sphere, we can compute the nonzero Riemann curvature components. They are (check this!):

$$R^{\theta}_{\ \phi\theta\phi} = \sin^2\theta, \qquad R^{\phi}_{\ \theta\phi\theta} = 1. \tag{5.38}$$

From the Riemann curvature, one defines the *Ricci curvature* by $R_{bd} := R^{a}_{\ bad}$ (recall that by Einstein's convention this means summing over the first and third indices).[7] For S^2 we have $R_{\theta\theta} = 1$ and $R_{\phi\phi} = \sin^2\theta$. Finally, we define the *Ricci scalar* or simply the *scalar curvature* by $R := g^{ab}R_{ab}$. Here, g^{ab} denotes the matrix element of the inverse matrix of (g_{ab}). For the 2-sphere we get $R = 2$, or $R = 2/r^2$ if the radius is r. (In two-dimensions, the Ricci scalar is always twice the so-called "Gaussian curvature".) This makes sense—a larger sphere has a smaller curvature. This is why locally on the surface of the Earth, we do not notice any curvature. R is a useful quantity because its value is independent of the choice of coordinates. This is in contrast, with say, Γ^a_{bc} or $R^{a}_{\ bcd}$, whose values depend on the coordinates. Other contractions of the curvature tensors like the *Kretschmann scalar* $R^{abcd}R_{abcd}$ is also coordinate independent.

Finally we are ready to say exactly what general relativity is. The *Einstein field equations* read

$$R_{ab} - \frac{1}{2}g_{ab}R + \Lambda g_{ab} = \frac{8\pi G}{c^4}T_{ab}, \tag{5.39}$$

where Λ is the *cosmological constant*. In Einstein's original work he did not include this,[8] so the combination $R_{ab} - \frac{1}{2}g_{ab}R$ is called the *Einstein tensor*. The quantity T_{ab}

---

[6] Riemann's famous Habilitation dissertation on this topic was in 1854. Just a few years later in 1857, another mathematician William Clifford (of the "Clifford algebra" fame) proposed the idea that the curvature of space can be continuously propagated from one point to another "after the manner of a wave". He envisioned that this could explain the motion of matter. Today we have instead gravitational waves that propagates in spacetime.

[7] For readers who have learned differential geometry or general relativity before, I think it is worth emphasizing that the Ricci curvature can be defined without any use of the metric. As long as a manifold has a connection defined, the Riemann curvature tensor and the Ricci curvature can be defined, without requiring the metric structure. (The connection coefficients are tied to the metric once it is chosen to be the Levi–Civita connection, but one could choose other connections.) Only the Ricci scalar needs the metric to be defined.

[8] Einstein introduced the cosmological constant so that the Universe described by general relativity can remain static, as he saw no evidence for a dynamically evolving cosmos at that time. More on this point later.

on the right-hand side is the *energy–momentum-(stress) tensor*, which describes the matter content of the physical system. It should be remarked that despite being highly regarded as one the most important achievements in the history of physics (and the whole of Science), Einstein was never satisfied with general relativity. He described the field equation as similar to a building, with one wing made of fine marble (referring to the geometrical left-hand side[9]), and the other wing of wood (the matter part). To Einstein, the T_{ab} should ideally be explained in a more fundamental theory. For the rest of his life, Einstein tried to formulate a unified field theory that can explain both gravity and matter.

In popular science and even textbooks, one may encounter statements along the line: gravity is *caused* by spacetime curvature. This is only a colloquial statement, it is technically *not* entirely correct. What we have in the Einstein field equations tell us that a certain combination of geometric quantities in terms of the metric and the curvature is equal to some mass-energy configuration. At this level we do not yet see how gravity comes into play. A comparison to Newtonian gravity can be made when we look at a metric that describes a spherically symmetric star, for example, from which we can extract information from the geometry that corresponds to what we would call gravity in Newtonian terms. But spacetime geometry is only a description of the phenomenon we call "gravity", there is no cause and effect involved.

In addition, we can consider the Newtonian limit of general relativity, via the "weak-field approximation", which allows us to treat gravitational fields as a perturbation of the flat Minkowski metric, as $g_{ab} = \eta_{ab} + h_{ab}$, where the components $|h_{ab}| \ll 1$. The Schwarzschild metric (see Section 9.1), which describes the gravitational field outside a static spherical body, can then be shown to reduce to the isotropic form

$$ds^2 = -\left(1 + \frac{2\Phi}{c^2}\right)c^2dt^2 + \left(1 - \frac{2\Phi}{c^2}\right)(dx^2 + dy^2 + dz^2). \tag{5.40}$$

In fact, this is how we know the value of κ in the Einstein field equations

$$R_{ab} - \frac{1}{2}g_{ab}R = \kappa T_{ab}. \tag{5.41}$$

General relativity itself does not tell us what κ is; we deduce what it *must* be by taking the limit of the theory into the regime Newtonian gravity holds. This is typically done using a perfect fluid for the energy-momentum tensor in the weak-field regime (in which the pressure terms are negligible compared to the energy density), so that one can relate the time-time component of the Ricci tensor to the energy density of the fluid, which turns out to satisfy[10]

[9] Hereinafter in this book, we denote left-hand side by LHS and the right-hand side by RHS.

[10] With $T_{00} = \rho c^2$.

$$R_{00} = \frac{1}{2}\kappa\rho c^2 = \frac{\nabla^2\Phi}{c^2}. \tag{5.42}$$

By comparing this to the Newtonian Poisson equation, $\nabla^2\Phi = 4\pi G\rho$, we can deduce that $\kappa = 8\pi G/c^4$. This derivation contains another important piece of information: the Newtonian gravitational field Φ can be derived from the time-time component of the Ricci tensor alone. This is why the usual "trampoline" analogy (with balls placed on it pressing down on the surface creating "curvature") in popular science textbooks and even in many science museums, are very misleading. What is important here is the *temporal curvature*. But does gravity *cause* time to be curved, or does time *cause* gravity? General relativity is silent on such an interpretative issue. More generally, general relativity does not even tell us *why* spacetime is curved in the presence of matter and energy. Why is there a coupling between objects and spacetime? The fact that stars can collapse to form a black hole suggests that matter can be transformed into spacetime curvature, but on the other hand, black holes can evaporate via Hawking radiation and thus spacetime curvature is transformed into matter. These suggest that there is an intimate connection between matter/energy and spacetime that is still not understood (or even appreciated). But we have carried the discussion too far into a more speculative regime and it is better to stop here for now.

Finally, some historical remarks are in order. We note that Equation (5.39) is actually a set of partial differential equations, so solving it in general (i.e., to find a metric g_{ab} that describes some T_{ab}) is highly nontrivial. It is customary in textbooks to first discuss the Schwarzschild solution, which was historically the first nontrivial equation found—surprisingly—in the same year Einstein published his theory of general relativity, by Karl Schwarzschild (1873–1916) who was serving in the military during the 1st World War[11]. Karl Schwarzschild wasn't the only one fascinated with general relativity amidst the war. The philosopher Rudolf Carnap (1891–1970) would try to explain the theory to his fellow soldiers in 1917 (Edmonds 2020). After the war ended that year, he would move to Berlin to study relativity formally (Einstein was a professor there at that time) and would go on to study the philosophy of space in general relativity. He is however, primarily known for his work in logical positivism, also known as logical empiricism (he was a member of the famous Vienna Circle), as well as language and semantics. I will defer the discussions of Schwarzschild's solution till the chapter on black holes (Chapter 9). Before we start introducing modern cosmology, we will make some remarks on the so-called "equivalence principle".

5.2 Equivalence Principle and Its Perils

In popular science books and even physics textbooks, one can often find the mention of the so-called "equivalence principle", which I personally do not like for many

[11] I always tell my students that if Schwarzschild can accomplish this with pen and papers while at war, surely they can do it in the comfort of their own room with the help of computer!

reasons. The first reason is a pragmatic one: one can do general relativity without it (differential geometry and the Einstein field equations are enough), so it is not essential. Of course the equivalence principle had its use in guiding Einstein towards general relativity (he had "the happiest thought" of imagining a man falling down from a roof, realizing that the person would not feel his weight during free fall), but to still use it today can be misleading especially to students, as I will explain below. There are a few versions of the equivalence principle, but the one most familiar to the readers is probably the statement that gravity is indistinguishable from acceleration, with the caveats that the measurements are done *locally* in a sufficiently small laboratory. It is easy to see why, on the surface of the Earth, if you were to drop two balls, both of them will drop pointing towards the center of the Earth. They do not drop parallel to the ground as they would have if there is no gravity, but instead the laboratory is accelerated upward in a rocket instead. Of course, the difference is tiny. Still, the point is that acceleration is not indistinguishable to gravity *in principle* unless the gravitational field is uniform, or unless by "local" we mean *at a single mathematical point*, which of course cannot possibly fit a laboratory! Mathematically, this version of the equivalence principle is already built into differential geometry: at each point in spacetime, one can always choose a coordinate system such that the metric tensor at that point takes the Minkowski form, and such that the Christoffel symbols vanish. This is known as the Riemannian normal coordinate system.

In the textbook on general relativity by John L. Synge in 1960, he wrote (Synge 1960):

> "When, in a relativistic discussion, I try to make things clearer by a spacetime diagram, the other participants look at it with polite detachment and, after a pause of embarrassment as if some childish indecency had been exhibited, resume the debate in their own terms. Perhaps they speak of the Principle of Equivalence. If so, it is my turn to have a blank mind, for I have never been able to understand this Principle. Does it mean that the signature[12] of the spacetime metric is $+2$ (or -2 if you prefer the other convention)? If so, it is important, but hardly a Principle. Does it mean that the effects of a gravitational field are indistinguishable from the effects of an observer's acceleration? If so, it is false. In Einstein's theory, either there is a gravitational field or there is none, according as the Riemann tensor does not or does vanish. This is an absolute property; it has nothing to do with any observer's world-line. Space-time is either flat or curved [...] The Principle of Equivalence performed the essential office of midwife at the birth of general relativity, but, as Einstein remarked, the infant would never have got beyond its long-clothes had it not been for Minkowski's concept. I suggest that the

[12] By signature he meant the *sum* of $\{-, +, +, +\}$, i.e. $-1 + 1 + 1 + 1 = 2$.

midwife be now buried with appropriate honors and the facts of absolute spacetime faced."

Now I turn to why the notion of the equivalence principle can lead to incorrect arguments and misconceptions. Sometimes an argument was made using the equivalence principle to deduce that in general relativity light rays can bend around massive objects like the Sun. The argument goes as follows: consider a small rocket ship in which a window allows light rays to enter. Since the ship is accelerating "upward", say, the trajectory of the rays as measured by the astronaut would appear to be bent. Since "acceleration = gravity", this argument is said to imply light bending by a gravitational field. However, recall that the equivalence principle is "local", so this cannot imply anything about actual light bending which occurs on a much larger scale. Furthermore, this argument did not make use of the Einstein field equations at all! So if it were to be correct, this would imply that light bending should be generically true in almost all theories of gravity (those theories that are formulated on a manifold and admit normal coordinate systems). However, this is not the case. Before general relativity, there was one attempt by Gunnar Nordström to formulate, essentially, the first example of a "metric theory" of gravity. Adriaan Fokker, together with Einstein, re-wrote Nordström theory into the form: $R = (24\pi G/c^4)T$, where T is a contraction of the energy-momentum tensor. Light, being an electromagnetic wave, is governed by the energy-momentum tensor of electromagnetism, which is trace-free, i.e., $T \equiv 0$. In such a theory then, electromagnetism does not gravitate and light cannot be bent by gravity. This is despite the fact that the manifold structure admits a Riemannian normal coordinate system!

In addition, the misleading statement "acceleration = gravity" is probably why there is a widespread misconception that special relativity cannot handle acceleration, and that the famous Twin Paradox cannot be resolved unless one appeals to general relativity. This is simply incorrect. Special relativity *can* handle acceleration just as well as Newtonian mechanics; after all, it *generalizes* the latter (i.e., Newtonian mechanics is a limit of special relativistic mechanics when $c \to \infty$). There is also no gravitational field involved in the paradox, and thus there is no need to involve general relativity.

5.3 Dynamical Universe: The Friedmann Equation

5.3.1 Cosmic Expansion and Acceleration

Cosmology is the study of the evolution of the entire cosmos. Einstein, like most people at that time, thought that our Universe is static, i.e., it has always been like this and will always be like this. He even tried to construct a static cosmological solution in general relativity, not realizing that such a universe is unstable. It is easy to complain in hindsight that Einstein should have known better, that he should have realized that a static universe is not stable (the mathematics to analyze stability was readily available and not difficult) and with that he could have predicted that the

Universe must be dynamical, as shown later by Georges Lemaître[13] (1894–1966), Edwin Hubble (1889–1953), and others. However, at the time there was simply no strong evidence that the Universe can be anything but static, so Einstein was just trying to describe what observations back then supported. It is often said that Einstein subsequently remarked that his introduction of the cosmological constant in order to construct a static universe is his "greatest blunder", but this may not be historically accurate (Barrow 2017).

It is worth mentioning that even after the expansion of the Universe was discovered, Einstein once considered the possibility that new matter continues to be formed in the Universe as it expands, so that the Universe looks overall the same (on average) at every instant of time. This is an earlier version of the "steady-state model" that will later be proposed again by Fred Hoyle (1915–2001), Thomas Gold (1920–2004), and Hermann Bondi (1919–2005) in 1948. Einstein's steady-state universe manuscript was probably written in early 1931. The difference between the later model of Hoyle–Gold–Thomas and Einstein's is that instead of a "creation field"[14] postulated by Hoyle, Gold, and Thomas, Einstein proposed that the cosmological constant assigns an energy to empty space that can be associated with the creation of matter. This is quite similar to some modern models in which dark energy can be converted into dark matter! However, Einstein did not publish his idea as the derivation, when correctly performed, would imply that such a universe would be devoid of matter. It is perhaps not surprising that Einstein would pursue such an idea in the first place, since a steady-state model is the "next closest" thing to a static universe that he initially envisioned. However, the failure of obtaining a working steady-state model eventually led him to fully embrace the dynamical expanding Universe picture. Einstein took the pragmatic approach in his study of cosmology, as he quickly abandoned the cosmological constant and the spatial curvature terms, as neither of them were required to explain observations (O'Raifeartaigh 2017).

How do we model the evolution of the Universe via general relativity? At the scale of the entire Universe, we can treat galaxy clusters as just "fluid particles"—in fact, matter is treated as "dust",[15] which is a pressureless fluid—so we can model them with the energy-momentum tensor of a perfect fluid ($T_{ab} = \mathrm{diag}(-\rho, p, p, p)$),

[13] In addition to being a physicist, Lemaître also served as a Catholic priest. He was the first to propose the idea of the hot Big Bang (he called it the "primeval atom") in 1927, which was then published in the paper titled *The Beginning of the World From the Point of View of Quantum Theory* in 1931 (Lemaître 1931). It is somewhat ironic that today some anti-science religious people think that the Big Bang is against the teaching of the Bible, given that a priest saw no contradiction between his science and religion. In the original draft of his 1931 paper, there was a paragraph that he decided to delete afterward: "I think that every one who believes in a supreme being supporting every being and every acting [sic.], believes also that God is essentially hidden and may be glad to see how present physics provides a veil hiding the creation" (Luminet 2011).

[14] "Simply" add a term C_{ab} to the RHS of the Einstein field equations!

[15] "To see a world in a grain of sand" as William Blake said.

Box 3: Units and Dimensions
It is customary to set the speed of light $c = 1$ in relativity. This is not only for convenience (since c occurs all over the place and we don't want to write them all), but it allows us to draw the light cone at 45 degrees. By setting $c = 1$ we are really treating time and space on equal footing as they would now have the same physical dimensions. Namely, time is now measured in units of length, say, meters. This may look strange, but humans used to measure time based on the changing shadow length of a sundial, thus "an inch of time worth one inch of gold" (一寸光阴一寸金) in Chinese idiom. Even today we often say the distance between two locations as "half an hour by walking", hence also measuring distance in time. So $c = 1$ is not that strange after all. We also often set the gravitational constant $G = 1$. With both $G = c = 1$, mass would also have a dimension of length, since GM/c^2 is a length.

Max Planck (1858–1947) came up with a system of units of measurement that are defined in terms of c, G, $\hbar$, and k_B, the latter two being the (reduced) Planck constant from quantum mechanics, and the Boltzmann constant from thermal physics and statistical mechanics. For example, the Planck length is defined by $\ell_\mathrm{P} = \sqrt{\hbar G/c^3}$ and the Planck mass by $m_\mathrm{P} = \sqrt{\hbar c/G}$. Since k_B is "only" a conversion factor between energy and temperature, much like c is the conversion between duration and length, we also often set $k_B = 1$. If $\hbar$ is also set to be 1, then everything is dimensionless. This is called the "Planck unit". Planck emphasized the universality of this unit system:

> "[...] it is possible to set up units for length, mass, time, and temperature, which are independent of special bodies or substances, necessarily retaining their meaning for all times and for all civilizations, including extraterrestrial and non-human ones, which can be called "natural units of measure".

The Planck mass and other "Planck quantities" are dimension-dependent. The Planck length in D-spacetime dimensions is

$$\ell_\mathrm{P}^{(D)} = \left(\frac{\hbar G}{c^3}\right)^{\frac{1}{D-2}}. \tag{5.43}$$

Evidently something strange happens in the $D = 2$ case, i.e., in $(1 + 1)$-dimensions. Similarly, in D-spacetime dimensions, the Planck time is $T_\mathrm{P}^{(D)} = \left(\frac{G\hbar}{c^{D+1}}\right)^{\frac{1}{D-2}}$. The fact that it does not work in $D = 2$ can be seen more clearly by noting that one cannot find a combination of x, y, z for the dimensional analysis $[G^x \hbar^y c^z] = \left(\frac{L}{MT^2}\right)^x \left(\frac{L^2 M}{T}\right)^y \left(\frac{L}{T}\right)^z = T$ to hold (here L is a length, M a mass, and T a time). Note that in the $(1 + 1)$-dimensional case, the gravitational constant G has units $LM^{-1}T^{-2}$, rather than the usual (3+1)-dimensional units of $L^3M^{-1}T^{-2}$. The specularity of lower dimensional spacetimes may explain why we live in $(3 + 1)$-dimensions. See also Section 11.1.

While there is nothing wrong with setting $G = c = \hbar = k_B = 1$, we must then be very careful about the numerical values that we assign to our system of interests. For example, many authors often set $M = 0.5$, 1, 5, 10 for black hole masses in their papers, but then this would actually correspond to highly quantum black holes around the Planck mass, and not large astrophysical black holes that the authors intended them to be (though often this does not change the qualitative results they are after). It is better to either leave G or $\hbar$ explicit and not set them to unity. The choice of keeping G is quite common in general relativity and the cosmology community. In this book, sometimes we use the Planck units, and sometimes we keep G explicit (especially when dealing with cosmology). Students need to be familiar with different conventions.

Before the introduction of Planck units in 1899, there were the "Stoney units", introduced by George Johnstone Stoney (1826–1911) in 1881, the physicist who also coined the name "electron". Both Planck units and Stoney units are examples of "natural units", which simply mean a set of units of measurement such that all physical constants can be defined from elements of the set. The Stoney units utilize: c, G, k_e, e, where k_e is the Coulomb constant ($k_e \equiv 1/4\pi\varepsilon_0$, where ε_0 is the vacuum permittivity) and e the charge of the electron. Thus, for example, a "Stoney length" is given by the combination

$$\ell_S = \sqrt{\frac{Ge^2}{4\pi\varepsilon_0 c^4}}, \tag{5.44}$$

while a "Stoney mass" is

$$m_S = \sqrt{\frac{e^2}{4\pi\varepsilon_0 G}}. \tag{5.45}$$

Just like the Planck units, one sets c, G, k_e and e to unity in the Stoney units. While there is no unit of charge included in Planck units, one can impose the additional requirement that $k_e = 1$, which is a common practice.

where ρ is the density and p the pressure. (We will set $c = 1$; see Box 3 above) The metric, obtained under the assumption that the Universe is homogeneous and isotropic at large scale, is known as the *Friedman-Lemaître-Robertson-Walker* (FLRW) universe, named after Alexander Friedmann (1888–1925), Georges Lemaître (1894–1966), Howard Robertson[16] (1903–1961), and Arthur Walker (1909–2001). In one of the commonly used coordinates, the metric reads

$$ds^2 = -dt^2 + a(t)^2\left[\frac{dr^2}{1 - kr^2} + r^2(d\theta^2 + \sin^2\theta d\phi^2)\right]. \tag{5.46}$$

If we choose the convention that the function $a(t)$ is dimensionless (it tells us how big the Universe is at any other time relative to its present size, $a(t_0)$, which is by this definition, equal to 1), and r carries the dimension of length, then k has dimension 1/length^2. Indeed k is the scalar curvature of the spatial geometry and can take any value. This is the convention we shall use. Of course, one can also work with the convention that $a(t)$ is dimensionful, and r dimensionless, in which case k is also dimensionless and thus customarily normalized to be either -1, 0, or $+1$. Confusingly, quite often in the literature people also set $k = -1, 0, +1$ even though they use the convention that $a(t)$ is dimensionless. This only makes sense if $k = -1, 0, +1$ are just representations of their respective classes

[16] Robertson led the so-called "Robertson Panel", a committee commissioned by the US Central Intelligence Agency in 1952 to review reports of Unidentified Flying Objects. See Section 13.5 for more details.

(positive, zero, and negative curvature), and are actually dimensionful but not clearly specified.

In both conventions, $k = 0$ means the spatial section is flat. This is the easiest case to study: we see that with $k = 0$, if $a(t) \equiv 1$ for all t, then this is just Minkowski spacetime in spherical coordinates. Thus the role of $a(t)$ is to re-scale the length measured by Pythagoras' theorem in the spatial geometry. Namely, given two galaxies marked by point A and B, their distance measured today would be different from the distance measured yesterday. Thus $a(t)$ is known as the *scale factor*. Notice that the coordinate distance between A and B does not change. For simplicity, in the spatial geometry, let us change to Cartesian coordinates and let A be at the origin (0, 0, 0) and B at (1, 1, 1), say. Then they will always be at those points and so their coordinate distance is always $\sqrt{1^2 + 1^2 + 1^2} = \sqrt{3}$. What changes is their actual, physical distance given by $a(t)\sqrt{3}$. In popular science parlance, one usually says that the galaxies[17] are not moving apart, but rather "new space" is being "created" in between the galaxies.

The velocity of recession of the galaxies is

$$\vec{v} = \frac{d}{dt}[a(t)\vec{r}] = \dot{a}(t)\vec{r} = \frac{\dot{a}(t)}{a(t)}\vec{R}, \tag{5.47}$$

where $R := a(t)r$ is the physical distance. (Note that in the second equality, we have used the fact that the coordinate distance r is fixed.) The quantity

$$H(t) := \frac{\dot{a}(t)}{a(t)} \tag{5.48}$$

is extremely important in cosmology. It is called the *Hubble parameter*, and its value at the current time t_0, denoted H_0 is called the *Hubble constant*. In fact, we recognize Equation (5.47) as the more general form of the Hubble–Lemaître law:[18] $v = H_0 D$, where D denotes the (physical) distance. Given that observation of galaxies back then could only be done with sufficiently nearby galaxies, H can be approximated as H_0 in the Hubble–Lemaître law. Clearly, an expanding Universe means H_0 is positive.

[17] Individual galaxies that are relatively close to each other can still be attracted by their mutual gravity and collide. In about 4.5 billion years our Milky Way Galaxy will likely collide with the Andromeda Galaxy; this is not inconsistent with the expansion of the cosmos at larger scales. In fact the Milky Way has assimilated various dwarf galaxies in the past. One of the major incidents was the *Gaia–Enceladus merger* some 8–11 billion years ago with a galaxy about 1/4 the size of the current Milky Way.

[18] This was previously known as the "Hubble law", but in 2018 the International Astronomical Union voted to rename it, since Lemaître did, in 1927, obtain the equation before Hubble in 1929 (let us also not forget the contribution of Milton Humason (1891–1972)—a "mule skinner" who delivered materials and equipment up the Mount Wilson Observatory during construction but later turned observational assistant and staff—who worked closely with Hubble to measure the redshift)! The realization that the Universe could be expanding was realized even earlier by Friedmann in 1922.

Observations have indicated that our Universe is consistent with $k = 0$ (of course observations can never tell us whether k is *exactly* zero due to error bars). Note that this does not necessarily mean that the Universe is infinitely large. If we relax the exactly isotropic assumption, the Universe may be flat but compact (for example, a flat 3-torus): imagine a box with opposite sites topologically identified, if you pass through any of the (non-solid) wall you will emerge from the opposite wall. For the $k = 1$ case, the spatial geometry is that of a sphere[19], though again we can also relax the isotropic assumption and consider, say, identifying the antipodal points. If we do this, instead of a 3-sphere S^3, we would have a real projective space $\mathbb{R}P^3$. In fact, when de Sitter first introduced what would later become known as de Sitter spacetime (which has a positive cosmological constant), he had followed Schwarzschild in imposing the spatial section to have $\mathbb{R}P^3$ topology instead of S^3, which is the assumption used today. Though the distinctions may be impossible to check in classical cosmology given the scale of the Universe, the distinctions between the two may be important in quantum cosmology (McInnes 2003; Ong & Yeom 2017). The $k = -1$ case corresponds to hyperbolic geometry. Just like using cubes to tile a flat space to get a toral universe, we can use dodecahedra to tile a hyperbolic space.[20] One manifold that can be constructed in this way is the "Seifert–Weber space". Another relatively famous example is the "Weeks manifold". In popular science books and some textbooks, the two-dimensional illustration for hyperbolic geometry is often that of a saddle. However, keen-eyed students would notice that the saddle surface (mathematically called a hyperbolic paraboloid) does *not* have a constant curvature throughout. This is because of Hilbert's theorem in differential geometry: there exists no complete regular surface of constant negative Gaussian curvature immersed in $\mathbb{R}^3$. Therefore such an illustration can be misleading if not carefully clarified.

The Einstein field equations (without Λ), along with the fact that T_{ab} satisfies[21] $\nabla^a T_{ab} = 0$ would give us two important equations. The first being the *Friedmann equation*

[19] To see why this is so, consider instead a two-dimensional sphere with radius L (and hence curvature $k = 1/L^2$). The claim is that its metric can be written as

$$ds^2 = \frac{dr^2}{1 - \frac{r^2}{L^2}} + r^2 d\phi^2. \tag{5.49}$$

To see this, note that if we embed $S^2 \hookrightarrow \mathbb{R}^3$, the sphere has equation $x^2 + y^2 + z^2 = L^2$. Then we can show that the Euclidean metric $dx^2 + dy^2 + dz^2$, restricted on the surface of the sphere, is exactly the form in Equation (5.49). This is left as an exercise for the student readers.

[20] If our Universe is indeed like this, it would make the ancient Greeks very happy, since the dodecahedron represents their celestial quintessence.

[21] ∇ denotes the covariant derivative; readers with no prior knowledge of differential geometry can just think of this as a generalization of differentiation in curved spacetimes. This equation is analogous to the conservation of energy, but not quite as we will soon discuss.

$$\left(\frac{\dot{a}}{a}\right)^2 = \frac{8\pi G\rho}{3} - \frac{k}{a^2}. \tag{5.50}$$

In terms of the Hubble parameter, the Friedmann equation is

$$H^2 = \frac{8\pi G\rho}{3} - \frac{k}{a^2}. \tag{5.51}$$

The second equation is the *fluid equation*

$$\dot{\rho} + \frac{3\dot{a}}{a}(\rho + p) = 0. \tag{5.52}$$

Both of these equations can be "derived" already in Newtonian physics (see, for example, the textbook by Liddle; Liddle 2015), but the physical interpretations are quite different which might cause confusions to beginners, so I prefer not to do so here. Instead, we will just accept that these equations are the results of general relativity and explore their consequences.

Note that the sign of k depends on whether ρ is greater or less than the *critical density* defined by

$$\rho_c(t) := \frac{3H^2(t)}{8\pi G}. \tag{5.53}$$

The density parameter is the dimensionless quantity defined by $\Omega := \rho/\rho_c$. If there are multiple components (say A and B) that we want to explicitly consider in the Universe, we can separate their contribution: $\Omega = \Omega_A + \Omega_B$.

By taking another time derivative of the Friedmann equation, and by substituting the fluid equation into the resulting equation, we could obtain the *acceleration equation*:

$$\frac{\ddot{a}}{a} = -\frac{4\pi G}{3}(\rho + 3p). \tag{5.54}$$

For normal matter like a perfect fluid (which we used as a model of galaxies, gases etc.), $\rho + 3p$ is positive, so $\ddot{a} < 0$. This means that the expansion of the Universe should be *decelerating*. It may be surprising to some readers to see that a larger pressure p would cause $\ddot{a}$ to be more negative. Our Newtonian intuition would be that positive pressure tends to push things apart, so why does it cause expansion to decelerate instead? The answer, essentially, is that pressure also contributes to the energy-momentum tensor, which curves spacetime in a manner that ends up with such an effect. Thus a Newtonian intuition can sometimes lead us astray in general relativity.

In 1998, our understanding of the Universe changed as observations indicated that the expansion is actually accelerating! The accelerated expansion can be explained if the cosmological constant is included in the Einstein field equations. The Friedmann equation now carries an additional term:

$$H^2 = \frac{8\pi G\rho}{3} - \frac{k}{a^2} + \frac{\Lambda}{3}. \tag{5.55}$$

The acceleration equation would then be modified into

$$\frac{\ddot{a}}{a} = -\frac{4\pi G}{3}(\rho + 3p) + \frac{\Lambda}{3}. \tag{5.56}$$

If $\Lambda > 0$ and is sufficiently large, the second positive term can overcome the first negative term to yield $\ddot{a} > 0$. It should be emphasized again that a positive pressure leads to a *deceleration* as can be seen in the equation above, which is contrary to our intuition that a positive pressure tends to "push things away". The cosmological constant term essentially acts as a *negative pressure* term. One often reads from popular science articles and books that it "came as a surprise" when observational evidence showed the expansion to be accelerating in 1998. As usual, the actual history is more complicated. It is certainly *not* the case that nobody cared about the cosmological constant before 1998. The truth is that the possibility of the cosmological constant had always been around, and had been used to explain various observations[22]. In fact, in 1995, J. P. Ostriker and Paul J. Steinhardt already gave the values of the matter density of about $\Omega_m \approx 0.3$ and the effective mass density of cosmological constant about $\Omega_\Lambda \approx 0.7$ (the values differ slightly when curvature term is included), based on observational constraints from nucleosynthesis and large-scale structures (Ostriker & Steinhardt 1995). Thus, the 1998 supernovae observations were not that much of a surprise. Nevertheless, they definitely provided a stronger evidence in support of a positive cosmological constant or, more generally, of dark energy (see Chapter 6).

Students may be confused why nowadays we use the cosmological constant in the context of *accelerated* expansion, whereas Einstein used it to make his universe *static*! This can be appreciated from the Friedmann equation. Equation (5.55) implies

$$\frac{da(t)}{dt} = a(t)\left[\frac{8\pi G\rho}{3} - \frac{k}{a^2} + \frac{\Lambda}{3}\right]^{\frac{1}{2}}. \tag{5.57}$$

[22] For example, Occhionero et al. discussed linear clustering of galaxies in a low density universe with $\Lambda > 0$ in 1980 (Occhionero et al. 1980); Efstathiou et al. argued for the need of a cosmological constant to explain large-scale structure in a *Nature* paper in 1990 (Efstathiou et al. 1990), and by 1995 Krauss and Turner wrote that "the observational case for a cosmological constant is so compelling today that it merits consideration in spite of its checkered history" (Krauss & Turner 1995). In fact, the Soviet physicist Matvei Bronstein (1906–1938) considered in 1933 a cosmological "constant" (more like modern dark energy) that can decay into matter and radiation. In the next chapter, we will see that Niels Bohr (1885–1962) once considered the possibility that energy conservation does not hold in nuclear processes. Since stars shine via nuclear processes, Bohr realized that this means energy emitted by stellar radiation might also not be conserved. Bronstein proposed that this non-conservation can be understood as being converted from the dynamical dark energy (Bronstein 1933). (Bronstein was the first to realize that quantizing gravity is not going to be straightforward as quantizing electrodynamics (Gorelik 2005). Unfortunately he was executed in 1938 under Stalin's order.)

The matter density ρ decreases at late times as the Universe expands and so does the curvature term. Hence even if $\Lambda > 0$ is very small, it eventually dominates the dynamic, and we have

$$\frac{da(t)}{dt} \sim a(t)\sqrt{\frac{\Lambda}{3}}, \tag{5.58}$$

which implies an exponentially growing scale factor $a(t)$. Of course, the Universe can be made static if Λ is chosen to cancel the matter and curvature terms so that[23] $\dot{a} = 0$. What Einstein did not realize was that this is *unstable*. Like balancing a pencil on its tip, a slight perturbation will disrupt the balance condition and make the Universe dynamical again.

Let us also remark that observations suggest that the Hubble parameter $H(t)$ is decreasing in time, i.e., $\dot{H} < 0$. However,

$$\dot{H} = \frac{d}{dt}\left(\frac{\dot{a}}{a}\right) = \frac{a\ddot{a} - \dot{a}^2}{a^2}, \tag{5.59}$$

which is not the same $\ddot{a}/a$ or $\ddot{a}$. Thus $\ddot{a}$ can be positive even if $H(t)$ is monotonically decreasing. In fact there is no physics in this paragraph, just elementary calculus.

5.3.2 Example: Matter-Dominated FLRW Universe and the Size of the Universe

Let us consider an example to see how we can model the evolution of the Universe (i.e., the behavior of the scale factor $a(t)$). Consider a Universe in which matter satisfies $\rho \gg p$. We will simplify the model by setting $p = 0$. Recall that such a perfect fluid is called "dust". From the fluid Equation (5.52) we can either separate the variables and integrate, or observe that it can be rewritten as

$$\frac{1}{a^3}\frac{d}{dt}(\rho a^3) = 0. \tag{5.60}$$

This implies ρa^3 is a constant in time, and thus $\rho \propto 1/a^3$. That is, the matter density dilutes inversely proportional to the physical volume, which in this case is in accordance to our intuition. Let $\rho(t) = \rho_0/a^3$ so that $a_0 = 1$ corresponds to the current era of the universe.

For a flat universe ($k = 0$), the Friedmann equation simplifies into:

$$\frac{\dot{a}^2}{a^2} = \frac{8\pi G\rho}{3} \Rightarrow H_0^2 = \frac{8\pi G\rho_0}{3}. \tag{5.61}$$

Consequently,

$$\frac{\dot{a}^2}{a^2} = \frac{8\pi G\rho_0}{3a^3} = \frac{H_0^2}{a^3}, \tag{5.62}$$

[23] We also need $\ddot{a} = 0$. From Equation (5.56) this would imply $\Lambda = 4\pi G\rho$ if pressure is negligible, which Equation (5.55) then implies that $k/a^2 = 12\pi G\rho/3 > 0$, i.e., Einstein's static universe must have a positively curved spatial section (usually taken to be a 3-sphere).

which implies

$$\dot{a}^2 = H_0^2 \cdot \frac{1}{a}. \tag{5.63}$$

Taking the square root yields

$$\frac{da}{dt} = \pm \frac{H_0}{\sqrt{a}}. \tag{5.64}$$

At this stage, note that both ± signs are mathematically correct. However, physically, we want to consider a universe that expands (as indicated by observations), hence we choose the + sign (we already implicitly did this in Equation (5.57)). The other solution would correspond to a universe that has always existed (no Big Bang) and then started to contract, a "time-reversed" version of the +-branch. Incidentally, in Lemaître's 1927 paper, he also considered the possibility that in the infinite past, a universe could already have a finite size, and gradually became larger (an eternally "loitering" phase). Such a universe has no singular beginning. If our Universe is like that, it would be hard to explain why after an infinitely long duration, the expansion rate suddenly increased.

Integrating the differential Equation (5.64) is straightforward, which gives

$$\frac{2}{3} a(t)^{\frac{3}{2}} = H_0 t + \text{constant}. \tag{5.65}$$

If we take $t = 0$ to correspond to the time of the Big Bang with $a = 0$, then the integration constant vanishes and

$$a(t) = \left(\frac{3H_0 t}{2}\right)^{\frac{2}{3}}. \tag{5.66}$$

We can check that the Hubble parameter goes like $1/t$, so the expansion is asymptotically slowing down to 0. This simple example is known as the "Einstein–de Sitter model" (O'Raifeartaigh et al. 2021).

Let us make a few remarks. First of all, for other values of k the exercise can be repeated *mutatis mutandis*, however the integration is more complicated. For example, for positive k (see the exercise section), we will find that $a(t)$ is a curve known as a "cycloid", which allows an infinitude of Big Bangs and Big Crunches. That is, in principle, mathematically the Universe can undergo a series of destruction and rebirth. Such a scenario is often referred to as the *cyclic universe*.

Next, this is a good time to return to the discussion of the size of the Universe. If $k = 0$ and our Universe is indeed infinitely large, how can it be "born" from a *point*? The answer is that, *it cannot*. Even if there is a topological identification so that a flat universe is compact, in the calculation above we can still choose the integration constant so that at $t = 0$, the scale factor is finite. Then such a compact universe would be born with a finite size; the mathematics does not dictate the initial size, and what we did in the calculation above is actually not quite correct. Crucially, if the universe is infinite *now*, it must *always* be infinite.[24] In particular, it must already be infinite at birth. Infinity is of course, rather tricky when it appears in our equations. Thus, for an infinite universe, instead of choosing the integration constant to be infinite, we can just take the coordinate grid to be the entire $\mathbb{R}^3$ so that the coordinate distance ℓ_x between the origin at an arbitrarily far point $(x, 0, 0)$, say, is unbounded. Then even at initial time the proper distance $\lim_{t\to 0}\lim_{x\to\infty}(a(t)\cdot\ell_x)$ is indeterminate, which is consistent with the size actually being infinite.

In this example, we mentioned that for dust, the energy density goes like $1/a^3$. What about other kind of matter? It turns out that,

$$\rho(t) = \frac{\rho_0}{a(t)^{3(1+w)}}, \tag{5.67}$$

where w satisfies the "equation of state" $p = w\rho$. To see this, consider a box of gas with volume V. We recall from Chapter 4 that the first law of thermodynamics is $dE = TdS - pdV$. In our model, galaxies are treated as dust particles and dark energy as a fluid, but there is a subtle difference compared to an actual "gas in a box" scenario. In the latter, if all the gas particles randomly—i.e., by chance—collected to only half of the box, or even to one corner of the box, this would temporarily lower the entropy because these are rare configurations. However, we don't actually permit this for galaxies and any fluid because the FLRW geometry presupposes homogeneity and isotropy. This should not be a surprise since in FLRW cosmology, we use a perfect fluid precisely because we want to preserve the Copernican principle—perfect fluids always have a constant entropy *by definition*; a perfect fluid has no heat conduction or dissipation due to viscosity, hence it does not generate any entropy within its flow. So as far as matter configuration is concerned, we can treat entropy to be constant[25] and thus $dS = 0$ and the first law reduces to $dE = -pdV$. On the other hand, the energy is $E = \rho c^2 V$, since ρ is the matter density. This means that (with $c = 1$)

$$\rho dV + Vd\rho = dE = -pdV, \tag{5.68}$$

and thus

$$Vd\rho = -(p + \rho)dV. \tag{5.69}$$

---

[24] One may ask: how can an infinite universe expand? A good analogy is the number line $\mathbb{R}$, for which we just map $x \mapsto 2x$ for all x, then it expands to twice its size, which is still the same infinity (technically, same cardinality). This is the same situation as the "Hilbert's Hotel" we mentioned in the beginning of Chapter 3.

[25] A great deal of confusion arises as to why this does not contradict the *increase* of entropy as per the expectation of the second law (our world is clearly not yet in a thermal equilibrium so entropy is expected to increase). Again, one has to be careful about the model used. At the large scale under discussion here, we do not take into account smaller scale phenomena like structure formation and star formation, etc. (This is similar to the FLRW cosmic expansion does not preclude galaxy mergers at local scales.) Entropy only increases if we consider those irreversible features.

Given that $p = w\rho$, we have

$$Vd\rho = -(1+w)\rho dV \Rightarrow \int \frac{d\rho}{\rho} = -(1+w)\int \frac{dV}{V}. \tag{5.70}$$

This then gives

$$\log \rho = -(1+w)\log V + \text{const.}, \tag{5.71}$$

or equivalently, with $V = Ca(t)^3$, where C is a constant (the *fixed* volume of the comoving box) we have

$$\rho(t) = \frac{\text{const.}}{a(t)^{3(1+w)}}. \tag{5.72}$$

With the requirement that $a_0 \equiv a(t_0) = 1$, we have, as claimed,

$$\rho(t) = \frac{\rho_0}{a(t)^{3(1+w)}}. \tag{5.73}$$

A matter-dominated universe (i.e., with dust) corresponds to $w = 0$. In the radiation-dominated case, where by radiation we mean electromagnetic radiation,[26] $w = 1/3$, and thus $\rho = 1/a^4$. Why is there an extra $1/a$ factor for radiation? One could appreciate this from the fact that as the Universe expands, not only is there a volume dilution $1/a^3$, but each individual photon wavelength gets stretched. This redshift results in a longer wavelength and therefore the photon is less energetic. This gives rise to another $1/a$ factor. A common question is what happens to the energy of the photons? What happens to energy conservation? This is actually a rather tricky question. The most straightforward answer is that energy is *not* always conserved. The equation $\nabla^a T_{ab} = 0$ cannot be interpreted as a true conservation law. In general relativity as well as quantum field theory (Noether's[27] theorem), energy conservation is not a given, but the consequence of time translation symmetry. In an expanding universe there is no time translation symmetry and hence no energy conservation. What the equation $\nabla^a T_{ab} = 0$ means is *not* that energy-momentum is conserved, but rather that energy-momentum can be transferred into the

[26] For readers well-versed in general relativity, here we are actually treating radiation as "photon gas". Since the FLRW metric is based on T_{ab} of a perfect fluid, we don't actually deal with full electromagnetism, but rather a fluid effective description of it. In particular, the Maxwell tensor of electromagnetism T^{EM} is trace-free: $g_{ab}T^{ab}_{\text{EM}} = 0$. Thus we require that for our fluid this is also satisfied. Using the perfect fluid energy-momentum tensor $T^{ab} = (\rho + p)u^a u^b + pg^{ab}$, it can be shown that $-\rho + 3p = 0$, i.e., $p = \rho/3$.

[27] Emmy Noether (1882–1935) was asked by Felix Klein (1849–1925) and David Hilbert to help clarify the conservation laws in the then new theory of general relativity. This resulted in the now celebrated Noether's theorem (there are actually two theorems: one that associates conserved quantities with global symmetries and the other concerns local symmetries; the latter is the mathematical foundation of modern gauge theories). However, her work was poorly understood between then and 1950. Some physicists would hear of her work and attempt to generalize her theorem (which was almost always presented in a simplified form), not realizing that her original version was superior and covered quite general settings! Readers who are interested in the history can consult the book by Yvette Kosmann-Schwarzbach and Bertram E. Schwarzbach (Kosmann-Schwarzbach & Schwarzbach 2011).

gravitational field itself. However, defining gravitational energy is another complicated issue, so it is best to just stick with energy being not conserved in general.

From the original acceleration equation without Λ, we see that $\ddot{a}$ is possible if and only if $\rho + 3p < 0$, or equivalently, $w < -1/3$. Depending on the range of w, this would correspond to different types of "dark energy", which we will return to in the next chapter. For now, let us note that the case $w = -1$ is special; it clearly satisfies the acceleration condition $w < -1/3$. On top of that, from Equation (5.67) we can see that $w = -1$ corresponds to ρ being constant for all t. This is the case of the cosmological *constant*, whose density remains unchanged even as the Universe expands. In this discussion, we are interpreting the cosmological constant as a fluid, just by moving the Λg_{ab} term in the Einstein field equations to the right-hand side. There are other more exotic equations of state. For example, the *Chaplygin gas* is a hypothetical perfect fluid that satisfies $p = -A/\rho$ for some constant A, and the *generalized Chaplygin gas* that satisfies $p = -A/\rho^{\alpha}$ for some $0 \leqslant \alpha \leqslant 1$. Both are named after Sergey Chaplygin[28] (1869–1942). An even more generalized version can be considered in phenomenological approaches. For example (Bouhmadi-Lopez et al. 2011),

$$\rho = \left(\frac{A}{a^{1+\beta}} + \frac{B}{a^{4(1+\alpha)}}\right)^{\frac{1}{1+\alpha}}, \qquad A > 0, B > 0, \tag{5.74}$$

where A, B, α, β are constants.

5.4 Modern Cosmology at a Glance

In the current understanding, the Universe began with a Big Bang. In his 1931 paper that started the idea of Big Bang (there is no equation in this short paper!), Lemaître envisioned that the Big Bang was the beginning of everything, including space and time:[29]

> "If the world has begun with a single quantum, the notions of space and time would altogether fail to have any meaning at the beginning; they would only begin to have a sensible meaning when the original quantum had been divided into a sufficient number of quanta."

Although the sentence does not make sense from the modern viewpoint insofar as what "quanta" means, the idea essentially agrees with the modern idea that space and time don't make sense at the initial singularity, and that only after that spacetime together with matter and radiation, as we know them, came into being. The initial stage is full of mysteries; if general relativity is correct all the way back to the Big Bang, then there is an initial spacetime singularity during which spacetime curvature diverges,

[28] Chaplygin introduced this fluid as a mathematical approximation for calculating the lifting force on a wing of an airplane in aerodynamics (Chaplygin 1904).

[29] Interestingly, the Chinese term for "Universe" is 宇宙, which is "spacetime": as defined in the text *Shizi* (尸子) written in c. 330 BCE by Shi Jiao (尸佼; c. 390–330 BCE), the character 宇 refers to "up down and four corners (all possible directions)" (上下四方曰宇), while 宙 is "the bygone past and the becoming of now" (往古來今曰宙).

and so nothing sensible can be said about it. There are speculations, of course, which either try to explain the beginning, or to construct a cosmology without a true beginning; but these will be left for a future chapter. For now, we will give a quick summary of what we know about the evolution of the Universe. We will not cover everything, instead we will just explain some important points.

First, we need to introduce some basic concepts about how distances are measured in astronomy and cosmology. Within the solar system as well as its neighborhood, the *astronomical unit* (AU) is often used; this is just the mean distance between the Sun and the Earth. The concept of a *light-year* is well known to most students of physics and astronomy, and it is just the distance that light travels in a year in vacuum. Another unit that is often used is *parsec*, abbreviated as "pc". It is about 3.26 light-years. A parsec is the distance that a object must lie from the Sun for its parallax angle to be exactly 1 arcsecond (one degree has 60 arcminutes, each of which has 60 arcseconds). Back in the days one of the main criticisms against heliocentrism was that if Earth is moving, we should see parallax in the stars, but this wasn't measurable by the technology back then because the amount was so tiny; the Universe was far larger and the stars were further than everyone had imagined. The closest star to us is Proxima Centauri, whose parallax angle is a mere 0.768 arcseconds. This translates into the distance of 1.3 parsecs or 4.24 light-years.

Another way to measure distance is by recognizing that the actual luminosity of stars ("absolute magnitude") is associated with their color. Using parallax we can determine the distances to the closest stars, and since luminosity drops as the inverse square of the distance, we can obtain the absolute luminosity from the observed brightness. We can then gather a set of data points {color, absolute magnitude}, and plot them in such a way that the horizontal axis runs from the hottest to the coolest stars from left to right (i.e., blue to red). Such a diagram is called a *Hertzsprung–Russell* (HR) diagram, named after Ejnar Hertzsprung (1873–1967) and Henry Norris Russell (1877–1957); see Figure 5.1. From the scatter plot of the available data, we would find that there are a few distinct families of stars—the "main sequence" stars like our Sun, the "giants" and "supergiants" that are near the end of their lifetime, and the stellar remnant "white dwarfs". The HR diagram can be used as an estimate of distance of stars that are too far to measure in terms of parallax. First, one observes its color as well as its *spectrum*.[30] The spectrum allows us to distinguish whether it belongs to the main sequence or other types of stars.

[30] Different elements absorb/emit light at a different frequency. The element helium was first found by analyzing the emission spectrum of the Sun, hence the name (*helios* is the Sun in Greek). In 1835 the French philosopher Auguste Comte (1798–1857) wrote that although we can possibly determine the shapes, the distances, the sizes, and the movements of the stars, we would never know how to study their chemical composition. He did not anticipate the advance of spectrography. However, Joseph von Fraunhofer (1787–1826) was already studying the spectrum of the Sun and a few bright stars in 1814 and 1823. Perhaps Comte wasn't aware of these works. For the interesting history of spectrography and its application in astrophysics, see Hearnshaw (2010). In cosmology, the spectra of distant galaxies, having passed through interstellar medium, will have a series of absorption lines due to the Lyα electron emission (electrons in a hydrogen atom jumping from the orbit $n = 2$ to $n = 1$). This is called the *Lyα forest*, which serves as a tool to study interstellar medium, such as the distribution and density of hydrogen clouds and their temperature.

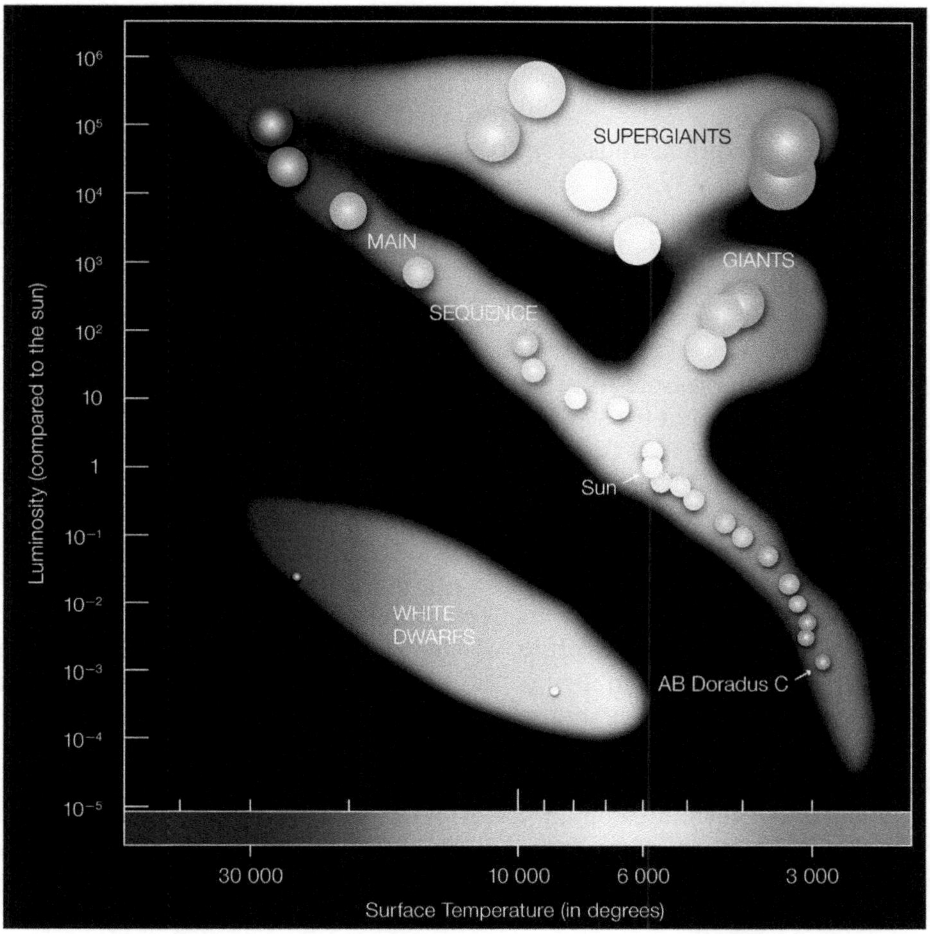

Figure 5.1. The Hertzprung–Russell diagram of stars, with surface temperatures plotted against their actual luminosities. Stars that burn hydrogen into helium lie on the diagonal branch, the so-called "main sequence". The lower end of the main sequence corresponds to faint cool stars called "red dwarfs". While more luminous stars like our Sun will one day exit the main sequence and becomes a giant or a supergiant, red dwarfs burn too little hydrogen and will never exit the main sequence. On the other hand, white dwarfs are a type of stellar remnants (see Section 8.4.2). *Credit:* ESO CC BY 4.0.

Then, we can roughly know its coordinates on the HR diagram and hence estimate its absolute magnitude. Comparing to its observed magnitude, we can use the inverse square law to estimate its distance.

To measure distances of faraway objects, we need the so-called *distance ladders.* The first rung of the ladder is the parallax measurement, which is only good for relatively close objects. The next rung is a kind of variable star that pulsates in both brightness and physical size, known as the *Cepheid variable.* Each such star has its own period of brightening and dimming. Cepheid variables are named[31] after its representative object Delta Cephei, discovered in 1784 by John Goodricke (1764–1786). However, the first observed was actually the star Eta Aquila, by Edward Pigott a month prior (1753–1825)! Henrietta Swan Leavitt (1868–1921) studied thousands of variable stars in the Magellanic Clouds (satellite galaxies of our Milky Way; visible to the naked eyes in the southern hemisphere). She found that the brighter a Cepheid variable star is, the longer its period, in a regular way. The distances to nearby Cepheid variables in our own galaxy can be measured using parallax[32]—the first rung of the ladder—and so this allows a calibration of the period–luminosity relation. For sufficiently closed galaxies for which Cepheid variables can be observed, although we don't know its true luminosity, we *can* measure its brightening period and from the period–luminosity relation, infer how bright the star is. We can also measure directly how bright it is as seen from Earth, and given that objects far away appear dimmer, we can calculate how far the star—and hence the host galaxy—must be from us. Cepheid variables were the first "standard candle" used in astronomy. A standard candle produced by the same factory has the same brightness because they burn from the same material, so by looking at how dim a candle is we can infer how far it is from us. The same goes for astronomical objects like the Cepheid variables; as long as there is a simple relation that governs their properties, we can use it to infer distances once they are calibrated (i.e., we need to figure out the absolute magnitude of the candle luminosity). The issue of calibration is not straightforward. In the case of the Cepheid variables, nearby ones turned out to be different from the ones further away because these stars have different metallicity (further stars that we can observe *now* are younger, and therefore contain less heavier elements[33].) The younger stars are brighter, and thus also further, than initially assumed. To measure galaxies that are even further away in which Cepheids are too dim to be seen, one uses the next rung of the ladder—Type Ia supernovae, which is the explosion caused by material from a companion star inspiralling onto a white dwarf. Again, one must first calibrate by first measuring the distances to closer galaxies that contain such supernovae using Cepheid variable stars. Each new rung of the ladder needs the previous rung.

Another way to indicate distance is by using the redshift of light, i.e., the fact that the wave is stretched by the expansion of the Universe. The redshift z is related to the scale factor of the FLRW metric via

$$\frac{a(t_{\mathrm{o}})}{a(t_{\mathrm{e}})} = 1 + z, \tag{5.75}$$

[31] The name was apparently coined by Agnes Mary Clerke (1842–1907), in her 1903 book *Problems in Astrophysics* (Clerke 1903).

[32] Direct measurements of parallax for "nearby" Cepheids are only possible relatively recently. Prior to this we had to use Cepheids that are in a cluster of main sequence stars, whose distance can be inferred from the HR diagram.

[33] In astrophysics, all elements except hydrogen and helium are called "metals"!

where t_o denotes the current observation time and t_e the time the light was emitted by the target. In terms of the wavelength, we have

$$\frac{a(t_o)}{a(t_e)} = \frac{\lambda_o}{\lambda_e}, \qquad z = \frac{\lambda_o - \lambda_e}{\lambda_e}. \tag{5.76}$$

Thus, at a given redshift z, the Universe was $1 + z$ times smaller than its current size. Specifically, $z = 1$ refers to an object that emitted the light we currently observe when the Universe was half its current size. Cosmological redshift is the most distant ladder we currently have. There are other notions of distance, such as the *luminosity distance* and *angular diameter distance*, which we will not get into here.

We now know that the Milk Way is a galaxy held together by the gravity of around 100 billion to 400 billion stars. The philosopher Immanuel Kant[34] (1724–1804) was right that there are other galaxies similar to our own (he called each galaxy an "island universe", and even considered that other stars have inhabited planets). We now know that there are at least 100 billion galaxies in the observable universe, and the number could be as high as 2 trillion. Galaxies form clusters, which then form superclusters. The Milky Way is part of the *Local Group*, which contains more than 54 galaxies, including our neighbor the Andromeda Galaxy. The Local Group belongs to the *Virgo Supercluster*, which is part of a much larger supercluster called the Laniakea Supercluster, defined only as recently as 2014. Laniakea, which means "immense heaven" in Hawaiian, contains a hundred thousand galaxies and has a diameter of roughly 520 million light-years. which *might* itself be a part of an even larger structure, known as the Shapley BoA (BoA stands for "basin of attraction"), to which our galaxy is being pulled towards (Valade et al. 2004). Even with the presence of these clusters, we usually assume in our mathematical model that the Universe is isotropic and homogeneous. This is the so-called "cosmological principle", or the "Copernican principle" (as Copernicus' heliocentrism taught us that we do not live in a special place in the cosmos; this is promoted to the philosophy that no part of the Universe is any more special than the rest.) The Universe looks homogeneous in the sense that at the largest scale, the statistical distribution of clusters seem to be uniform[35]. The FLRW model then tells us how the Universe evolves as it expands.

In the previous subsection we mentioned that the energy density of radiation evolves like $1/a^4$, whereas matter evolves like $1/a^3$ (curvature decays like $1/a^2$, this

[34] Kant is famous for his three volumes of "critiques", namely *Critique of Pure Reason*, *Critique of Practial Reason*, and *Critique of Judgement*. These works are hard to read, yet Albert Einstein read them all by the age of 16. But then Einstein also mastered differential and integral calculus by 14.

[35] There are debates concerning whether this is really true. For example, recent discoveries of unexpected structures nicknamed "giant arc" (discovered in 2021; Lopez et al. 2022) and the "big ring" (found in 2024; Lopez et al. 2024) both seem to challenge this assumption (Aluri et al. 2023; Lopez et al. 2025). However, even a uniform distribution can admit local "large" structures so it is unclear if these are statistically significant.

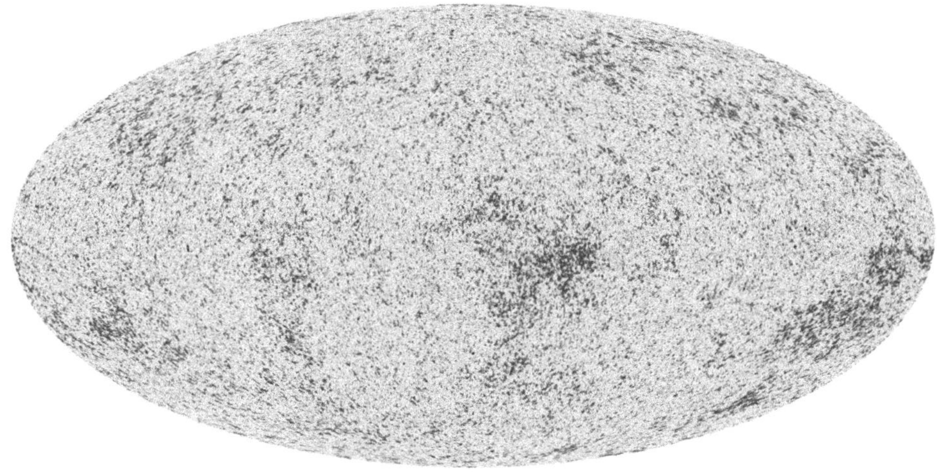

Figure 5.2. The cosmic microwave background (CMB) image based on data from the Planck Legacy release, the mission's final data release, published in July 2018. The image shows anisotropies in the CMB; some places are colder while some places hotter. Nevertheless, the temperature difference is very small; the temperature is uniform to roughly 1 part in 100,000. Like the surface of the Earth, we cannot make a two-dimensional map of the CMB sky without distorting some features. Usually one utilizes the Mollweide projection (also called the Babinet or elliptical projection), named after Karl Brandan Mollweide (1774–1825), which is sometimes also used for the Earth's map. This projection trades accuracy of angle and shape for accuracy of proportion in area. *Credit:* ESA and the Planck Collaboration.

makes sense because curvature of a sphere, for example, is $1/R^2$ where R is its radius; the larger the space is, the smaller the curvature). Thus in the sufficiently early Universe, radiation will dominate over matter. This is to be expected: the early Universe is much hotter than the late Universe. As the Universe expands the temperature becomes colder, and at some point matter becomes dominant over radiation, the cross-over is called "matter-radiation transition". Before that the temperature is too high for atoms to stay stable, so in the highly ionized environment photons get absorbed and re-emitted all the time. Without free-streaming photons, the Universe was opaque, as if in a thick fog. Only after the matter-radiation transition, stable atoms formed,[36] and photons can travel freely for the first time (technically we say that the photons "decoupled"). These are the photons of the cosmic microwave radiation (CMB). The CMB map is shown in Figure 5.2. Since light takes time to reach us, looking at far away objects means that we are looking back in time. The moment of matter-radiation transition is like when a veil was lifted, that veil is the "last scattering surface" from which the CMB photons were emitted. Another often-used analogy would be a cloud: we can only

[36] Somewhat confusingly this era is also referred to as the epoch of "recombination", during which electrons combined with protons to form hydrogen atoms *for the first time ever*!

see the bottom "surface" of the cloud; the interior of the cloud is like the earlier opaque Universe in which photons are scattered frequently and in random directions. Since the photons are stretched by the cosmic expansion, although what we observe as CMB temperature today is only at 2.726 Kelvin, it was about as hot as the surface of the Sun back then. The fact that there should be a "relic temperature" from the Big Bang was predicted by Ralph Asher Alpher (1921–2007) and Robert Herman (1914–1997) in the 1920s before it was discovered in 1964 by Arno Penzias and Robert Wilson. The predicted temperature was 5 K, quite close to the measured value (Alpher 2014), though some have expected even higher temperatures of $O(10)$ K.

We can explain a bit more about what the CMB represents. Essentially, the temperature fluctuation seen in a given direction in the CMB sky (determined by the angles on a sphere) is given in terms of spherical harmonic decomposition:

$$\delta T(\theta, \phi) = \sum_{\ell=1}^{\infty} \sum_{m=-\ell}^{\ell} a_{\ell m} Y_{\ell m}(\theta, \phi), \qquad a_{\ell m} \in \mathbb{C}. \tag{5.77}$$

Essentially, the number ℓ is the "multipole moment", which determines the "wavelength" of the mode, i.e., the number of waves along a meridian. It describes "angular resolution" in the sense that for each ℓ, the sky is partitioned into $2\pi/(\ell + 1)$ parts. The "monopole" mode $\ell = 0$ covers the whole sky (full 360 degrees); the "dipole" $\ell = 1$ mode would have two features that are separated by a half-turn and so on. The number m, on the other hand, determines the "shape" of the mode. The *power spectrum* is defined to be

$$C_\ell = \frac{1}{2\ell + 1} \sum_{m=-\ell}^{\ell} |a_{\ell m}|^2. \tag{5.78}$$

Usually the plot, as in Figure 5.3 below, is in terms of a normalized power spectrum defined by

$$D_\ell \equiv \frac{\ell(\ell + 1)}{2\pi} C_\ell. \tag{5.79}$$

Referring now to Figure 5.3, the reason the temperature is colder than average for small ℓ was because the photons had to "climb out" of the gravitational potential well at the time of last scattering, and therefore lost some energy. This gravitational redshift is called the *Sachs–Wolfe effect*, and part of the graph before it starts to climb toward the first peak is sometimes called the "Sachs–Wolfe plateau". The wavy form of the curve is due to the imprints of sound waves moving through the ionized hydrogen gas. This is known as the *baryon acoustic oscillation* (BAO). Back then the temperature was high enough that photons couldn't freely propagate in the ionized plasma, thus the regions where the density of protons and

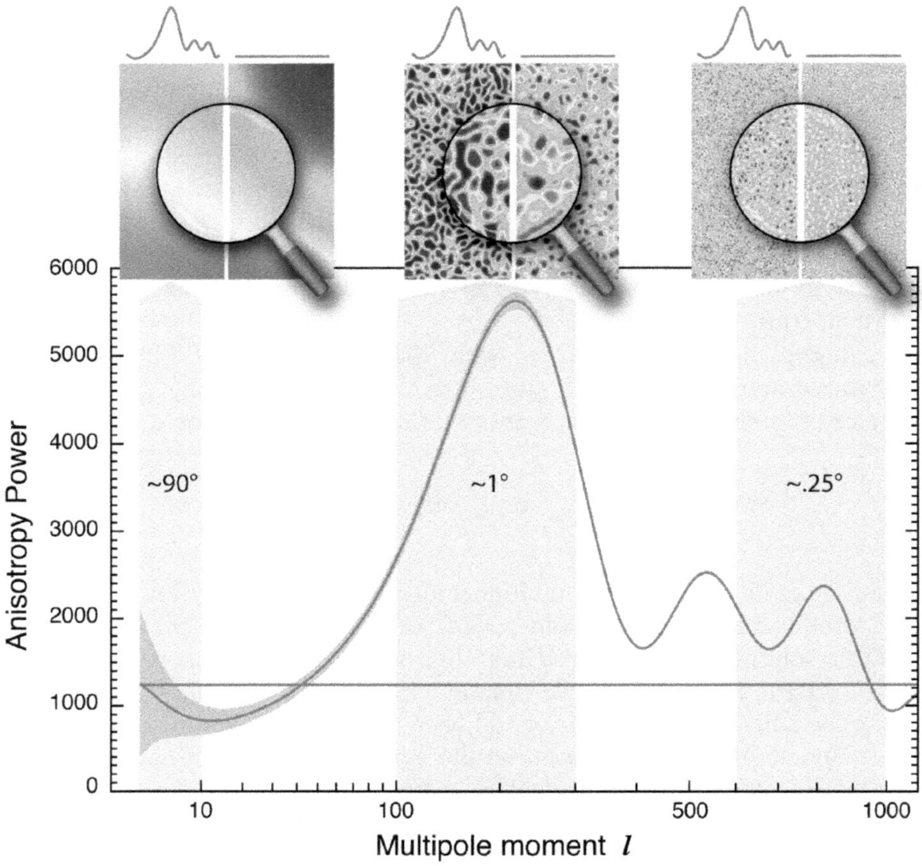

Figure 5.3. This graph shows the sky's temperature (colored patches) when measured at different angular scales across the sky, as observed by the Wilkinson Microwave Anisotropy Probe (WMAP) satellite. Angular scale decreases with increasing multipole moment. The wiggly line is the actual sky measurement. The straight line is the same measurement of sky energy if there were a true uniform distribution of temperature, fixed by the monopole mode, this is essentially fixed by the variations in the strength of gravity at the time of last scattering. The sky images above three separate portions of the graph show their appearance under these two scenarios. *Credit:* NASA/WMAP Science Team/Theophilus B Griswold.

electrons was higher, were also the same regions where there were more photons. The different densities therefore created hotter or colder regions as tracked by the CMB photons. The peaks in the graph provide important information about how much matter—including dark matter—was present in the early Universe. The CMB is a treasure trove of information. For example, if the Universe has a nontrivial topology, we may be able to see multiple images of the same star that have circumnavigated the Universe[37]. This is difficult in practice since we need to analyze the sky statistically; and depending on the size of the topological identification (the "fundamental domain"), the signals may not even be observable. Similar analysis can be done via the CMB, although so far there is no evidence for this.

Note that the larger ℓ is, the smaller the region it corresponds to. Generally $\ell < 50$ is considered to contain information of the fluctuations generated by inflation, if such a phase indeed happened in the past (see Section 13.3). If inflation occurred, the fluctuations are expected to be Gaussian (the deviation is called "non-Gaussianity", denoted by f_{NL}; this can be the effect of a multi-field inflation, for example), and the power spectrum should be slightly "red-tilted" in the sense that large scale perturbation has more power. The C_ℓ's can be expressed as an integral over the Fourier space function $P(k)$, called the "primordial curvature power spectrum" (sometimes denoted $P_{\mathscr{R}}(k)$):

$$C_\ell = 4\pi \int d^3k\ P(k)\ \nabla_\ell\ (k)^2, \tag{5.80}$$

where $\nabla_\ell(k)$ is a certain "transfer function", which describes various physics like photon–baryon interactions. It is the shape of ∇_ℓ that determines the CMB peaks. Inflation typically predicts a power-law form

$$P(k) = A_s\left(\frac{k}{k_*}\right)^{n_s - 1}, \tag{5.81}$$

where k_* is a chosen pivot scale (usually 0.05 Mpc^{-1}). Here A_s is called the amplitude and n_s is the *spectral index*. If $n_s > 1$ we have a blue tilt, but if $n_s < 1$ as predicted by standard inflationary models, we have a red tilt. If $n_s = 1$ exactly, the spectrum is scale-invariant (the same amplitude of fluctuations at all scales).

---

[37] In Italo Calvino's *Cosmicomics*, there is a story "Games Without End", which describes a topologically nontrivial universe:

> "In any case, I preferred to look ahead, and what did I see? Pfwfp, whom my eyes had just left behind me, was speeding on his galaxy directly in front of me. 'Ah!' I cried, 'now it's my turn to chase you!'
>
> 'What?' Pfwfp said, from before me or behind me, I'm not really sure which, 'I'm the one who's chasing you!'
>
> I turned around: there was Pfwfp, still at my heels. I looked ahead again: and he was there, racing off with his back turned to me. But as I looked more closely, I saw that in front of this galaxy of his that was preceding me there was another, and that the other galaxy was mine, because there I was on it, unmistakable even though seen from behind. And I turned towards Pfwfp following me and narrowed my eyes: I saw that his galaxy was being chased by another, mine, with me on top of it, turning at that same time to look back."

The results from the Planck satellite in 2015 gave $n_s = 0.9649 \pm 0.0042$ (Alpher 2014). If n_s is not a constant but a function of k, one usually parametrizes the running of n_s as a power series of $\ln(k/k_*)$:

$$n_s(k) = n_s(k_*) + \alpha_s \ln\left(\frac{k}{k_*}\right) + \frac{1}{2}\beta_s \ln^2\left(\frac{k}{k_*}\right) + \cdots. \quad (5.82)$$

The coefficient of the linear logarithmic term, α_s, is called the *running spectral index* and is given by

$$\alpha_s := \frac{dn_s}{d\ln(k)}. \quad (5.83)$$

If $\alpha_s < 0$, n_s decreases at larger k, that means less power for smaller scales, hence an evolution towards red tilt. Similarly, we have a blue tilt evolution if $\alpha_s > 0$. Planck's result is $\alpha_s = -0.0045 \pm 0.0067$ (Alpher 2014).

No matter which direction we look into the sky, we will always detect almost the same temperature of the CMB. Tiny variations exist of course, but these are at the parts per million level! These variations are caused by tiny quantum and thermal fluctuations, but they are extremely important. If the Universe is perfect with no variation whatever, then nothing interesting would ever happen. Tiny fluctuations mean that matter distribution is not completely uniform, and as such, gravity will clump regions with more matter concentration. Over time, this gave rise to structure formation which consequently resulted in galaxies, planets, and life forms like us. We are here because the Universe was not a perfect place!

Today the CMB is often accepted as the "smoking-gun" evidence for the reality of the Big Bang. However, the history is not so straightforward. Even with the discovery of the CMB, one of the proponents of the steady-state theory, Fred Hoyle, did not accept the Big Bang theory. He maintained that the CMB temperature was *not* explained by the Big Bang cosmology, since whatever the temperature was measured, say, 27 K instead of 2.7 K, that would just as well be accepted.[38] It is true that there are free parameters in the Big Bang theory, but most physicists today think that there is indeed a hot Big Bang phase (though not necessarily the initial singularity). This belief is founded not just upon the existence of the CMB, but also other consistencies, for example, the abundances of primordial hydrogen and helium. The early Universe made equal number of protons and neutrons, but as the temperature cooled down when the Universe expanded, the ratio changed because the rest mass of the proton is smaller than that of the neutron (so at some

[38] Before the discovery of Penzias and Wilson, there was also the discovery by Andrew McKellar (1910–1960) who, way back in 1941 (23 years before Penzias and Wilson), deduced from spectra analysis of organic cyano radical (CN) in interstellar space that ambient space has temperature of about 2.3 K. Hoyle had used this as an argument *against* the Big Bang proposal, as the proponents of the Big Bang actually expected temperature above 3 K, and even about 20 K Hoyle (2025). The discovery was well known but nobody realized its true significance until the "re-discovery" of Penzias and Wilson. In fact, in a textbook by Herzberg in 1950, on p. 496 Herzberg (1950), one finds his comment that the 2.3 K temperature "has of course only a very restricted meaning".

point there won't be enough energy to create new neutrons, but new protons could still be produced); and furthermore free neutrons are not stable and would decay with a mean lifetime of about 877.75 seconds (Castelvecchi 2021). The consequence is that the proton-to-neutron ratio changed to about 8 to 1 in the first few minutes after the Big Bang. As we will now show, this can be argued via a rather elementary (but not easy!) calculation, and it agrees with observations!

The ratio of the number density between neutrons and protons can be described by the Maxwell–Boltzmann distribution (see Section 4.2):

$$\frac{N_n}{N_p} = \left(\frac{m_n}{m_p}\right)^{\frac{3}{2}} \exp\left[-\frac{(m_n - m_p)c^2}{k_B T}\right], \tag{5.84}$$

where m_n, m_p are the masses of a neutron and proton, respectively. We have restored c and k_B for clarity. In terms of MeV, $(m_n - m_p)c^2 \approx 1.3$ MeV. Since $m_n \approx m_p$, when the temperature is very high, the exponential suppressing term implies that the ratio is almost 1 at the beginning. Initially neutrons can be converted into protons and vice versa via the process $n + \nu_e \leftrightarrow p + e^-$, where ν_e is the electron neutrino (see Section 6.1). However, at around $k_B T = 0.8$ MeV, this process can no longer occur (because the weak interaction rate falls below the Hubble expansion rate; we skipped lots of steps here!), and the ratio of neutrons to protons was fixed to about $N_n/N_p \approx 1/5$, because

$$\frac{N_n}{N_p} \approx \exp\left(-\frac{1.3}{0.8}\right) \approx 0.2 = 1{:}\ 5. \tag{5.85}$$

We say that neutron has "frozen out". This is *not* yet the 1:8 ratio we mentioned before. To get there we need some knowledge of atomic physics, specifically the reactions in nucleosynthesis. Essentially, protons and neutrons can combine to form deuterium d via the interaction $p + n \leftrightarrow d + \gamma$, where γ denotes a photon. Then $d + p$ and $d + d$ would give helium-3 and helium-4, respectively. These processes can go both ways initially, but not when the temperature has dropped low enough for the environmental photons to disrupt the stability of deuterium (the technical term is "photo-dissociation"). This happens at $k_B T \approx 0.1$ MeV. As you have no doubt noticed, lots of concept are involved; cosmology and particle astrophysics can be quite messy!

Next, for a radiation-dominated FLRW universe, its scale factor goes like[39] $a(t) \sim \sqrt{t}$. Since temperature is inversely proportional to the scale factor, we thus have

$$\frac{T}{T_{\rm eq}} = \left(\frac{t_{\rm eq}}{t}\right)^{\frac{1}{2}}, \tag{5.86}$$

[39] This can be obtained in the same way we computed Equation (5.66), by replacing $\rho = \rho_0/a^3$ for matter-dominated Universe to $\rho = \rho_0/a^4$ for radiation-dominated Universe.

where the subscript “eq” denotes the time when matter and radiation are in equilibrium. Plugging in the numbers ($T_{\text{eq}} \approx 16,\,000$ K and $t_{\text{eq}} \approx 50,\,000$ years; which are obtained from the Big Bang model) we will finally obtain

$$\left(\frac{1\ \text{sec}}{t}\right)^{\frac{1}{2}} \approx \frac{k_B T}{2\ \text{MeV}}. \tag{5.87}$$

Thus, with $kT \sim 0.1$ MeV, we get $t \approx 400$ s. This is the time after the Big Bang when deuterium and thus helium became stable. Therefore, from the time of neutron freeze-out to $t \sim 400$ s, the ratio N_n/N_p would have again changed due to neutron decays:

$$\frac{N_n}{N_p} \approx \frac{1}{5}\exp\left[-\frac{400}{877.75}\right] \approx 0.1268 \approx 1{:}\ 8, \tag{5.88}$$

where 877.75 seconds being the lifetime of a free neutron. Note that if we use the *half-life* $\tau = 877.75 \times \ln 2$ instead, then the exponential factor would have the more familiar form of the “half-life equation”[40]$\exp[-\ln 2 \times 400/\tau]$. where the factor $-\ln 2$ comes from the radioactive exponential decay. Thus, for each neutron we should have about 8 protons.

Knowing the n-to-p ratio gives us the helium-to-hydrogen ratio that can be compared with observations. Finally, recall that a hydrogen atom has one proton and no neutron, while the stable isotope helium-4 has two protons and two neutrons. All neutrons either end up in helium-4 or they eventually decay. The final mass fraction of helium is thus

$$\frac{N_{\text{He}}}{N_n + N_p} = \frac{2N_n}{N_n + N_p} = \frac{2}{1 + \frac{N_p}{N_n}} = \frac{2}{1+8} = \frac{2}{9} \approx 0.22. \tag{5.89}$$

In other words, we expect the early Universe to have around 22% of helium-4 and 78% of hydrogen. The estimate can vary depending on how accurate we want the model to be (recall that in mathematical modeling, we can always add more complexities), but the helium ratio lies between 23% and 26% (73% to 75% hydrogen and the rest being other heavier elements). The theoretical work that predicted the production of hydrogen, helium, and other heavier elements in the early Universe was carried out in 1948, in the famous “Alpher–Bethe–Gamow” paper[41] (Alpher et al. 1948). Subsequent research, however, found that except for lithium, heavier elements could not be formed this way—elements up to iron were the product of nuclear fusion inside stars, and others (like gold) were produced either by exploding stars (supernovae) or during collisions of neutron stars. The primordial production

[40] The decay equation or half-life equation is $N(t) = N(0)e^{-\lambda t}$, where $\lambda = 1/\tau$ is the decay constant, where τ is the mean lifetime. In terms of the half-life $t_{0.5}$, we have $\lambda = \ln 2/t_{0.5}$.

[41] Ralph Alpher was then a PhD student under the supervision of George Gamow (1904–1968). Hans Bethe’s (1906–2005) name was added by Gamow so that the paper author list would sound like “$\alpha\beta\gamma$”, much to the dismay of Alpher.

of deuterium, hydrogen, helium, and lithium is known as *Big Bang nucleosynthesis* (BBN). BBN cannot produce heavier elements because both the density of matter and the temperature is cooling down as the Universe expands, this is not a favorable condition for nuclear fusion to occur.

Returning to the CMB, it actually allows us to test the prediction of BBN. The predicted abundance of deuterium, helium, and lithium actually depends on the density of ordinary matter in the early universe, which can be treated as a free parameter. The ratio of these elements with respect to hydrogen can then be plotted as a function of the density of ordinary matter (i.e., not including dark matter) relative to photons. The latter can be extracted from the observation of the CMB such as the Wilkinson Microwave Anisotropy Probe (WMAP) satellite. As shown in Figure 5.4, the WMAP observation of the ordinary matter density ($4.6\% \pm 0.2\%$) fixed a line through all the different curves, the intersections of which are the ratios that can be compared to other direct observations. The fact that the predicted abundances shown by the circles in the graph are in good agreement with observed abundances obtained by other means is strong evidence that the theory is correct, at least in principle. Strictly speaking, there *is* a mismatch for the lithium abundance: observations of lithium from old stars in our galaxy is only $1/3$ of the predicted amount. While the status of the problem remains open, it is plausible that it is our understanding of the astrophysics of stars that may account for the mismatch (Korn 2024). Readers who are interested in the missing lithium problem may consult the specialized book *Lithium Across the Universe* by Eduardo Martin (Martin 2023).

Another important fact is that the CMB matches the Planck's blackbody spectrum. This means that matter and radiation were at thermal equilibrium. However, thermal equilibrium should imply a high entropy state, in fact, a maximum entropy state. Yet, as we have argued, this cannot be the case, otherwise nothing interesting would happen since then except for tiny fluctuations now and then. By the second law of thermodynamics, we expect the early Universe to have a much smaller entropy than today. So what gives? A likely explanation is that we need to take into account the gravitational degrees of freedom (whatever that means!). This is easier said than done, since much like the energy of a gravitational field, the entropy of a gravitational field is also very tricky to define. However, we do know that if there is a strong gravitational field in a box filled with gas, the gas will be attracted towards the source of the gravity and clump together. The arrow of time in this case moves in the direction of structure formation. On the other hand, without gravity, an initial clump of gaseous particles would disperse and uniformly fill the box. So what constitutes as a high entropy state cannot be decided *by appearance* alone. The CMB may indicate that the matter-radiation sector was in thermal equilibrium, but the blackbody spectrum does not tell us any information about gravity. The whole system was not at thermal equilibrium. The arrow of time suggests that the entropy is lower the earlier we go back in time. Therefore in some sense the Big Bang must be in a very special state. What is the initial condition of the Universe? Was there a physical law that necessitates a low entropic initial condition?

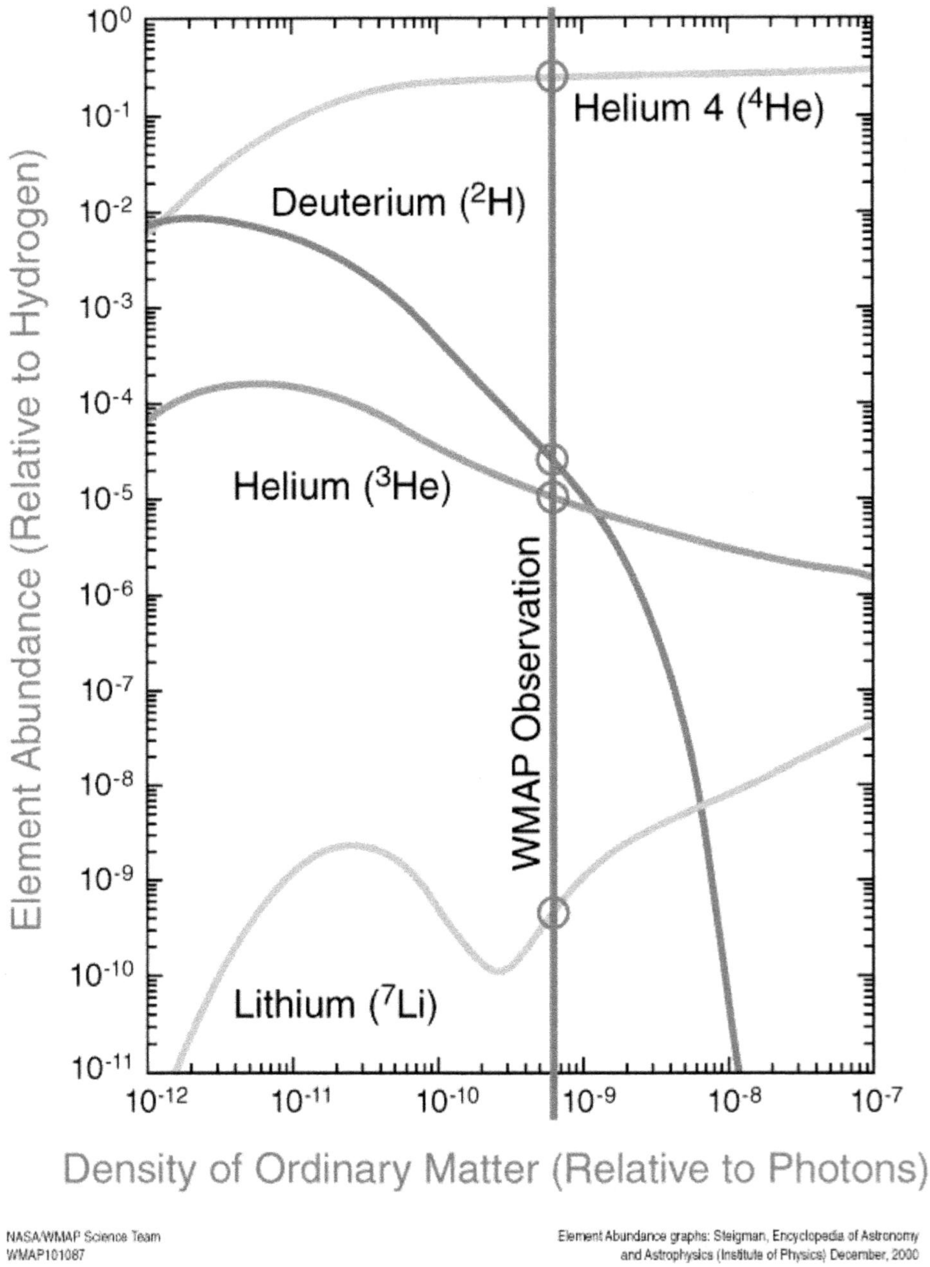

Figure 5.4. The curves are the theoretically obtained abundance of deuterium, helium, and lithium (relative to hydrogen) produced by the hot Big Bang. The WMAP observation fixed the points on the various curves that can be checked with other observations. About 24% of the ordinary matter in the Universe is helium. *Credit:* NASA/WMAP Science Team.

In relation to this, is the mainstream view that right after the Big Bang[42] there was a brief period during which the Universe underwent an exponential rate of expansion called "inflation". Did inflation really occur? Given that these are not yet established physics (though most cosmologists have high confidence in the inflationary paradigm), we defer the discussion to Section 13.3.

In Figure 5.5, we show the history of the Universe, to the best of our knowledge today. In the upper part, the illustration shows how the observable Universe started at the Big Bang and inflated. In the subsequent evolution, the expansion rate is not noticeable at the scale relative to the inflated size. The fact that there is anything at all in the Universe is due to the quantum fluctuations during inflation, which seeded structural formation as the fluctuations were magnified by inflation. The inserts below show a zoomed-in view on some of the microscopic processes taking place during cosmic history, which also shows the cosmic redshifting of photons as they lose energy when the Universe gradually expands.

For now, one mystery regarding matter is worth mentioning: when a matter particle is emitted from the "vacuum", it is accompanied by its antimatter particle. For example, the antimatter of an electron is a positron (e^+), which has a positive charge but is otherwise completely identical to the electron (hence the mass of e^+ is also the same as that of an electron: m_e). Some particles are their own antiparticles, for example, the photon. When a particle meets its antiparticle, they annihilate. One may ask if two photons can annihilate each other. The answer is yes, but we don't usually call such an event "annihilation". Usually by annihilation we mean two massive particles converting their mass–energy into massless particles. For example, an electron annihilates a positron and emits two photons. The time-reversed process is two colliding photons converting their energy into an electron–positron pair, which is equivalent to "photon–antiphoton annihilation". The photons must have enough energy to create an electron–positron pair.[43]

Anyway, here comes a big unsolved puzzle: if there is an exact symmetry between the number of particles and antiparticles, why didn't all of them annihilate each other in the first fractions of a second after the Big Bang? To explain the fact that there are still particles (and us!) today, there must have been a matter-antimatter asymmetry, so that the particle-to-antiparticle ratio is not exactly 1: 1 but $10^6 + 1$: 10^6, i.e., about one particle per billion survived the annihilation. The reason for this asymmetry is yet to be established conclusively.

We emphasize again that the cosmological constant, no matter how small it is, will always eventually dominate at late time since its density is constant. It just needs to wait till both radiation and matter are sufficiently diluted. However, we still do not know whether cosmological constant is really responsible for the accelerated

[42] Sometimes one reads that the Big Bang happened *after* inflation. This confusion is not that of physics but of semantics. Here by Big Bang we mean the initial singularity. However, in some texts, "Big Bang" refers to the (very hot) era right after the initial singularity during which matter and radiation are created. In fact the original Big Bang theory before inflation was suggested is often called the "hot Big Bang model". In inflationary model, this hot era happens after the end of inflation. This is called the "re-heating" of the Universe.

[43] The energy in the center-of-momentum frame must be at least $2m_ec^2$.

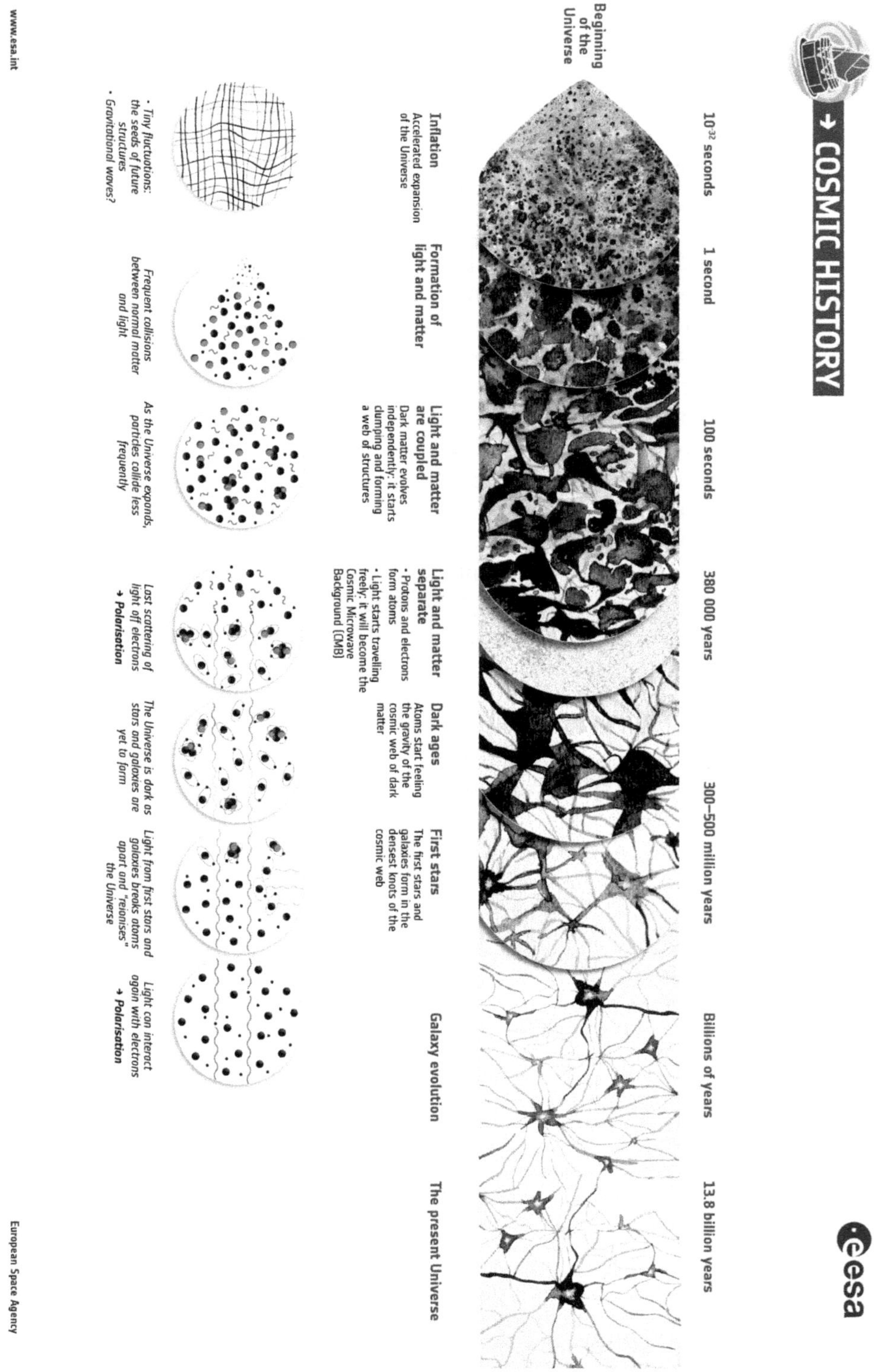

Figure 5.5. A summary of the 13.8 billion year history of the Universe, to the best of our understanding today. See texts for more description. *Credit:* ESO.

expansion of the Universe. Thus we usually refer to the cause of said acceleration as "dark energy". Similarly, there is also an unknown extra source of gravity that we could not account for, which we refer to as the "dark matter". Roughly speaking, the mass–energy content of the Universe currently consists of about 68% dark energy, 27% dark matter, and only 5% ordinary matter (including radiation). Dark matter and dark energy are collectively referred to as the "dark sector". In the next chapter, we shall discuss in detail the physics (and speculations) of the dark sector.

There are two major issues in cosmology today that might indicate our model of the Universe may need to be revised, or that some techniques used in the relevant observations contain as-yet-undiscovered problems. The *Hubble tension* refers to the mismatch between the locally measured expansion rate and the inferred rate via the cosmic microwave background. Historically, for many decades the astronomy community thought that the Hubble constant is between 50 and 100 km s^{-1} Mpc^{-1}. (Hubble's original estimate, based on inaccurate distance estimates, was about 500 km s^{-1} Mpc^{-1}!) As such, the "little h" was defined via

$$H_0 = h \times 100 \text{ km s}^{-1}\text{ Mpc}^{-1}. \tag{5.90}$$

The current value for H_0 based on "late time cosmology" by directly measuring the expansion rate via distance ladder techniques (observed by telescopes including the Hubble telescope and the latest JWST), converged to a value around 73 km s^{-1} Mpc^{-1}, whereas those derived from the CMB gave a value of around 67 km s^{-1} Mpc^{-1}, hence the tension. This is shown in Figure 5.6. In 2025, the Dark Energy Spectroscopic

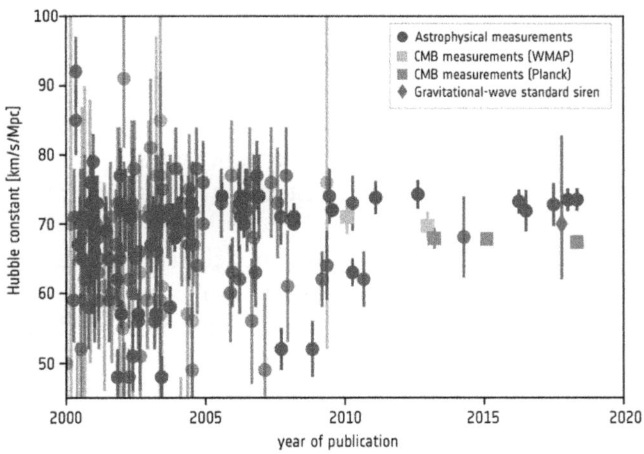

Figure 5.6. Various measurements of the Hubble constant over 20 years. Astrophysical measurements gave a value of $H_0 = 73.5$ km s^{-1} Mpc^{-1}, with an uncertainty of only 2%. These measurements are shown in blue. On the other hand the value estimated from our standard cosmological model utilizing the data of the early cosmos from the CMB gave a value of around 67.4 km s^{-1} Mpc^{-1}, with uncertainty of less than a percent. While it is remarkable that these two very different techniques gave very close values for H_0, the problem is that they don't agree *within the error bars*. The detection of gravitational waves with an electromagnetic counterpart (the collision of two neutron stars detected in 2017) enabled the first Hubble Constant H_0 measurement through the "standard siren" method. *Credit:* ESA/Planck Collaboration.

Instrument (DESI) that mapped 5000 galaxies, combining with BBN, inferred a value of 68.51 km s^{-1} Mpc^{-1} (DESI Collaboration 2025), which is closer to the CMB measurements. The recent DESI data also suggested that dark energy is not a constant DESI Collaboration (2025, 2025, 2025, 2025); more on this in the next chapter.

The second problem is the $\sigma 8$ *tension*, which concerns the measurement of galaxy clusters on a scale of 8 h^{-1} Mpc (via weak gravitational lensing; more on light bending and gravitational lensing in Chapter 7). Analysis revealed that matter has not clumped as much as expected assuming the concordance ΛCDM cosmology and its parameters constraints given by the CMB data. The $\sigma 8$ tension is also referred to as the *S8 tension*, where the quantity $S8$ is related to $\sigma 8$ via

$$S_8 \equiv \sigma_8 \left(\frac{\Omega_m}{0.3}\right)^{\frac{1}{2}}. \tag{5.91}$$

5.5 Exercises

(1) Understand how the Euler–Lagrange equation can be used to derive Newton's second law $F = ma$.

(2) Consider the metric

$$ds^2 = \frac{dx^2 + dy^2}{y^2}, \qquad y > 0. \tag{5.92}$$

This is the Poincaré half-plane model for hyperbolic geometry.

(a) What are the geodesics in this geometry?

(b) Show that the length of a vertical curve with end points being a fixed point (x_0, y_0) and (x_0, ε) tends to infinity as $\varepsilon \to 0$. That is to say, what appears like a finite line segment is actually infinite according to someone who lives in the hyperbolic world.

(c) Show that the scalar curvature at every point is constant and negative.

Additional note: The half-plane model can be transformed into the Poincaré disk model via a conformal transformation. The geometry on the disk model is well illustrated by some paintings of M.C. Escher's (1898–1972), e.g., the angels and demons in *Circle Limit IV* (*Heaven and Hell*) all have the same size, despite the fact that they appear to our eyes as becoming smaller and smaller towards the boundary circle (which is infinitely far away from the perspective of the inhabitants inside) (Dunham 2012). A good understanding of the hyperbolic plane and the Poincaré disk, as well as their higher dimensional generalizations, are crucial if you are interested in gauge/gravity duality, more traditionally called "AdS/CFT" (short for anti-de Sitter/conformal field theory) correspondence.

(3) This question is for students who are familiar with general relativity. Assume that we have a connection abstractly denoted by Γ, whose connection coefficients are Γ^a_{bc}. Furthermore assume that the connection

is defined independently of the metric tensor. The connection need not be torsion-free[44] or even metric compatible.[45] Yet we will show that something that looks like Einstein field equations can be derived. Consider the *Palatini action* defined by[46]

$$S(g, \Gamma) = \int d^4x \sqrt{-g}\, R(\Gamma), \quad \text{where } R(\Gamma) := g^{ab} R_{ab}(\Gamma), \tag{5.93}$$

and R_{ab} is the Ricci tensor. Show that the variation $\delta S/\delta g^{ab} = 0$ implies

$$R_{ab}(\Gamma) - \frac{1}{2} g_{ab}(\Gamma) = 0. \tag{5.94}$$

You may use the fact that

$$\frac{1}{\sqrt{-g}} \frac{\delta\sqrt{-g}}{\delta g^{ab}} = -\frac{1}{2} g_{ab} \tag{5.95}$$

without proof. Note that despite the appearance, Equation (5.94) is not yet the Einstein field equation because Γ is not necessarily the Levi–Civita connection. Try to prove that if we impose the torsion-free condition, then given the equation of motion for Γ, the connection is necessarily metric compatible, and hence by the fundamental theorem of (semi-)Riemannian geometry, it is uniquely the Levi–Civita connection, thus recovering general relativity.

Additional note: As pointed out in Ferraris et al. (1982), the original Palatini paper of 1919 was rather far from what is usually meant by Palatini's method discussed above, which was actually due to Einstein in 1925.

(4) In this chapter we discussed the Einstein–de Sitter model (FLRW cosmology with $k = 0$ and filled with dust). Try to solve the $k = 1$ case. To do this, we note that the Friedmann equation (with $c = 1$ and $k = 1$), can be written in the form

$$\dot{a}^2 = \frac{A^2 - a}{a}, \tag{5.96}$$

where

$$A^2 \equiv \frac{8\pi G \rho_0}{3}, \tag{5.97}$$

for some ρ_0 as initial condition (dust evolves as $\rho(t) = \rho_0/a^3$). Solving for t gives

[44] Torsion-free means that $\Gamma^a_{\ bc} = \Gamma^a_{\ cb}$.

[45] Metric compatibility means that $\nabla_c g_{ab} = 0$.

[46] See Section 6.3 for more explanation regarding the factor $\sqrt{-g} \equiv \sqrt{-\det(g_{ab})}$.

$$t = \int_0^a \left(\frac{a'}{A^2 - a'} \right)^{\frac{1}{2}} da'. \tag{5.98}$$

This integral can be performed analytically by setting $a = A^2 \sin^2(\theta/2)$. Check that your result is

$$t = \frac{A^2}{2}(\theta - \sin\theta). \tag{5.99}$$

Show that

$$a = \frac{A^2}{2}(1 - \cos\theta). \tag{5.100}$$

These are parametric equations that define the *cycloid* curve, with the Big Bang at the beginning and the Big Crunch at the end. It is possible that the Universe "repeats" in a continuous cycle of bangs and crunches. The cycloid curve suitably rotated is the same as the Brachistochrone curve (curve of fastest descent); it also appears in the context of the tautochrone problem. Both of these are interesting mathematical problems worth finding out more about.

References

Alpher, V. S. 2014, AsJPh, 23, 17

Alpher, R. A., Bethe, H., & Gamow, G. 1948, PhRv, 73, 803

Aluri, P. K., Cea, P., Chingangbam, P., et al. 2023, CQGra, 40, 094001

Barrow, J. D. 2017, The Philosophy of Cosmology, ed. K. Chamcham, J. Silk, J. D. Barrow, & S. Saunders (Cambridge: Cambridge Univ. Press) 85

Bouhmadi-López, M., Chen, P., & Liu, Y.-W. 2011, PhRvD, 84, 023505

Bronstein, M. 1933, PhyZS, 3, 73

Castelvecchi, D. 2021, Natur, 598, 549

Chaplygin, S. 1904, SMMUMP, 21, 1

Clerke, A. M. 1903, Problems in Astrophysics (London: A&C Black)

DESI Collaboration 2025, [arXiv:2503.14738 [astro-ph.CO]] 2025arXiv250314738D

DESI Collaboration 2025, [arXiv:2503.14739 [astro-ph.CO]] 2025arXiv250314739D

DESI Collaboration 2025, [arXiv:2503.14743 [astro-ph.CO]] 2025arXiv250314743LD

DESI Collaboration 2025, JCAP, 02, 021

Dunham, D. 2012, Mathematics and Modern Art, Springer Proceedings in Mathematics Vol. 18, ed. C. Bruter (Berlin: Springer)

Edmonds, D. 2020, The Murder of Professor Schlick: The Rise and Fall of the Vienna Circle (Princeton, NJ: Princeton Univ. Press)

Efstathiou, G., Sutherland, W. J., & Maddox, S. J. 1990, Natur, 348, 705

Ferraris, M., Francaviglia, M., & Reina, C. 1982, GReGr, 14, 243

Gorelik, G. E. 2005, Ph-Usp, 48, 1039

Hearnshaw, J. 2010, JAHH, 13, 90

Herzberg, G. 1950, Molecular Spectra and Molecular Structure I. Spectra of Diatomic Molecules (2nd ed.; Princeton, NJ: Van Nostrand)

Hoyle, F. 1988, A Century of Ideas, Fundamental Theories of Physics ed. B. G. Sidharth, (Dordrecht: Springer) 149
Korn, A. J. 2024, EPJWC, 297, 01007
Kosmann-Schwarzbach, Y., & Schwarzbach, B. E. 2011, The Noether Theorems: Invariance and Conservation Laws in the Twentieth Century (Berlin: Springer)
Krauss, L. M., & Turner, M. S. 1995, GReGr, 27, 1137
Lemaître, G. 1931, Natur, 127, 706
Liddle, A. 2015, An Introduction to Modern Cosmology (3rd ed; New York: Wiley)
Lopez, A. M., Clowes, R. G., & Williger, G. M. 2022, MNRAS, 516, 1557
Lopez, A. M., Clowes, R. G., & Williger, G. M. 2024, JCAP, 07, 055
Lopez, A. M., Clowes, R. G., & Williger, G. M. 2025, RSPTA, 383, 20240029
Luminet, J.-P. 2011, GReGr, 43, 2911
Manin, Y. 2009, NAMS, 56, 1268
Martin, E. 2023, Lithium Across the Universe (Bristol: IOP Publishing)
McInnes, B. 2003, JHEP, 09, 009
Occhionero, F., Vittorio, N., Carnevali, P., & Santangelo, P. 1980, AA, 86, 212
Ong, Y. C., & Yeom, D.-h. 2017, JCAP, 04, 040
O'Raifeartaigh, C., O'Keeffe, M., & Mitton, S. 2021, EPJH, 46, 4
O'Raifeartaigh, C. 2017, The Philosophy of Cosmology, ed. K. Chamcham, J. Silk, J. D. Barrow, & S. Saunders (Cambridge: Cambridge Univ. Press) 467
Ostriker, J. P., & Steinhardt, P. J. 1995, Natur, 377, 600
Planck Collaboration 2020, A&A, 641, A10
Synge, J. L. 1960, Relativity: The General Theory (Amsterdam: North-Holland Publishing Company)
Valade, A., Libeskind, N. I., Pomarède, D., et al. 2004, NatAs, 8, 1610

Chapter 6

The Dark Sector

> *"The Dark Arts are many, varied, ever-changing, and eternal. Fighting them is like fighting a many-headed monster, which, each time a neck is severed, sprouts a head even fiercer and cleverer than before. You are fighting that which is unfixed, mutating, indestructible."*—Severus Snape (J. K. Rowling, 'Harry Potter and the Half-Blood Prince'.)

The best model for the cosmos that fits observational data today is the ΛCDM model, which in addition to normal matter, also contains two major ingredients: the positive cosmological constant Λ, and cold dark matter (CDM). Together they make up about 95% of the matter-energy budget of the Universe. Nevertheless, there are signs that the accelerated expansion of the Universe is not due to a simple Λ, but rather some peculiar form of dynamical dark energy. Similarly, the nature of the CDM remains unknown despite decades of theoretical speculations and experimental searches. In this chapter we review the dark side of the Universe. In addition, we also discuss whether the dark sector is actually "real", or whether it is the theory of general relativity that needs to be modified at cosmological scale. We should pay attention to some lessons from history.

One of the biggest problems in modern cosmology is probably the nature of dark matter and dark energy, which has eluded physicists and astronomers for decades. The essential problem is simply that observational data do not seem to agree with how the Universe is expected to behave assuming the equations given by general relativity are correct, *unless* there exist hitherto unknown substances that primarily only interact gravitationally, which comprise about a quarter of the mass–energy budget of the Universe; as well as unknown fields that resulted in the accelerated expansion of the Universe. The former is called "dark matter" and the latter "dark energy". Together, they comprise the "dark sector" of the Universe.

doi:10.1088/2514-3433/ade963ch6

However, while reading this chapter it is important to always remind oneself that these names are just place holders for our ignorance. In his book *What Do You Care What Other People Think?*, Richard P. Feynman wrote:

> "You can know the name of a bird in all the languages of the world, but when you're finished, you'll know absolutely nothing whatever about the bird [...] So let's look at the bird and see what it's doing—that's what counts. I learned very early the difference between knowing the name of something and knowing something."

We know the effects caused by dark matter and dark energy, but we don't even know if they are indeed new kind of matter or energy.

There are alternative proposals in which instead of dark matter and dark energy, the discrepancies between theory and observation are attributed to our theory of gravity being not entirely correct. This gave rise to an entire field of research called "modified gravity", which we will explain in this chapter. We will also take a look at historical examples in which anomalies are sometimes successfully explained by modifying the theory, but sometimes by positing the existence of "new matter" that was later discovered. It is important to pursue different possibilities as we can only know the correct answer once we have it.

6.1 Dark Matter

Consider a spiral galaxy like our own Milky Way. Stars in the galaxy go around, "orbiting" the galactic center. If what we know about gravity is correct (in this scenario, we are in the so-called "weak field regime" so that general relativity reduces to Newtonian gravity), then as the distance of the star from the galactic center increases, the speed should slow down. This can be seen readily by setting the magnitude of Newtonian gravity equal to that of the centripetal force:

$$\frac{GM(r)m}{r^2} = \frac{mv^2}{r} \Longrightarrow v(r) = \sqrt{\frac{GM(r)}{r}}, \tag{6.1}$$

where $M(r)$ is the mass enclosed within radius r of the galaxy, and m the mass of a specific star. Clearly $v(r)$ is a monotonically decreasing function of r. This formula fails at the origin $r = 0$ where the divergence is not physical; this is just due to the simplified model used to illustrate the problem; actual realistic models are more complicated.

Far out at the edge of the galaxy, stars should move with a relatively slow speed. However, this is not what observations see. Time after time, galaxy after galaxy, astronomers measured the speed of these stars at the far edge of the galaxies to be traveling much faster than expected. We can plot the graph of the speed of the stars vs their distance from the galactic center. This is the famous "galactic rotation curve". An example is shown in Figure 6.1. Instead of decreasing, the speed is often seen to tend to some (relatively large) constant value. Now, the speed is determined by how much mass is contained within the galaxy. Therefore one possible explanation is that there exists more matter than the eyes (and all the telescopes) can see. Hence the name "dark matter".

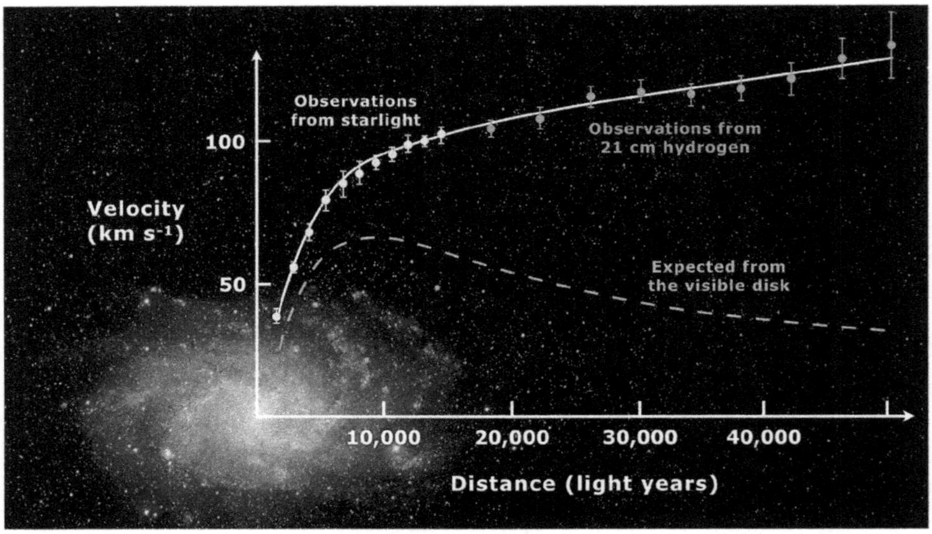

Figure 6.1. Rotation curve of spiral galaxy Messier 33, otherwise known as the Triangulum Galaxy, as determined by the data represented by yellow and blue points with error bars, and a predicted one from distribution of the visible matter according to Newtonian gravity (gray line). The data and the model predictions are from Corbelli & Salucci (2000). The discrepancy between the two curves can be accounted for by adding a dark matter halo surrounding the galaxy. *Credit:* This [Rotation curve of spiral galaxy Messier 33 (Triangulum)] image has been obtained by the author(s) from the Wikimedia website where it was made available by [Accrama] under a CC BY 4.0 licence. It is included within this book on that basis. It is attributed to [Mario De Leo].

Before proceeding it is good to have some background knowledge about particle physics. The idea that everyday stuff is made of small indivisible constituents dated back to ancient Greece. Called "atoms", such basic ingredients were believed to be hard and uniform. Some believed that they have extended shapes like hooks so they can attach to each other—the precursor of attractive forces and molecular bonds of today's chemistry. Yet what we call atoms today are not indivisible. An atom consists of the atomic nucleus and the electrons that orbit it. As far as we know the electrons are indeed indivisible (though all particles are quantum objects and we should not think of them as little hard spheres). The nucleus consists of a collection of protons and neutrons, which are in turn made of the quarks. There are six types of quarks. In general, quarks combine to form "hadrons", which are further categorized into baryons and mesons. Baryons, which include the protons and neutrons, are made of three quarks; whereas mesons are made of one quark and one anti-quark. Quarks carry a different sort of charge; instead of electric charge, it is called a "color charge". This should not be confused with the ordinary colors that we see; they are just names. There are three colors: red, blue, and green; but each has a corresponding anti-color. Hadrons are color-neutral. For example, in the baryons, each quark must be of a different color so that the combination is white; whereas the mesons must have one color matched with its anti-color. The quarks form three families: the up and down (which make up the protons and neutrons), the charm and strange, and the top and bottom. The six types are referred to as the "flavors".

Each family of quarks is matched with a corresponding electron-type particle. For the up and down quarks, it is just the ordinary electron; for the charm and strange quarks, the "muon"; and for the top and bottom quarks, the "tau". The muons and taus behave like electrons but they are more massive and unstable. Finally, the electron, muon and tau all have a corresponding neutrino (more on this later). The neutrinos, together with the electron, muon and tau, are collectively called the "leptons". The masses of these particles get heavier as we go from the first to the third family, but no one knows why there are three families. The Nobel Laureate of Physics of 1944, Isidor Isaac Rabi (1898–1988), famously remarked "who ordered all of that?". Thus far we have only discussed the matter particles, but there are also the force carriers ("gauge bosons"): in quantum field theory, elementary forces are mediated by particles too. The force carriers for electromagnetic, strong nuclear, and weak nuclear forces are, respectively, the photon, the gluons (there are 8 types), and the W-bosons (W^+ and W^-) together with the Z-bosons. In addition, there is the scalar particle known as the Higgs boson, which was the latest to be discovered (in 2012 at the Large Hadron Collider, or LHC for short). It is the only fundamental scalar particle known so far (admittedly there are still some models that argue for the Higgs particle to be composite). These particles form what is known as the *standard model of particle physics*, summarized in Figure 6.2.

It is remarkable that all known matter in the Universe can be explained with a collection of relatively small number of particles. As Hermann Weyl wrote in *Philosophy of Mathematics and Natural Science* (Weyl 2009),

> "The composition of the material world out of one or a few units, existing in a huge number of completely alike specimens, must surely be looked upon as one of the most fundamental features in the nature of the Universe; and one that is most profoundly in need of interpretation."

Unfortunately, this explanation remains to be found.

The common point of view by particle physicists and high energy physicists is that dark matter is a kind (or maybe there is more than one kind) of new particle beyond the standard model of particle physics. For example, the idea of "supersymmetry" (SUSY) is that for each particle there is a "super-partner". A "selectron" is the super-partner of an electron, and a "Higgsino" is the superpartner of the Higgs particle, etc. Some of these supersymmetric particles, or "sparticles", could serve as dark matter candidates. SUSY was proposed not for solving the dark matter problem, but rather mainly as part of a larger scheme to unify the forces. We know that the electromagnetic force can be unified with the weak nuclear force to form a single "electroweak" force at sufficiently high energy. Using supersymmetry, the electroweak force can further unify with the strong nuclear force at an even higher energy scale. Such a theory is referred to as the Grand Unification Theory (GUT). The hope was that one could go even one step further and unify this putative GUT force with gravity, and obtain a "Theory of Everything" (TOE). Of course, Nature does not demand such a unification, but it is philosophically pleasing if we can

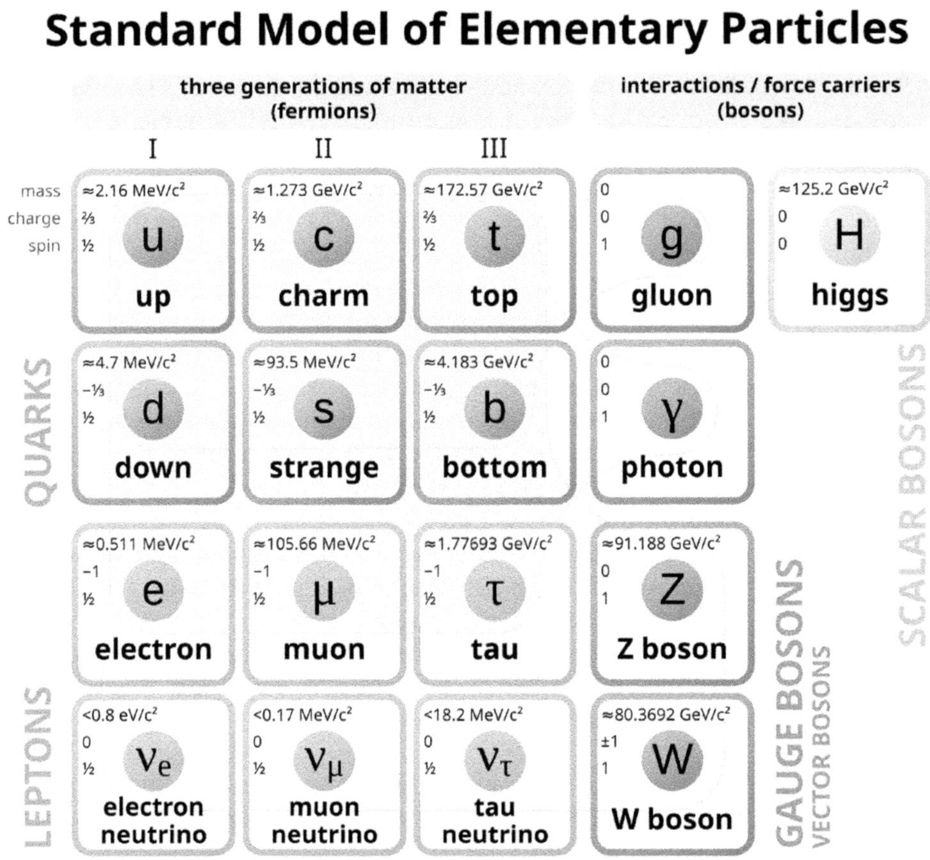

Figure 6.2. Standard model of elementary particles: the 12 fundamental fermions and 5 fundamental bosons. The faint loops indicate which bosons (orange-red) couple to which fermions (purple and green). *Credit:* This [Standard Model of Elementary Particles] has been obtained by the author(s) from the Wikimedia website, where it is stated to have been released into the public domain. It is included within this article on that basis.

combine all forces into one.[1] However, almost two decades after the LHC started running, no sign of SUSY has yet to be glimpsed. The non-discovery does not rule out the idea of SUSY entirely, but in my personal opinion it seems increasingly unlikely that dark matter can be explained by SUSY.

[1] In Western Medieval alchemical thoughts, it is believed that the four elements (fire, water, earth, wind), represented as a square, can be combined into a higher unity represented by a circle (hence “squaring the circle”). The end result is sometimes referred to as “quintessence”, the fifth element that represents the soul and the cosmos. Similarly, in the *Daodejing* (《道德经》) of Taoism in ancient China, we read that: “The Way gives birth to the One; The One gives birth to two; Two give birth to three; And three give birth to all things”. Also in Indian philosophy, one finds the idea of *Prakṛti*, the ultimate cause of all material existence, or the primordial nature or matter. Thus we see that the philosophical urge to unify Nature has always been with us. The philosophy that everything is fundamentally made of the same stuff (like “atoms”) is just one example. Now that we know there are elementary particles that are not the same (the leptons and the baryons), one tries to unify the forces instead at a higher energy scale. In this view, only one “superforce” existed, which split into the forces we know as the temperature of the Universe decreased.

Another popular candidate is the "axion", a particle first proposed independently by Frank Wilczek (1951–current) and Steven Weinberg (1933–2021) in 1978 as the Goldstone boson of the Peccei–Quinn theory, which was in turn introduced by Roberto Peccei (1942–2020) and Helen Quinn (1943–present) in 1977 to solve the strong CP problem in the theory of quantum chromodynamics (QCD; the theory of the quarks and nuclear forces). It is too technical for us to explain this problem here, so it suffices to say that if axions exist in the right range of mass, it might serve as a dark matter candidate. Peccei–Quinn theory, which introduces a scalar field and a new symmetry into the standard model, in hindsight, naturally predicts a new particle (the axion), but Peccei and Quinn did not make this association.

Most of the proposed dark matter particles belong to a class called WIMPs or "weakly interacting massive particles". They do not interact via the electromagnetic force. We can primarily only see their gravitational effects. From the perspective of experimental searches, it is fortunate that depending on the type of the particle proposed, dark matter can interact with normal matter though only very rarely. This is a lot like the neutrino, which was first proposed by Wolfgang Pauli[2] (1913–1993) in 1930 to explain the apparent violation of energy and momentum during beta decay. This is a radioactive process during which a neutron decays into a proton, plus an electron and an anti-electron neutrino (as mentioned, there are various species of neutrinos. The one associated with electrons is called the "electron neutrino", ν_e. Its antiparticle is denoted $\overline{\nu}_e$.) At the time of Pauli, physicists had no idea of the existence of neutrinos; all they saw was the decay into protons and electrons. But such a process is in conflict with energy and momentum conservation[3]. Pauli proposed that a new particle must exist to balance the equation, though he was reluctant to do so. He even said that he had done a "terrible thing": "I have postulated a particle that cannot be detected". It would take two more decades before neutrinos were detected experimentally. This is because neutrinos hardly interact with anything. At this moment, trillions of neutrinos shoot right past your body (10^{11} neutrinos pass through your thumbnail every second, day and night), and in fact the entire Earth, as if it is not even there. However, there is a chance of about

[2] Pauli once sought therapy from the psychoanalyst Carl Jung. Since Jung was interested in the collective unconscious, he was intrigued by Pauli's ability to vividly recall his dreams. Jung would go on and compile some 1300 of Pauli's dreams (how many dreams can *you* remember?) and use them for his research. Pauli continued to discuss with Jung even after he was no longer a patient. Central to their discussions was Jung's idea of *archetypes*, symbolic images that represent common human experiences and motivations that can appear in dreams. Privately, Pauli attributed his work on CPT symmetry to the inspirations from his dreams that they discussed. Pauli wrote: "Every time I have talked to Mr Jung (about the synchronistic phenomenon and such) a certain spiritual fertilization takes place." (Miller 2024)

[3] In the previous chapter we mentioned that energy is not really conserved in an expanding cosmos, but of course at a local scale in a laboratory on Earth, there is no doubt that we can apply energy conservation. This is of course what we have learned nowadays; back in the day there were serious physicists, including Niels Bohr (1885–1962), who considered the possibility that energy may not be conserved in nuclear processes. If *you* were a physicist in those days, would you choose to believe in a new particle that no one has ever seen and probably will never be seen, or choose to give up on energy conservation?

1 in 1 trillion trillion that a neutrino would interact with an atomic nucleus. To improve the odds, modern neutrino detectors utilize either a huge body of water or ice. For example, the IceCube detector in Antarctica, and the Super-Kamiokande Neutrino Telescope in Japan that contains 50,000 tonnes of water, which allows it to detect 2.4 neutrino interactions per day. The reason why this is called a *telescope* is because astrophysical processes like supernovae would release a large amount of neutrinos, about 10^{58} of them (in fact these carry way more energy than the photons and other particles in the explosion), and since they barely interact with the gases in the stellar vicinity, they can reach Earth even before the photons (note that they are not traveling faster than light, just that the photons are trapped momentarily in the hot plasma of the explosion). During the event of SN 1987A, neutrino detectors on Earth—including Kamiokande II, the predecessor of Super-Kamiokande—detected a total of 25 neutrino-events over a period of about 13 seconds. The light from the supernova only arrived 3 hours later. It was a blue supergiant star called Sanduleak-69 202 in the constellation of Dorado that had exploded.

We have side-tracked somewhat from dark matter, but the point is that dark matter is not the first "possibly new particle" that was proposed based on something that *was not seen* in observations. In the case of the neutrino, Pauli proposed the existence of neutrinos to save the conservation laws, whereas in the case of dark matter, it was Fritz Zwicky (1898–1974) who proposed it in the 1930s (calling it *Dunkle Materie*) in order to explain the peculiar speed of the individual galaxies in the Coma Cluster. In his 1937 paper (Zwicky 1937), he estimated the mass of the Coma Cluster using the Virial theorem (he was probably the first to use the virial theorem to determine the mass of a galaxy cluster; see Section 4.2 for an introduction to Virial theorem). The idea, essentially, is that $\langle v^2 \rangle \sim GM/r$, where r is the radius of the visible cluster (about 2×10^6 light-years). This yields at least 4.5×10^{13} solar masses for the cluster and 4.5×10^{10} solar masses for each galaxy on average (since the cluster contains about a 1000 galaxies). Zwicky wrote:

> "This result is somewhat unexpected, in view of the fact that the luminosity of an average nebula is equal to that of about 8.5×10^7 suns."

Zwicky deduced that this implied that the mass-to-light ratio is about 500—there is more mass than the amount of light indicated, i.e., dark matter that contributes to the mass without increasing the overall luminosity. It took a while to rule out the "missing mass" being due to massive collapsed objects like black holes, dwarf stars (which are not so luminous), and intracluster gas or dust.

Some 40 years later, Vera Rubin (1928–2016) established the galaxy rotation curve for stars in various galaxies, showing that the flat rotation curve feature is common, and therefore dark matter not only affects motion of galaxies within a cluster, but also of stars around a galaxy. How far out do rotation curves remain flat? If galaxies are surrounded by a physically realistic halo of dark matter, once we go far enough we should see the curve go down again. However, in 2024, a new paper utilized a weak gravitational lensing technique claimed that the curves may remain flat *indefinitely* far away, as they saw no clear hint of any decline even out to 1 Mpc away, well beyond the expected virial radii of dark matter halos (Mistele et al. 2024). This suggests that either dark matter halos are more extended than previously thought, or perhaps we need to rethink our theory of gravity.

Observational evidence also prefers dark matter particles that do not move too fast. From statistical mechanics and thermodynamics, we know that fast

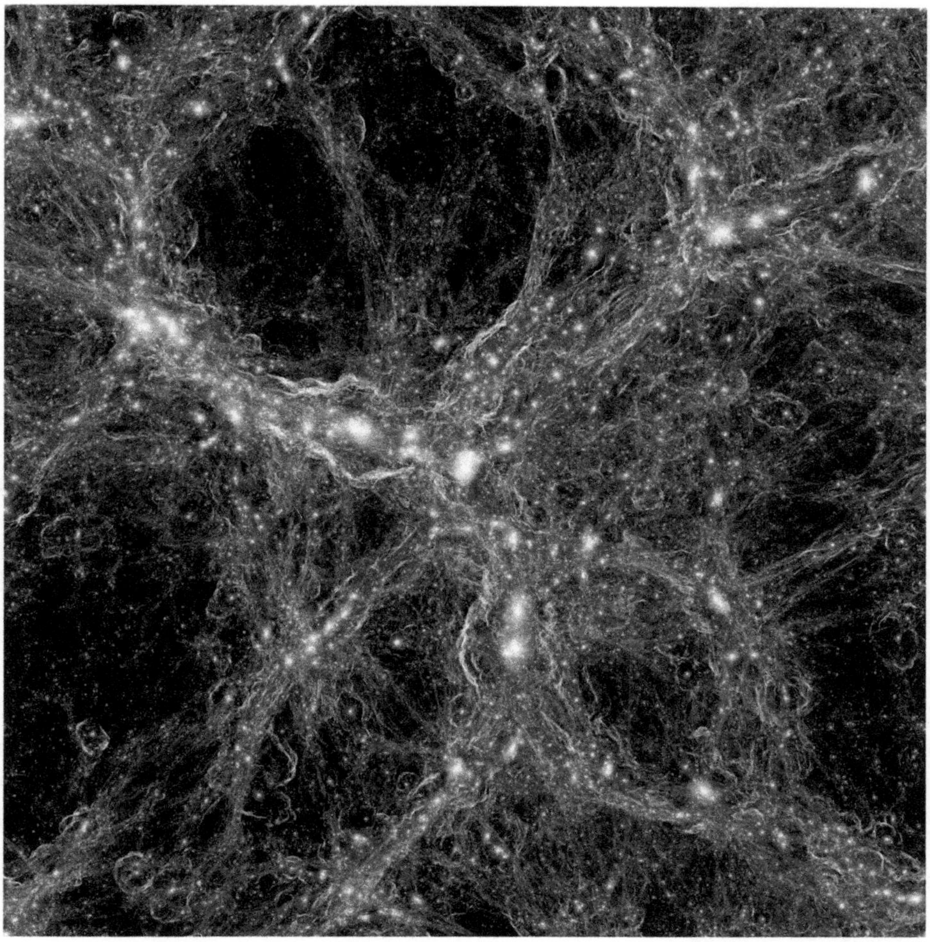

Figure 6.3. In ΛCDM, dark matter plays an important role in structure formation, as dark matter organized itself into filaments, which in turn served as a scaffolding for visible matter to form galaxies. This has been "verified" via many computer simulations. That is, by incorporating dark matter into our model, the simulations produced large-scale structures of galaxy clusters that look very similar to what actual observations indicate. Shown here is a composite image (of the full TNG100-1 box; i.e., the simulation box size is 100 megaparsecs across) which overlays a projection of the dark matter density with the output of on-the-fly cosmological shock finder, here used to derive the average Mach number of shocks along each line of sight. All the gravitationally collapsed structures (in orange/white) are surrounded by successive shock surfaces (blue) which encode their formation histories. *Credit:* Reprinted with permission from the TNG collaboration. (c) 2023 The TNG Collaboration.

moving particles in a container gives rise to a higher temperature, so we refer to dark matter particles with relatively small speed as "cold dark matter". The reason why dark matter cannot be too warm is that otherwise it will have enough kinetic energy to disrupt structure formations (see Figure 6.3) in the early Universe and then the distribution and morphology of galaxies today will be quite different.

Of course, calling what we don't know "dark matter" does not necessarily mean that it is a kind a new particle. Again, we can make a comparison with the history of neutrino. As we have mentioned, back then, Neils Bohr proposed another possible explanation for the beta decay observations: maybe energy and momentum are *not* absolutely conserved but only statistically so (Walther Nernst also considered such a possibility already in some other contexts). If true, this would be similar to the second law of thermodynamics, which Boltzmann had shown to be statistical in nature. Thus, overall entropy is non-decreasing, but it *can* and *will*, from time to time decrease just because of fluctuations. However, it seems difficult to explain why *every time* the beta decay experiment is performed the conservation laws are violated, even if the statistical nature of these laws are such that they are satisfied overall in the macroscopic world, but at the scale of the atomic nuclei they are not. Again, this would be similar to the second law of thermodynamics: it is much less likely that all the air particles in a room suddenly, by chance, gather at one corner and suffocate almost everyone, but this probability increases as the room size decreases. Still, if the violation to energy and momentum conservation is observed *every time*, even this explanation seems to be unlikely. It turned out that Pauli was right, and conservation laws are not statistical. So what about dark matter? We only deduced its existence because we have a great deal of confidence in our description of gravity. However, what if general relativity is not quite correct at the galactic or cosmological scale? We will discuss this issue soon, but let us first introduce dark energy properly.

6.2 Dark Energy

In Chapter 5, we have mentioned that the expansion of the Universe is accelerating, and this can be explained with a cosmological constant (see Carroll 2001 for a useful overview). Of course, there are other possibilities. Recall that the energy density of a perfect fluid satisfies

$$\rho(t) = \frac{\rho_0}{a(t)^{3(1+w)}}, \tag{6.2}$$

where $w < -1/3$ for the equation of state is required to have an accelerated expansion. Thus, given that the energy density is positive (usually required for stability purpose), dark energy must have a *negative pressure.* The cosmological constant, when considered as a perfect fluid, satisfies $w \equiv -1$ at all time, with energy density given by

$$\rho_\Lambda = \frac{\Lambda}{8\pi G}. \tag{6.3}$$

On the other hand, we can consider a fluid with w that varies in time[4] (we don't want it to vary in *space* because in the FLRW metric we assume the Universe to be homogeneous at large scale. Having said that, we can also relax this assumption; see later discussion). If the fluid satisfies $-1 < w < 0$, it is called a "quintessence", which is not to be confused with the Medieval notion of the quintessence, but the latter represents the cosmos, so the name is somewhat fitting. The simplest type of quintessence is a scalar field ϕ with some suitable potential V. In this case, the parameter w is given by

$$w = \frac{\frac{1}{2}\dot{\phi}^2 - V^2}{\frac{1}{2}\dot{\phi}^2 + V^2}. \tag{6.4}$$

With an appropriate form of V, the field can evolve such that at late times (i.e., the current era) the value of w goes below $-1/3$, and in fact tends to -1. This latter requirement is because the "concordance model" of cosmology with a cosmological constant and cold dark matter (ΛCDM) explains observational data rather well, so even if dark energy is not a constant, its w should be close to -1 at late time. It is sometimes stated in the literature that the main reason for favoring quintessence over a cosmological constant is that, if there is indeed a cosmological constant, then it is likely a vacuum energy (since its energy density is unchanged no matter how large the Universe is), but then quantum field theory would estimate that the most natural scale of vacuum energy is about 10^{122} times the required value in cosmology. Why is the cosmological constant so unexpectedly small yet nonzero? Most physicists therefore consider the cosmological constant as "unnatural". However, using a quintessence to describe dark energy is just sweeping the unnaturalness somewhere else: the mass scale of the scalar field has to be unnaturally small, much smaller than any other known mass scales in particle physics. Of course there are more complicated models that attempted to be more "natural", but it is difficult to come up with a natural and convincing model.

A fluid with $w < -1$ is called "phantom energy". It will lead to a *Big Rip* singularity in the future, in which every structure, and eventually even the Universe itself, tears apart (Caldwell et al. 2003). Let us take a closer look at how this fate comes about. As per usual, we assume that the Universe is spatially flat ($k = 0$). When the phantom energy dominates over other matter fields, we can consider the Friedmann equation to be

$$H^2 = \frac{8\pi G}{3}\rho_p, \tag{6.5}$$

[4] The idea that dark energy can be time-varying is not new; Einstein himself realized that his cosmological constant can be generalized to be a function of time but decided against it. It was in 1917 that Schrödinger suggested that the cosmological constant term could be placed on the RHS of the Einstein field equations and therefore be viewed as a kind of matter field rather than a geometric property. Einstein responded that if the cosmological constant is indeed a constant, putting it on either side of the equation is equivalent. If not a constant, however, Einstein thought that this was too much of a speculation. He wrote: "The course taken by Herr Schrödinger does not appear passable to me because it leads too deeply into the thicket of hypotheses." (Einstein 1918) As we have seen many times, Einstein was a pragmatic physicist who didn't like to consider "what if" scenarios unless forced upon by observations or strong theoretical motivations.

in which the subscript denotes the phantom field. Divided by the Hubble constant H_0^2 throughout and using the definition that $H(t) = \dot{a}(t)/a(t)$, we have

$$\frac{\left(\frac{da}{dt}\right)^2}{a^2}\frac{1}{H_0^2} = \frac{8\pi G}{3H_0^2}\frac{\rho_{p,\,0}}{a^{3(1+w_p)}}, \tag{6.6}$$

where $\rho_{p,\,0}$ is the current energy density of the phantom field, and $w_p < -1$. Let $\rho_c := 3H_0^2/8\pi G$, which is known as the "critical density", and let $\Omega_{p,\,0} := \rho_{p,\,0}/\rho_c$ be the density parameter. Together with the density parameter of other matter fields $\Omega_{m,0} := \rho_{m,0}/\rho_c$, we have $\Omega_{p,\,0} + \Omega_{m,0} = 1$. The differential equation can now be written as

$$\frac{da^2/a^2}{H_0^2\, dt^2} = (1 - \Omega_{m,0})a^{-3(1+w_p)}. \tag{6.7}$$

Integrating from the current age of the universe (t_0) to some future time t', we get

$$H_0 \int_{t_0}^{t'} dt = (1 - \Omega_{m,0})^{-\frac{1}{2}} \int_{a_0}^{a'} a^{\frac{1+3w_p}{2}} da. \tag{6.8}$$

In our convention, $a_0 = 1$, thus it can be shown that this gives

$$H_0(t' - t_0) = (1 - \Omega_{m,0})^{-\frac{1}{2}}\left(\frac{2}{3}\frac{\left|a'^{3(1+w_p)/2} - 1\right|}{\left|1 + w_p\right|}\right). \tag{6.9}$$

We now notice that $a'_{\text{rip}} \to \infty$ occurs at a *finite* time t_{rip} since the equation at that moment is (recall that $w_p < -1$)

$$H_0(t_{\text{rip}} - t_0) = (1 - \Omega_{m,0})^{-\frac{1}{2}}\left(\frac{2}{3}\frac{1}{\left|1 + w_p\right|}\right) < \infty. \tag{6.10}$$

As the scale factor goes to infinity, the expansion rate tends to infinity, thus the "rip".

Note that the existence of a quintessence or a phantom does not mean that we do not have a cosmological constant. Any two or all three components can be present simultaneously. This would even allow the cosmological constant to be negative. A negative cosmological constant would lead to an eventual contraction of the Universe and a Big Crunch. However, if the current era is still dominated by the quintessence, it can exhibit a temporary phase in which the expansion is accelerating. Speaking of a negative cosmological constant, it is also possible that the sign of the cosmological constant can change, such that the very early cosmos has a negative cosmological constant, which at some point switches its sign. This is known as the Λ_sCDM model (Akarsu et al. 2021). It has been proposed that such a sign change could help to ameliorate the Hubble tension. In the Λ_sCDM model the flip of sign was not explained and thus seems somewhat ad hoc. An alternative possibility is to consider a scenario in which both quintessence and phantom are present. We can view their combined effect as an "effective dark energy". This gives rise to the possibility that the overall sign of the effective "Λ" can change. Models of dark energy that allow w to cross from $w < -1$ to $w > -1$ or vice versa is said to cross a "phantom divide", and this behavior is called "phantom crossing".

In the previous chapter, we have mentioned that data from the Dark Energy Spectroscopic Instrument (DESI) suggested that dark energy is not a cosmological constant (DESI Collaboration 2025, 2025, 2025, 2025). In particular, if we parametrize the dark energy equation of state as

$$w(a) = w_0 + w_a(1 - a), \tag{6.11}$$

where a is the expansion factor, then if ΛCDM is correct, i.e., if dark energy is indeed the cosmological constant, we would expect that $w_0 = -1$ and $w_a = 0$. However, the best fit of DESI data together with CMB and the Dark Energy Survey Year 5 (DESY5) data combined gave

$$w_0 = -0.752 \pm 0.057, \tag{6.12}$$

and

$$w_a = -0.86^{+0.23}_{-0.20}. \tag{6.13}$$

Figure 6.4 shows the constraint plot along with some annotated explanations. In fact, DESI data favors a phantom crossing (DESI Collaboration 2025). However, since the parametrization using Equation (6.11) is only the first two terms of a Taylor series expansion of w at $a = 1$, some have criticized the validity of this approach (Nesseris et al. 2025).

Now, let us discuss another possibility: the need for Λ seems to arise from the assumption of the FLRW metric, which in turn assumes large-scale homogeneity. While this is often motivated by the "Copernican principle", which states that we do not live in a special place in the Universe, there is a danger that we have replaced the dogma of "the Earth is in the center of the Cosmos" to the other extreme, but maybe equally dogmatic notion that "nowhere is special". Clearly the latter is not exactly true—we know that almost all the mass in the solar system is in the Sun, not equally distributed among all the planets. Likewise, stellar density in a globular cluster is different from that of a diffused cluster. At the very large scale, superclusters of galaxies are connected via web-like structures, separated by regions of space that are devoid of stars, called voids. The Universe is only homogeneous once we take some kind of statistical average. (See also Section 5.4.) What if we are located inside a relatively under-dense region? Then nearby galaxies would be gravitationally attracted to the denser region outside, and this motion on top of the overall expansion would look like an acceleration to us.

Such a possibility, with many variants, have been considered in the literature, the simplest is the Lemaître–Tolman–Bondi (LTB) metric, which takes the form

$$ds^2 = -dt^2 + X^2(r, t)dr^2 + A^2(r, t)(d\theta^2 + \sin^2\theta d\phi^2), \tag{6.14}$$

where the functions A and X can vary not only in time, but in space. Another lesser known model is Stephanie cosmology, which describes a universe filled with a

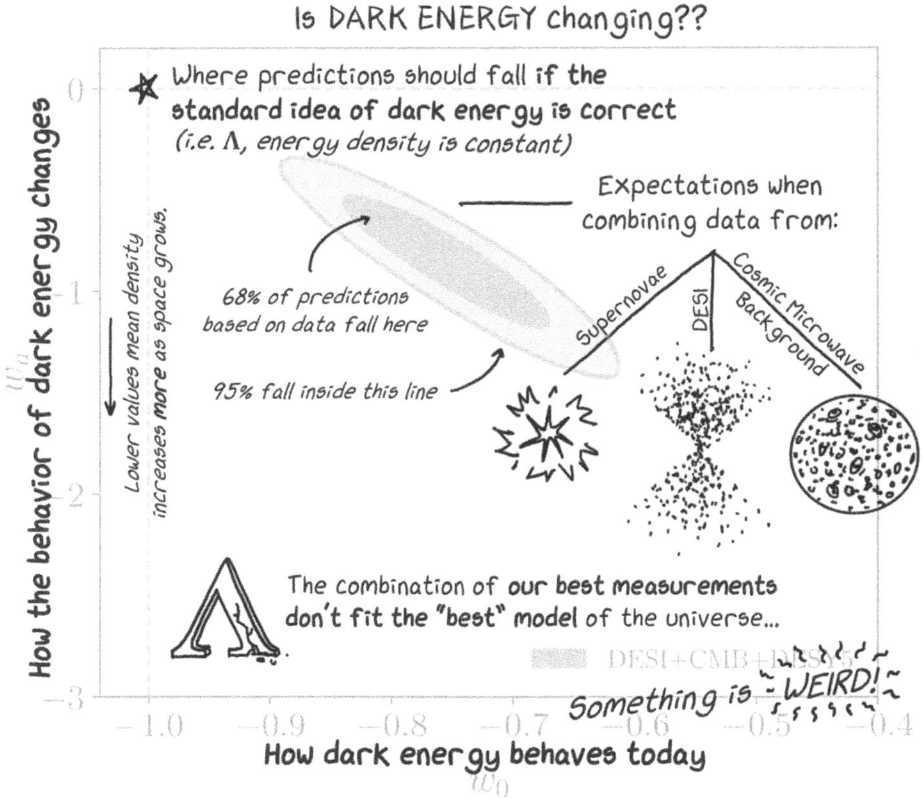

Figure 6.4. The green contour plot shows the constraints on parameters of an evolving dark energy, based on the combined DESI+CMB+DESY5 data. DESI is based on galaxy survey, while DESY5 is based on supernovae data. The horizontal axis is w_0 and the vertical axis is w_a. If dark energy is the cosmological constant, we should have $w_0 = -1$, $w_a = 0$. Yet the best fit contour does not include this point. This suggests that dark energy is evolving in time, and hence not a cosmological constant. Given that this is a very recent finding at the time of writing, whether this is really the case awaits future research. *Credit:* Claire Lamman / DESI Survey

homogeneous energy density but supporting an inhomogeneous pressure. A Stephani cosmological model is defined by a metric of the form

$$ds^2 = -\frac{dt^2}{V^2(t, r)} + \frac{a^2(t)}{V^2(t, r)}[dr^2 + r^2(d\theta^2 + \sin^2\theta d\phi^2)], \tag{6.15}$$

where $V(r, t) = 1 - \beta a(t)r^2$ and $k(t) = -4\beta a(t)$, in which β is a free parameter. A flat FLRW cosmology is recovered in the limit $\beta \to 0$. In a Stephani model, the spatial scalar curvature is a function of time,[5] $k = k(t)$. In fact it can change sign throughout the cosmic history.

[5] This is why normalizing k to be either -1, 0 or 1 in the standard FLRW geometry—as we discussed in Section 5.4—can give rise to confusions, as this could give the wrong impression that k cannot vary smoothly and has to "jump" to other discrete values, which is of course not correct. Having said that, there is still a topology change when the sign changes. Whether this is "drastic" is a matter of personal taste.

However, the general consensus is that inhomogeneous models alone cannot explain all the observational data, and some kind of dark energy is still preferred. Like dark matter, however, "dark energy" is just a placeholder for an unexplained phenomenon. It is likewise possible that a different theory of gravity can explain the accelerated expansion without the need for dark energy. In the next section, we will explain what it means to modify gravity, and the various difficulties that come from doing so. Before ending this section, it is worth noting that the dark sector could be a lot more complicated than expected. In the ordinary matter sector, a mere 5% of the universe mass–energy content, there is already so much interesting physics; is it too presumptuous of us to think that the rest, the 95% or so, "merely" consists of a few types of rather boring dark matter and dark energy? There could be many types of dark matter and dark energy, with nontrivial interactions between them. Figure 6.5 summarizes the energy-matter budget of the Universe.

Indeed, some interactions between the dark sectors have been proposed, for example, models that allow dark energy to be converted into dark matter and vice versa. Such models can also potentially explain the so-called "coincidence problem" or the "why now" problem (see Box 4 below for more discussions): why do we live in

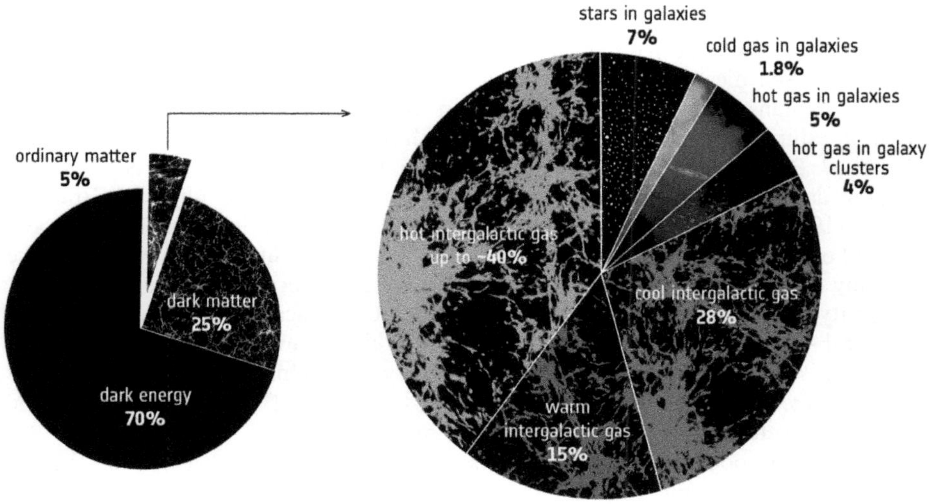

Figure 6.5. The energy and mass contents of the Universe today. About 70% of the cosmos is thought to be dark energy, 25% dark matter, and 5% normal matter. Stars in galaxies across the Universe only make up about 7% of all ordinary matter. Not shown here are particles like photons and neutrinos, which make up about 0.01% and 0.1% of the normal matter, respectively. The percentages evolve over time. When stable atoms first formed some 380,000 years ago, photons made up 15% of the entire mass–energy budget of the Universe, while neutrinos make up about 10%. Dark matter was about 63% and the other baryonic matter about 12%. The percentage of dark energy was negligible back then. This is simply due to dark energy being constant over time, while radiation decays like $1/a^4$ as opposed to non-relativistic matter that decays like $1/a^3$, where $a = a(t)$ is the scale factor of our Universe, modeled by the FLRW geometry. Neutrinos are relativistic and therefore behave effectively like radiation. *Credit:* European Space Agency (ESA).

the era that dark matter density is of the same order of magnitude as the dark energy? The interacting dark energy model can be modeled by coupling the fluid equation for the dark energy and dark matter together:

$$\begin{cases} \dot{\rho}_{\rm DE} + 3H(\rho_{\rm DE} + p_{\rm DE}) = -\gamma\rho_{\rm DE}\rho_{\rm DM}, \\ \dot{\rho}_{\rm DM} + 3H(\rho_{\rm DM} + p_{\rm DM}) = \gamma\rho_{\rm DE}\rho_{\rm DM}. \end{cases} \tag{6.16}$$

If $\gamma > 0$, we have dark energy converting into dark matter. Likewise, dark matter can be converted into dark energy with $\gamma < 0$. This is similar to the form of the Lotka–Volterra equations, Equation (1.20), in Section 1.5. In fact, it can be written in the same form if H is a slow-varying function (and hence treated as a constant) (Aydiner 2018; Cruz et al. 2019; Ong 2023). Other form of interaction terms are possible.

Box 4: From Coincidences to The Doomsday Argument

It is not clear if the "coincidence problem" regarding the densities of dark matter and dark energy is really a problem. Sometimes coincidences do happen. For example, we are living in a privileged time on Earth for which a total solar eclipse is possible. Earth is also the only place in the solar system from which this phenomenon can be observed. The Moon is moving further away from Earth at a rate of about 3.8 cm per year, so eventually it won't be large enough to cover the Sun "just right" during a solar eclipse. The last totality is expected to occur some 1.2 billion years from now. This is still way before the Sun dies, which would occur about 5 billion years from now. The first modern humans appeared some 300,000 years ago. So, why do *we* find ourselves living in the age during which totality is possible, which consists of just about 1/5 of the age of the Sun? Perhaps humanity does not survive that long? This is related to the "Doomsday argument" in philosophy, first proposed by the physicist Brandon Carter (1942–present) in 1983. Essentially, if humanity lasts a long time and so there are trillions of people that will eventually exist, why do we find ourselves born in the first tiny fraction of all people? If we are to be a typical observer, this would imply that humanity cannot survive for very long. This is the Copernican principle applied in the temporal sense, instead of spatially. How valid is this argument is debatable, since the Universe is *not* "homogeneous in time"; the very early Universe is not the same as the late stage Universe. Also, what do we mean by a "typical" observer? Is there some kind of measure for the probability to make sense?

Having no good answer for these questions, I shall end with the quote by the mathematician and philosopher Blaise Pascal (1623–1662) from his 1670 work *Pensées*:

> "When I consider the short duration of my life, swallowed up in the eternity before and after, the little space which I fill, and even can see, engulfed in the infinite immensity of spaces of which I am ignorant, and which know me not, I am frightened, and am astonished at being here rather than there; for there is no reason why here rather than there, why now rather than then. Who has put me here? By whose order and direction have this place and time been allotted to me? *Memoria hospitis unius diei prætereuntis*. ["the remembrance of a guest of one day that passeth by"]"

6.3 Modified Gravity

Laymen (including many "crackpots"), somehow have the impression that physicists take general relativity as "sacred" and would defend it to their deaths. Nothing can be further from the truth: a quick glance at the preprint arXiv server would show that every day there are discussions on different approaches of "modified gravity", and new proposals are made all the time. Indeed many physicists are not satisfied with general relativity and seek to "go beyond it". Some physicists are bothered by the singularities in the theory, but partly the interests in modified gravity is also based on the hope that we can somehow explain the dark sector without actually appealing to new particles or a new form of fields.

To modify gravity usually means to modify the theory of general relativity. There are many ways one can do this, too many in fact to give a proper review. Still, we can give some representative examples. The Einstein field equations are essentially the "equations of motion" of the theory, and just like we can derive equations of motion from varying the Lagrangian or action in classical mechanics,[6] we can also obtain Einstein field equations by varying the *Hilbert action* (or the Hilbert–Einstein action), given by

$$S = \int d^4x \sqrt{-g} R. \tag{6.17}$$

The factor $\sqrt{-g} \equiv \sqrt{-\det(g_{ab})}$ that involves the determinant of the metric is just the proper term to define a volume form in curved geometry. It is essentially the *Jacobian* students first encountered in multivariable calculus. Essentially, it describes how the coordinate grids have been deformed under a coordinate transformation.

To see how this term arises, consider the change from a coordinate system (x^i) to another $(\tilde{x}^j)$, say the standard Cartesian system to the polar system in two spatial dimensions of the flat plane. The Jacobian is, as one has probably seen in a calculus textbook,

$$J^i_j := \left(\frac{\partial x^i}{\partial \tilde{x}^j}\right) = \begin{pmatrix} \dfrac{\partial x}{\partial r} & \dfrac{\partial x}{\partial \theta} \\ \dfrac{\partial y}{\partial r} & \dfrac{\partial y}{\partial \theta} \end{pmatrix} = \begin{pmatrix} \cos\theta & -r\sin\theta \\ \sin\theta & r\cos\theta \end{pmatrix}, \tag{6.18}$$

since $x = r\cos\theta$ and $y = r\sin\theta$. The metric tensor ds^2 is invariant under coordinate transformation, but in the new coordinates, the components $\tilde{g}_{ij}$ will be given by

$$\tilde{g}\left(\frac{\partial}{\partial \tilde{x}^i}, \frac{\partial}{\partial \tilde{x}^j}\right) = \tilde{g}\left(\frac{\partial x^k}{\partial \tilde{x}^i}\frac{\partial}{\partial x^k}, \frac{\partial x^l}{\partial \tilde{x}^j}\frac{\partial}{\partial x^l}\right) = \frac{\partial x^k}{\partial \tilde{x}^i}\frac{\partial x^l}{\partial \tilde{x}^j} g\left(\frac{\partial}{\partial x^k}, \frac{\partial}{\partial x^l}\right), \tag{6.19}$$

[6] A quick recap: for many classical mechanic systems, its Lagrangian turns out to be $L = T - V$, the difference between kinetic and potential energy. The action is defined as $S = \int L\, dt$.

where we have made use of the linearity of the metric tensor. This equation is just the transformation behavior of the metric tensor under coordinate transformation. However, the last equation can be written in terms of the Jacobian, so that we have

$$\tilde{g}_{ij} = g_{kl} J_i^k J_j^l. \tag{6.20}$$

Viewing these as the components of matrices, and using the property of the determinant that $\det(AB) = \det(A)\det(B)$, we have

$$\det(\tilde{g}_{ij}) = (\det J)^2 \det(g_{ij}). \tag{6.21}$$

Since the original metric (g_{ij}) is just the Cartesian one, its matrix determinant is 1. Thus we have the determinant of the Jacobian being the square root of the determinant of the metric in the new coordinates:

$$\det J = \sqrt{\det(\tilde{g}_{ij})}. \tag{6.22}$$

In the Lorentzian case, because the determinant is negative, an extra minus sign is required to make the determinant a real function. Things are of course more complicated for curved manifolds (as we need to first define what integration on a manifold means), but the essential idea is the same.

Note the simplicity of the Hilbert action. The scalar curvature R is the natural scalar quantity associated with the Riemann curvature tensor, and how beautiful that just R, and not other quantities, is needed to obtain the Einstein field equation from the variational principle.

The first method to modify gravity is simply to consider a different action. For example, instead of R, we can consider an arbitrary function

$$S = \int d^4x \sqrt{-g} f(R). \tag{6.23}$$

This is known as $f(R)$ gravity. Of course R has a dimension of $[\text{length}]^{-2}$, so we need extra care interpreting the function $f(R)$. We can write in terms of power series expansion

$$f(R) = R(1 + c_1 R + c_2 R^2 + \cdots), \tag{6.24}$$

where the c_n's are dimensionful, or we can also write $f(R) = RF(R/R_*)$ for some dimensionless function F, where R_* is some scale that determines when the modification becomes important. Because of the powers in R, we expect that such a modification would only be important when curvature is large enough. It is important that whatever modification we make, the resulting theory essentially reduces to general relativity in the regime that the latter has been well-tested; just like general relativity can reduce to Newtonian gravity.

However, due to the higher curvature terms in $f(R)$ gravity the equations of motion involve fourth order differential equations, which are hard to deal with. This is expected due to a result known as *Lovelock's theorem*, proved by David Lovelock

(1938–present) in 1971, which says that general relativity is unique provided that a few assumptions are made:

(1) We still work in the same geometric framework of Riemannian geometry;[7]
(2) Spacetime has four dimensions;
(3) Whatever tensor combinations that replace the Einstein tensor on the left-hand side, say D_{ab}, such that $D_{ab} = T_{ab}$, must satisfy $\nabla^a D_{ab} = 0$;
(4) The field equations only contain the derivatives of the metric up to second order.

Specifically, Lovelock theorem says that if these requirements are made, then the tensor D_{ab} *must* be of the form

$$D_{ab} = AG_{ab} + Bg_{ab}, \tag{6.25}$$

where A and B are constants, and G_{ab} the Einstein tensors. The constants can be fixed by considering suitable physical systems and requiring that its weak field non-relativistic limit recovers Newtonian results. This uniquely fixes general relativity.

Hence it is not a surprise that in $f(R)$ gravity, assumption (4) is violated—in order to modify gravity, we must give up at least one of these premises of the theorem. While assumption (4) is not strictly speaking a necessary property a gravity theory must have, it is a nice property. Higher order derivatives are hard to deal with, not only because solving them is mathematically challenging, but also in terms of initial value problems, if we want to know how a physical system evolves in time, we need to know what the initial conditions are. For a second-order equation of the metric, we only need to know the metric at initial time, as well as the value of its first derivative. For an kth order equation, however, we would need to know up to the $(k-1)$th order derivatives.[8] This is not only mathematically complicated but physically challenging (this would be like dealing with a strange mechanics problem in which we need to measure not just position, velocity, but also acceleration, jerk, and all the higher order quantities). In addition, higher order field equations usually lead to instability. This is also well-known from classical mechanics, known as the *Ostrogradsky's theorem* proved in 1850 by Mikhail Vasilevich Ostrogradsky (1801–1862). The theorem does not imply that all solutions of higher-derivative theories are unstable, but it would be best to avoid the possibility of such an issue from ever arising if possible. This is the reason for imposing assumption (4).

Instead of considering higher orders in the scalar curvature R, one can also consider constructing other scalar quantities from the Riemann curvature tensor.

[7] Technically it is a Lorentzian geometry, since the temporal dimension implies a causal structure, which is absent from ordinary Riemannian geometry. Mathematicians usually reserve the term "Riemannian" for geometry that deals with spatial dimensions only.

[8] The existence and uniqueness of solutions involve the Cauchy–Kovalevskaya theorem, proved by Sofya Kovalevskaya (1850–1891). A special case was proven by Augustin Cauchy (1789–1857).

For example, we can replace R in the Hilbert action with the specific combination called the *Gauss–Bonnet term*

$$R^2 - 4R^{ab}R_{ab} + R^{abcd}R_{abcd}. \tag{6.26}$$

The resulting theory of gravity is called Gauss–Bonnet gravity, and its action is given by

$$S = \int d^Dx\sqrt{-g}\left[R + \frac{\alpha}{D-4}(R^2 - 4R^{ab}R_{ab} + R^{abcd}R_{abcd})\right], \tag{6.27}$$

where α is a constant.

This theory is only good for higher dimensions ($D \geqslant 5$), because in four dimensions the Gauss–Bonnet part does not contribute since it yields a topological term,[9] and thus the theory reduces to general relativity, as we expect from Lovelock's theorem. However, one can still introduce the Gauss–Bonnet term in a nontrivial way in four dimensions by introducing another scalar field ϕ that couples to gravity via some coupling function f:

$$S = \int d^4x\sqrt{-g}[R + f(\phi)(R^2 - 4R^{ab}R_{ab} + R^{abcd}R_{abcd})]. \tag{6.28}$$

Since the theory involves not only the metric but a scalar field, Lovelock's theorem does not apply. This does not mean that the theory is completely viable since any modified theory of gravity needs to pass numerous observational tests, as well as be mathematically consistent with no pathological properties.

Another possibility is to consider a completely different geometry. In general relativity, as in Riemannian geometry, there is only curvature. Technically speaking, curvature is a property *not* of spacetime itself, but of a structure called *connection* imposed on it. A connection essentially tells us how to compare vectors at different points of a curved spacetime, and leads to the concept of covariant derivatives ∇^a that we have seen in the last chapter. In general relativity the connection is called *Levi–Civita connection*. It has the special property that it is *torsion-free*. For our purpose, there is no need to understand what torsion is in detail; it suffices to know that it is a geometric quantity distinct from curvature. For a general connection, both curvature and torsion can be nonzero. In fact, the fact that general relativity works so well without torsion is rather surprising—why should Nature choose among all the generic connections, one that has no torsion[10]? In an extension of general relativity called Einstein–Cartan theory, or the Einstein–Cartan–Sciama–Kibble theory, torsion is coupled to the spin of matter. This effect only becomes noticeable when the spin matter density is sufficiently large, say right after the

[9] For a similar reason, general relativity is trivial in two dimensions, since the Hilbert action becomes a topological term. In the case that we have a two-dimensional compact surface Σ (with both dimensions being spatial), the Gauss–Bonnet theorem says that integrating the scalar curvature over the surface gives essentially the Euler characteristics of the surface up to a constant: $\int_\Sigma d^2x \; R = 4\pi\chi(\Sigma)$.

[10] Could it be that torsion is just too small to have a noticeable effect? Could the smallness of torsion be related to the smallness of the cosmological constant?

Big Bang, otherwise the theory is completely indistinguishable from general relativity. Thus, such a theory cannot explain dark energy or dark matter. It is also worth noting that torsion does not propagate in the Einstein–Cartan theory, its effect is completely localized inside the matter source. However, perhaps other torsion-based theories can?

Surprisingly, it is also possible to construct a gravity theory that utilizes only torsion, where curvature identically vanishes. This is the so-called "teleparallel" class of theories. Surprisingly, one can construct a "teleparallel equivalent of general relativity" (TEGR). As the name implies, it is equivalent to general relativity at the action level. This is essentially because a "torsion scalar" can be constructed in this theory that differs from the scalar curvature of general relativity by a total divergence term, which under the integral of the action, gives no contribution to the field equations. The theory is, however, quite unnatural because test particles do not follow geodesics defined from the teleparallel connection (called the *Weitzenböck connection*), so TEGR is not as geometrically elegant as general relativity. One could go a step further by considering $f(T)$ theory. Note that although TEGR is equivalent to general relativity, $f(T)$ is not equivalent to $f(R)$ due to the aforementioned total divergence term now contained inside $f(T)$ that cannot be integrated out. It is easier to handle $f(T)$ theory in some sense because the field equations are only second order. However, the structure of the theory is a lot more difficult to understand than $f(R)$; even the question of how many degrees of freedom the theory has is still debatable after a decade since the theory's conception. Thus one has traded a difficulty with more tricky ones. In addition to curvature and torsion, there is also another quantity called *non-metricity*, which is given by a nonzero $Q_{cab} \equiv \nabla_c g_{ab}$. In such a geometry, the length of a vector can change as we parallel translate it along a curve. Again, it is possible to construct a special class of non-metricity theory such that it is equivalent to general relativity; and then also to extend it to a modified gravity called $f(Q)$ gravity. The status of the theory remains unclear. Much like $f(T)$, there are debates concerning the number of degrees of freedom in $f(Q)$.

Another possibility is to consider adding a mass to graviton. A graviton is a hypothetical force particle that mediates gravitational interaction, much like a photon mediates electromagnetic interaction. No graviton has yet been detected, but it is a natural description from a field theory point of view. In the standard model of particle physics, neutrinos are massless. However we know that this cannot be true for all neutrino species, otherwise we cannot explain the fact that they "oscillate" (see Section 8.4.3): as they move through space, an electron neutrino can change into a muon neutrino or a tau neutrino. In fact, these three types of neutrinos can change into each other with some probability. The graviton is massless because the "speed of gravity" is the same as the speed of light. This means that if the Sun suddenly disappears, the orbit of Earth would only react to its absence some 8 minutes later, just when visually we see that the Sun disappeared (the last ray of light emitted 8 minutes ago having just reached us at the same time). But what if gravitons, like some neutrinos, actually have a tiny mass? Such a theory is known as "massive gravity",[11] which has a long history. The first attempt was the Fierz–Pauli theory in 1939. In 1970, it was shown that the massless limit of the theory does not recover general relativity, which is now called the van Dam–Veltman–Zakharov (vDVZ) discontinuity. In fact, the light-bending effect would be off by a whopping 25%. The so-called *Vainshtein mechanism*, proposed two years afterwards, can force the theory to recover the correct limit that matches linearized general relativity. However, in the same year, David Boulware and Stanley Deser showed that the non-linearity introduced by Vainshtein mechanism excites a new, sixth degree of freedom, which is a "ghost mode" that can cause instabilities in the theory. A ghost-free version can be constructed by introducing even more non-linearities. This is known as the dRGT massive gravity theory, short for the authors Claudia de Rham, Gregory Gabadadze, and Andrew Tolley. Yet the theory is not without complications; not only is it plagued by some propagations that are superluminal, but also instabilities in cosmological solutions.

Indeed, for a modified theory of gravity, in principle we need to check two things: (1) it does not contradict any observation, and (2) its mathematical structure is healthy. The first requirement of course should be updated as and when new observational data becomes available. Thus a theory that is viable today may cease to be tomorrow. In fact, when gravitational wave was finally detected, it allows us to put a tight bound on the speed of gravitational wave, which ruled out a good number of modified gravity theories. The second requirement is more difficult. It is easy to modify general relativity, but to do so in the "right manner" without giving rise to problems is rather difficult. Actually, as we will discuss in greater detail in the next section, modified gravity is not a new idea—the Hall–Newcomb theory in the 1890s, named after Asaph Hall[12] (1829–1907) and Simon Newcomb (1835–1909), proposed to modify Newton's inverse square law into the phenomenological law

$$F = -\frac{GMm}{r^{2.0000001612}}, \tag{6.29}$$

in order to explain the precession of the Mercury perihelion. However, this would mess up the orbit of other planetary bodies which were already consistent with data. Indeed, de Sitter showed in 1913 that this force law is inconsistent with the motion of the Moon's perigee (Turner 1913; Vaynberg 2024). So it is with modified gravity in general—new unexpected problems tend to appear somewhere.

A rather common problem that plagues various modified gravity theories—including $f(T)$ and $f(Q)$ gravity, as well as Gauss–Bonnet type gravity—is that they are not predictive. For a healthy theory, given an initial condition, we want to be able to compute what happens in the future. Mathematically, we say that the Cauchy problem should be well-posed. This is true in general relativity, though it does not mean such calculations are easy! It took a lot of efforts from the numerical relativity community to find suitable ways to perform calculations that can model

[11] Due to the graviton being a spin-2 particle, a massive version would have $2s + 1 = 5$ degrees of freedom.
[12] He is best known for having discovered the two moons of Mars, Deimos and Phobos, in 1877.

black hole collisions and mergers, and also for mathematicians to prove the well-posedness. The point is that if a theory is ill-posed, then it is quite useless for physics. This loss of predictability can happen if the equations of motion cannot be *uniquely* solved. Often this may be accompanied by superluminal propagation in the theory, or even the appearance of local closed causal curves in some examples.

The modern point of view is that gravity theories like general relativity and these modified theories of gravity are only "effective theories", which means that they are to be replaced by other theories when energy scale or curvature is large enough. Therefore, problems that plague the effective theory may no longer arise in the full theory. For example, Gauss–Bonnet gravity suffers from acausal pathologies unless the coupling constant is sufficiently small. But Gauss–Bonnet modification is only a low energy leading correction, if embedded into string theory, such that towers of higher spin particles are included, then the pathologies no longer arise (Camanho et al. 2016). But then this defeats the purpose, I think; the whole point is to look for a healthy effective theory so that one could do sensible physics without all the nitty-gritty of the more complicated full theory. If the effective theory is pathological, we have to be careful about its predictions.

6.4 Some Lessons From History

Let us start with the precession of the Mercury perihelion. Kepler's second law, which is a consequence of the inverse square law of gravity, says that planetary orbits are elliptical. But if we measure the orbit of the Mercury closely, we will notice that every turn of the orbit does not close up to a nice ellipse, instead the ellipse "moves" about. Equivalently, the closest point of the orbit to the Sun—the perihelion—does not stay fix but precesses in a peculiar manner. In total, there are 565 seconds of arc per Earth century. Taking into account the effects of other planets on the orbit of Mercury, most of these can be explained. However, there remained 43 arcsecond per century that cannot be accounted for. One possible explanation was proposed: there could be another planet lurking closer to the Sun than Mercury, and such an "intra-Mercurial object" would provide extra perturbations on the orbit of Mercury.

Such an idea is quite reasonable. In fact, the planet Neptune was discovered in much the same way. Astronomers realized that the orbit of Uranus did not quite fit the prediction of Newton's theory, and they conjectured that a new planet exists further out that is responsible for the orbit perturbation. A careful mathematical analysis allowed them to deduce the rough position of this putative planet. A thorough telescopic search then found the predicted planet Neptune not far from the predicted patch of sky. Because of this, Neptune can be considered as the first planet whose existence was first predicted from mathematical calculations. With such a success, it seems likely that an intra-Mercurial planet is responsible for the anomalous precession of the perihelion of Mercury. This planet was given the name "Vulcan" by the French astronomer Urbain Le Verrier (1811–1877), one of the astronomers who calculated the position of Neptune prior to its discovery (the other one was the British astronomer John Couch Adams (1819–1892); not unexpectedly a dispute erupted concerning who was the first to do so[13]). Despite some claims of observation, no concrete convincing evidence was ever obtained for its existence. In addition to a hypothetical planet, Newcomb proposed in 1895 that the extra gravitational perturbation may originate from a ring of fine material—instead of a single body—either situated in between the Sun and Mercury, or between Mercury and Venus. The zodiacal light, which is a faint glow of light caused by sunlight scattered by interplanetary dust, was thought to be the evidence for such a material. This proposal was found to be not viable either (Turner 1913).

Then a new theory of gravitation appeared—general relativity. Because the Sun is massive, spacetime curvature around it is strong enough that we need to consider general relativistic correction on the orbit of Mercury. A nontrivial calculation shows that it gives the extra 43 arcsecond per century value required to explain the precession of perihelion.[14] Thus in this case, there is no new planet, instead a new theory is required to solve the mystery. Some readers may have heard that there was another theory before general relativity that gave the same perihelion shift—a theory by Paul Gerber (1854–1909). This is true. However, he reached the right result via incorrect reasoning, i.e., the formula doesn't follow from his assumed premises.

Returning to the issue of dark matter and dark energy, are they like Neptune or Vulcan? We simply don't know just yet. The dark matter search has been ongoing for decades with no success. This does not mean it does not exist, after all there are so many possible kinds of dark matter, and most of them require a different experiment dedicated to its detection. In the decades of non-detection, for each model, we can rule out some parameter space, but it is difficult to say for sure if a certain candidate is *definitely* ruled out. On the modified gravity side, there are also many different theories to check, and so far none of them seem to be promising in solving the dark matter or dark energy issue without facing some kind of criticism themselves. In addition, modified gravity needs to be able to fit the CMB data (as well as other constraints) as well as dark matter. Science needs to progress by allowing different people to pursue different possibilities, and hopefully one of them will work. However, we should keep Hermann Weyl's quote in Chapter 1 in mind, and avoid "cheap and easy generalizing for its own sake". It is easy to consider some kind of new theory and even combinations of different theories and calculate something to publish. This is even understandable in today's "publish or perish" culture worsened by the academic job market and universities' obsession for better rankings, but I honestly wish that at the very least students set their target higher at the beginning of their physics career; we only have a finite amount of time and energy, it would be wise to focus our efforts on investigating physics questions that are well-motivated with some clear purpose in mind.

Incidentally, we have mentioned that a simple generalization like the Newcomb–Hall theory didn't work to explain the anomalous precession of the Mercury

[13] Especially given the historical relations between France and Britain!

[14] The calculation can be found in most general relativity textbooks and will not be covered here.

perihelion. Most modified gravity theories today that try to extend general relativity are analogous to Newcomb–Hall type modification to Newtonian gravity. Would it be possible that we need an entire different kind of theory, a new paradigm change that completely goes beyond the language of general relativity (but still reduces to general relativity in a certain limit)? Maybe. However, if this is indeed the case, such a theory is still out of our reach. (Note that such a theory need not necessarily be a theory of quantum gravity.)

One of the themes of this book is to appreciate that the ways humans think have always been similar,[15] the fact that modified gravity was not so new is one such piece of evidence. In fact, recall that most physicists are not comfortable with the smallness of the cosmological constant, and so many other alternative proposals have been made:

(1) The void model, or the "Swiss-Cheese" model, in which we are located in an underdense region of the Universe.
(2) Modified gravity.
(3) Some other fields like quintessence, phantom, quintom, K-essence, etc.

In addition, it has been argued that vacuum energy does not gravitate, i.e., the large vacuum energy density predicted by quantum field theory does not contribute to the energy density of the universe. If so it would be a mistake to equate vacuum energy with a cosmological constant. In fact ideas similar to these have been proposed in another context a long time ago: Newtonian cosmology.

As discussed in Section 3.2, Newtonian gravity is plagued by a serious problem when applied to an infinite cosmos uniformly filled with matter, namely that some physical quantities diverge. For example, consider the mass density ρ, which is mass divided by volume, so $\int \rho r dr \sim \int M/r^2 dr$, which gives the potential $\Phi \sim M/r$. However, the integral $\int \rho r dr$ is divergent if ρ is constant and the Universe is infinite. To avoid infinitely large quantities, Föppl argued that perhaps some matter possesses negative masses, which can regulate the integrals to be finite. Even in modern cosmology, negative masses have been considered to resolve the dark matter and dark energy puzzle in one way or another. Then we have the idea of Charlier and Selety that the matter distribution is not uniform but "hierachical" in a way that renders the aforementioned integral finite. This idea is analogous to the void model. In a different but similar problem, Lord Kelvin had proposed that the æether that fills our Universe simply does not gravitate. This is very similar to the idea that vacuum energy may indeed permeate space, but does not gravitate. Of course, modified Newtonian gravity was also considered back then. Let us take a closer look at the last option.

---

[15] In cosmology, we also note that ancient people had various creation myths about how the world came to be. In fact, both the possibility of an eternal universe and a universe with a beginning had been considered. The cyclic universe scenario is also rather natural since Nature already has so many cycles, so it is not surprising to encounter beliefs that involve the destruction and rebirth of the cosmos; e.g., in Hindu and Mayan worldviews.

We first note that the Newcomb–Hall type modification $1/r^{2+\varepsilon}$ would not work, since in that case the force integral[16] becomes (Norton 1999)

$$\int \rho dr \longrightarrow \int \frac{\rho}{r^{\varepsilon}} dr, \tag{6.30}$$

which still diverges in the limit $r \to \infty$ if ρ is constant (since $\varepsilon \ll 1$). The divergence can be avoided if we modify the force so that it decays much faster at large r. What better way to do this than to multiply the Newton inverse square force by an exponentially decaying factor? This was precisely what Hugo von Seeliger (1849–1924) proposed in 1895:

$$F = -\frac{GMm}{r^2} \longrightarrow -\frac{GMm}{r^2} e^{-\lambda r}, \tag{6.31}$$

where $\lambda > 0$ is a constant with dimension $[\text{length}]^{-1}$. Sometimes ideas in theoretical physics are not so complicated!

Another modification was proposed a year later by Seeliger together with Neumann. In this scenario, the *potential*, not the force per se, that was multiplied with an exponentially decaying factor (with a different λ):

$$\Phi(r) = -\frac{GM}{r} e^{-\sqrt{\lambda} r}. \tag{6.32}$$

This is of the form now known as the *Yukawa potential*, since it was later proposed in the context of particle physics in 1935 by Hideki Yukawa (1907–1981). More specifically, the Yukawa potential is

$$V_{\text{Yukawa}}(r) = -g^2 \frac{e^{-\alpha m r}}{r}, \tag{6.33}$$

where m is the particle mass, and $1/\alpha m$ sets the length scale of the range of the associated force. Interpreted in a similar way, the Neumann–Seeliger gravity amounts to an early example of massive gravity, where λ is related to the graviton mass.

Newtonian gravity can be expressed in terms of the Poisson equation as

$$\nabla^2 \Phi = 4\pi G \rho. \tag{6.34}$$

It is not difficult to show that for the Neumann–Seeliger gravity, the Poisson equation has an additional term

$$\nabla^2 \Phi - \lambda \Phi = 4\pi G \rho. \tag{6.35}$$

Historically, Einstein mistakenly claimed that the Newtonian limit of general relativity with a cosmological constant term introduces an exponential cutoff.

[16] Since $\Phi \sim \int \rho r \, dr$, and force is the derivative of the potential, we have $F \sim \rho r$, and so for ρ constant, $F \sim \int \rho \, dr$.

This led to a lot of confusions in the literature Pitts (2020). This is not true, as said limit actually yields

$$\nabla^2\Phi + \Lambda = 4\pi G\rho. \tag{6.36}$$

Instead of just *less* gravitational attraction at large r, there is a *repulsion*! This is actually not surprising since Einstein introduced Λ to make his cosmological model static; the repulsion is used to balance gravitational attraction. This is clear when we write down the potential and force expressions:

$$\Phi = -\frac{GM}{r} - \frac{\Lambda r^2}{6}; \quad F = -\frac{GMm}{r^2} + \frac{\Lambda mr}{3}. \tag{6.37}$$

Incidentally, note that this is a sum of an inverse square law and a linear law. This is the only combination that satisfies the converse of the famous (exterior) Newton's shell theorem. The shell theorem says that a spherically symmetric body of mass M has a gravitational field identical to that of a point particle of the same mass located at its center. The converse asks: given that a force F between two masses M and m separated by a distance r is of the form $F = Mmf(r)$, such that a spherically symmetric body affects external objects gravitationally as though it is a point mass concentrated at its center, what can we say about the function f? It turns out that it has to be an inverse square or linear force, or a combination thereof.

Before we end this section, it is worth commenting on another attempt to modify Newtonian gravity. Just like planet Mercury, the Moon's orbit also precesses. More precisely we are interested in the Moon's apsidal motion. This once raised a problem because the observed value was twice the amount predicted by Newtonian gravity, according to the calculations of many experts back then (Bodenmann 2010): including Leonhard Euler (1707–1783), Jean le Rond d'Alembert (1717–1783), and Alexis Clairaut (1713–1765). D'Alembert postulated an additional lunar magnetic force to explain the mismatch. On the other hand, Clairaut proposed in 1747 that Newton's inverse square law should be modified, by including a higher order correction (c.f. the idea of $f(R)$ gravity and the like), into (Klyve 2014)

$$F = -\frac{GMm}{r^2} + \frac{C}{r^4}, \tag{6.38}$$

where C is a constant to be determined. Two years later, although even Euler still thought that Newton's law was in need of modification, Clairaut resolved the problem. He realized that Newton's law is still correct, the problem was that some higher-order terms in the perturbation equation that were previously thought to be unimportant are actually essential.

I think this piece of history is interesting for at least three reasons: (1) modified gravity with a higher-order correction was already considered at least as far back as 1747 in astrophysics, (2) higher-order terms can be crucial, and most importantly, (3) Clairaut *changed his mind and gave up his own theory* when he realized his mistake! This is a hallmark of a good scientist.

6.5 Exercises

(1) One of the earliest attempts to explain the flat galaxy rotation curves is the Modified Newtonian dynamics (MOND) theory proposed by Mordehai Milgrom (1946–present) in 1983 (Milgrom 1983). The idea is to replace $F = ma$ by

$$F = ma\mu, \tag{6.39}$$

where $\mu = \mu(a/a_0)$ is a function that depends on the acceleration. It is called an "interpolating function". The characteristic scale a_0 dictates the transition between the Newtonian regime and the MOND regime. When a is large, i.e., near the galactic center, we need $\mu \to 1$ to recover the Newtonian behavior. On the other hand, if we were to reproduce the galaxy rotation curve ($v \to$const. at large r), show that we need $\mu \to a/a_0$ for small a.

(2) The challenge for MOND is to construct a relativistic version of it. Early attempts include the "TeVeS" (Tensor–vector–scalar gravity) theory proposed in 2004 by Jacob Bekenstein (2004). Another theory is simply called "RMOND" (relativistic MOND), which seems to be consistent with CMB physics (Skordis & Zlosnik 2021; Kashfi & Roshan 2022). Learn more about such theories and compare them to dark matter theory like WIMPS. In which aspects do MOND-like theories perform better than traditional dark matter models? In which aspects do they still face major challenges?

(3) If dark energy is just the cosmological constant, show that as it becomes the dominant term at later times, the expansion of the Universe is exponential. (We have done this in Section 5.3, but you can try to be more rigorous. For example, how "late" is late?) How does the Hubble constant relate to the cosmological constant?

(4) As an example of modified gravity based on an extra dimension, consider the Dvali–Gabadadze–Porrati (DGP) theory proposed in 2000. In this model our world is a brane in a five-dimensional Minkowski spacetime (Dvali et al. 2000). Gravity would appear to behave as usual at short distances,[17] but when $r \gg r_c$ it behaves differently. The cross-over scale r_c is a new length scale that is related to the five-dimensional gravitational constant. The four-dimensional Friedmann equation in the flat universe ($k = 0$) receives an extra correction term due to the higher dimensional effect and becomes (Lue et al. 2004)

$$H^2 = \frac{8\pi G\rho}{3} + \frac{H}{r_c}. \tag{6.40}$$

Show that the term H/r_c is equivalent to adding a dark energy whose equation-of-state parameter evolves to $w = -1$ at late time as matter density dilutes. If you are interested, find out more about the current state of this theory.

Hint: If ρ_q represents the energy density of an effective dark energy, we can write

[17] This is not to say that it reduces to general relativity. In fact, it does not, a new ultra-light gravitational scalar would appear, alongside the usual gravitons.

$$H^2 = \frac{8\pi G}{3}(\rho + \rho_q), \tag{6.41}$$

and thus identify

$$\frac{8\pi G\rho_q}{3} = \frac{H}{r_c}. \tag{6.42}$$

References

Akarsu, O., Kumar, S., Ozulker, E., & Vazquez, J. A. 2021, PhRvD, 104, 123512
Aydiner, E. 2018, NatSR, 8, 721
Bekenstein, J. D. 2004, PhRvD, 70, 083509 Erratum 2005, , 71 069901
Bodenmann, S. 2010, PhT, 63, 27
Caldwell, R. R., Kamionkowski, M., & Weinberg, N. N. 2003, PhRvL, 91, 071301
Camanho, X. O., Edelstein, J. D., Maldacena, J., & Zhiboedov, A. 2016, JHEP, 2016, 20
Carroll, S. M. 2001, LRR, 4, 1
Corbelli, E., & Salucci, P. 2000, MNRAS, 311, 441
Cruz, M., Lepe, S., & Morales-Navarrete, G. 2019, NuPhB, 943, 114623
DESI Collaboration 2025, JCAP, 02, 021
DESI Collaboration 2025, [arXiv:2503.14743 [astro-ph.CO]] 2025arXiv250314743L
DESI Collaboration 2025, [arXiv:2503.14738 [astro-ph.CO]] 2025arXiv250314738D
DESI Collaboration 2025, PhRvD, 112, 083514
Dvali, G., Gabadadze, G., & Porrati, M. 2000, PhLB, 485, 208
Einstein, A. 1918, PhyZ, 19, 165
Kashfi, T., & Roshan, M. 2022, JCAP, 10, 029
Klyve, D. 2014, ed. E. Sandifer How Euler Did Even More (Washington, DC: Mathematical Association of America) 199
Lue, A., Scoccimarro, R., & Starkman, G. D. 2004, PhRvD, 69, 124015
Milgrom, M. 1983, ApJ, 270, 365
Miller, A. I. 2024, NatRP, 6, 289
Mistele, T., McGaugh, S., Lelli, F., Schombert, J., & Li, P. 2024, ApJL, 969, L3
Nesseris, S., Akrami, Y., & Starkman, G. D. 2025, [arXiv:2503.22529 [astro-ph.CO]] 2025arXiv250322529N
Norton, J. 1999, The Expanding Worlds of General Relativity: Einstein Studies Vol. 7, (Boston, MA: Birkháuser)
Ong, Y. C. 2023, Univ, 9, 437
Pitts, J. B. 2020, ed. A. S. Blum, R. Lalli, & J. Renn The Renaissance of General Relativity in Context, Einstein Studies Vol. 16 (Cham: Birkháuser)
Skordis, C., & Zlosnik, T. 2021, PhRvL, 127, 161302
Turner, A. B. 1913, JRASC, 7, 276
Vaynberg, E. 2024, Sy, 204, 82
Weyl, H. 2009, Philosophy of Mathematics and Natural Science (Princeton Univ. Press: Princeton, NJ)
Zwicky, F. 1937, ApJ, 86, 217

Chapter 7

Light: Particle, Wave, and Both

> *'Light was once the sight of God. As the gaze of Ra*[1] *spanned space... In the night sky, planets and stars once played host to gods and angels, who in turn passed their gift of light to man."* —Arthur Zajonc, "Catching the Light: The Entwined History of Light and Mind".

Light: ubiquitous yet a surprisingly complicated phenomenon. Questions raised regarding the nature of light has led to many paradigm shifts in physics. We will examine the bending of light in Newtonian gravity, which is only possible if light is a particle with a nonzero mass. General relativity would predict twice the value of the Newtonian result, and since then gravitational lensing has been used in astrophysics and cosmology on a regular basis. We also discuss electromagnetism and special relativity, as well as how the æther worldview struggled and was eventually abandoned. It is particularly important to see how scientists tried to save the idea of the æther. Just like the æther, some of us may be working on completely bogus ideas today without knowing when to give up. Finally, we discuss how emission lines, atomic models and thermodynamics led to the birth of quantum physics. Some well-known stories like the "ultraviolet catastrophe" and the "plum-pudding model" are pointed out as being over-simplified historical fictions.

Imagine that you were an ancient human, holding a torch at night wandering around. As your fire flickered in the dark, you saw two greenish-yellow points of light, emanated from the eyes of a cat.[2] If you were a natural philosopher, how would you go about explaining why cats can see so well in the dark while we cannot?

---

[1] Egyptian Sun god.

[2] Today we know that the "eyeshine" of some animals is due to the reflection from their eye structure called *tapetum lucidum*. This layer of tissue behind the retina reflects some light back through the retina, thus increasing the light available to the photoreceptors, improving night vision.

doi:10.1088/2514-3433/ade963ch7

A natural guess based on this observation would be that cat's eyes emit some kind of light. If this is true for cats, it probably also holds for other animals with eyes, including us. The difference could be that cat's eyes somehow emit more or "better" light than us. Indeed, for a long time the theory of vision is precisely that: visual perception is accomplished by eye beams emitted by the eyes. This is known as the "extramission" theory, as opposed to the eyes only acting as a receiver of light, which is known as the "intromission" theory. Even today, some people believe in the "evil eye", i.e., that a malevolent glare can cast a curse on other people. Thus based on the understanding back then, the eyes being capable of emitting some kind of rays was not such a far-fetched hypothesis (think about it: how would you try to falsify this claim?). The intromission theory eventually won when the structure of the eyes were well-studied and how images formed on the retina became better understood. Among the first to contribute to the understanding of the visual system was Ibn al-Haytham (c. 965–c. 1040), whose name was Latinized in the Western world as Alhazen. He wrote the *Kitāb al-Manāzir*, or the "Book of Optics", which described the theory of vision and the camera obscura. An illustration from his work is shown below in Figure 7.1.

After the theory of vision, one still has to figure out what light actually *is*. This simple question turned out to lead to an epic journey, which revolutionized our worldview twice over by the conception of special relativity and quantum mechanics. Albert Einstein, for example, imagined chasing after a beam of light in the vacuum of space when he was only 16 years old. This question remained with him and subsequently led to his discovery of the theory of special relativity. In this chapter we will discuss the debate that started all this: whether light is a wave or a particle. We then discuss how the wave theory was seemingly verified by the realization that light is an electromagnetic wave in accordance to Maxwell's theory of electromagnetism. This, in turn, gradually led to the question of the nature of the luminiferous æther, the hypothetical medium that light waves propagate in. It should be emphasized, however, that the luminiferous æther was *not* incompatible with Maxwell's theory, as some students seem to have misunderstood. Eventually special relativity removed the need for such an æther. Finally we discuss how light has both the properties of wave and of particle, a remarkable fact that led to modern quantum mechanics. There are many interesting ideas to be learned throughout this convoluted journey, which in my opinion is one of the greatest stories in the history of science.

7.1 The Nature of Light

Light and darkness, day and night. These dichotomies are familiar to our ancestors. Despite light being everywhere, what exactly *is* light turned out to be a much more difficult question to answer. When the mechanical view of the cosmos became popular, it was natural to think that light beams consist of tiny particles, the "corpuscles of light". Sir Isaac Newton, well known for splitting white light into the spectrum of a rainbow using a prism, was among those who believed that the constituents of light are corpuscular.

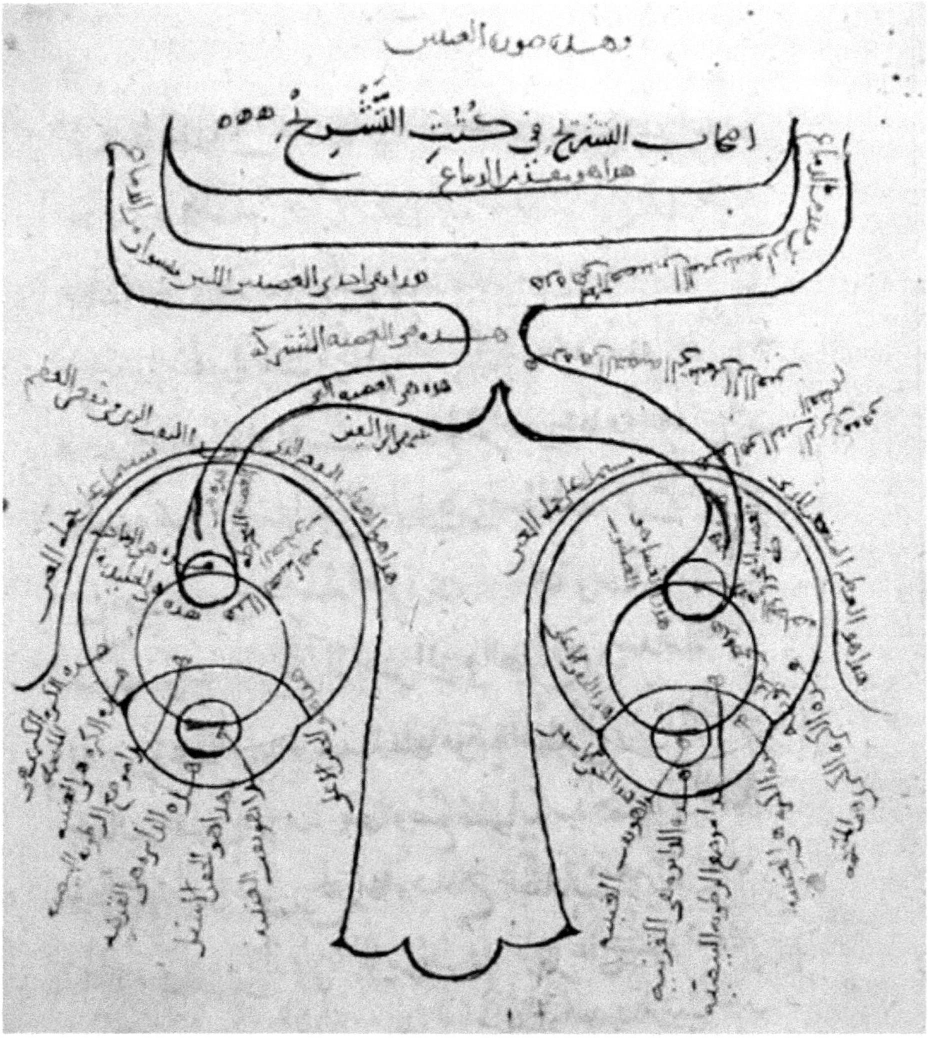

Figure 7.1. Illustration of the visual system by Hasan Ibn al-Haytham's in his "Book of Optics". This image comes from an 11th century copy of al-Haytham's manuscript held in the Süleymaniye Library in Istanbul, Turkey (MS Fatih 3212, vol. 1, fol. 81b, Süleymaniye Mosque Library, Istanbul). *Credit:* This [Alhazen1652] has been obtained by the author(s) from the Wikimedia website, where it is stated to have been released into the public domain. It is included within this article on that basis.

The first person who seriously proposed that light is a wave seems to be Christiaan Huygens (1629–1695) in 1678, though even before that Francesco Maria Grimaldi (1618–1663) observed that light can diffract, which is indicative of a wave property. Grimaldi's work *Physico-Mathesis de Lumine* was posthumously published in 1665. Christiaan Huygens wrote in his work *Traite de la Lumièere* of 1690 that [as quoted by Ernst Mach, "On the Principle of the Conservation of Energy" in Popular Scientific Lectures (1895)]:

> "There can be no doubt that light consists of the motion of a certain substance. For if we examine its production, we find that here on earth it is principally fire and flame which engender it, both of which contain beyond doubt bodies which are in rapid movement, since they dissolve and destroy many other bodies more solid than they: while if we regard its effects, we see that when light is accumulated, say by concave mirrors, it has the property of combustion just as fire has, that is to say, it disunites the parts of bodies, which is assuredly a proof of *motion*, at least in the *true philosophy*, in which the causes of all natural effects are conceived as *mechanical* causes. Which in my judgment must be accomplished or all hope of ever understanding physics is renounced."

Newton eventually became aware of the phenomenon of light diffraction, though he still tried to explain it within the framework of the corpuscular theory. One possibility, as Newton proposed, was that there exists an "æthereal medium" that permeates space, which could transmit vibrations faster than light. As the corpuscles were overtook by these "vibrational modes", they gave rise to what we perceive as refraction or diffraction. Thus, the "æther" was also a useful concept in the corpuscular view of light, not just for the wave theory that became more popular later on, though for a very different reason.

The wave theory became more and more convincing afterward. In 1802, Thomas Young (1773–1829) demonstrated with the now famous double-slit experiment that light passing through two slits give rise to interference fringes. Later on in 1817, from the interference pattern Young was able to deduce that the wavelength of light is no more than 1 micrometer or so. Another physicist who made important contributions to the wave theory of light was Augustin-Jean Fresnel (1788–1827), who by carefully studying diffraction patterns under various experimental settings, devised a mathematical theory for diffraction. In 1819, the French Academy of Sciences organized a competition on the research of diffraction, and so Fresnel submitted his work for the prize. The judges were François Arago (1786–1853), Siméon Poisson (1781–1840), and Pierre-Simon Laplace (1749–1827), all of whom supported the corpuscular theory. One might think that they would have prejudices, but Fresnel's work was too impressive to be ignored. In fact, Poisson predicted—on the basis of Fresnel's theory—that if the wave theory is correct, then parallel light rays that hit a non-transparent disk would nevertheless give rise to a bright spot in the center of the shadow of the disk. Poisson thought that the existence of such a spot would be quite absurd. Yet its existence was experimentally confirmed soon after by François Arago. This is now called the "Poisson spot", and its existence became a strong evidence that light is a wave.

The wave theory became firmly established after James Clerk Maxwell (1831–1879) published his mathematical theory of electromagnetism. The modern version of Maxwell's equations read (with bold-faced quantities being vectorial):

$$\nabla \cdot \mathbf{E} = \frac{\rho}{\varepsilon_0}, \quad \text{(Gauss's Law)} \tag{7.1}$$

$$\nabla \cdot \mathbf{B} = 0, \quad \text{(Gauss's Law for Magnetism)} \tag{7.2}$$

$$\nabla \times \mathbf{E} = -\frac{\partial \mathbf{B}}{\partial t}, \quad \text{(Faraday's Law)} \tag{7.3}$$

$$\nabla \times \mathbf{B} = \mu_0 \left(\mathbf{J} + \varepsilon_0 \frac{\partial \mathbf{E}}{\partial t} \right), \quad \text{(Ampère–Maxwell Law)} \tag{7.4}$$

where **E** and **B** are the electric and magnetic fields, respectively; ρ is the electric charge density, **J** the current density, while ε_0 and μ_0 are the vacuum permittivity and the vacuum permeability, respectively.

The history of electromagnetism is interesting on its own, but we shall not get into too much detail. It is nevertheless worth mentioning that the fact that a conducting wire can cause the needle of a compass to deflect was noted by the Danish chemist and physicist Hans Christian Ørsted (1777–1851) already in 1820. Unlike many of his contemporaries, he had suspected that electricity and magnetism are closely related. The 19th century, the century of electricity, was also the age of *romanticism*, a counter-movement to the Enlightenment of the 18th century. This intellectual movement emphasized individual creativity and also posited that the world around us, as revealed by natural philosophy, is highly connected to ourselves. Thus we should not just seek deductive science, but also need to place emphasis on intuition and embrace a more holistic understanding of the world around us. The Romanticism movement was not anti-science, it just took a different approach to science. Indeed the science of electricity became a source of wonder in the 19th century society (Figure 7.2), from public lectures to demonstrations in salons, and even inspired literary works like *Frankenstein* (1818) by Mary Shelley (1797–1851). Ørsted himself, heavily influenced by the philosophical and aesthetics ideas of the romantic era (Lynning & Jacobsen 2011), was perhaps more open to discover the secret link between electricity and magnetism. (Ørsted was also interested in literature and poetry. He was a close friend of Hans Christian Andersen, the author famous for his fairy tales.)

Before the publication of Maxwell's theory in 1865, there were already hints that light might be related to electromagnetic phenomena. Notably, in 1857, Gustav Kirchhoff (1824–1887) showed that the speed of a signal along an electric wire was equal to the speed of light.[3] A year prior, Wilhelm Eduard Weber (1804–1891) and Rudolf Kohlrausch (1809–1858) measured the numerical value of the ratio of the electrostatic unit of charge to the electromagnetic unit of charge, and found that it is the speed of light. It cannot be overstated how much of a surprise this was—light and electromagnetic phenomena seem so distinct but are related. The ancients were familiar with lightning and magnets, as well as light. To them, all three seemed unrelated. Then, the investigations by Michael Faraday and other physicists gradually led to the realization that magnetism is closely related to electricity.

[3] The speed of light was first measured back in 1676, by the Danish astronomer Ole Rømer. The measurement was based on an ingenious method: timing eclipses of Jupiter's moon Io. See Box 1 in Section 3.3.

Figure 7.2. A painting either by, or in the style of, Paul Lelong (1799–1846), which depicts some condenser jars (also known as Leyden jars, early capacitors that store electricity), an electrostatic generator, and other equipment, as well as a vase with flowers. This still life painting is a testament to the fact that experimentation with electricity played an important part in the life of 19th century intellectuals. Perhaps it also contained the romantic message that the effort to understand Nature is itself at least as beautiful as flowers in a finely designed vase. *Credit:* Electricity: condenser jars, an electro-static generator, and a vase with flowers. Gouache painting. Wellcome Collection. *Source*: https://wellcomecollection.org/works/wm8ne5nu/images?id=kwcnjh2x

Finally, Maxwell unified magnetism with electricity into a theory of electromagnetism. Then, as if by magic, one realized that this theory admits an *electromagnetic wave* whose speed is the speed of light! Unexpectedly, light too can be unified into the picture.

To see this, we set $\rho = 0$ and $\mathbf{J} = 0$ in the Maxwell's equations and consider the "curl of curl identity" for a vector field, applied to the electric field:

$$\nabla \times (\nabla \times \mathbf{E}) = \nabla (\nabla \cdot \mathbf{E}) - \nabla^2 (\mathbf{E}). \tag{7.5}$$

By the source-free Gauss law, the term $\nabla(\nabla \cdot \mathbf{E})$ is zero. On the other hand, by Faraday's law,

$$\nabla \times (\nabla \times \mathbf{E}) = \nabla \times \left(-\frac{\partial \mathbf{B}}{\partial t}\right). \tag{7.6}$$

Because of the symmetry of partial derivatives (Schwarz's theorem), the time derivative can be "factored out" and we get

$$\nabla \times (\nabla \times \mathbf{E}) = -\frac{\partial}{\partial t}(\nabla \times \mathbf{B}). \tag{7.7}$$

The Ampère–Maxwell Law then leads to the *wave equation* for the electric field:

$$\mu_0 \varepsilon_0 \frac{\partial^2 E}{\partial t^2} - \nabla^2 E = 0, \tag{7.8}$$

whose propagation speed is

$$c = \frac{1}{\sqrt{\mu_0 \varepsilon_0}}. \tag{7.9}$$

This is also the same for the magnetic field. The known values of μ_0 and ε_0 then give c, which is close to the known speed of light measured back then.

Maxwell wrote:

> "That light is not itself a substance may be proved from the phenomenon of interference. A beam of light from a single source is divided by certain optical methods into two parts, and these, after traveling by different paths, are made to reunite and fall upon a screen. If either half of the beam is stopped, the other falls on the screen and illuminates it, but if both are allowed to pass, the screen in certain places becomes dark, and thus shows that the two portions of light have destroyed each other. Now, we cannot suppose that two bodies when put together can annihilate each other; therefore light cannot be a substance. [...] What we have proved is that one portion of light can be the exact opposite of another portion [...] Such quantities are the measures, not of substances, but always of processes taking place in a substance. We therefore conclude that light is [...] a process going on in a substance [...] so that when the two portions [of light] are combined no process goes on at all. [...] the light is extinguished when the difference of the length of the paths is an odd multiple of [...] a half wave-length. [...] we see on the screen a set of fringes consisting of dark lines at equal intervals, with bright bands of graduated intensity between them. [...] if the two rays are polarized [...] when the two planes of polarization are parallel the phenomena of interference appear as above [...] As the plane turns [...] light bands become less distinct [...] at right angles [...] illumination of the screen becomes uniform, and no trace of interference can be discovered. [...] The process may, however, be an electromagnetic one [...] the electric displacement and the magnetic disturbance are perpendicular to each other, either [...] supposed to be in the plane of polarization." [From "Ether, or Æther," Encyclopædia Britannica, Ninth Edition (1878)]

The first experimental evidence for the existence of the electromagnetic waves was provided by Heinrich Rudolf Hertz (1857–1894) in late 1886, utilizing two "spark gaps". He found that when a spark was generated in his apparatus, another spark was generated in the spark gap of (what amounts to) an antenna located at the other side of the room, which was not connected to electrical source. An electromagnetic wave has *propagated across* the room. This is the discovery of radio waves that would kick-start a new chapter of technological development in communication. Hertz was also able to, in a subsequent experiment, measure the speed of the wave and showed that it is indeed the speed of light. Nowadays, we can perform astronomical observations at various wavelength of the electromagnetic wave spectrum, each of which can reveal different physics, as shown in Figure 7.3. For example, infrared light can penetrate dust cloud and allows us to see more stars that are otherwise hidden.

The Maxwell equations we know today consist of a set of four differential equations, but this was not the original formulation of Maxwell, which consisted of 20 equations of 20 variables (which would surely horrify today's students who already struggle with the modern formulation!). The simplification was due to Oliver Heaviside (1850–1925), a self-taught physicist. Heaviside once remarked: "Shall I

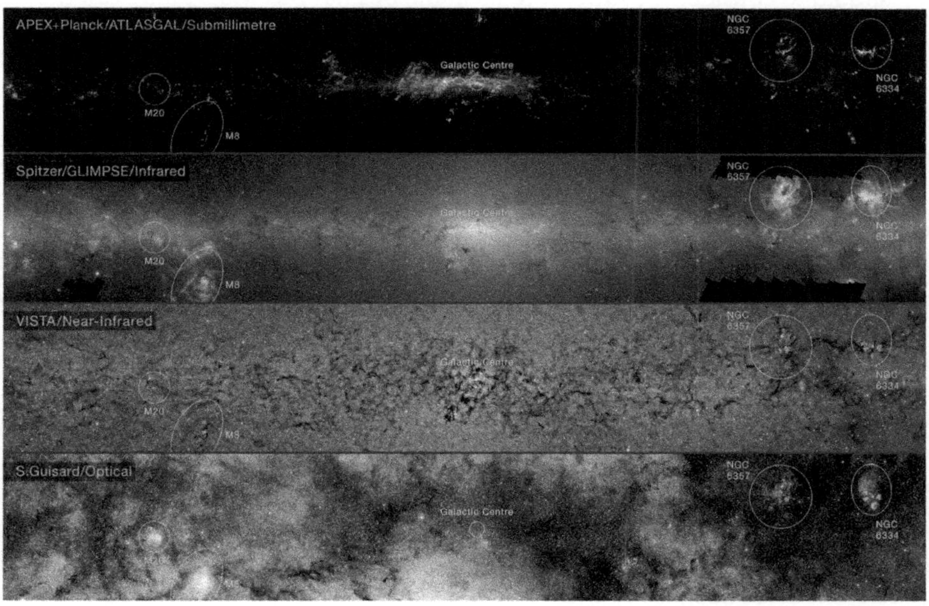

Figure 7.3. The central regions of the Milky Way observed at different wavelengths. Note that the significance of the colors varies from image to image and they cannot be directly compared. ATLASGAL is an acronym for APEX Telescope Large Area Survey of the Galaxy, located at 5100 m altitude on Chajnantor, Chile. APEX itself stands for "The Atacama Pathfinder Experiment". VISTA (Visible and Infrared Survey Telescope for Astronomy) is an infrared survey telescope at the Paranal Observatory in Chile. *Credit:* ESO/ATLASGAL consortium/NASA/GLIMPSE consortium/VVV Survey/ESA/Planck/D. Minniti/S. Guisard. Acknowledgement: Ignacio Toledo, Martin Kornmesser, Mahdi Zamani. CC BY 4.0

refuse my dinner because I do not fully understand the process of digestion?" when criticized for his use of operators before they could be rigorously formulated. Today there are many different units of electromagnetism, one of them being the Heaviside–Lorentz unit. In this unit Coulomb's law that describes the force between two charges q_1 and q_2 reads

$$F = \frac{q_1 q_2}{4\pi\varepsilon_0 r^2}, \tag{7.10}$$

where ε_0 is the vacuum permittivity. The factor of $4\pi\varepsilon_0$ does not appear in the Gaussian units. The electromagnetic Lagrangian in the Heaviside–Lorentz unit, however, does not have any factor of 4π. The Maxwell equations are also free of 4π's in this unit. In fact, Heaviside wanted to get rid of as many 4π as possible. As he wrote in 1892,

> "The excrescene 4π is a mere question of units, and need not be discussed here. The 4π's are particularly obnoxious and misleading in the theory of magnetism. Privately I use units which get rid of them completely, and then, for publication, literally season with 4π's to suit the taste of B.A. unit-fed readers. Of course, if it comes to numerical comparisons we should to consider the ratio of units in the ordinary to what I may call the rational system. Sometimes it is $\sqrt{4\pi}$, sometimes $(4\pi)^{-\frac{1}{2}}$, sometimes 4π, sometimes unity, but in the mere algebra it is simply a matter of putting in 4π's here and there in translating from rational to ordinary units."

We do this nowadays too with setting the speed of light or the gravitational constant to unity. However, restoring them in the final step "to ordinary units" is often quite tedious. In any case, the value 4π is not accidental, it is the consequence of spherical symmetry: a surface of constant electric field strength that is generated by a point charge is spherical, and a sphere of radius r—in 3 spatial dimensions—has area $4\pi r^2$.

7.2 Light Bending from Newtonian Gravity to Cosmology

The corpuscular viewpoint of light allows it to be subject to gravitational pull, and thus there would also be light bending in Newtonian gravity. Unlike the general relativistic case, which is discussed in detail in many textbooks (and so will be omitted here), Newtonian light bending was only mentioned in passing, if any. Therefore, I think it is interesting to include the Newtonian result here and see how it can be computed. See also Soares (2025).

Suppose a light corpuscle has a mass m, whose orbit around the Sun is given by $r = r(\theta)$. We have in mind a light particle that goes very close to the Sun and gets deflected (a "Sun-grazing orbit"). The velocity is given by $v^2 = \dot{r}^2 + r^2\dot{\theta}^2$, where the dot denotes time derivative. At the turning point of the orbit, which is at some distance $r = R$ from the Sun, we have $\dot{r} = 0$ and $v = R\dot{\theta}$. At this point the particle would have kinetic energy $mv^2/2$ and gravitational potential energy $-GMm/R$.

At spatial infinity, gravitational potential tends to zero and we let the kinetic energy be denoted by $mc^2/2$. The conservation of energy then requires that

$$\frac{1}{2}mc^2 = \frac{1}{2}mv^2 - \frac{GMm}{R} \Longrightarrow v^2 = c^2 + \frac{2GM}{R} = R^2\dot{\theta}^2. \tag{7.11}$$

Note that this result is *independent* of m, but we do require $m \neq 0$ in the first place (otherwise it experiences no gravity). Still, in principle m can be arbitrarily small.[4]

In the Newtonian picture, the velocity of light near gravitating bodies is larger than the speed of light in "free space". From this equation one immediately obtains

$$\dot{\theta}^2 = \frac{1}{R^2}\left(c^2 + \frac{2GM}{R}\right). \tag{7.12}$$

The Cartesian coordinates (x,y) relate to the polar coordinates (r, θ) via $(x, y) = (r\cos\theta, r\sin\theta)$. Differentiating with respect to time twice yields

$$(\ddot{x}, \ddot{y}) = (\ddot{r} - r\dot{\theta}^2)(\cos\theta, \sin\theta) + (r\ddot{\theta} + 2\dot{r}\dot{\theta})(-\sin\theta, \cos\theta). \tag{7.13}$$

We therefore observe that the centripetal acceleration is

$$a = -\frac{GM}{R^2} = \ddot{r} - r\dot{\theta}^2. \tag{7.14}$$

On the other hand, we know that Newtonian orbits are conic sections. Thus r must take the general form of the conics:

$$r = \frac{l}{1 + e\cos\theta}, \tag{7.15}$$

where e is the eccentricity of the orbit, and l is a constant known as the "semi-latus rectum". Writing this as $r + er\cos\theta = l$ and differentiate it twice, we can show that at $r = R$ and $\theta = 0$ during the closest approach to the Sun (we can always align the coordinate axes such that this is true), we have

$$\ddot{r} + e\ddot{r} - er\dot{\theta}^2 = 0. \tag{7.16}$$

By combining this with Equation (7.12), we can obtain

$$\ddot{r} = \frac{c^2}{R} + \frac{GM}{R^2}. \tag{7.17}$$

On the other hand, Equation (7.16) can be written as

$$\ddot{r} = e(r\dot{\theta}^2 - \ddot{r}), \tag{7.18}$$

[4] In modern field theory a photon can also be massive if we consider the Proca equation. Instead of a Maxwell field (whose quantization leads to massless photon), we refer to this field as the Proca field, named after Romanian physicist Alexandru Proca (1897–1955).

which by the centripetal acceleration relation, is equivalent to

$$\ddot{r} = \frac{eGM}{R^2}. \tag{7.19}$$

The two equations, Equations (7.17) and (7.19) must match. This allows us to solve for the eccentricity:

$$e = 1 + \frac{c^2/R}{GM/R^2}. \tag{7.20}$$

If we substitute in the values of the various constants,[5] we will find that $e \sim O(10^5) \gg 1$, i.e., the orbit is that of a hyperbola. With elementary algebra, we can re-write the conic into the more familiar Cartesian form (this is left as an exercise for the students):

$$\left(x - \frac{el}{e^2 - 1}\right)^2 - \frac{y^2}{e^2 - 1} = \frac{l^2}{(e^2 - 1)^2}. \tag{7.21}$$

The hyperbola has gradients (its asymptotes) given by $\pm\sqrt{e^2 - 1} \approx \pm e$.

Finally the deflection angle α is the angle subtended by the two asymptote lines; see Figure 7.4. Two lines intersecting at a point would define two angles. The deflection angle is the one that is smaller (the two angles can of course be equal), so that in the limit that there is no deflection, the deflection angle would be zero instead of 180 degrees. Using elementary trigonometry, we have

$$\tan\left(\frac{\pi}{2} - \frac{\alpha}{2}\right) \approx e. \tag{7.22}$$

This is equivalent to $\cot(\alpha/2) = e$, which is also equivalent to $\tan(\alpha/2) = 1/e$. For a small angle, we can approximate this as

$$\tan\left(\frac{\alpha}{2}\right) \approx \frac{\alpha}{2} = \frac{1}{e} \Longrightarrow \alpha \approx \frac{2}{e}. \tag{7.23}$$

The numerical value, once converted into degrees (or rather arc-seconds), is $0.875''$. This is *half* the value that general relativity gives. To compute light bending around the Sun using general relativity, one has to work out the geodesic equation for massless particle and analyze the behavior of the solution. The first observation of light bending by the Sun was conducted in 1919 during a solar eclipse (that's the only time we could see stars near to the Sun!). This event was big news; Figure 7.5 shows a newspaper report from that time.

Historically, Albert Einstein first proposed that light can be bent by gravitational field in 1911, before the theory of general relativity was finalized. Even in the

[5] The speed of light $c = 3 \times 10^8$ m s^{-1}, the gravitational constant $G = 6.67 \times 10^{-11}$ m^3 kg^{-1} s^{-2}, the solar mass $M = 1.99 \times 10^{30}$ kg, and R being very close to the Sun can be approximated by the solar radius, which is $R \approx 6.95 \times 10^8$ m.

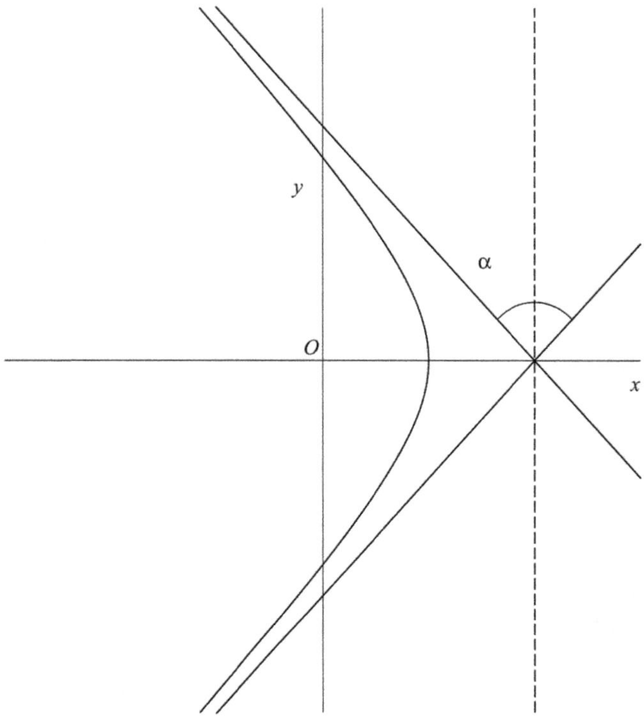

Figure 7.4. This figure is an exaggerated and not-to-scale illustration of Newtonian light bending around the Sun, whose center is located at the origin O. The two straight lines have gradients close to $+e$ and $-e$, respectively, where e is the eccentricity of the hyperbolic orbit of the corpuscle of light. The angle α subtended by the asymptotes is the angle of light bending. The vertical dashed line is just the angle bisector for easy reference.

"Entwurf" (meaning "outline") paper of 1913,[6] his result for light bending was still the same as the one in 1911, namely the light bending amount was identical to the Newtonian result. Only in the final general relativity result the deflection angle became $1.75''$, which was the value that matched the observation. In a sense it was fortunate for Einstein that observational support[7] did not happen until after the correct prediction was made. If observations had been made in 1912[8], it would have observed twice the amount of deflection predicted by Einstein. In that case, even if Einstein would finally find the right theory afterward, its impact might have been less, in the sense that the physics community might feel that Einstein "tweaked" his

[6] In full the title is *Entwurf einer verallgemeinerten Relativitátstheorie und eine Theorie der Gravitation*, or "Outline of a Generalized Theory of Relativity and of a Theory of Gravitation".

[7] By which we meant observation from measuring stars' apparent position during a solar eclipse; the expedition was led by the famous astronomer Sir Arthur Eddington (1882–1944). The fact that a famous British scientist helped to validate a theory proposed by Einstein, a German (he renounced his German citizenship in 1896 and became a Swiss citizen in 1901), is often perceived as a testament that science can transcend national boundaries, even in the midst of war.

[8] Such expeditions were indeed sent to Brazil (Crispino 2011). Another attempt was made in 1914. Neither was successful. Erwin Finlay-Freundlic (1885–1964) who led the 1914 expedition, being a German, was arrested as a spy by the Russian soldiers, after World War II broke out during his travel to conduct the observations.

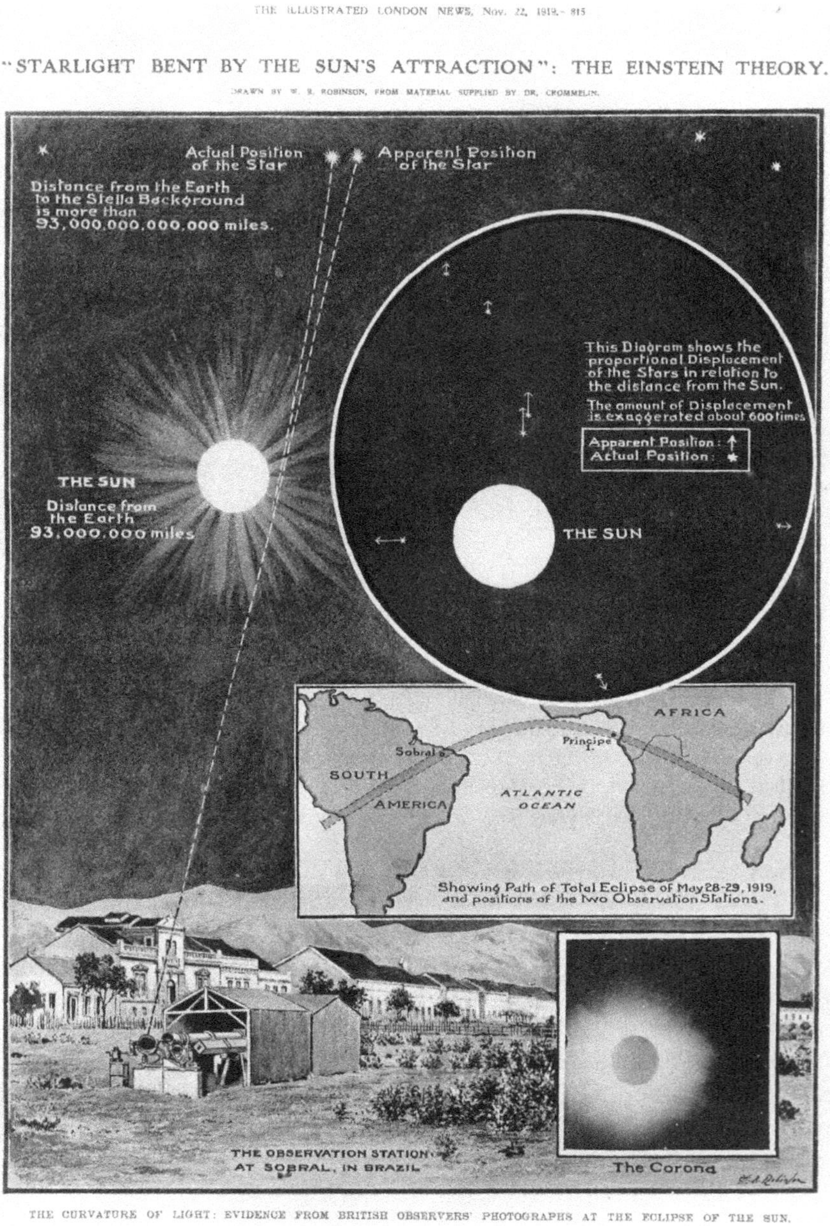

Figure 7.5. Illustration of the solar eclipse of 1919 and how light bending worked, from *Illustrated London News* dated 22 November 1919. The artist is W. B. Robinson, from material supplied by Dr. Crommelin. *Credit:* This [1919 Eclipse expedition to test relativity] has been obtained by the author(s) from the Wikimedia website, where it is stated to have been released into the public domain. It is included within this article on that basis.

theory to accommodate the observed value. Although in such an alternative history, this would still be the correct scientific method, it would likely not have the same impact as a theory that gives the correct prediction at the onset.

There is, however, one mystery that remains. The technology and technique to test for light bending around the Sun during a solar eclipse[9] had been available for quite some time. Why didn't anyone try to do it before Einstein's suggestion that light can be bent by gravity? Surely even in the context of Newtonian gravity, it is also a way to test for the corpuscular nature of light against the wave nature of light, as the latter would have no mass to be affected by Newtonian gravity? I couldn't find an answer to this question; perhaps there is one but I am not a historian of science and can only provide a *guess*. Perhaps the wave nature of light had become so well accepted that once the technology to test for light bending became available, it no longer occurred to anyone that one could (or even need to) test whether there is any light bending at all.

Gravitational light bending, or *gravitational lensing*, is now used on a regular basis in astrophysics and cosmology. Suppose a star is blocked by another foreground object. If the object is sufficiently massive, it can bend the light of the background star. This can produce several effects. For example, it can form multiple images of the same background star, where each image is from photons that take a different curved path to hit our eyes. When the background star is precisely behind the foreground object, light rays from the source star would pass on all sides equally to reach us, thus forming a ring shaped image known as an "Einstein ring". When four images are formed around the lensing object roughly symmetrically, it is called an "Einstein cross". One of the applications in astrophysics is "microlensing", in which a background light source is magnified by the gravity of a foreground object. This allows us to detect exoplanets, among other applications. In cosmology, light rays can be bent in the presence of dark matter. This helps us to map the distribution of dark matter even though we cannot see it directly by other means.

On the cosmological scale, light from distant galaxies can be gravitationally lensed along different paths so much so that the time it takes for photons traveling on one path can be vastly longer (years and even decades) than the other paths. One of the most remarkable examples of this phenomenon is the gravitational lensing by the galaxy cluster MACS1149, as shown in Figure 7.6 below. This lensing event produced three amplified images of a background spiral galaxy. (Galaxy-cluster gravitational lenses can magnify background galaxies up to 50 times or so.) The spiral galaxy itself hosts a supernova nicknamed "Refsdal", which was itself split into four images by a cluster galaxy. Refsdal was discovered in November 2014, and since the host galaxy has been lensed, the image of the supernova was duplicated in other images. In one of these the image of the supernova appeared *a year* later. This lensing configuration allows a team of astronomers to measure the Hubble constant from the time delay (Grillo et al. 2020). They also suggested that one could also

[9] Ancient records of solar eclipses also helped scientists to determine how much the Earth's rotation has slowed down throughout the millennia. If the rotation had remained the same, we could use computers to calculate where the eclipse would have been in the years mentioned in the ancient records, and we would find that they wouldn't have been able to observe the eclipse, for the shadow of the Moon would have fallen somewhere else on Earth. Adjusting for the location discrepancy, it was found that the Earth's rotation is slowing down by about 1.8 milliseconds per century on average (Stephenson et al. 2016).

Figure 7.6. This image shows the huge galaxy cluster MACS J1149+2223, whose light took over 5 billion years to reach us. Close to the center are multiple distorted images of the same background galaxy, which hosts a supernova, which is *itself* lensed into 4 images in an Einstein cross configuration. A close-up of the Einstein cross is shown in the inset. *Credit:* NASA, ESA, S. Rodney (John Hopkins University, USA) and the FrontierSN team; T. Treu (University of California Los Angeles, USA), P. Kelly (University of California Berkeley, USA) and the GLASS team; J. Lotz (STScI) and the Frontier Fields team; M. Postman (STScI) and the CLASH team; and Z. Levay (STScI). CC BY 4.0

measure the value of the equation-of-state parameter w of the dark energy density, currently with a 30% uncertainty. Therefore, time-delay cluster lenses have the potential to become a cosmological probe.

In 2021, another team of astronomers reported the discovery of a multiple imaged supernova, nicknamed "Requiem",[10] that was lensed by a foreground galaxy cluster, shown in Figure 7.7. By looking through the archival Hubble Space Telescope

[10] Requiem means "rest" in Latin. It is often used to refer to a Catholic mass or musical composition for the dead, but more generally also refers to any token of remembrance. Supernova is of course, the death of a star.

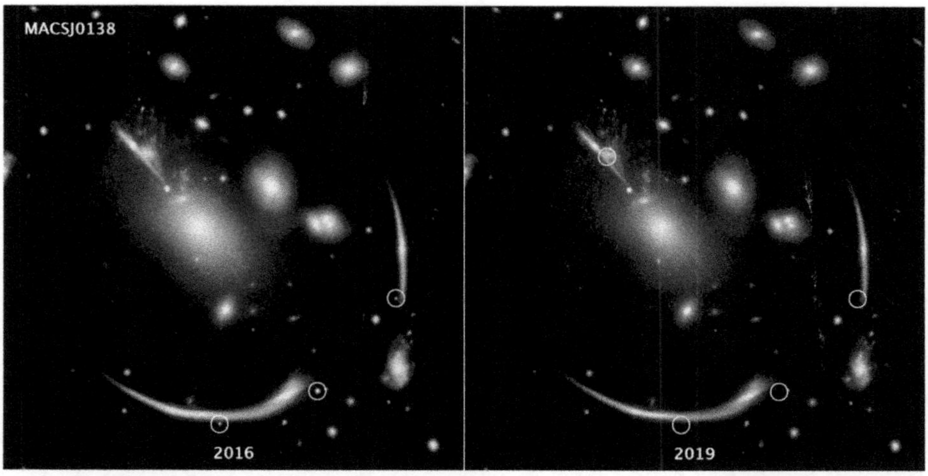

Figure 7.7. These images are a composite of separate exposures acquired by the WFC3/IR instrument on the Hubble Space Telescope. Several filters were used to sample narrow wavelength ranges. The same supernova appear three times in the 2016 image on the left, which already disappeared by 2019, as shown in the right image. Notice that the shape of the host galaxy has been distorted by gravitational lensing. A fourth image is predicted to appear in the year 2037 $\pm$ 2, in the position highlighted by the yellow circle at top left of the right image. The light from Supernova Requiem needed an estimated 10 billion years for its journey, based on the distance of its host galaxy. The light that Hubble captured from the cluster, MACS J0138.0-2155, took about 4 billion years to reach Earth. *Credit:* Steve A. Rodney (University of South Carolina), Gabriel Brammer (Cosmic Dawn Center/Niels Bohr Institute/University of Copenhagen); Image Processing: Joseph DePasquale (STScI)

images, they found three lensed images of the supernova, which appeared with time delays of less than 200 days. They then *predicted* that a fourth image would appear in the year 2037 $\pm$ 2 (Rodney et al. 2021). If this prediction comes to be, it would be a remarkable test for both general relativity and our understanding of the galactic environment (the matter distribution of which dictates the lensing path).

In 2024, based on the observation data of JWST (James Webb Space Telescope, finally launched on December 25, 2021, after many years of delay), the system J1721 +8842 was found to show a distant quasar being *doubly lensed*, instead of a lensed dual quasar as previously thought. A quasar is short for "quasi-stellar object", named for its star-like appearance but is extremely luminous. It is now understood to be an active galactic nucleus, i.e., a central supermassive black hole is actively feeding in a distant galaxy, so the gas in the accretion disk heats up brightly. The six lensed images of the quasar are produced by two gravitational lenses: the first at redshift of $z = 2.382$ and a second one at redshift $z = 0.184$. (Recall that a redshift of 2 means that the light from the object has been stretched to twice its original wavelength due to the expansion of the Universe. Thus, the larger the redshift, the further away an object is. See Section 5.4 for more details.) The double-lensing causes the light to take a zig-zag path across space. This is the first example of an Einstein zig-zag lens ever discovered (Dux et al. 2025).

Figure 7.8. The "Molten Ring" Galaxy, one of the most amazing examples of gravitational lensing in action. *Credit:* NASA, ESA, Saurabh Jha (Rutgers, The State University of New Jersey). Acknowledgment: Leo Shatz.

Another beautiful example of gravitational lensing is nicknamed the "Molten Ring" Galaxy, formally designated GAL-CLUS-022058s, located in the southern hemisphere constellation of Fornax. The image Figure 7.8 shows a remote galaxy being magnified by a factor of 20 and distorted via gravitational lensing by an elliptical galaxy (center of the image). It is an example of "Einstein ring", though not a perfect one. The galaxy is located at 9.4 billion light-years away. This means that we are observing the galaxy as it was during the era of peak star formation in the history of the Universe.

Finally, I would like to quote Hermann Weyl a paragraph from his book *Philosophy of Mathematics and Natural Science*, originally made in the context of Einstein's static universe, which has the topology $\mathbb{R} \times S^3$. Being spatially closed it allows light to circumnavigate around the cosmos, the effect of which is that we will see multiple images of the same star, if the radius of the S^3 is small enough). Still, what he said remains applicable in the context of gravitational lensing, in the sense that we have explained, namely that light travels around different paths can take vastly different time to reach us. Weyl wrote (Weyl 2009):

> "An observer should therefore see infinitely many images of a star [...], showing him the star in states between which an 'eon" has elapsed, the time needed by the light to travel around the sphere of the world. The present would be shot through with the ghosts of the long ago."

The ghosts of the long ago! How poetic.

7.3 Electromagnetism, Special Relativity, Field and the Æther

The history of special relativity has been treated in great detail in many books, and so I would rather not spend too much time on it. Still, the influence of electromagnetism on the formulation of special relativity cannot be overstated. In Newtonian mechanics, there is always a frame of reference (with the inertial frame defined by Newton's First Law of Motion[11]). Laws of motions in Newtonian mechanics satisfy Galilean transformation. For example, if a frame S' is moving with a constant speed v towards the positive x direction with respect to a static frame S, then the point x in S would be measured in S' as $x' = x - vt$. The problem is that when we derive the speed of light of an electromagnetic wave from Maxwell equations, in which frame is this speed relative to? A natural conjecture was that it should be the frame of the luminiferous æther, the "substance" that permeates the universe in which the electromagnetic wave propagates in (after all, it was thought back then, a wave needs to propagate in *something*). However, if this is the case, and if Galilean transformation applies to electromagnetism, then an experiment performed in a laboratory with a nonzero speed relative to the æther should find a directionally dependent speed of light. But this was never measured in any experiment of the Michelson–Morley type. In other words, electromagnetism is incompatible with Newtonian mechanics' worldview. How does one deal with such an incompatibility?

Most physicists are, by our professional training, very careful and not very radical. We tend to keep as many ingredients of the old understanding as possible, only modifying things here and there. So it is not a surprise that physicists first attempted to modify either the properties of æther, or the properties of matter themselves. For example, the æther must now be static ("immobile"), otherwise the speed of light would vary together with any changes to the æther. However, the æther must also satisfy a bunch of other requirements in order to be feasible. For example, light is a transverse wave because its components vibrate perpendicular to the direction of propagation. However, transverse waves generally do not propagate in *fluids*, only in *solids*. But how can the æther be a solid yet we can pass *through* it without interacting with it? Indeed, Newton rejected Descartes' idea of vortices[12] on the ground that planetary bodies passing through the vortices would be slowed down by drag, and so the planetary orbits would be different from what was observed,

---

[11] Note that gravity messes this up. Newton's first law of motion essentially defines inertial frame as one that is not subject to any force; but gravity, being an inverse-square force, is always present in the Universe, no matter how small. Furthermore, gravitational force is *universal* so everything feels it. How then can one still define an inertial frame? Of course in practice one simply ignores gravity when it is weak; but strictly speaking there is no such thing as an inertial frame. One could also say that perhaps this already hinted that gravity should not be treated as a force. In fact, Newton's theory of gravity can be geometrized into Newton–Cartan theory, in which gravity becomes an effect of the curvature of *space*. One advantage in this formulation is that the factor of 2 in the Coriolis force naturally arises from the symmetry of the geodesic equation.

[12] The idea of vortices did not die out completely. Maxwell devised an electromagnetic æther with "molecular vortices" in 1862—in its original conception of electromagnetism, light travels through molecular vortices in the æther. In our own time, a class of dark matter candidate is the "superfluid dark matter", which can also lead to vortices.

which was well explained by the inverse-square law with no surrounding fluid of any kind. The æther, if it exists, would be subject to the same criticism, so its viscosity must be extremely small. This seems to be contradictory to it being some kind of a solid. Newton, however, did consider the possibility that the æther is not uniform, but with density that decreases closer to massive objects like planets. Such a model would reduce the drag to an arbitrarily small amount. The same possibility was also raised by the great mathematician Leonhard Euler later on, who also tried to *explain gravity* as the effect of the denser æther in space *pushing down* on us. Euler's remarkable idea was published anonymously[13] in 1743 in an essay *De Causa Gravitatis*. (Thomas Young, however, considered that æther is denser inside matter, thinning out continuously outside of matter.) Another possibility is that the constituents of the æther are so fine that they can pass through other matter unhindered; other matter would simply be "porous" to them.

It is interesting to compare some of these ideas to modern ideas in other problems in physics. For example, George Gabriel Stokes (1819–1903; of the Stokes' theorem fame[14]), proposed that the æther might be "dilatant", so that to anything that passes through it, it behaves like a fluid if the velocity is slow, but behaves as almost solid at high velocity. This is similar to the non-Newtonian fluid, which is a type of fluid whose viscosity changes when a force is applied. Thus material objects like planetary bodies can pass through it with very little resistance but for light it would behave like a solid as required. Such a model incorporates different aspects to make æther more viable. Indeed one finds similar idea today in, for example, the "Vainshtein mechanism" mentioned in Chapter 4. This mechanism, otherwise known as "k-mouflage", allows a theory of gravity to have more degrees of freedom than general relativity at a cosmological scale (in order to explain dark sectors), but suppresses those degrees of freedom near a gravitating source like the Earth, so as to match the well-tested general relativity results. Similarly a candidate of dark energy called a "chameleon" field gives rise to a fifth force whose range is very small in regions of high density but much larger in low-density interstellar space, thus accounting for the accelerated expansion of the Universe. Also worth noting is the idea of the mathematician Augustin-Louis Cauchy (1789–1857), who in 1839, proposed that æther might have a *negative compressibility*, which was aimed to explain the lack of any longitudinal wave mode. Lord Kelvin would further support this idea in 1888 and call it "labile æther" (also called "contractile æther"). Compressibility is a measure of the instantaneous relative volume change of a fluid or solid as a response to a pressure:

$$\text{Compressibility} = -\frac{1}{V}\frac{\partial V}{\partial p}. \tag{7.24}$$

By definition, because of the overall negative sign, if an increase in pressure induces a reduction in volume, the compressibility is positive. Thus, Cauchy's negative

[13] Possibly because he thought the idea too radical and did not want to defend the idea publicly.

[14] He also made some contributions in the field of biology (Ceredig 2020).

compressibility corresponds to the peculiar case that an increase in volume resulted from an increase in pressure. Equivalently, a negative pressure in such a substance gives rise to volume increase.

But how does one explain the *non-detection* of the variation in the speed of light? Even if the æther is completely at rest, the Earth is not, so a relative motion must exist. Hendrik Lorentz (1853–1928) and George FitzGerald (1851–1901) independently[15] proposed the idea that perhaps the length of any object *shortens* in the direction of its motion. With the appropriate amount of shortening, this could explain why the speed of light appears to be independent of direction even if there is an æther. The now familiar length contraction by the "Lorentz factor" $\sqrt{1 - v^2/c^2}$ was established by Larmor in 1897 and also by Lorentz in 1904. Note that in order for length contraction to affect all sorts of materials moving in the æther, it was thought to be necessary that one should have an understanding not just of electromagnetic force, but forces that govern molecules and atoms. It was not clear back then if there are other forces that bind matter together apart from electromagnetism. If other forces exist, do they transform also like electromagnetic force? And if not, how can length contraction works for different material so universally? Even eminent mathematicians like Henri Poincaré were involved in trying to answer these questions. However, these models appear quite ad hoc to the modern eyes with hindsight.

It was Einstein who took a completely different and extremely radical approach: *just take the constancy of the speed of light in all inertial frames as an axiom*[16]! With this assumption the theory of special relativity was born. Subsequent geometric formulation of the theory by Hermann Minkowski further demonstrated that the speed of light defines the causal structure of spacetime. The Lorentz transformation preserves the metric $ds^2 = \eta_{ab}dx^a dx^b$, much like how Galilean transformation preserves the usual length given by the Pythagoras theorem. One way to appreciate this is to consider the Lorentz transformation of a particle moving in the x-direction with a constant speed v:

$$\begin{cases} x' = \gamma(x - vt), \\ t' = \gamma(t - vx/c^2), \end{cases} \tag{7.25}$$

where γ denotes the Lorentz factor. By introducing the dimensionless quantity $\beta := v/c$ we can write this in a more symmetric form

[15] Lorentz's paper was published in 1892, and FitzGerald's in 1889. Lorentz was not aware of FitzGerald's work until 1894. He then wrote to FitzGerald, who replied that he remembered that he did send the article (a single paragraph of "letter" without any equation (FitzGerald 1889) to *Science* but he didn't know if it was published!

[16] Another assumption is the Principle of Relativity, which predated Einstein. This says that experiments cannot distinguish between inertial frames that are static and one that is in constant-speed motion. Galileo was perhaps the first to propose a universal principle of relativity (which is *not* limited to frames that are invariant under what we now call "Galilean transformation" (Browne 2020)). Interestingly, Galileo used a ship traveling at constant velocity on a smooth sea as an example to illustrate the principle. A similar line of thinking can be found in ancient Chinese texts. ("地恒动不止而人不知，譬如人在大舟中，闭牖而坐，舟行而人不觉也。"–《尚书纬· 考灵曜》.)

$$\begin{cases} x' = \gamma(x - \beta ct), \\ ct' = \gamma(ct - \beta x). \end{cases} \tag{7.26}$$

In terms of the *rapidity* θ (see Box 5 below for more discussions), defined via $\beta = \tanh\theta$, we can write the aforementioned Lorentz transformation in matrix notation as

$$\begin{pmatrix} ct' \\ x' \end{pmatrix} = \begin{pmatrix} \cosh\theta & -\sinh\theta \\ -\sinh\theta & \cosh\theta \end{pmatrix} \begin{pmatrix} ct \\ x \end{pmatrix}. \tag{7.27}$$

The Lorentz transformation matrix

$$\begin{pmatrix} \cosh\theta & -\sinh\theta \\ -\sinh\theta & \cosh\theta \end{pmatrix} \tag{7.28}$$

looks very much like the more familiar rotation matrix

$$\begin{pmatrix} \cos\theta & -\sin\theta \\ \sin\theta & \cos\theta \end{pmatrix}, \tag{7.29}$$

but with the trigonometric (or "circular") functions replaced by hyperbolic functions. The Lorentz transformation is thus a "rotation" in hyperbolic geometry that is inherent to Minkowski spacetime.

Box 5: Why Rapidity?
Why do we bother with introducing the concept of rapidity? One mathematical reason is as follows: we learned in special relativity that velocities do not add linearly, instead of $v_1 + v_2 = v$, we have

$$v = \frac{v_1 + v_2}{1 + \frac{v_1 v_2}{c^2}}. \tag{7.30}$$

From the Lorentz transformation in matrix form

$$\begin{pmatrix} ct' \\ x' \end{pmatrix} = \begin{pmatrix} \gamma & -\gamma\beta \\ -\gamma\beta & \gamma \end{pmatrix} \begin{pmatrix} ct \\ x \end{pmatrix}, \tag{7.31}$$

we can define the transformation matrix

$$\Lambda(\beta) := \begin{pmatrix} \gamma & -\gamma\beta \\ -\gamma\beta & \gamma \end{pmatrix}. \tag{7.32}$$

We can check that indeed $\Lambda(\beta_1)\Lambda(\beta_2) = \Lambda\left(\frac{\beta_1 + \beta_2}{1 + \beta_1\beta_2}\right)$. It is often useful to have a linear relation instead of a nonlinear one. Thus, one seeks a function $f(\theta) = \beta$ which is one-to-one and satisfying the homomorphism

$$\Lambda[f(\theta_1)]\Lambda[f(\theta_2)] = \Lambda[f(\theta_1 + \theta_2)]. \tag{7.33}$$

This is equivalent to finding a function that satisfies

$$f(\theta_1 + \theta_2) = \frac{f(\theta_1) + f(\theta_2)}{1 + f(\theta_1)f(\theta_2)}, \tag{7.34}$$

and $f(\theta) = \tanh\theta$ fits the bill. It is in fact, the unique choice. See Exercise 1.3.15 in (Naber 2012). In conclusion, rapidity is useful because it is additive unlike velocity. Furthermore, rapidity so defined is inherently related to the geometry of Minkowski spacetime and thus a natural quantity.

The idea of the æther, once considered one of the most important questions in physics[17], whose existence became harder to justify amidst non-detection by various experiments[18], finally became obsolete in this special relativity worldview. Why didn't other physicists give up æther prior to Einstein? Why did they rather propose Lorentz transformation to render the æther undetectable rather than just say that the æther does not exist? I think there were several reasons. First of all, it wasn't obvious that the æther is not detectable by other means, even if one cannot reveal its existence via relative motion. It was also true that the idea of æther was too deeply ingrained to consider giving it up (how could light, a wave, propagate without a medium? Today we just accept this as a fact, no longer bothered by the strangeness of such an idea!). However, I think more importantly, was the fixation on trying to explain how light and matter interact with the æther so as to render the speed of light constant in all inertial frames. To emphasize, this is a very difficult problem, given that one has to understand the forces that govern not just electromagnetism and also the then unclear constituents of matter. In some sense, in the eyes of the critics, Einstein just assumed what the others were trying to prove. Still, Einstein did more by doing less (no need of complicated models of interactions and æther)—special relativity is more elegant and it gives rise to an entire paradigm shift of how we view space and time. In this sense, special theory is "special" because Einstein's way of thinking is special; he was able to view the problems from a completely new angle.

Why this is an immense revolution can only be appreciated by looking back into the literature of that time. As mentioned, most physicists were trying to understand how the æther works, and how light and matter interact with it. Thus the inner structure of an electron was also part of the investigation. In other words, in order to explain how an electron moves in the aether, one needs a model that incorporates

[17] Oliver Lodge wrote in his *Modern Views of Electricity* (1889) that the nature of the æther "is *the* question of the physical world at the present time. But it is not unanswerable. It is in my belief, not far from being answered" (emphasis in original). This statement remained in the second edition of 1892, but removed by the third edition in 1907, after the publication of special relativity (Goldberg 1949).

[18] Joseph Larmor (1857–1942) would even argue that the æther is undetectable *in principle*; Ebenezer Cunningham (1881–1977) contemplated that the æther is only a "mental image" one constructs, and indeed each inertial observer can construct an æther. There would therefore be an infinite number of æther frames (Hunt 1986). But just like we can treat every point as a "center" in a Universe without a center, there simply isn't a need for an æther anymore if you can define any number of them.

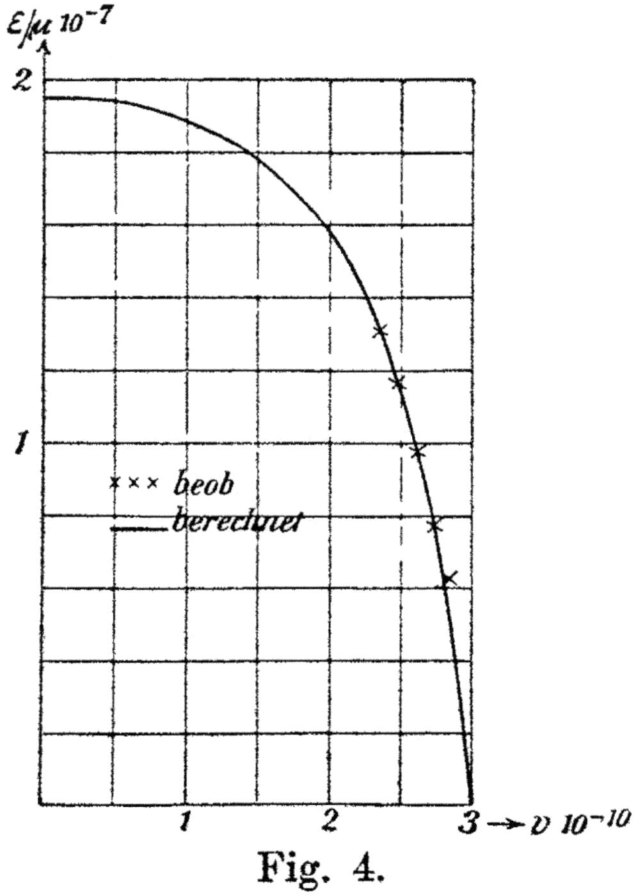

Figure 7.9. Drawing from Walter Kaufmann's article "The magnetic and electrical deflectability of Becquerel rays and the apparent mass of electrons", Göttinger Nachrichten, 1901. The vertical axis is essentially the charge-to-mass ratio, while the horizontal axis the velocity. The charge-to-mass ratio drops quickly towards zero, indicating that the effective mass (the relativistic) mass of the electron increases as $v \to c$. *Credit:* This [Göttinger Nachrichten 1901 Fig 1-4] has been obtained by the author(s) from the Wikimedia website, where it is stated to have been released into the public domain. It is included within this book on that basis.

their interaction, as well as the models for the electron structure and that of the æther. This quickly became a mess as there were various predictions for how the charge-to-mass ratio for an electron changes as its speed increases. In 1901, four years before special relativity, Walter Kaufmann (1871–1947) measured that the charge-to-mass ratio did indeed decrease when the electron is moving faster. In other words, Kaufmann found that "mass" (what we today call the *relativistic mass*) increases as the function of v/c, and seems to become infinite as $v \to c$, as shown in Figure 7.9. Kaufmann's original experiment did not have the accuracy required to test Einstein's special relativity (whose formula for the relativistic mass increase is independent of the particle structure) and the various other theories at that time. It also contained some mistakes in the analysis. Still, various other experiments, as well as future repeats of Kaufmann's experiment, eventually ruled in Einstein's favor.

Thus, Einstein's simplicity won, and the messy literature concerning particles interacting with the æther became obsolete.

Yet Einstein wasn't done; he would go on to revolutionize physics again by proposing that light has *both* the property of wave and that of particle. As we will see later, conversely, matter would also have a wave property. In quantum mechanics, one eventually has to question what we mean by "real". We end this section by a quote of Poincaré who, despite criticizing Einstein "resolution" of the constancy of speed of light problem, and despite him believing in the existence of an absolute æther frame, did at one point, not only considered the possibility that æther does not exist, but also that material objects "do not exist" (though he quickly backtracked on that thought):

> "It matters little whether the æther really exists; that is the affair of the metaphysicians. The essential thing for us is that everything happens as if it existed, and that this hypothesis is convenient for the explanation of phenomena. After all, have we any other reason to believe in the existence of material objects? That, too, is only a convenient hypothesis; only this will never cease to be so, whereas, no doubt, some day the æther will be thrown aside as useless."

Did we really completely forget about the æther though? Not completely, at least in spirit we have something similar that permeates all space, though it serves a completely different purpose than æther: *field*. Maxwell theory of electromagnetism is the first *field* theory. Nowadays the existence of say, magnetic field, is illustrated in schools by sprinkling iron powder around a magnet. We are so used to this now, but back in the days it must have been difficult for people to believe that there is a mysterious "field" surrounding a conducting wire, since electricity flows through the wire, not around it in *empty space*. In fact, magnets were hard to explain since they could move a piece of iron without touching it, seemingly acting *across distance*. This is totally against the "mechanical view" of the cosmos, as we discussed in Chapter 2. Of course, magnets were well known even in ancient times. Aristotle did not attempt to explain magnetic force, though he mentioned the idea proposed by Thales of Miletus that magnets must have a "soul".[19] Note that in most cases we do not directly measure a "field", be it an electric field or gravitational field. Rather, we measure the force experienced by a test particle in said field. Hermann Weyl, in his book *Philosophy of Mathematics and Natural Science* emphasized this point (with P denoting a point in space) with the electric field as an example (the emphasis is his original) (Weyl 2009):

> "Employing various test particles we find that the force [...] may be split up into two factors:

[19] There were plenty of strange beliefs surrounding magnets. In *Credulities Past and Present* (1880), William Jones mentioned an ancient belief that a magnet washed in spring water would be seen to breathe, and would reply to questions in "a voice like that of a sucking child". This belief is especially strange since it is easily verifiable to be false by conducting an experiment.

$$\vec{F}(P) = e \cdot \vec{E}(P),$$

where the vectorial factor $\vec{E}(P)$, the "electric field strength", is a point function independent of the state of the test particle, while the scalar factor e, the "charge'[20] of the test particle, is determined exclusively by the inner state of the particle, depending neither on its position nor on the conductors, and is thus found to be the same no matter into what electric field we may place the particle. Here we start from the force as the given thing; but the facts outlined lead us to conceive of an electric field, mathematically described by the vectorial point function $\vec{E}(P)$, which surrounds the conductors and which *exists, no matter whether the force it exerts on a test particle be ascertained or not.* The test particle serves merely to render the field accessible to observation and measurement. [...] Now the equation $\vec{F}(P) = e \cdot \vec{E}(P)$ is no longer to be looked upon as a definition of $\vec{E}$ but as a *law of nature* (to be corrected if circumstances warrant it) determining the ponderomotoric force which an electric field $\vec{E}$ exerts on a point charge e. [...] A systematic presentation will introduce $\vec{E}$, the electric field strength, in a purely "symbolic" way without explanations and then lay down the laws it satisfies (for instance, that the line integral of $\vec{E}$ extended over a closed curve in space is zero) as well as the laws according to which ponderomotoric forces are connected with it. If forces are considered observable, the link between our symbols and experience will thus have been established."

[20] Why the quotation mark for a charge? Because what *is* a charge is not at all clear. Weyl wrote: "In line with Leibniz's ideas, the material particle, although imbedded in a spatial environment from which its field effects take their start, would itself than be a *monad* existing beyond space and time. Hence one may not say, 'Here *is* a charge," but only, "This closed surface within the field *surrounds* a charge". Italic is his original. Here he probably had in mind the idealization in the abstraction had treated physical particles as an unphysical geometric point. One could of course be more precise and define a size for particles like an electron. Even then, what is a charge other than a quantum number that causes a particle to react with an electromagnetic field?

[21] Likewise, we may introduce the concept of "mass" first when teaching physics, but mass has to be measured by some means. As pointed out by Weyl, Galileo defined momentum first before mass, not the other way round. For Galileo, two bodies have the same mass m if at equal velocities v, they have the same momentum. To introduce mass first and then to define momentum as mv is to treat the mass already as an abstract concept, albeit one that is distilled through experiences and observations. The quantity mv is also interesting: in the 14th century, there was an idea that motion is due to *virtus derelica.* Nicolas Bonet (c.1280–1343) wrote that:

> "in a violent motion some nonpermanent and transient form is impressed in the mobile so that motion in a void is possible as long as this form endures; but when it disappears that motion ceases."

The "nonpermanent and transient form" was called the *impetus.* John Buridan (1301–1358) argued that the measure of how much impetus is given by the speed and the quantity of matter of the moving body, which is the same as the momentum defined later. For Buridan, the celestial realm has no resistance, and thus the original impetus impressed upon the celestial bodies or the celestial spheres will remain indefinitely, thus there's no need for an external divine force to constantly push them around.

In other words, the field is an abstract construction we inferred, and although it can only be detected if there is a test particle that couples to it, we infer the existence of the field independently of test particles, and distill the laws of nature that the field obeys[21]. For pedagogical purposes however, students are taught the other way around: start with the field as a given physical entity, and compute what it does to test particles. Coulomb's law, for example, discuss the electric field surrounding a charged point particle, *even in the absence of test particles.* What it really means is that *if* there is a test particle coming into the scene, it will behave in a certain way in response to the field. This is not saying that field is not "real". It is a mathematical construct that is rather abstract, but it does describe reality. A skeptical person might argue that we can only detect the field when a test particle is around, so is it really there when there is no test particle? This is akin to asking whether a tree really falls in the forest if no one hears it. We define a field to describe the behavior of particle interactions, and there is no point asking what happens when there is no particle to warrant such definition in the first place. *However*, the equation $\vec{F}(P) = e \cdot \vec{E}(P)$ *suggests* that we can treat the field as an entity that exists by itself. Eventually the idea will be developed further in quantum field theory, in which the field is treated as the fundamental entity, whose quantization leads to what we perceive as particles. An electron, for example, is the "excitation" within the *electron field* (and that's why all electrons are identical). Coming back to the Coulomb's law example without test particles, we are used to the way of thinking that an electric field (not to be confused with the quantum electron field) is *produced* by an electrical charge, like an electron. Yet again, the theory of electromagnetism is silent on interpretative issues like this much like in general relativity. We know that the field coexists with the charge, but is it really correct to say that the charge *manifests* the field?

There is one crucial difference between fields and the æther. Questions like how objects move within the æther unimpeded do not arise for fields, because fields exist ultimately—even in quantum field theory—"only" as a useful mathematical construct used to describe particle interactions, unlike aether that was treated as a *physical substance* like a fluid, but one that is increasingly endowed with more and more fantastical properties to explain its "non-detection".[22] When special relativity became a successful theory, the notion of æther gradually faded away rather than being "disproved"; æther in general cannot be ruled out. Einstein gave a lecture *Æther and the Theory of Relativity* in 1920 (thus after the birth of general relativity), in which he wrote:

> "It may be added that the whole change in the conception of the æther which the special theory of relativity brought about, consisted in taking away from the æther its last mechanical quality, namely, its immobility.

[22] Fields are, by construction, Lorentz invariant. As for the æther, whether it is Lorentz invariant or not depends on what properties the æther is supposed to possess. Before the idea of Lorentz transformation was introduced, the aether was not Lorentz invariant and would therefore, if existed, be detected by the Michelson-Morley experiment. Lorentz transformation was introduced *precisely for the purpose* of constructing a (in some sense) "Lorentz-invariant æther theory".

> [...] the special theory of relativity does not compel us to deny æther. We may assume the existence of an æther; only we must give up ascribing a definite state of motion to it, i.e., we must by abstraction take from it the last mechanical characteristic which Lorentz had still left it. [...] according to the general theory of relativity space is endowed with physical qualities; in this sense, therefore, there exists an æther. According to the general theory of relativity space without æther is unthinkable; for in such space there not only would be no propagation of light, but also no possibility of existence for standards of space and time (measuring-rods and clocks), nor therefore any spacetime intervals in the physical sense. But this æther may not be thought of as endowed with the quality characteristic of ponderable media, as consisting of parts which may be tracked through time. The idea of motion may not be applied to it."

In other words, it is up to us to *define* what æther means. At some point, it is best to do away with the term to avoid confusions. Thus we speak of metric tensor, and fields,[23] but no longer of æther, though in modern times, the term æther still survives in a class of modified gravity theories in which there is a preferred reference frame, and involves both the metric tensor and an additional timelike vector field referred to as the "æther". Such a theory is called "Einstein-æther theory" (Jacobson & Mattingly 2004).

Plenty of advancements in science are like that: new, better ideas take over and the old ones are gradually forgotten.[24] Sometimes we feel that in modern theoretical physics there are too many wild ideas and there seems to be no real progress. However, there has always been no shortage of weird ideas throughout history, it's just that we no longer teach the failed ones today. This often gives the impression that science steadily marches on and are currently stuck. Progress may be slower during some era (maybe including now), but getting stuck is rather the norm.

Finally, let us mention some attempts at generalizing electromagnetism. One of the obvious shortcomings of electromagnetism is similar to the one in general relativity: matter is posited but not a part of the theory itself. In other words, electromagnetism describes how a charged particle like an electron moves in the presence of an electromagnetic field, but it doesn't tell us how an electron came to be. (Modern quantum field theory view is that an electron is an excitation of an *electron field*, which is distinct from the electromagnetic field; but then why is there an electron field?) Likewise, in general relativity matter interacts with spacetime geometry, but the theory doesn't tell us how matter field came to be. Gustav Mie (1868–1957) attempted to formulate a theory in the 1910s in which elementary

[23] We emphasize again that fields, though they permeate space, are quite different from the traditional æther which is a substantial medium.

[24] "A new scientific truth does not triumph by convincing its opponents and making them see the light, but rather because its opponents eventually die, and a new generation grows up that is familiar with it."—Max Planck, *Scientific Autobiography and Other Papers*.

particles are knots in the æther (see also Section 7.4). The idea is to devise a theory in which stable particles are solutions to the field equation. Furthermore, Mie wanted the theory to reduce to Maxwell's equations in a certain weak field limit. In addition, Mie hoped that gravitation would also emerge naturally from such a unified field theory (Smeenk & Martin 2007). To obtain a stable charge configurations such as an electron, Mie had to introduce nonlinear terms into the Maxwell equations[25]. Unfortunately Mie was not successful in building the theory; he couldn't find a suitable Lagrangian that would work; though he did try to illustrate his ideas with simple toy model. Consider for example the Lagrangian density

$$\mathscr{L} = -\frac{1}{4}F_{ab}F^{ab} + \alpha(A_c A^c)^3, \tag{7.35}$$

where F_{ab} is the electromagnetic tensor given by $F_{ab} = \partial_a A_b - \partial_b A_a$, where A^c is the component of the so-called "4-vector potential". When the coupling constant α vanishes, we recover the original electromagnetic Lagrangian density. To preserve Lorentz invariant, other types of terms that can be used are limited. Pauli noted in 1921 that one term was missed by Mie: $\frac{1}{4}F^*_{ab}F^{ab}$, where F^*_{ab} is the Hodge dual of F_{ab} (Smeenk & Martin 2007).

There are many other theories of nonlinear electrodynamics. For example, in the 1930s, Max Born (1882–1970) and Leopold Infeld (1938–1950) proposed the Born–Infeld model. This was introduced to remove the divergence of the electron's self-energy in classical electrodynamics by introducing an upper bound of the electric field at the origin. The Born–Infeld Lagrangian takes the form

$$\mathscr{L} = -s^2\left[-\det\left(\eta_{ab} + \frac{F_{ab}}{s}\right)\right]^{\frac{1}{2}} + s^2, \tag{7.36}$$

where s is a scale parameter, and η_{ab} is the matrix representation of the Minkowski metric. The scale parameter provides a cutoff for the value of the electric field to regulate the self-energy of point charges. For small values of the electric and magnetic fields, the theory reduces to the original Maxwell theory. Another nonlinear extension is known as the "Power–Maxwell electrodynamic", which aims to make electrodynamics conformally invariant even in higher dimensions. In this theory, the contraction $F_{ab}F^{ab}$ in the Lagrangian density is promoted to a power form $(F_{ab}F^{ab})^p$, hence the name. Conformal invariance means that under the conformal transformation the spacetime metric $g_{ab} \longmapsto \Omega^2 g_{ab}$ for some function Ω (as well as $A^c \longmapsto A^c$), the Lagrangian remains unchanged. The energy-momentum tensor for the modified field will also be traceless ($T^a_a = 0$) in all dimensions. This implies that the power depends on the number of spatial dimension d via the relation $p = (d + 1)/4$. In three spatial dimensions this reduces to Maxwell electrodynamics (Hassaine & Martinez 2007).

[25] Though he was not the first to consider possible generalizations of Maxwell equations. Einstein had also considered higher order or nonlinear extensions.

7.4 From the Dual Nature of Light to Atomic Models

In 1885, a mathematics teacher Johann Jakob Balmer (1825–1898) studied the visible spectral lines of the hydrogen atom, which consist of six distinct lines of different brightness. Among these is a deep-red color line with a wavelength of about 656 nm, called the "Hydrogen-alpha" or simply the Hα line. This is followed by a light blue line of wavelength 486 nm and so on. Actually Balmer was initially only aware of the first four that are obvious to the naked eyes (the last two have a wavelength below 400 nm, thus considered as ultraviolet, though they are still barely visible to the naked eyes).

Balmer soon noticed that he could recognize a pattern for the wavelength:

$$\lambda = \mathfrak{h} \cdot \frac{n^2}{n^2 - 4}, \tag{7.37}$$

where $n = 3, 4, 5, 6, \cdots$. The prefactor $\mathfrak{h}$ is about 364.56 nm, now called the Balmer constant. With this empirical formula he substituted in $n = 7$ and obtained a prediction for the wavelength of the fifth line in the series: 397 nm, which he later learned that this was exactly what was detected by Anders Jonas Ångström (1814–1874). The series is therefore known as the Balmer series.

A few other similar series were subsequently investigated, and in 1888 Johannes Rydberg (1854–1919) came up with the more general empirical formula

$$\frac{1}{\lambda} = R_H\left(\frac{1}{m^2} - \frac{1}{n^2}\right), \quad n > m, \tag{7.38}$$

where $R_H = 4/\mathfrak{h}$ is now called the Rydberg constant. This formula (the Rydberg formula) reduces to the Balmer series when $m = 2$.

Such a result created a profound mystery: what does the hydrogen emission line have to do with the integers? More generally, why are there emission lines at all, which are discrete, instead of a continuum of spectra? The explanation would have to wait for a new revolution in the history of science—the birth of quantum mechanics. This occurred only two years later in 1900, when Max Planck introduced a new constant that bears his name, in the effort to explain the thermal distribution of radiation emitted by a blackbody.

One often reads from textbooks that the problem facing Planck and the community of physicists back then was the so-called "ultraviolet catastrophe", that is, the Rayleigh–Jeans law predicted that the distribution of the energy (or intensity) emitted by a blackbody across the electromagnetic spectrum would be infinite as the wavelength becomes shorter and shorter (or equivalently, as the frequency becomes higher and higher). This is a historical fiction (Bomark & Renstr 2024), but for now let us continue. No divergence was seen in experiments. In fact, observational data did not indicate any huge amount of energy at all in the limit of zero wavelength (or infinite frequency)—instead it tends towards *zero*. We sometimes read that the "prediction" of quantum field theory for the value of the cosmological constant is the worst in the history of physics, with a value that is 10^{122} times larger than what observations indicated. Yet, surely the ultraviolet catastrophe[26] is infinitely worse! Thankfully, as we will soon explain, this story is not historically accurate! Planck managed to come up with a distribution, now called *Planck distribution* for the radiation that could fit the data. The energy density at frequency ν is, according to Planck, given by the formula

$$u_\nu(\nu, T) = \frac{8\pi h v^3}{c^3} \frac{1}{\exp\left(\frac{h\nu}{k_B T}\right) - 1}, \tag{7.39}$$

where k_B is the Boltzmann constant, and h the Planck constant. In terms of wavelength, we have

$$u_\lambda(\lambda, T) = \frac{8\pi hc}{\lambda^5} \frac{1}{\exp\left(\frac{hc}{k_B T\lambda}\right) - 1}, \tag{7.40}$$

Now back to the actual history: the name "ultraviolet catastrophe" was given by Paul Ehrenfest much later in 1911, and in fact Planck did not think along this line for he was not aware of any catastrophe; he was simply seeking a better distribution. Several years back, in 1896, Wilhelm Wien (1864–1928) proposed the radiation law of the form

$$u_\lambda(\lambda, T) = c_1 \frac{T}{\lambda^5} e^{-\frac{c_2}{\lambda T}}, \tag{7.41}$$

which seemed to fit the data. But as more experiments were performed, deviations were found at the large wavelength regime. Planck therefore attempted to devise a better formula. Note that the deviation found was for the long wavelength limit, not the short wavelength limit that would correspond to the ultraviolet catastrophe. Planck's distribution worked, and it was presented during a meeting at the Prussian Academy in 1900. Rayleigh's proposal was presented at the same time but didn't gain much attention. Rayleigh tried to improve on Wien's result in the long wavelength regime by applying the equipartition law in statistical mechanics and obtained an expression of the form

$$u_\lambda(\lambda, T) = c_3 \frac{T}{\lambda^4}. \tag{7.42}$$

This was *not* meant to describe the radiation energy density at all wavelengths, and he was aware that one cannot justify the use of the equipartition law in general. Of course, it is trivially obvious that this expression will lead to "ultraviolet catastrophe", and he did also considered introducing an exponential cutoff (Rayleigh 1900) which

[26] Ultraviolet, or simply UV, has since become a synonym for "high energy", especially in particle physics, quantum field theory, and high energy physics in general. Its opposite is of course IR, or infrared.

would have avoided this issue (again, a natural thing to do, just like in Seeliger's modified Newtonian gravity; see Section 6.4), so that

$$u_\lambda(\lambda, T) = c_3 \frac{T}{\lambda^4} e^{-\frac{c_4}{\lambda T}}. \tag{7.43}$$

It is not fair to present the straw man Equation (7.42) and claimed as if this led to a serious issue, while even back then this was well understood and alternatives were available. Somehow the history became confused in books after years of re-telling.

Anyway, back to our main narrative. In deriving his formula, Planck first modeled a blackbody with a cavity with perfectly reflective walls (which is still the way we teach undergraduates today), filled with finitely many distinct yet identical resonant oscillators with various frequencies ν. For each frequency there corresponds some number of oscillators with said characteristic frequency, and the energy at that frequency gets distributed to each oscillator in an integer multiple of a "unit of energy" ε.

In other words, Planck had made the assumption that a unit of energy ε is proportional to the frequency, namely $\varepsilon = h\nu$. Today we recognize this as the "quantization of energy", that energy comes in discrete packets instead of a continuum. Yet Planck's opinion was that this was merely a mathematical technicality and didn't think too much of its possible physical ramification. Modern readers may think that this is rather strange; how was the fact that energy is now discrete not worth a further thought by such a great physicist like Planck? Of course, it's true that Planck was then more concerned with "getting the right answer", as he once said

> "I considered the [quantum hypothesis] a purely formal assumption, and I did not give it much thought except for this: that I had obtained a positive result under any circumstances and at whatever cost."

That said, there was another reason: such an idea was not entirely new. He was inspired by what Boltzmann did when dealing with distributions of energy in a given system (though as we shall see in Chapter 10, he was at one point, rather skeptical of the probabilistic methods of Boltzmann for understanding entropy). In the probabilistic treatment of statistical mechanics, to help with the calculations, it is often useful to introduce some cutoff scale δ in the intermediate step, and then to take $\delta \to 0$ at the end to get the right answer. This scale has no physical significance; it is only a mathematical technique[27]. Thus, as soon as the obtained energy distribution matched the experimental data, Planck thought that was all to be done. Though, from hindsight, we see that the occurrence of h in the Planck distribution is quite

[27] We do similar things in quantum field theory with "dimensional regularization" when we consider the dimension of spacetime to be $4 - \delta$. We don't seriously think that this is anything physical!

unlike the arbitrary cutoffs in earlier physical considerations: we cannot get rid of it at the end by taking $h \to 0$. In fact, h has a definite value $6.62607015 \times 10^{-34}$ J Hz^{-1}. Nowadays we often work with the reduced Planck constant $\hbar = h/2\pi$ instead.

It was Einstein who first took the Planck constant h seriously as "new physics" when he used the assumption of discrete energy quanta to explain the photoelectric effect (the observation that when light beams of sufficient energy shine onto a metal surface, electrons can be emitted from the material) in 1905. This earned him the 1921 Nobel Prize in Physics (he never won the Nobel Prize for either special relativity or general relativity). Textbooks often portrayed this episode in history as convincing evidence that quantum mechanics is correct, asserting that classical (i.e., non-quantum) physics couldn't explain all the features of the photoelectric effect convincingly. However, although by now there is no doubt that quantum mechanics is indeed correct, we should be careful of what "quantum" really means in this context. While it is true that by assuming that light comes in packets of energy, now called *photons*, Einstein could explain the photoelectric effect, there are also other ways to get the right answer. Already back in 1969, Willis E. Lamb, Jr. and Marlan O. Scully wrote:

> "A misconception which most physicists acquire in their formative years is that the photoelectric effect requires the quantization of the electromagnetic field for its explanation. [...] In fact we shall see that the photoelectric effect may be completely explained without invoking the concept of 'light quanta'."

We now know that photons do exist, from other modern experiments, but the typical photoelectric effect does not necessitate the existence of photons. The trick to explain the photoelectric effect without photons is to consider a "semi-classical" theory in which the material atoms are quantized (or more precisely, electrons in these atoms can only take some specific orbits, as in the *Bohr model* that we will soon discuss), but that the incoming light is still treated classically as an electromagnetic wave. In such an approach, light would still be a wave and not a particle. Still, Einstein's insight that light can be a wave yet sometimes behave like being made of particles was revolutionary.

We now know that the energy of a photon is $E = h\nu$, and its momentum is $p = E/c = h/\lambda$. The relativistic energy for any particle satisfies

$$E^2 = (pc)^2 + (mc^2)^2. \tag{7.44}$$

Thus for a massless particle, the second term vanishes and we recover $E = pc$. Meanwhile, for a massive particle, we can show that this expression reduces to

$$E = \gamma mc^2, \tag{7.45}$$

where γ is the Lorentz factor.

In 1924, Louis de Broglie (1892–1987) would turn the idea the other way round in his PhD thesis: if waves can exhibit properties of particles, can particles also behave like a wave? For a massive object in Newtonian mechanics, $p = mv$, so he proposed that for such a body, one should associate with it a wavelength

$$\lambda = \frac{h}{p} = \frac{h}{mv}. \tag{7.46}$$

which is now called the *de Broglie wavelength*. Since macroscopic bodies have relatively large mass, the de Broglie wavelength is tiny, thus explaining why we do not see particles behaving like waves—a ball does not "diffract" as it goes through a door, for example. For small objects like an electron, the effect becomes pronounced. The famous double-slit experiment that shows light behaving as a wave can be performed using electrons instead (Figure 7.10). This was carried out in 1927 by Clinton Davisson and Lester Germer, and independently by George Paget Thomson and his research student Alexander Reid. Instead of leaving just two prominent strips on the screen as the electrons pass through the two slits, an interference pattern can be observed. Even more bizarre was the fact that the same interference pattern occurs even if we shoot out one electron at a time. If an experiment is devised to detect exactly which slit an electron passes through, however, the interference pattern would not appear; there would be only two strips. Variation upon variation of the double-slit

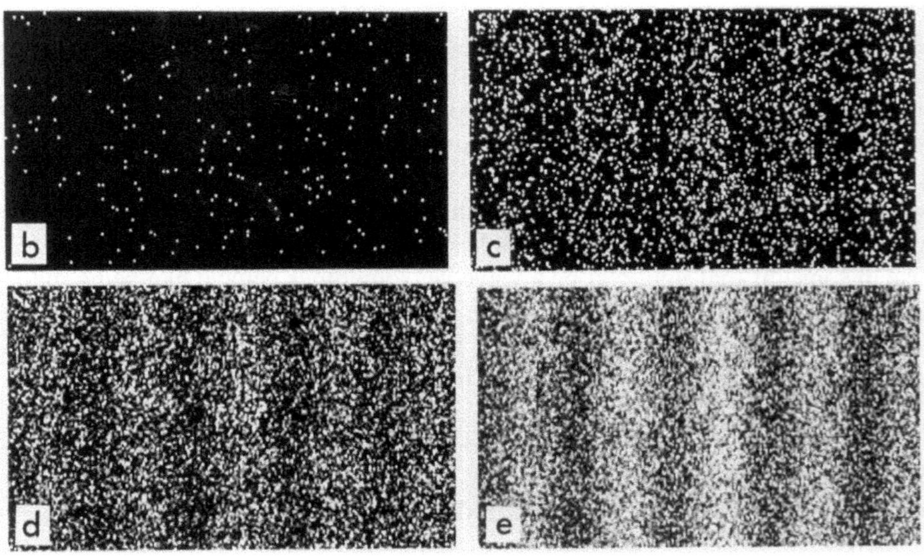

Figure 7.10. A double-slit experiment conducted with electrons. As more and more electrons are fired through the slits, a wave interference pattern gradually forms on the screen. This remains the case even if only one electron is fired at a single time. Particles can therefore exhibit wave behaviors, illustrating the "wave-particle duality" of the quantum world. *Credit:* This [Double-slit experiment results Tanamura four] image has been obtained by the author(s) from the Wikimedia website where it was made available by [Cantons-de-l'Est] under a CC BY-SA 3.0 licence. It is included within this book on that basis. It is attributed to [Dr. Tonomura and Belsazar].

experiment have been carried out since then, revealing many more strange properties of the quantum world.

All this came from the realization of de Broglie that we should treat particles and waves on equal footing. In other words, they are different facets of the same thing; two sides of the same coin. Recognizing this, Niels Bohr proposed the "principle of complementarity", in the sense that particle and wave behaviors are both necessary and they complement each other[28].

Things are obvious in hindsight. However, from the idea of photons in 1905 to 1924, almost 20 years passed before the idea occurred to de Broglie. *Why wasn't it obvious?* One of the reasons is of course: no one had ever seen a particle behaving like a wave, so no one had any reason to think along this line. However, theoretical physicists are often very creative people, and predictions for hitherto unobserved phenomena are not rare in physics, so this cannot be the only reason. Though I have no strong evidence for this claim, I think it probably has to do with the fact that the nature of light has always been "in the limelight" (pun intended!), with debates about whether it is a wave or particle throughout the history of physics, so when the idea of photons appeared, it brought with it new discussions concerning the nature of light and subsequently also of matter. Still, in Bohr's model, atoms and electrons are still particle, only the electron orbits are "quantized", there were no consideration that the particles themselves can also be a wave. Indeed, unlike light, the nature of matter was not as fiercely debated. It is true that there were debates concerning whether atoms are physically real or just mathematical quantities, but compared to light, matter is such tangible stuff that almost no one would seriously contemplate the possibility that it can be "wavelike". How could a solid tangible object be a wave? In other words, the possibility of particle-wave duality was simply too far-fetched.

We end this chapter with some discussions on the Bohr model. The structure of atoms was another important episode in the history of physics, but unfortunately it would deviate too much from our main discussion to go into greater detail. There is some widely accepted historical fiction here that should be clarified, however. It is commonly mentioned that Ernest Rutherford (1871–1937) proposed that the atom consisted of a central nucleus surrounded by orbiting electrons. This is not true.

---

[28] Niels Bohr was fascinated with the Chinese philosophy of *Yin* and *Yang*, with two opposite qualities containing a bit of each other and complementing one another in a harmony (c.f. the notion of *coincidentia oppositorum* proposed by Nicholas of Cusa). When the King of Denmark Frederick IX decided to confer the Order of the Elephant on Bohr in 1947, Bohr designed his own coat of arms incorporating the *Taiji* symbol (though with red color replacing the traditional white for *Yin*), and a Latin motto *contraria sunt complementa*, "opposites are complementary". It has been argued that Bohr was also inspired by the art form known as "cubism" (he bought the cubism painting *La Femme au Cheva*, "Woman with a Horse", by Jean Metzinger (1883–1956)). Just like motion is represented by showing snapshots of different moments of time in a single Cubism painting, an electron is an entity that has both wave and particle properties simultaneously (Schinckus 2016). Cubism was itself inspired by viewing time as the fourth dimension, as its founder Pablo Picasso (1881–1973) was made aware of Poincaré's book *Science and Hypothesis* via a mathematically trained insurance actuary Maurice Princent (1875–1973), who explained to him the idea of a fourth dimension. It is interesting to note that Cubism was developed also around 1905, the same time Einstein discovered special relativity (Brush 2011).

Rutherford's famous gold foil experiment led him to the conclusion that the atom has a very dense core (the nucleus), but he had nothing to say about how the electrons were distributed. The orbiting electron picture has a big problem that people at that time, being well-versed in the physics of Maxwell's electromagneticism, would be aware of: a point charge in orbit is necessarily accelerating, and so it emits electromagnetic radiation. The electrons would thus very quickly spiral into the nucleus as it loses energy and cannot maintain its orbit. It is often said that Bohr's model, based on the rudimentary understanding of quantum physics back then (strictly speaking, quantum *mechanics* wasn't yet born), improved on this model by demanding that electrons can only stay on some discrete fixed orbit, unless they absorb or emit enough energy to "jump" to other orbits. This is true, but even before Bohr, J.J. Thomson (1856–1940) had attempted to address the stability problem by devising complicated models of electrons in systems of rings to render the whole configuration stable. The textbook's "plum-pudding model" in which the electrons are embedded into a diffused positively charged ball is a modern fiction that simply doesn't fairly depict Thomson's sophisticated thinking.[29]

There are of course also questions one should raise regarding Bohr's model. For example, why don't the electrons radiate electromagnetic waves and lose energy when it stays on the permissible orbit? Well, he simply *postulated* that it must be so. This is not very satisfying, but the model did give us some insights on why only certain orbits are allowed.

If we treat the hydrogen atom as a point charge electron orbiting a nucleus like a planet orbiting the Sun classically, then the centripetal force of the orbiting electron is provided by the electrostatic force

$$\frac{mv^2}{r} = \frac{e^2}{4\pi\varepsilon_0 r^2} \Longrightarrow r = \frac{e^2}{4\pi\varepsilon_0 m v^2}. \tag{7.47}$$

where e is the charge of the electron. The quantization rule for the angular momentum was known to be $L = mvr = n\hbar$, where n is a positive integer. This gives

$$v = \frac{n\hbar}{mr}. \tag{7.48}$$

Upon substituting this into the preceding equation, we can obtain

$$r = \frac{4\pi\varepsilon_0 \hbar^2 n^2}{me^2}. \tag{7.49}$$

[29] See Hon and Goldstein's work (Hon & Goldstein 2013) for an investigation into how this myth of plum-pudding model came from; see also Renstrøm & Bomark (2025). Essentially, it was popular science gone wrong. Popularization of scientific ideas often tries to draw analogies that were not accurate and can be misleading. This is true nowadays and also true back then. After a while, the inaccurate image was confused with the original model and got introduced back into technical literature as such. This reminds me of the history of unidentified flying objects or UFOs (now officially called UAP—unidentified anomalous phenomena; see Section 13.5). In 1947, the pilot Kenneth Arnold saw a few strange objects which he later described as "they flew erratic, like a saucer if you skip it across the water." This description of their motion was later confused with their shape. In the years after that, many sightings of "flying saucers" were reported.

Setting $n = 1$ gives $r \approx 5.3 \times 10^{-11}$ m, a length scale known as the Bohr's radius, which is essentially the size of hydrogen atom. Note that Equation (7.48) for $n = 1$ is essentially equivalent to $\lambda = h/p$ (modulo a factor of 2π), but of course Bohr did not know about particle-wave duality back then.

Using the Bohr model, the Balmer series can be explained. As an electron jumps to the second energy level closest to the nucleus (this is the $m = 2$ in the Rydberg formula) from more distant levels, a photon with a certain energy (and hence wavelength) is emitted. For example, the $3 \to 2$ transition produces the Hα line.

The Bohr model is *not* the final picture of atomic structure, which will require proper quantum mechanics in the modern sense, in which we have to solve the Schrödinger equation. But let us mention that there were other models that were not usually mentioned in modern textbooks, not even as historical tidbits. One of the early pioneers was Joseph Larmor,[30] who already proposed a "solar system" model of atoms back in 1897, but he did not have the idea of a nucleus back then (he did anticipate a positive charge constituent since overall an atom is neutral). However, Larmor had another idea that was actually much more intricate and interesting (though ultimately incorrect). Before introducing his idea, it was worth mentioning that Lord Kelvin proposed that atoms are vortices in the æther in 1867. Peter Guthrie Tait (1831–1901), who worked on the vortex theory, pioneered what would become knot theory in mathematics.[31] From modern eyes, this is similar to the idea that deep down at the fundamental level, there is no distinction between matter and geometry ("geometrodynamics"), or the more recent idea of "it from bits" that everything is fundamentally made of quantum information. Even in the more mundane (by modern standard) context of quantum field theory, fundamentally there are only quantum fields, as what we call particles are just excitations of their corresponding field.

Larmor proposed the idea that electrons resulted from a rotational strain in the æther, i.e., "defects" in the sea of æther that were somehow endowed with a charge. In modern times, we find the concept of defects in certain quantum field theories (e.g., grand unified theories), in the context of phase transitions that can give rise to "topological defects" like monopoles and cosmic strings in the early Universe, for example. Larmor contemplated the possibility that an atom is just a complicated structure made of charged particles. Of course, Larmor was the first to calculate the rate of energy radiation from an accelerating electron ("Larmor radiation" and "Larmor power"), so he was well aware that an accelerating charge gives off electromagnetic waves and loses energy in the process. Thus, he was faced with the same stability issue we discussed before. Let us take a look at how he thought about this problem in 1894:

> "It may be objected that a rapidly revolving system of electrons is effectively a vibrator, and would be subject to intense radiation of its

[30] Larmor was also the first to propose time dilation in 1897, and with it obtained the full Lorentz transformation.

[31] A failed physical theory can give rise to interesting and valid mathematics!

> energy. That however does not seem to be the case. We may on the contrary propound the general principle that whenever the motion of any dynamical system is determined by imposed conditions at its boundaries or elsewhere, which are of a steady character, a steady motion of the system will usually correspond, after the preliminary oscillations, if any, have disappeared by radiation or viscosity. A system of electrons moving steadily across the medium, or rotating steadily round a center, would thus carry a steady configuration of strain along with it; and no radiation will be propagated away except when this steady state of motion is disturbed."

So, like Bohr almost 20 years later, and before the era of quantum physics, Larmor proposed essentially the same ad hoc claim that in certain "steady motion", accelerating charges simply do not radiate! Note that this is really not an explanation, but merely stating what is observed: since the observed atoms are stable, there cannot be electromagnetic radiation that renders it unstable. This even has the feel of what we today call the (weak) *anthropic principle*: the Universe is as it is, because otherwise we cannot be here to ask question about why it is like this. The feeling of tautology leaves us unsatiated.

7.5 Exercises

(1) We mentioned that for a massive particle, the equation

$$E^2 = (pc)^2 + (mc^2)^2. \tag{7.50}$$

is equivalent to

$$E = \gamma mc^2, \tag{7.51}$$

where γ is the Lorentz factor. Prove that this is indeed the case.

(2) There is another atomic model that is quite often neglected in most textbooks: the "Saturnian model" proposed by Japanese physicist Hantaro Nagaoka (1865–1950) in 1904. In this model, the positive charge is carried by a sphere, while the negative charged electrons form a ring that orbits around it. Find out more about this model and why it doesn't work. Why don't the actual rings of Saturn face the same problems? (Saturn's "ring" actually consist of many distinct rings, hence the plural.)

(3) In the Bohr model, an electron orbits the nucleus of an atom under a central force (the electrostatic force). Balancing the centripetal force with the electrostatic force as we have done in Equation (7.47), and by imposing the quantization rule onto the angular momentum, we derived the Bohr radius:

$$r = \frac{4\pi\varepsilon_0\hbar^2 n^2}{me^2}. \tag{7.52}$$

Compare and contrast this derivation of the Bohr radius with Feynman's derivation in *Feynman Lectures of Physics*, Vol. 3 (Chapter 2, Sections 2–4), available online at https://www.feynmanlectures.caltech.edu/III_02.html#Ch2S4. Are these two derivations equivalent? Why?

Hint: Did Feynman consider the angular momentum at all? If not, why?

Additional note: By the way, do not be too bothered by constants like 2 and π, since the calculation is only meant to be heuristic to capture the essence of the physics. Beginner students often get too troubled by getting every single detail right, but it is more important to get the *idea* correct first and foremost—the detail can wait. As Richard Hamming (1915–1998) once wrote in his *Numerical Methods for Scientists and Engineers* (1962): "The purpose of computing is insight, not numbers".

References

Bomark, N.-E., & Renstrøm, R. 2024, EPSCHEP (EPS-HEP2023), 449, 660
Brush, S. G. 2011, PhT, 54, 49
Browne, K. M. 2020, AmJPh, 88, 207
Ceredig, R. 2020, RSPTA, 378, 20200105
Crispino, L. C. B. 2011, IJMPD, 29, 2041001
Dux, F., Millon, M., Lemon, C., et al. 2025, A&A, 694, A300
FitzGerald, G. F. 1889, Sci., 13, 390
Goldberg, S. 1949, Chy, 2, 89
Grillo, C., Rosati, P., Suyu, S. H., et al. 2020, ApJ, 898, 87
Hassaine, M., & Martinez, C. 2007, PhRvD, 75, 027502
Hon, G., & Goldstein, B. R. 2013, AnP, 525, A129
Hunt, B. 1986, HSPBS, 16, 111
Jacobson, T., & Mattingly, D. 2004, PhRvD, 70, 024003
Lynning, K. H., & Jacobsen, A. S. 2011, SHPS, 42, 45
Naber, G. L. 2012, The Geometry of Minkowski Spacetime: An Introduction to the Mathematics of the Special Theory of Relativity (2nd ed.; New York: Springer)
Schinckus, C. 2016, FoS, 22, 709
Smeenk, C., & Martin, C. 2007, The Genesis of General Relativity, Boston Studies in the Philosophy of Science, Vol. 250, ed. M. Janssen, et al. (Berlin: Springer)
Soares, D. S. L. 2009, [arXiv:physics/0508030 [physics.gen-ph]] 2005physics...8030S
Stephenson, F. R., Morrison, L. V., & Hohenkerk, C. Y. 2016, PRSA, 472, 20160404
Rayleigh, L. 1900, LEDPMJS, 49, 539
Renstrøm, R., & Bomark, N.-E. 2022, [arXiv:2212.08572 [physics.ed-ph]] 2022arXiv221208572R
Rodney, S. A., Brammer, G. B., Pierel, J. D. R., et al. 2021, NatAs, 5, 1118
Weyl, H. 2009, Philosophy of Mathematics and Natural Science (Princeton, NJ: Princeton Univ. Press)

Chapter 8

From Quantum Mechanics to Imaginary Time

> *"It was about ten days before Christmas in the winter of 1925. [...] But this was no ordinary vacation. Leaving his wife at home, Schrödinger booked a villa in the Swiss Alps for two and a half weeks, taking with him his notebooks, two pearls, and an old Viennese girlfriend. [...] The Viennese-born physicist placed a pearl in each ear to screen out any distracting noises."*—Leon Lederman, 'The God Particle'.

Having introduced the birth of quantum physics at the end of the last chapter, we now formally explain some aspects of quantum mechanics. Instead of the usual material one can find in standard textbooks, we will start by motivating the standard "trick" of Wick rotation via the heat equation, and the Black–Scholes equation in finance, which gives us the Schrödinger equation. We then explore how Wick rotation in a path integral works, and how temperature of a field theory is essentially cyclic imaginary time. Later on we will discuss quantum entanglement and the Bell inequality, as well as how the cosmos provides an arena to close a certain loophole in the Bell test. We also discuss how quantum mechanics can explain various phenomena in astrophysics and cosmology, such as: how the Sun shines; how white dwarfs and neutron stars are supported by degenerative pressure; how neutrino oscillation works and what is a Majorana particle; and what is "quantum cosmology".

If there is a field of physics that is the most misused by cranks and charlatans, it is probably quantum mechanics; you have probably heard of many dubious claims like quantum healing, quantum water or other equally nonsensical products. Some students are also especially interested in quantum mechanics given that many of its results seem so counterintuitive it is almost magical. I guess we all yearn for something extraordinary in our otherwise mundane lives. Philosophically speaking,

doi:10.1088/2514-3433/ade963ch8

there are also a lot of unresolved issues concerning how we interpret quantum mechanics. The "Many-World interpretation", for example, could be especially intriguing to laymen and students alike, since it reminds them of science fiction stories that involve parallel universes or the "multiverse". I personally feel that many people are attracted to the idea of parallel universes because we cannot help but to contemplate what kind of life we would be living if we have made different choices at various points. It's the same sentiment expressed in *Invisible Cities* by Italo Calvino:

> "Marco enters a city; he sees someone in a square living a life or an instant that could be his; he could now be in that man's place, if he had stopped in time, long ago; or if, long ago, at a crossroads, instead of taking one road he had taken the opposite one, and after long wandering he had come to be in the place of that man in that square. By now, from that real or hypothetical past of his, he is excluded; he cannot stop; he must go on to another city, where another of his pasts awaits him, or something perhaps that had been a possible future of his and is now someone else's present. Future not achieved are only branches of the past: dead branches."

Even if parallel universes exist "out there", do they provide any solace knowing that each and every possible life choice is somehow lived out, *yet we can only live out one of them?* As Calvino also wrote, concerning all the possible futures, "one contains what is accepted as necessary when it is not yet so; the others, what is imagined as possible and, a moment later, is possible no longer".

Perhaps the greatest allure of quantum mechanics is that it implies that reality is not at all like how we perceive it to be. As Niels Bohr once purportedly[1] said:

> "Everything we call real is made of things that cannot be regarded as real. If quantum mechanics hasn't profoundly shocked you, you haven't understood it yet."

In this chapter, we won't have much to say about philosophical issues despite my strong interests in them. In particular I will not discuss the various interpretations of quantum mechanics in length. Instead I would like to focus on the physics. In addition to the usual topics like the Schrödinger equation (though there will be no examples given on how to solve it—those are readily available in standard textbooks) and path integrals, we will also briefly discuss quantum entanglement and Bell inequality (though not in full technical detail), a topic that is rarely discussed in cosmology textbooks. The reason I include it here is because the various tests of Bell inequality, known collectively as the "Bell test", has eventually *put the cosmos to use* in the attempt to close a certain loophole. Using an extremely large scale to test for the laws that govern the very small, is too good a story not to be

[1] I could not find the original source.

included! Lastly we shall discuss the applications of quantum mechanics in astrophysics and cosmology.

8.1 The Schrödinger Equation and Imaginary Time

If there is a "matter wave" as suggested by de Broglie, what is the wave equation that governs its dynamics? This question was raised by Peter Debye[2] (1884–1966), and the answer was found by Erwin Schrödinger. The equation is

$$i\hbar\frac{\partial\psi}{\partial t} = -\frac{\hbar^2}{2m}\frac{\partial^2\psi}{\partial x^2} + V(x)\psi, \tag{8.1}$$

where ψ is the *wavefunction* and $V(x)$ is a potential term, with $i = \sqrt{-1}$ the imaginary unit.

Now, just like Newton's laws of motion, the Schrödinger equation cannot be derived from a more fundamental principle. However, we can try to motivate it. First, we note that this is very different from the classical wave equation that propagates at speed v:

$$\frac{\partial^2 y}{\partial t^2} - v^2\frac{\partial^2 y}{\partial x^2} = 0. \tag{8.2}$$

It is straightforward to check that any function of the form $y = f(x - vt)$ is a solution to Equation (8.2). So why can't this wave equation describe the quantum matter wave? Well, let us try an ansatz of the form

$$y = A\cos(kx - \omega t), \tag{8.3}$$

where ω is the angular velocity and k the wave number. The cosine function (or a sine function) is an obvious choice to model a wave. However, we know that energy is quantized, so $E = h\nu = \hbar\omega$. Similarly, the wave number is related to the quantization of momentum via $p = \hbar k$. If we use this quantization relation the ansatz would become

$$y = A\cos\left(\frac{px - Et}{\hbar}\right). \tag{8.4}$$

Substituting this into the wave equation Equation (8.2), we would obtain a simple relation between E and p, which is $E = pv$. However, we know that this cannot be true unless $v = c$. This can be deduced from the equation $E^2 = p^2c^2 + m^2c^4$ from Einstein's special relativity. Yet we want to have a wave equation for *matter*, with nonzero mass. So the wave equation is not suitable. One might object that maybe with a different, more complicated ansatz, the wave equation could work. However, this is not the case. The fundamental problem here is that for non-relativistic matter, say, a free particle of mass m, its kinetic energy is

[2] Debye purportedly said to Schrödinger, "Schrödinger, you are not working right now on very important problems anyway. Why don't you tell us some time about that thesis of de Broglie, which seems to have attracted some attention."

$$E = \frac{p^2}{2m}, \tag{8.5}$$

which implies that for the quantum particle,

$$\omega = \frac{\hbar}{2m}k^2. \tag{8.6}$$

The fact that ω is not proportional to k but rather to k^2 suggests that we should not treat the time derivative on an equal footing with spatial derivative, as was the case in Equation (8.2). Instead, we should have an equation for which

$$\frac{\partial y}{\partial t} = C\frac{\partial^2 y}{\partial x^2}, \tag{8.7}$$

for some proportionality constant C. However, even then we cannot have a simple ansatz as in Equation (8.4) since the first derivative would result in a sine function, whereas the second derivative would give a cosine function. What works is a plane-wave type solution which contains both cosine and sine combination, namely

$$y = A\exp[i(kx - \omega t)]. \tag{8.8}$$

We can now substitute this new ansatz into Equation (8.7). The readers can check that this fixes the constant C and we finally end up with the Schrödinger equation for free particles:

$$\frac{\partial f}{\partial t} = i\frac{\hbar}{2m}\frac{\partial^2 f}{\partial x^2}. \tag{8.9}$$

Equivalently, upon replacing f by the notation for wavefunction ψ, we have

$$i\hbar\frac{\partial \psi}{\partial t} = -\frac{\hbar^2}{2m}\frac{\partial^2 \psi}{\partial x^2}. \tag{8.10}$$

In three-dimensional space, we simply replace $\frac{\partial^2}{\partial x^2}$ with $\frac{\partial^2}{\partial x^2} + \frac{\partial^2}{\partial y^2} + \frac{\partial^2}{\partial z^2}$. With this motivation we see that the imaginary unit i is not as mysterious as one may think. It appears quite naturally due to Euler's famous identity $e^{ix} = \cos x + i\sin x$ and its application in representing waves.

What is the *physical interpretation* of ψ? This is an extremely difficult problem that gave rise to an entire branch of philosophy of science—the question of the interpretation of quantum mechanics, which we will not get into. The usual textbook interpretation is that ψ is a *probability wave.* Recall the double slit experiment performed with electrons. We can compute the probability of the electron being detected at location x at time t by computing the probability as

$$\text{Prob}(x, t) \equiv |\psi(x, t)|^2 = \psi(x, t)^*\psi(x, t), \tag{8.11}$$

where $\psi(x, t)^*$ denotes the complex conjugate of $\psi(x, t)$. In the standard interpretation this is just a rule with no derivation; we accept it simply because it works! We call it the *Born rule*. The wavefunction can be normalized by using the fact that the probability of finding the electron *somewhere* at time t must be 1. That is to say,

$$\int_{-\infty}^{+\infty} |\psi(x, t)|^2 dx = 1. \tag{8.12}$$

Nothing material is actually waving in this interpretation, it is like talking about a "crime wave"—a crime wave means that there are more crimes statistically speaking.

This inherent probabilistic nature of quantum mechanics is a revolution. Even though in classical physics, statistical methods are used when we are dealing with a system with many particles (as in the case of statistical mechanics), fundamentally everything is still deterministic. In quantum mechanics however, even if there is only *one* particle, there is still an inherent uncertainty built into the theory. In the single-electron double slit experiment, the electrons only become a definite existence when detected at the screen or detector. Between the source and the detector, we cannot say what happens to the electron; there is no electron per se to be spoken of, only a probability wave. We cannot even predict exactly where a given electron will hit the screen; we can only determine the distribution of a collection of electrons after the experiment has run for a while. John Wheeler once metaphorically described the state of a photon in between its source and the detector as a "great smoky dragon":

> "The mouth of the dragon is sharp, where it bites the counter [the detector]. The tail of the dragon is sharp, where the photon starts. What the dragon does or looks like in between we have no right to speak."

This is why in modern chemistry, we no longer have definite orbits for electrons around a nucleus. There cannot be. We can only compute the probability of the electrons appearing at some location. Thus the modern depiction of atomic structure shows only "electron clouds". This raises serious philosophical issues. If the wavefunction is in a superposition of states before the system is observed, does that mean that the Moon is not really there when no one is looking at it? (This is a question once raised by Einstein[3].) The famous half-alive-half-dead Schrödinger's cat was devised to highlight such an ontological issue. The common consensus (at least in the interpretation of quantum mechanics that accepts wavefunction collapse—the so-called "Copenhagen interpretation", which itself comes in many sub-varieties) is that one indeed cannot talk about what happens when the system is not observed/measured. However, *why* do a quantum system's many possible states collapse into a single definite state when measured? This is known as the "measurement problem". Note that most physicists would not think that wavefunction

[3] To which Pauli's thought was: "One should no more rack one's brain about the problem of whether something one cannot know anything about exists all the same, than about the ancient question of how many angels are able to sit on the point of a needle. But it seems to me that Einstein's questions are ultimately always of this kind." (Mermin 2025)

collapse requires a *conscious* observer.[4] Sometimes one reads the argument that since a system is always interacting with its environment (a cosmic ray or neutrino hitting the system, for example), there will be *decoherence*, a process that causes the system to lose its quantum properties. Gravity acting between the Earth and Moon, for example, *may* decohere the Moon wavefunction and therefore the Moon remains a classical object even when no one is looking at it. However, decoherence cannot completely solve the measurement problem because it cannot explain the origin of Born's rule. One has to postulate the Born rule even in the decoherence picture.

Finally, let us take a look again at the Schrödinger equation and pay attention to the asymmetry in the time and spatial derivative. There is another famous differential equation in physics that behaves like that—the heat equation, which we already introduced in the beginning of Chapter 4. In one-dimensional space, we have

$$\frac{\partial u}{\partial t} - \alpha \frac{\partial^2 u}{\partial x^2} = 0, \tag{8.13}$$

where α is called the *diffusion constant*. Indeed, the heat equation describes a diffusion process: imagine a box of gas in which initially some molecules in the center have higher kinetic energy (a hotter spot); their motion will eventually transfer and spread out their kinetic energy to other molecules so that in the end we have a uniform temperature distribution. If we compare this with the Schrödinger equation, Equation (8.10), we see that they are very similar. In fact, if we let $t = -i\hbar\tau$, then the Schrödinger equation becomes

$$\frac{\partial \psi}{\partial \tau} - \frac{\hbar^2}{2m} \frac{\partial^2 \psi}{\partial x^2} = 0, \tag{8.14}$$

which is the heat equation with diffusion constant given by $\alpha = \hbar^2/2m$.

Another more involved example is to consider the Black–Scholes equation in financial mathematics:

$$\frac{\partial V}{\partial t} + rS\frac{\partial V}{\partial S} + \frac{1}{2}\sigma^2 S^2 \frac{\partial^2 V}{\partial S^2} = rV, \tag{8.15}$$

where S is the price of a stock, σ is the *volatility* of the stock, and r is the annualized risk-free interest rate, continuously compounded. By completing the square (see Exercise at the end of this section) this can be put into the form[5]

$$\frac{\partial V}{\partial t} = -\frac{\sigma^2}{2}\left[S\frac{\partial}{\partial S} - \frac{1}{2}\left(1 - \frac{2r}{\sigma^2}\right)\right]^2 V + \frac{\sigma^2}{8}\left(1 + \frac{2r}{\sigma^2}\right)^2 V. \tag{8.16}$$

Then, define $x = \ln S$, and introduce two new quantities

[4] To distinguish the possible role of consciousness, by the Copenhagen interpretation we usually mean the one without the need of a conscious observer; whereas the "Von Neumann–Wigner interpretation" is used to refer to the more controversial school of thought that "consciousness causes collapse".

[5] Part of the calculations below are based on the discussion in Physics Stack Exchange (2025).

$$p := -i\sigma\left[\frac{\partial}{\partial x} - \frac{1}{2}\left(1 - \frac{2r}{\sigma^2}\right)\right], \tag{8.17}$$

and

$$U := \frac{\sigma^2}{8}\left(1 + \frac{2r}{\sigma^2}\right)^2. \tag{8.18}$$

This allows Equation (8.16) to be reduced into the form

$$\frac{\partial V}{\partial t} = \left(\frac{p^2}{2} + U\right)V. \tag{8.19}$$

Compare this with the Schrödinger equation of a unit-mass particle under some potential U after the transformation of $t = -i\hbar\tau$, and momentum operator $p = -i\hbar\partial_x$ (with ψ replaced by V):

$$\frac{\partial V}{\partial \tau} = \left(-\frac{p^2}{2} - U\right)V. \tag{8.20}$$

This differs from the Black–Scholes equation of the form Equation (8.19) by a minus sign. However, in financial markets, the Black–Scholes equation is typically determined by the *final condition* (more often called "terminal condition") instead of an initial condition. In fact, the equation is ill-posed if one insists on solving it forward in time, in the sense that numerically a tiny difference in the initial data could lead to a vastly different solution at a later time. This is why in finance, we cannot predict future values from the current worth of the portfolio. See Chapter 8 of *Introduction to Partial Differential Equations* by Peter J. Olver (2013).

In other words, the Black–Scholes equation provides the dynamics for the option values backwards in time, while the Schrödinger equation propagates a particle forward in (now imaginary) time. Thus to make a fair comparison we let $\tau \mapsto -\tau$ in the Black–Scholes equation. Then we see that the two are in fact equivalent.

The fact that the Schrödinger equation in "imaginary time" is equivalent to a diffusion process is not a coincidence, as we will discuss in the next section.

8.2 Path Integral and Wick Rotation

Consider again the double slit experiment. A source emits electrons one at a time, passing through a plate with two slits (say, A and B) and hitting the screen at the back. Standard quantum mechanics via the Born rule then allows us to compute the probability of the electron being detected at some given location on the screen. Quantum mechanically, the wavefunction will be the sum of two possible states: one that passes through slit A and one that passes through slit B, and the wavefunction is in a superposition of states. Of course there is nothing special about *two* slits; we can add more slits. In fact, we can also add more plates in between the first plate and the screen. Thus, we may have a configuration of plates, each with some number of slits. To compute the probability, it would be more convenient to employ the "path integral" method, in which we "sum up" all the possible paths the particle can take, say, the ith hole on the jth screen. Now imagine that we increase the number of plates and the number of slits and continue to do so in the limit towards infinity. But a plate with infinitely many slits is as good as having no plate.[6] Thus we can conclude that even in the absence of plates, we still have to sum up all possible paths when we consider the propagation of an electron from a source to the detector—remember that Wheeler's smoke dragon only has a definite tail and head, and we have no right to speak of what happens in between.

The quantum probability amplitude K that describes all the possible paths is called a *propagator,* also known as the "evolution kernel", or the "transition amplitude". In quantum mechanics, the time-dependent Schrödinger equation is usually written as

$$i\hbar\frac{d}{dt}\,|\psi(t)\rangle = \hat{H}\,|\psi(t)\rangle. \tag{8.21}$$

We say that the Hamiltonian $\hat{H}$ *generates the time evolution* of quantum states. Thus we define

$$K(x_f, x_i; t) = \left\langle x_f \middle| e^{-\frac{it\hat{H}}{\hbar}} \middle| x_i \right\rangle, \tag{8.22}$$

where x_i and x_f denote the initial and final positions of the particle, respectively.

Though developed and made popular by Richard Feynman, the idea of path integrals was already introduced by Paul Dirac before that, who gave the amplitude of a particle that propagates from a point q_i to another point q_f in time $t = t_f - t_i$ as

$$\left\langle q_f \,\middle|\, e^{-\frac{i\hat{H}t}{\hbar}} \,\middle|\, q_i \right\rangle = \int Dq(t) e^{\frac{i}{\hbar}\int_0^T dtL(q,\, \dot{q})}, \tag{8.23}$$

where $S = \int_0^T dtL(q, \dot{q})$ is the classical action (integral of the Lagrangian). This will look very obscure for readers who have not learned path integrals before. But even without any prior knowledge, one can try to appreciate the equation somewhat. The LHS is just a short-hand notation for the probability amplitude from q_i to q_f, while the RHS is an "integral", thus a sum, over all the possible paths. The wave nature of quantum particles is crucial since we need the constructive interference to produce the actual classical path. Since $\hbar$ is very small, $\exp(iS/\hbar)$ is highly oscillatory. Therefore nearby paths tend to cancel, except around the classical path (since the first variation of the action δS vanishes on the classical path[7]). Thus it is not quite

[6] Recall from Chapter 3 that Nicholas of Cusa once described God as being everywhere and nowhere; and also the old saying that an infinite Universe is like a sphere with a center everywhere, and hence has no center.

[7] The classical path $x_{\rm cl}$ is the solution to the functional variation of the classical action; so

$$\frac{\delta S}{\delta x(t)} = 0. \tag{8.24}$$

This is the "Hamilton's principle of least action" in classical mechanics.

correct to say that Nature chooses one of the infinitely many paths; not every path is on equal footing—the classical path is what "survives" the cancellation.

Whenever one writes down a notation like $\int Dq(t)$, it is tempting to interpret it as an integral; in fact this is why we call it a "path integral". However, we should take this as a suggestive name only, because the path integral is *not* rigorously defined. Consider the "integral" $\int_\Gamma D\gamma$, where γ: $[t_i, t_f] \to \mathbb{R}^d$ is any path connecting the endpoints $\gamma(t_i) = q_i$ and $\gamma(t_f) = q_f$, and Γ is the space of such paths. Then we would like $D\gamma$ to be a Lebesgue-type measure on the space Γ of paths. Unfortunately, there is no such measure because a (nontrivial) translational invariance Lebesgue-type measure cannot be defined on infinite dimensional Hilbert spaces. This is a theorem in functional analysis. Therefore, despite the name path "integral", and despite the fact that it works (i.e., produces results that match experimental data), it is not an integral in the traditional sense. As the mathematician E. J. McShane (1904–1989), who specialized in calculus of variation and integration theory, once lamented,

> "There are in this world optimists who feel that any symbol that starts off with an integral sign must necessarily denote something that will have every property that they should like an integral to possess. This is of course quite annoying to us rigorous mathematicians; what is even more annoying is that by doing so they often come up with the right answer."

Even ignoring the issue of mathematical rigor, highly oscillatory terms are tricky to deal with in actual calculations. Fortunately, by Wick rotating into imaginary time, we can convert $\exp(iS/\hbar)$ into the form "$\exp(-S/\hbar)$", as we will discuss in detail soon. The main advantage of doing so is that an exponentially decaying function is much simpler to handle! The Wick-rotated path integral turns out to be well-defined mathematically, it is exactly the same type of integral developed to deal with diffusion processes like Brownian motion. The associated measure is called the *Wiener measure* (which is *not* translational invariant), introduced in 1923 by Norbert Wiener (1894–1964), a mathematician and also a philosopher.[8] It is true that the "path space" is still infinitely large, but in Brownian motion we know that the path is not smooth between intervals. Intuitively, the infinity that arises from the derivative of a non-smooth path—i.e., the $\dot{\gamma}$ in $\exp(-S/\hbar) = \exp\left(-\int_0^T |\dot{\gamma}|^2 |/\hbar\right)$—"cancels" with the divergence of the infinite path space, thus rendering the measure well-defined. This, however, cannot be done when the integral is highly oscillatory before the Wick rotation. Readers who are interested in the connection between quantum mechanic and Brownian motions can refer to Favella (1967).

Let us now consider the Wick-rotated version of the propagator and look at more detail. Just like the usual integral can be seen as the limit of the Riemann sum, the

[8] He wrote an essay titled "The Theory of Ignorance" when he was 10 years old, in which he argued that human knowledge can only be an approximation, and fundamentally we cannot be "certain of anything".

idea is to split the propagator into many small intervals. First, recall that $t = -i\hbar\tau$ after Wick rotation, so we have $e^{-it\hat{H}/\hbar} = e^{\tau\hat{H}} = e^{-N\cdot\frac{\tau\hat{H}}{N}}$, so that[9]

$$K(q_i, q_f; \tau) = \left\langle q_f \left| \left(e^{-\frac{\tau\hat{H}}{N}}\right)^N \right| q_i \right\rangle \tag{8.25}$$

$$= \langle q_f | \underbrace{e^{-\varepsilon\hat{H}} \cdots e^{-\varepsilon\hat{H}}}_{N \text{ times}} | q_i \rangle; \quad \varepsilon \equiv \frac{T}{N}. \tag{8.26}$$

Take $\hat{H} = \frac{\hat{p}^2}{2m} + \hat{V}$, where V is the potential term.

We can now insert $N - 1$ copies of the "completeness relation"[10]

$$\int_{\mathbb{R}} dq_i |q_i\rangle \langle q_i| = 1 \tag{8.27}$$

into the propagator and obtain

$$K(q_i, q_f; t) = \int_{\mathbb{R}} dq_{N-1} \cdots \int_{\mathbb{R}} dq_1 \langle q_f | e^{-\varepsilon\hat{H}} | q_{N-1}\rangle \cdots \langle q_1 | e^{-\varepsilon\hat{H}} | q_i\rangle. \tag{8.28}$$

Now we may evaluate, for each factor, using the relation that $\int_{\mathbb{R}} dp|p\rangle\langle p| = 1$,

$$\langle q_{i+1} | e^{-\varepsilon\hat{H}} | q_i\rangle = \int_{\mathbb{R}} dp \langle q_{i+1} | e^{-\varepsilon\hat{H}} | p\rangle\langle p | q_i\rangle \tag{8.29}$$

$$= e^{-\varepsilon V(q_i)} \int_{\mathbb{R}} dp\ e^{-\frac{\varepsilon p^2}{2m}} \left[\frac{e^{\frac{ip(q_{i+1} - q_i)}{\hbar}}}{2\pi\hbar}\right] \tag{8.30}$$

$$= \frac{e^{-\varepsilon V(q_i)}}{2\pi\hbar} \left[\sqrt{\frac{2\pi m}{\varepsilon}}\, e^{-\frac{(q_{i+1} - q_i)^2}{(2\varepsilon/m)\hbar^2}}\right] \tag{8.31}$$

$$= \frac{1}{2\pi\hbar}\sqrt{\frac{2\pi m}{\varepsilon}}\, e^{-\frac{m}{2\varepsilon\hbar^2}(q_{i+1} - q_i)^2} e^{-\varepsilon V(q_i)} \tag{8.32}$$

$$= N(\varepsilon) e^{-\varepsilon L}, \tag{8.33}$$

[9] In the following I will be brave and try to include $\hbar$ explicitly instead of setting it to unity as is often done in the literature. This is because we are learning quantum mechanics, so it should be important to see where $\hbar$ appears. Also, setting $\hbar = 1$ could miss some subtleties in the calculations. In any case you should not trust my calculations as I am sure there would be mistakes somewhere; spotting errors is in any case good training for any physics student. I attempted this calculation before in my Springer thesis (Ong 2016), but there are many mistakes therein, and I now take the opportunity to correct as many as possible. (Factors of $\hbar$ are considered "typos" or minor mistakes and not really worth issuing an errata for the book.)

[10] Any quantum state can be regarded as a vector, which can be expanded in a certain basis $\{|q_i\rangle\}$. For a discrete basis set, the completeness relation is just the statement that the sum of the outer product $\sum_i |q_i\rangle\langle q_i| = 1$, where 1 here denotes an identity operator. In the continuous version, replace the sum by an integral.

where we have defined the prefactor of the exponential term as

$$N(\varepsilon) := \frac{1}{\hbar}\sqrt{\frac{m}{2\pi\varepsilon}}, \tag{8.34}$$

and the "Lagrangian" is

$$L = \frac{m}{2}\left(\frac{q_{i+1} - q_i}{\varepsilon\hbar}\right)^2 + \frac{1}{2}[V(x_{i+1}) + V(x_i)], \tag{8.35}$$

where we have used the *mid-point prescription* to the potential term discretization. In the derivation above, to go from Equations (8.29) to (8.30), we have used the fact that[11]

$$\langle x|p\rangle = \frac{1}{\sqrt{2\pi\hbar}}e^{\frac{ipx}{\hbar}}, \tag{8.36}$$

while from Equations (8.30) to (8.31) we have evaluated the Gaussian-type integral via the standard formula

$$\int_{\mathbb{R}} (e^{-\frac{1}{2}ax^2 + iJx})\, dx = \sqrt{\frac{2\pi}{a}}\, e^{-\frac{J^2}{2a}}. \tag{8.37}$$

Therefore, with $q_0 = q_i$ and $q_N = q_f$, we can show that we have

$$K(q_i, q_f; \tau) = \int_{\mathbb{R}} \prod_{n=1}^{N-1} dq_n \prod_{n=0}^{N-1} \langle q_{n+1}|e^{-\varepsilon\hat{H}}|q_n\rangle \tag{8.38}$$

$$= \int_{\mathbb{R}} \prod_{n=1}^{N-1} dq_n \prod_{n=0}^{N-1} N(\varepsilon)e^{-\varepsilon L(q_{n+1}, q_n)} \tag{8.39}$$

$$= \int_{\mathbb{R}} \prod_{n=1}^{N-1} dq_n\left(\frac{m}{2\pi\varepsilon\hbar^2}\right)^{\frac{N}{2}} e^{-\varepsilon\sum_{n=0}^{N-1} L(q_{n+1}, q_n)}. \tag{8.40}$$

Thus,

$$K(q_i, q_f; \tau) \equiv \int Dq\; e^{-S} \tag{8.41}$$

where

$$\int Dq \equiv \int_{\mathbb{R}} \prod_{n=1}^{N-1} dq_n\left(\frac{m}{2\pi\varepsilon\hbar^2}\right)^{\frac{N}{2}}, \tag{8.42}$$

[11] The normalization of the completeness relations can be chosen such that $\int_{\mathbb{R}} dq|q\rangle\langle q| = 1$, and $\int_{\mathbb{R}} dp|p\rangle\langle p| = B$, and $\langle x|p\rangle = Ae^{ipx/\hbar}$, with the constraint $2\pi|A|^2\hbar = B$. By choosing $B=1$, we have $A = 1/\sqrt{2\pi\hbar}$.

and

$$S = S_E = \varepsilon \sum_{n=0}^{N-1} \frac{m}{2}\left(\frac{q_{n+1} - q_n}{\varepsilon\hbar}\right)^2 + \varepsilon \sum_{n=0}^{N-1} V(q_n). \tag{8.43}$$

We have emphasized with a subscript E that here S is the Euclidean action. The occurrence of $\varepsilon\hbar$ instead of just ε in the ratio $(q_{n+1} - q_n)/\varepsilon\hbar$ is because ε is not dimensionless, it has the same dimension as $\tau = t/\hbar$. In the literature and some textbooks, some authors work with a dimensionless ε, in which case the various expressions would be somewhat different. We may now take the formal continuum limit $\varepsilon \longrightarrow 0$. Then the action becomes

$$S \to \int_0^{T} \left[\frac{1}{\hbar^2} \cdot \frac{m}{2}\left(\frac{dq}{d\tau'}\right)^2 + V(q)\right] d\tau'. \tag{8.44}$$

We can now reverse the Wick rotation back to Minkowski time and obtain the Minkowskian propagator

$$K = \int Dq \, e^{iS/\hbar}; \quad S = \int_0^{t} \left[\frac{m}{2}\left(\frac{dq}{dt'}\right)^2 - V(q)\right] dt'. \tag{8.45}$$

Note that in quantum mechanics, two operators $\hat{A}$ and $\hat{B}$ are generically non-commutative. Their *commutator* is defined by $[\hat{A}, \hat{B}] := \hat{A}\hat{B} - \hat{B}\hat{A}$. As a result, we don't usually have $e^{\hat{X}}e^{\hat{Y}} = e^{\hat{X}+\hat{Y}}$. Instead, $e^{\hat{X}}e^{\hat{Y}} = e^{\hat{Z}}$, where

$$\hat{Z} = \hat{X} + \hat{Y} + \frac{1}{2}[\hat{X}, \hat{Y}] + \frac{1}{12}[\hat{X}, [\hat{X}, \hat{Y}]] - \frac{1}{12}[\hat{Y}, [\hat{X}, \hat{Y}]] + \cdots, \tag{8.46}$$

which is known as the Baker–Campbell–Hausdorff formula. This means that generically $[\hat{p}^2, \hat{V}] \neq 0$; it is this non-commutativity that makes quantum mechanics non-trivial. In the calculation above we have ignored this fact when we separated out the kinetic term and the potential term from the Hamiltonian. This is justifiable because the correction terms due to the commutators is of order $O(\varepsilon^2)$ that does not contribute in the continuum limit (both $\hat{X} = -\varepsilon\hat{p}^2/2m$ and $\hat{Y} = -\varepsilon\hat{V}$ in our calculation already contain one power of ε).

Next, we shall discuss a profound relation between quantum field theory and statistical mechanics (and hence thermodynamics). Indeed, this is why Wick rotation is very useful for us—we can use it to compute the Hawking temperature of a black hole in a later chapter. We will also see that the Hawking temperature is intimately related to the geometry of the Wick-rotated manifold, so that in a sense, general relativity being a geometric theory, already somehow knows about Hawking temperature, which is a quantum effect. This probably hints at something deeper and more profound connections existing between quantum field theory and general relativity. For now let us establish the relation between thermodynamics and Wick rotation.

As mentioned in Tsvelik (2003), which also sums up what we have said regarding the path integral "sidestepping" the complications of the Baker–Campbell–Hausdorff formula,

> "*The reason why the terms 'quantum field theory' and 'statistical mechanics' are used together so often is related to the essential equivalence between these two disciplines. Namely, a quantum field theory of a D-dimensional system can be formulated as a statistical mechanics theory of a (D+1)-dimensional system. This equivalence is a real godsend for anyone studying these subjects. Indeed, it allows one to get rid of non-commuting operators and to forget about time ordering, which seem to be characteristic properties of quantum mechanics. Instead one has a way of formulating the quantum field theory in terms of ordinary commuting functions, more or less conventional integrals, etc.*".

For a quantum mechanical system at a finite nonzero temperature T, let us consider its partition function Z. In the canonical ensemble, Z is a function of T. With the Boltzmann constant set to unity, we have

$$Z(T) = \mathrm{Tr}[e^{-\beta\hat{H}}], \tag{8.47}$$

where $\beta = 1/T$, and $\hat{H}$ is the Hamiltonian of the system. Recall from Chapter 4 that Z occurs in the Gibbs distribution

$$P_\lambda = \frac{e^{-\beta E_\lambda}}{Z}, \tag{8.48}$$

which gives the probability of a system being in some particular state $|\lambda\rangle$ with energy E_λ at temperature $T = 1/\beta$. In this case, $Z = \sum_\lambda e^{-\beta E_\lambda}$, which is required as a normalization factor to guarantee that the probability sums to one: $\sum_\lambda P_\lambda = 1$. Indeed, as mentioned in Chapter 4, Z stands for "Zustandsumme", a German word that literally means "sum over states".

From Z, we can compute many useful quantities such as

(1) Entropy, $S = -\frac{\partial F}{\partial T} = \log Z + T\frac{\partial(\log Z)}{\partial T}$,

(2) Free energy, $F = -T\log Z$,

(3) Helmholtz energy, $E = \frac{1}{Z}\mathrm{Tr}[\hat{H}e^{-\beta\hat{H}}]$.

The claim is that Z can be related to the path integral. To see this, we write[12]

[12] To appreciate this formula, consider the trace of a diagonal $n \times n$ matrix diag$(a_{11}, a_{22}, \cdots a_{nn})$, which is just the sum $\sum_{i=1}^n a_{ii}$. The continuous version is an integral, and the subscript "ii" corresponds to the two x's in the integrand. This is why in this case the propagator does not go from x to some other y but both are x. One could also say that the initial and final positions are the same. Equivalently, instead of integrating over a line interval, we are integrating over what is topologically a *circle*.

$$Z = \mathrm{Tr}[e^{-\beta\hat{H}}] = \int_{-\infty}^{+\infty} dx \,\langle x|e^{-\beta\hat{H}}|x\rangle \tag{8.49}$$

$$= \int_{-\infty}^{+\infty} dx \,\langle xg|(e^{-\varepsilon\hat{H}})^N g|x\rangle, \tag{8.50}$$

where $N := \beta/\varepsilon$. Like what we have done before, this can be put into the path integral form

$$Z = \int Dx \, e^{-S_E}, \tag{8.51}$$

where S_E is the Euclidean action of the system. For a simple system whose Lagrangian is the difference between its kinetic and potential energy, we therefore have

$$S_E = \int_0^\beta \left[\frac{m}{2\hbar^2}\left(\frac{dq}{d\tau}\right)^2 + V(q)\right] d\tau. \tag{8.52}$$

The factor of $\hbar$ is not so convenient in the integrand; it would be simpler if given a Lorentzian action we can immediately replace t by τ and flip the sign of the potential term. This can be done by absorbing the $\hbar^2$ into a re-definition of τ:

$$S_E = \int_0^\beta \left[\frac{m}{2\hbar^2}\left(\frac{dq}{d\tau}\right)^2 + V(q)\right] d\tau \tag{8.53}$$

$$= \int_0^\beta \left[\frac{m}{2}\left(\frac{dq}{d(\hbar\tau)}\right)^2 + V(q)\right] d\tau \tag{8.54}$$

$$= \int_0^{\beta\hbar} \left[\frac{m}{2}\left(\frac{dq}{dy}\right)^2 + V(q)\right] dy \cdot \frac{1}{\hbar}, \tag{8.55}$$

where $y \equiv \hbar\tau$ is a new integration variable. Thus the Euclidean action is of the form

$$e^{-S_E} = \exp\left\{-\int \left[\frac{m}{2}\left(\frac{dq}{dy}\right)^2 + V(q)\right]\frac{dy}{\hbar}\right\}. \tag{8.56}$$

Now we notice that this is almost like the Lorentzian action

$$e^{iS/\hbar} = \exp\left\{i\int \left[\frac{m}{2}\left(\frac{dq}{dt}\right)^2 - V(q)\right] dt \Big/ \hbar\right\}, \tag{8.57}$$

if we can just replace t with y and flip the potential sign (and replace minus sign with i in front of the integral). So this is what we shall do by re-defining y as τ! This is equivalent to setting $\hbar = 1$, which was done without batting an eye in the literature. Thus in the literature, we find that what is often called the Euclidean action is

$$\tilde{S}_E = \int_0^{\beta\hbar} \left[\frac{m}{2}\left(\frac{dq}{dt}\right)^2 + V(q) \right] d\tau, \tag{8.58}$$

and the action is now $e^{-\tilde{S}_E/\hbar}$ instead of e^{-S_E} previously. (Check that the action in both cases are dimensionless!)

Since the propagator is from x back to x, we see that $\beta\hbar$ can be viewed as the period of a cycle. Because $\beta = 1/T$, we can say that temperature corresponds to "cyclic imaginary time". As A. Zee wrote (Zee 2003),

> "Surely you would hit it big with mystical types if you were to tell them that temperature is equivalent to cyclic imaginary time. [...] Some physicists, myself included, feel that there may be something profound here that we have not quite understood."

We should emphasize that no matter what formulation one uses for quantum mechanics, the theory is inherently probabilistic. This is a feature of Nature that is unavoidable, and we have to learn to live with uncertainty (Figure 8.1).

8.3 Quantum Entanglement, Bell Inequality and Its Cosmic Test

Consider a polarizer, which is an optical filter that allows light waves of a specific polarization to pass through. The basic idea is simple, and is based on the fact that light, being a wave, is vibrating. Most light sources in Nature give rise to unpolarized light, i.e., its plane of vibration changes randomly as it traverses through space. A polarized light, on the other hand, has a fixed plane of vibration. A vertical polarizer therefore only allows light waves that are vibrating in that direction to pass through. If we then add another horizontal polarizer in the path of the light that has passed through the first filter, we will see that no light is able to pass through. So far this is quite easy to understand, since only the vertically polarized light is present, it does not have any component that is horizontally vibrating, and hence it gets blocked completely by the second filter. However, if we put a *third* filter, which is aligned to say, 45 degrees, *in between* the two filters, then suddenly some light is able to pass through! This is absolutely strange and counterintuitive: if two filters can completely block all light, then why would *adding* an additional filter now does not? This is like—imagine a zombie apocalypse—having two lines of defense can completely stop any zombie from passing through, but adding another barrier in between now allows some zombies to pass through!

The answer lies in quantum mechanics. Briefly, the direction of polarization is associated with the oscillations of the electric field within the light wave. Only the component of the electric field $\overrightarrow{E}$ that is parallel to the polarizer filter can pass through, namely the fraction of the electric field vector that passes through is $\cos\theta$,

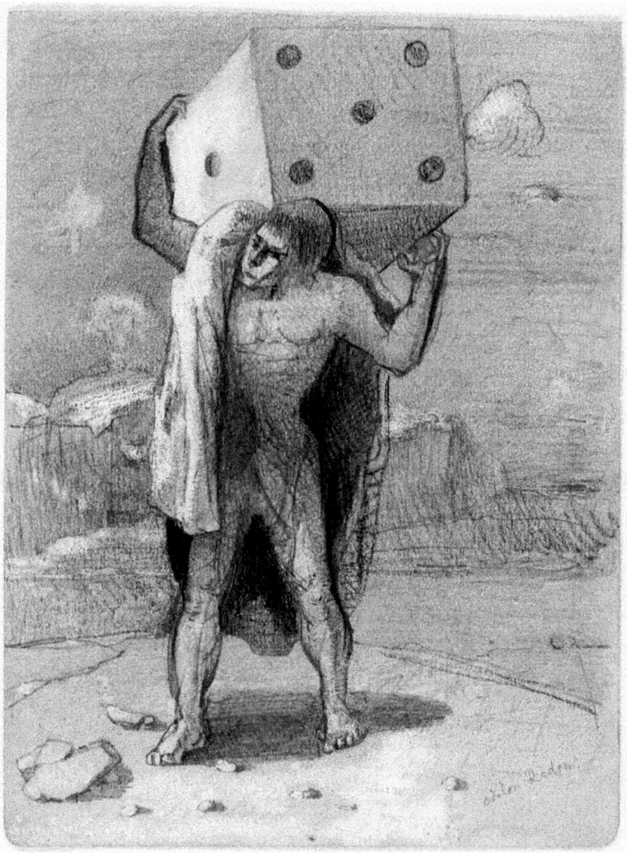

Figure 8.1. *The Die*, also called *The Weight of Passions*, by Odilon Redon (1840–1916). We often make choices in life based on probability, by weighting the chance of success or failure of various options. The fact that quantum mechanics is fundamentally probabilistic was—and perhaps still is—difficult for some physicists to accept. Once there was hope that although we cannot live our lives with certainty (is that even a good thing?), at least the laws of nature is deterministic. Quantum mechanics tells us that this is not the case; uncertainty is built into the very fabric of our world. The path integral picture suggests that Nature somehow chooses the most "optimal" path among all possible paths, which is reminiscent of Leibniz's idea of "the best of all possible worlds" (namely the belief that despite there being sufferings in this world, it is already the best that could be. Perhaps by changing one thing here to lessen someone's suffering, it creates a worse disaster elsewhere. This was Leibniz's attempt to solve the "problem of evil" in theology: if God is good and all powerful, why does he allow so much unnecessary suffering?) But even if we consider each possibility as being realized in some other world as in the "many-world interpretation" of quantum mechanics, the theory is still probabilistic since we cannot know for sure which "branch" of the wavefunction we will end up with after a measurement is made. Life is fundamentally full of uncertainty. *The Die* reminds me of the Greek mythological Atlas, carrying the world on his shoulders, but now the world is quantum mechanical. *Credit:* Margaret Day Blake Endowment. Source: https://www.artic.edu/artworks/147063/the-die-also-called-the-weight-of-passions

where θ is the angle between the original electric field vector and the polarizer. Quantum mechanically, this means that the probability of a photon passing through the filter is given by $(\cos\theta)^2$. Now suppose we have three polarizers; the first one is horizontally polarized, the third vertically, and the second one has its axis aligned in some fixed direction defined by the vector $|d\rangle$. Before passing through the first polarizer, the light is assumed to be unpolarized. Thus the probability of it passing through the first polarizer is $1/2$ (like flipping a coin for a random distribution). Let the state of being horizontally polarized be denoted by $|h\rangle$ and that of being vertically polarized by $|v\rangle$. The photon that has passed through the second polarizer would have a superposition $|d\rangle = \cos\theta|h\rangle + \sin\theta|v\rangle$. After passing through the first polarizer, the light is already horizontally polarized. So the probability that it goes through the second filter is $\cos^2(\theta)$ as mentioned (also the inner product $\langle v|h\rangle = 0$). Finally, the probability of a $|d\rangle$-polarized photon passing through the vertically polarized third polarizer is $\sin^2\theta$. The probability for a photon to pass through all three polarizers is therefore $\cos^2\theta \sin^2\theta/2$. Clearly this can be nonzero.

The results do not make sense to our classical intuition; somehow *less* light is blocked by having *more* filters. Since the beginning of quantum mechanics, the fact that we can only compute probabilities makes a lot of physicists, including Einstein, uncomfortable. One possibility that has been suggested is that there is a "hidden variable" that perhaps our theory still is incomplete to know, or even unknowable in principle but there regardless. This philosophy is sometimes called quantum "realism"—some properties are there even when we do not perform measurement. Thus instead of "just" having some probability to go through a filter, a photon may have a *hidden variable* that dictates whether it can go through a certain polarizer or not *definitely*. However, if this is the case, we can use a Venn diagram to analyze the three-filter scenario. Let us label the polarizers as A, B, C, with B situated in between the first filter A and the last filter C. Denote the number of photons that pass through filter X but not Y by $N(X, \overline{Y})$. Then we can count $N(A, \overline{C})$, $N(A, \overline{B})$, $N(B, \overline{C})$. This should satisfy the inequality

$$N(A, \overline{C}) \leqslant N(A, \overline{B}) + N(B, \overline{C}). \tag{8.59}$$

You can convince yourself that this inequality is true by just drawing a Venn diagram. In a more familiar set theoretic notation, this is just a special case for the following result: for any three sets S_1, S_2, S_3, we must have

$$N(S_1 \cap S_2) \leqslant N(S_1 \cap S_3) + N(\overline{S_3} \cap S_2). \tag{8.60}$$

(Of course this also works if we replace the number N by the probability P.) This set theoretic result has a fancy name: *Bell's inequality*, named after John Bell (1928–1990) who published his results in 1964 in a short-lived obscure journal *Physics*. Anyway, the upshot is that for the three-filter experiment, Bell's equality is *violated*. How can a set-theoretical result be violated? This is a *reductio ad absurdum*—it must be that we *cannot* treat quantum objects classically in the sense that they have definite properties as the hidden variable hypothesis asserts.

However, using three subsequent polarizers like this means there is a loophole that can still allow for a hidden variable. Perhaps the hidden variable also includes information about the interactions with the polarizers, so that passing through one filter changes how the photon will later interact with the subsequent filters. In other words, maybe whether a photon can pass through a filter or not is not a fixed information as we have previously assumed. To get rid of this loophole, we can utilize *quantum entanglement*. Before we continue with the improved experimental set-up, let us give a quick non-technical introduction to quantum entanglement,[13] a feature of quantum mechanics that, according to Erwin Schrödinger, is *the* trait that distinguishes it from classical physics. The usual example is to consider particles with spin, say an electron that can be either spin-up or spin-down. Spin is a quantum mechanical property and should not be thought of as classical rotation. If two electrons are quantum entangled, then the act of measuring the spin of one would instantly determine the spin of another, no matter how far away they are separated. Thus, if we measure the electron in our lab to be spinning "up", then its partner, which was sent to some far away place (say, the Andromeda Galaxy) prior to the measurement, must be spinning "down". Despite what Einstein called "spooky action at a distance", quantum entanglement does not allow communication (i.e., transmitting of *information*) faster than light. This is because information transfer requires us to control the state of the particle, but whatever result ("up" or "down") appears in our lab (and hence far away) is fundamentally random.

Now, back to the experiments that test the Bell inequality. Any such experiment goes by the name "Bell test".[14] The simplest set-up is essentially to consider two entangled photons, in the sense that if we send them through two identical filters, both of them either pass through or both get blocked, no matter how far they are separated in space. This separation means that neither of the photons can influence the other by passing information between them without violating special relativity. Now, suppose we have a polarizer at location R_1 and R_2, separated sufficiently far apart. We send two entangled photons, one to each location. At each location, the polarizer is randomly rotated so that they allow photons that are either vertically polarized (A), 45-degrees polarized (B), or horizontally polarized (C), through. This is repeated many times, and we would have a table that records whether the photons at both locations pass through the polarizer or not. Though more complicated than the sequential three-filter set-up, we can also draw a Venn diagram and check that indeed Bell's inequality is still violated. Since the entangled photons cannot send information to each other, the loophole seems to be closed, and we have to conclude that there cannot be a hidden variable if quantum mechanics is *local*. Locality here simply means that a particle cannot affect another particle that is far away beyond

---

[13] For readers who are well-versed with quantum mechanics, we can explain a little bit more. Suppose we have two particles labeled P and Q, then the state of each particle is described by a vector (or more precisely, due to the need or normalization, a *ray*) in a Hilbert space. Let $\mathscr{H}_P$ and $\mathscr{H}_Q$ denote the respective single-particle Hilbert spaces. The combined system of two particles is then described by the tensor product space $\mathscr{H} = \mathscr{H}_P \otimes \mathscr{H}_Q$. If P is in a state $|u\rangle$, and Q is in a state $|v\rangle$, then $|u\rangle \otimes |v\rangle \in \mathscr{H}$. However, there are states in $\mathscr{H}$ that *cannot* be written in the form $|\psi_P\rangle \otimes |\psi_Q\rangle$ for any $|\psi_P\rangle \in \mathscr{H}_P$ and $|\psi_Q\rangle \in \mathscr{H}_Q$. Such a state is said to be entangled.

[14] In practice, it is easier to test the so-called Clauser–Horne–Shimony–Holt (CHSH) inequality, whose violation also indicates inconsistency with a local hidden-variable theory. We will explain the meaning of "local" soon.

the limit set by special relativity. Hidden variables are still possible in a *nonlocal* theory. Note the important distinction: quantum entanglement does not imply nonlocality; what quantum entanglement implies is that there are *correlations* that cannot be explained by a local classical theory.

There is still another loophole however—how can we be sure that the polarizers at R_1 and R_2 are chosen *truly* at random? Sure, we can use high-quality random number generators, but they could *in principle* still be influenced by some very recent event in their common past. In any case, random number generators are based on some algorithm and thus also not truly random. In other words, there is a possibility that some hidden, classical variable might influence the measurements that we perform on the entangled particle, thus making the results look as if there is a quantum correlation when there is none. To fully close this so-called "freedom-of-choice loophole" requires the use of independent sources (as far as this can be guaranteed) of randomness that are available to us. This is where astronomy comes to the rescue. In 2017, a research team set up two telescopes and pointed them at different stars in the Milky Way. Each photon received can be of a different color;[15] this was set during emission at random. Using this in place of the random number generators built on Earth, the researchers still measured violation of Bell's inequality. If there is a classical correlation that somehow conspires to "fake" quantum correlation, this would have to be set in motion some 600 years ago (Handsteiner et al. 2017). A subsequent experiment in 2018 utilized quasars far outside our galaxy, the time for such a local influence was pushed back to about 7.8 Gyr ago, which seems to be highly implausible (Rauch et al. 2018).

There is another loophole, called the "fair sampling" loophole. Since photons are hard to detect, the table we built that records whether a photon and its partner passes through the polarizer or not does not contain the data from *all* of the photons. In other words, not all entangled pairs are measured. If the sampling isn't fair, this could introduce a bias in the data that looks like a Bell violation but it could be a false positive. To close this loophole requires photon detectors that can operate at high enough efficiency. This was also achieved in 2018 by a Chinese research team (Li et al. 2018). Combining all the experiments performed so far, it seems that we have convincingly ruled out local hidden-variable theories of quantum mechanics: the Universe fundamentally cannot have *local realism*.

Finally, it is worth noting that quantum entanglement has entered the forefront of technological development, not just pure theoretical research. For example, it could be utilized in secure communication via quantum key distribution (QKD), a cryptographic protocol that uses the principles of quantum mechanics to securely exchange cryptographic keys between two parties. It is in principle more secure than the traditional "public key cryptography" (such as the "RSA algorithm"), which relies on number theoretic computational difficulty to crack the code (essentially the difficulty to factorize large numbers into their prime components). Thus, someone

[15] The color of a star is determined by the peak of its blackbody radiation, but individual photons can have different energy.

with large computational power can theoretically, potentially break the code. In QKD, on the other hand, utilizing quantum entanglement, one can detect if an eavesdropper is trying to intercept the communication.

China launched a quantum entanglement satellite in 2016 called Micius, or 墨子 (Mozi[16]) in Mandarin, to study the feasibility of a future global communication network. At the same time, the project will provide opportunities to test and further refine technologies that would have applications in various areas, such as space laser aiming, space weak light detection, space high precision synchronization, small satellite high-precision attitude maneuvers, and high-speed single photon detection. Micius is a part of the international project Quantum Experiments at Space Scale (QUESS), with ground facilities in China and receiving stations in Austria. In late 2025 or early 2026, the European Space Agency is set to launch the similar Eagle-1 quantum key distribution satellite.

8.4 Quantum Mechanics in Astrophysics and Cosmology

Quantum mechanics has revolutionized our civilization in terms of technological advancement and scientific understanding in the last century. If there is no quantum mechanics, much of our modern society would not be the same. Though quantum mechanics governs physics at the smallest scales, its effects extend to astronomically large objects. Without quantum mechanics, many phenomena in the cosmos would remain inexplicable. In this section we will take a look at a few examples: (1) how the stars, including our Sun, shine; (2) how some stars can resist gravitational collapse even after its nuclear fuel ran out; (3) how a type of neutrino changes into another (neutrino oscillation), and briefly, (4) quantum cosmology.

8.4.1 How Does the Sun Shine?

It requires no further elaboration that sunshine is crucial to most life forms on Earth—we need the warmth and the energy. But what is the mechanism behind the burning heat? The answer is nuclear fusion. This is well known to most sufficiently educated persons today, but how did people try to explain the heat source before we know about nuclear fusion? Again, one finds creativity in the proposed solutions in this episode of the history of science. Some proposals seem almost too far-fetched today and were no longer mentioned in textbooks. For example, Julius Mayer (1814–1878) and Lord Kelvin both proposed that the heat is generated by meteors or asteroids bombarding the Sun. Another more plausible proposal, often referred to as the Helmholtz–Thomson or Kelvin–Helmholtz theory, explained the heat as the result of gravitational contraction. The basic idea is simple enough to go into some details here.

[16] Mozi (c.470–c.391 BCE) founded the Mohist school of thought, which in addition to the traditional subjects that other Chinese schools were interested in (such as ethics, politics, and moral philosophy), also pursued logic and scientific thoughts, which include geometry, mechanics, and optics, such as the principle of *camera obscura*. Unfortunately Mohism fell into decline under the Qin dynasty.

The gravitational potential energy of the Sun, assuming a spherical cloud of uniform density for simplicity (with mass M and radius R), is given by the integral

$$E_{\text{grav}} = -\int_0^R \frac{Gm(r)\rho(r)4\pi r^2}{r}dr \tag{8.61}$$

$$=-\int_0^R \frac{G\left(\int_0^r \rho(r')4\pi r'^2 dr'\right)\rho(r)4\pi r^2}{r}dr \tag{8.62}$$

$$=-\rho^2\int_0^R \frac{16\pi^2 G r^4}{3}dr \quad (\text{since } \rho\text{=const.}) \tag{8.63}$$

$$=-\rho^2 R^5\frac{16\pi^2 G}{15}. \tag{8.64}$$

In the calculations above $m(r)$ and $\rho(r)$ denote the mass and density inside a ball of radius $r < R$. Since

$$\rho = \frac{M}{\frac{4}{3}\pi R^3}, \tag{8.65}$$

we get for the final expression of the gravitational potential energy:

$$E_{\text{grav}} = -\frac{3}{5}\frac{GM^2}{R}. \tag{8.66}$$

As it turns out, due to a result known as the Virial theorem (see Section 4.2), half of the gravitational potential energy is released during the collapse. The luminosity of the Sun L, which can be measured, tells us how much energy it radiates. Therefore there is a characteristic timescale ("Kelvin–Helmholtz timescale" or simple "thermal timescale") set by

$$t_{\text{KH}} = \frac{|E_{\text{grav}}|}{L} \sim \frac{GM^2}{RL}. \tag{8.67}$$

For the Sun this works out to be around 10^7 years. This age is, however, too small since we have rocks that dated back to geological eras far older than that; for now the oldest rock ever dated (in 1999) was about 4 billion years old (the Earth itself is thought to be about 4.54 billion years old), part of the Acasta Gneiss of the Slave Craton in northwestern Canada. To our best understanding, the Earth cannot form *before* the Sun—the planets took some hundreds of millions of years to form from the protoplanetary disk that surrounded the newly formed Sun. Therefore we conclude that the heat of the Sun cannot be explained by gravitational contraction alone. Gravitational contraction is, however, crucial, as it was what gave birth to our Sun and other stars in the beginning. Readers can consult Box 6 below for some details on this process.

Box 6: Jeans Mass and Gravitational Collapse

Gravitational collapse can happen when a cloud of gas of radius R (assumed to have constant density) has a gravitational potential energy whose magnitude exceeds that of two times the kinetic energy, as dictated by the Virial theorem. This condition is called the *Jeans criterion*, and the phenomenon is referred to as *Jeans instability*, named after Sir James Jeans (1877–1946). As a mathematical inequality, the Jeans criterion reads

$$|E_{\text{grav}}| = \frac{3}{5}\frac{GM^2}{R} > 2 \times \frac{3}{2}Nk_BT, \tag{8.68}$$

where we have used the equipartition theorem (Section 4.2) for the average kinetic energy of the gas. Now, in terms of the density

$$\rho = \frac{M}{\frac{4}{3}\pi R^3}, \tag{8.69}$$

this inequality is equivalent to

$$M^2 > 5Nk_BT \cdot \frac{R}{G} = \frac{5Nk_BT}{G}\left(\frac{\frac{4}{3}\pi\rho}{M}\right)^{-\frac{1}{3}}. \tag{8.70}$$

Letting m be the mass of a single gas particle, we have $M = Nm$. This implies the inequality

$$M > \frac{5k_BT}{mG}\left(\frac{\frac{4}{3}\pi\rho}{M}\right)^{-\frac{1}{3}}, \tag{8.71}$$

or

$$M > M_J := \left(\frac{5k_BT}{mG}\right)^{\frac{3}{2}}\left(\frac{3}{4\pi\rho}\right)^{\frac{1}{2}}. \tag{8.72}$$

The quantity M_J is called the *Jeans mass*. If μ denotes the mass ratio m/m_H (i.e., μ is the molecular weight), where m_H is the mass of the hydrogen atom, then we can write

$$M_J = \left(\frac{5k_BT}{G\mu m_H}\right)^{\frac{3}{2}}\left(\frac{3}{4\pi\rho}\right)^{\frac{1}{2}}. \tag{8.73}$$

Note that this is independent of R, i.e., independent of the initial size of the gas cloud. When a gas cloud locally satisfies the Jeans criterion, it begins to collapse in patches. Thus multiple protostars can form from a single collection of gas. This explains why around 85% of stars are part of binary or multiple star systems. Single stars like our Sun is comparatively rare. (It is also possible that our Sun was also born as a binary system, but its partner was somehow ejected out of the system.)

We can also define the *Jeans length*, R_J, which is the radius of a uniform gas cloud whose mass is M_J. It is given by

$$R_J = \left(\frac{15k_BT}{4\pi G\mu m_H\rho}\right)^{\frac{1}{2}}. \tag{8.74}$$

We now know that the Sun and other stable stars mainly consist of hydrogen. As a ball of gas of hydrogen contracts, it heats up. If there is enough mass to begin with, this heat can kick start thermonuclear fusion that combines four hydrogen atoms into one helium atom (more specifically, four protons into a helium-4 nucleus). The details of the fusion process are rather complicated, but essentially we can compute the fractional energy change of the process:

$$\frac{4m_p c^2 - m_{\text{He}}c^2}{4m_p c^2} \approx 0.007. \tag{8.75}$$

The nuclear timescale of the Sun is thus $0.007Mc^2/L$, which is about 10^{11} years, four orders of magnitude older than the thermal timescale, and enough to explain geological evidence.

Nuclear fusion requires quantum mechanics in a non-obvious manner. In short, in order to fuse protons, we need to overcome a potential barrier (protons are positively charged and so they naturally repel each other). Classically there is not enough heat energy to overcome this potential effectively. However, *quantum tunneling* can do the job. Consider non-relativistic particles whose total energy is $E = p^2/2\,m + V$, where V is a potential term. In general, the time-dependent Schrödinger equation is

$$i\hbar\frac{\partial}{\partial t}\,|\psi(t,x)\rangle = \hat{H}\,|\psi(t,x)\rangle. \tag{8.76}$$

Consider for simplicity that the system is time-independent. Then we have instead

$$\hat{H}\,|\psi(t,x)\rangle = E\,|\psi(x)\rangle. \tag{8.77}$$

Then, we have

$$E\psi(x) = -\frac{\hbar^2}{2m}\frac{d^2\psi(x)}{dx^2} + V(x)\psi(x). \tag{8.78}$$

For E=const., we have

$$\frac{d^2\psi(x)}{dx^2} = -\frac{2m}{\hbar^2}[E - V(x)]\psi(x). \tag{8.79}$$

Let us simplify things further by considering $V =$ const. This is not realistic, but typically in physics we can learn about the *essence* by working out the simplest examples. As we have emphasized repeatedly, in model-building, it is often helpful to start with the simplest model, understand it, and only then add more complications. The essential physics of the equation now depends on the sign of $E - V$. If $E > V$, then by defining k via

$$k^2 := \frac{2m}{\hbar^2}(E - V), \tag{8.80}$$

we have an ordinary differential equation of the form

$$\frac{d^2\psi(x)}{dx^2} = -k^2\psi(x), \tag{8.81}$$

whose general solution is a plane wave:

$$\psi(x) = A\sin(kx) + B\cos(kx). \tag{8.82}$$

Plane wave solutions over the whole of $\mathbb{R}$ cannot be normalized. That is, no values of A and B can satisfy $\int_{\mathbb{R}} |\psi(x)|^2 dx = 1$. The reason is because the fixed amplitude wave cannot give a convergent integral. Therefore such a solution only exists if we truncate off the wave outside a bounded region. This is the "particle in a box" scenario one encounters in most quantum mechanics textbooks.

On the other hand, if $V > E$, then the ODE would differ by a sign on the RHS:

$$\frac{d^2\psi(x)}{dx^2} = \kappa^2\psi(x), \tag{8.83}$$

where

$$\kappa^2 := \frac{2m}{\hbar^2}(V - E). \tag{8.84}$$

The solution is a linear combination of exponential terms:

$$\psi(x) = Ce^{\kappa x} + De^{-\kappa x}. \tag{8.85}$$

Tunneling happens when there is a potential wall that classically the particle cannot go over ($V > E$). However, quantum mechanically, the particle started out as a free particle represented by an oscillating wavefunction. As it encounters the wall, there is an exponential decay into the wall. Then as it emerges from the other side of the potential wall, the solution again changes back into an oscillating wave, now with reduced amplitude. Of course the thicker the width of the wall, the smaller the tunneling rate.

Much of the same scenario happens in the Sun, though of course the mathematical model would need to be more complicated: it is three-dimensional and the wavefunction should be considered as sums of spherical harmonics, and the potential is not constant but the Coulomb barrier $V \propto 1/r$, but one gets the idea. Without quantum tunneling, the Sun needs to be around 10^{10} K to overcome the potential barrier to sustain thermonuclear fusion. This is far above the mean temperature of the Sun ($\sim 10^6$ K) and even its core temperature ($\sim 10^7$ K).

8.4.2 Degenerate Pressure Inside White Dwarfs and Neutron Stars

After a star burns away all of its hydrogen fuel, it can continue to burn helium and then other gases down the periodic table, until it hits iron. No energy can be produced by iron fusion, so the star is "dead". The outward radiation pressure generated by nuclear fusion is what keeps a star stable. When this process stops, the star can no longer withstand gravitational collapse. What happens next depends on the mass of the star. For stars whose mass is more or less the same as our Sun, it will end as a "white dwarf", which is stabilized by "electron degeneracy pressure", which we will explain shortly. Like its name suggests, a white dwarf is relatively small—it packs the mass of a Sun into the size of our planet Earth, and its density is about 10^9 kg m^{-3}. Though no longer producing energy, it still glows white hot from the remaining heat. While the exact mass range is still debatable, for stars that are about 1.4 to 3 solar masses, the electrons and protons will become so tightly squeezed that they fuse into a collection of neutrons, hence the name "neutron star" (as the density becomes larger towards the core, the neutron star can exhibit various exotic states of nuclear matter). A neutron star is even more compact, as it packs the mass of our Sun into the size of a sphere of 10 km radius; its density is about 10^{17} kg m^{-3}. A rather macabre way to appreciate this is as follows: there are roughly 8 billion people on Earth (as in 2024). Take the average weight per person as 60 kg as a crude approximate. Setting

$$10^{17} = \frac{8 \times 10^9 \times 60}{L^3} \tag{8.86}$$

yields $L \approx 0.017$ m. Thus, the density of a neutron star is roughly the density of all humans being squeezed into a box whose side length is 2 cm! For a white dwarf, the box is considerably larger: $L \approx 0.78$ m, which would still be very uncomfortable...

There are many models available and so far we do not know for sure what the interior of a neutron star looks like. A neutron star is supported by neutron degenerative pressure. For a star that is heavier than 3 solar masses[17] it is generally believed that the end result would be a black hole.

In this section we will focus on the white dwarf case and heuristically derive the upper bound of its mass, known as the *Chandrasekhar limit*. The pressure of a classical ideal gas in a container satisfies

$$P = k_B \frac{NT}{V}, \tag{8.87}$$

where we have restored the Boltzmann constant k_B for clarity; N, T, V are the number of particles, the temperature, and the volume, respectively. At the microscopic level we can understand pressure as a consequence of the collision between the particles, thus increasing the number of particles has the same effect as decreasing the volume—the rate of collision goes up. Increasing the temperature also has the same effect because it amounts to increasing the mean kinetic energy of the particles. For "degenerate matter" inside a white dwarf, however, the pressure is not as sensitive to temperature, and persists even at zero temperature! For Fermi gas at relatively low density, the pressure is given by

[17] In a realistic situation we also need to consider mass loss during the final phases, so this mass refers not to the initial mass when the star is stable, but the final mass near the end of its life after it went through various mass loss channels.

$$P = \frac{(3\pi^2)^{\frac{2}{3}}\hbar^2}{5m}\left(\frac{N}{V}\right)^{\frac{5}{3}}, \tag{8.88}$$

where m is the mass of the individual particle. At sufficient high density, the motion of the particles becomes relativistic and $P \propto (N/V)^{4/3}$. The proportionality constant depends on the properties of the particles making up the gas. We have actually seen these behaviors in Exercise 4.4.(2), in which we found that the adiabatic index γ in PV^γ=const. takes the value $5/3$ for the non-relativistic case, and the value $4/3$ for the relativistic case.

For simplicity, let us focus on the non-relativistic case. As we shall see, the radius of a white dwarf is proportional to the inverse of its volume, or

$$R \propto \frac{\hbar^2}{GM^{\frac{1}{3}}}. \tag{8.89}$$

The proportionality constant again depends on the particle content. Note that unlike non-degenerate stars, whose size is larger if it contains more mass, the situation is completely opposite for degenerate stars: more massive stars are *smaller*. This equation suggests that as $M \to \infty$, the radius $R \to 0$. However this does not happen because when the star is sufficiently small, there is little room for the particles to move around and so their motion becomes relativistic and thus this simplified model fails. More realistically, one would find that incorporating relativistic effects, there is an upper mass limit, known as the *Chandrasekhar limit*, which still corresponds to a white dwarf of a finite size (in fact we know that a black hole will form if the star becomes too small for a given mass). The Chandrasekhar limit is given by

$$M_{\rm Ch} = K\left(\frac{\hbar c}{G}\right)^{\frac{3}{2}}\frac{1}{(\mu_e m_{\rm H})^2}, \tag{8.90}$$

where K is a constant to be specified later (it cannot be obtained via the heuristic approach), μ_e is the average molecular weight per electron, and m_H is the mass of the hydrogen atom. We have restored the speed of light c explicitly for clarity.

We can derive the mass–radius relation Equation (8.89) and the Chandrasekhar mass Equation (8.90) heuristically[18] by considering another famous quantum mechanical result: the *Heisenberg's uncertainty principle* $\Delta x \Delta p \gtrsim \hbar$, which essentially says that position and momentum cannot both be measured exactly.

To this aim, we will consider the white dwarf as an "electron star", so that the total mass M is given by Nm_e. This is a very crude toy model; the white dwarf cannot possibly contain only electrons, there would also be protons around. Also, the total

[18] A more rigorous treatment is accomplished via the Lane–Emden equation, which is the Poisson equation with polytropic fluid. The constant K in Equation (8.90) is a constant related to the solution to the Lane–Emden equation.

mass needs to take into account gravitational binding energy, so M is not simply the mass of all the electrons.[19] Nevertheless, this should give us more or less the correct result if we only care about the qualitative features (because such a crude informal estimate is often quite short, it is also referred to as "back-of-the-envelope" calculation). Let us consider the total kinetic energy of a non-relativistic white dwarf:

$$E_k = \frac{N\Delta p^2}{2m_e} \sim \frac{N\hbar^2}{(\Delta x)^2 2m_e}, \tag{8.91}$$

where m_e is the electron mass. We have used the estimate $\Delta x \Delta p \sim \hbar$. Let the volume of the star be V. We shall assume the density to be homogeneous. Define the number density by $n := N/V = M/(m_e V)$. We have $\Delta x \sim n^{-1/3} = (V/N)^{1/3}$. This is roughly the room an electron can move while being squeezed together. Their motion gives rise to a huge Δp which in turn is responsible for the degenerative pressure. With the relation $V \sim R^3$, where R is the radius of the star, we can rewrite Equation (8.91) as:

$$E_k = \frac{N\hbar^2 n^{\frac{2}{3}}}{2m_e} = \frac{N\hbar^2}{2m_e}\left(M^{\frac{2}{3}} m_e^{-\frac{2}{3}} R^{-2}\right) = \frac{M^{\frac{5}{3}}\hbar^2}{2m_e^{\frac{8}{3}} R^2}. \tag{8.92}$$

In order for the star to be at equilibrium, i.e., to withstand gravitational collapse, we must balance the kinetic energy with gravitational binding energy. For a spherical mass of uniform density, the gravitational binding energy is just $E_{\text{grav}} = -3GM^2/5R$ given in Equation (8.66). Thus

$$|E_{\text{grav}}| \sim \frac{GM^2}{R} = E_k, \tag{8.93}$$

which leads to[20]

$$R \sim \frac{\hbar^2}{2m_e^{\frac{8}{3}} G M^{\frac{1}{3}}}. \tag{8.94}$$

Thus we have obtained the correct mass–radius relation $R \propto M^{-\frac{1}{3}}$ for a non-relativistic white dwarf, Equation (8.89). For the relativistic regime, we have to repeat the calculation with relativistic kinetic energy

$$E_k = N(\gamma - 1)m_e c^2, \quad p = \gamma m_e v. \tag{8.95}$$

With $v \sim c$, the kinetic energy is essentially

[19] This approximation was also made in Box 6 in deriving the Jeans criterion. However, since in that context the gas cloud has yet to contract, the effect from gravity is considerably weaker than in a bound white dwarf.

[20] A more realistic model is to consider the number of electrons to be $N_e = MZ/Am_p$, where A is the atomic mass (number of nucleons per atom), and Z is the number of electrons per atom, while m_p is the proton mass. Then $m_e^{8/3}$ is replaced by $m_e \cdot m_p^{5/3}$.

$$\Delta pc - m_e c^2 \sim \frac{\hbar c}{\Delta x} - m_e c^2 \sim \left(\frac{M}{m_e V}\right)^{\frac{1}{3}} \hbar c - m_e c^2. \tag{8.96}$$

For Δp large enough to withstand gravitational collapse, the rest mass term is negligible and one obtains

$$\frac{GM^2}{R} \sim N\left[\left(\frac{M}{m_e V}\right)^{\frac{1}{3}} \hbar c\right] \sim \frac{M^{\frac{4}{3}} \hbar c}{m_e^{\frac{4}{3}} R}. \tag{8.97}$$

We note that the R-dependence cancels from both sides of the equation, and we are left with an expression for M, the Chandrasekhar limit in Equation (8.90):

$$M_{\rm Ch} \sim \frac{1}{m_e^2}\left(\frac{\hbar c}{G}\right)^{\frac{3}{2}}. \tag{8.98}$$

The actual result is (assuming the star is made of an ideal Fermi gas)

$$M_{\rm Ch} \sim \frac{\omega_3^0 \sqrt{3\pi}}{2}\left(\frac{\hbar c}{G}\right)^{\frac{3}{2}} \frac{1}{(\mu_e m_H)^2}, \tag{8.99}$$

where $\omega_3^0 \approx 2.018$ is related to the solution of the Lane–Emden equation. Numerically, $M_{\rm Ch}$ is about 1.44 times the mass of our Sun.

We mentioned above that for non-relativistic white dwarfs, $P \propto \rho^{5/3}$, whereas for the relativistic case, $P \propto \rho^{4/3}$. In a more realistic situation, when a star collapses, γ is not a constant. Intuitively, when the star becomes smaller under gravitational collapse, the motion of the particles inside it would become more and more agitated, leading to their velocity becoming more and more relativistic. We therefore should expect that γ should decrease towards 4/3. One often uses an expression that interpolates between the two cases, which is good enough for most modeling purposes:

$$P \approx \left[\left(K_1 \rho^{\frac{5}{3}}\right)^{-2} + \left(K_2 \rho^{\frac{4}{3}}\right)^{-2}\right]^{-\frac{1}{2}}, \tag{8.100}$$

where K_1 and K_2 are some dimensionful proportionality constants. Other interpolating functions are possible, depending on the complications that our model building requires.

The Chandrasekhar limit was named after Subrahmanyan Chandrasekhar (1910–1995), who came upon the idea during his boat trip to England to pursue his PhD study at University of Cambridge. The idea was heavily criticized by Arthur Eddington and the two entered a fierce dispute. Eddington never retracted his attack on Chandrasekhar's theory till his death. In 1983, Chandrasekhar received the Nobel Prize for Physics, along with William A. Fowler (1911–1995) who discovered how various elements can be formed by the nuclear reactions in stars, for their theoretical studies of the physical processes of importance to the structure and evolution of the stars.

8.4.3 Neutrino Oscillation, Dirac Equation, and Majorana Neutrino

We have introduced neutrinos in Section 6.1. To recap, neutrinos come with three "flavors": electron neutrino ν_e, muon neutrino ν_μ, and tau neutrino ν_τ. In the 1960s, the Homestake Experiment (or the Brookhaven Solar Neutrino Experiment) headed by astrophysicist Raymond Davis, Jr. (1914–2006) and John N. Bahcall (1934–2005), found that the number of electron neutrinos (the only type produced in the Sun) detected did not match theoretical prediction based on the luminosity of the Sun. The number deficit was between one half and two thirds. This became known as the "solar neutrino problem". Though initially skeptical (maybe there was an unknown problem with the experiment), the results became gradually accepted when combined with other, different, neutrino detection experiments (e.g., the Super-Kamiokande detector). Raymond Davis Jr. and Masatoshi Koshiba (1926–2020; one of the designers of the Kamiokande experiment) received the Nobel Prize in Physics in 2002 "for pioneering contributions to astrophysics, in particular for the detection of cosmic neutrinos".

Bruno Pontecorvo[21] (1913–1993) proposed the later accepted explanation for the missing neutrino puzzle—the flavor of neutrinos can change during its travel from the source to the detector. This, however, requires the standard model of particle physics to be slightly modified—neutrinos cannot *all* be massless. A layperson might interpret this as the statement that, for example, ν_e, ν_μ and ν_τ all have possibly different but definite mass, just like we know what is the mass of an electron or a muon. However, the situation is more complicated. In fact, there are *mass eigenstates* ν_1, ν_2, ν_3, in addition to the flavors. Thus, an electron neutrino is actually in a superposition of all the mass eigenstates:

$$|\nu_e\rangle = U_{e_1}|\nu_1\rangle + U_{e_2}|\nu_2\rangle + U_{e_3}|\nu_3\rangle. \tag{8.101}$$

Similarly for ν_μ and ν_τ. Thus in general, we have, for the neutrinos with a flavor v_α, the superposition of the mass eigenstates ν_i's:

$$|\nu_\alpha\rangle = \sum_i U_{\alpha i}|\nu_i\rangle. \tag{8.102}$$

The $U_{\alpha i}$'s are elements of the Pontecorvo–Maki–Nakagawa–Sakata (PMNS) matrix. Since there are three neutrino flavors we have a 3×3 PMNS matrix.

We say that there is a *mixing* between the flavor and mass eigenstates of neutrinos. Because the masses are different, the quantum mechanical phases of the mass states also evolve differently during propagation. The result is that although a source can only emit ν_e, after a while we will have a mixture of all the flavors. There is a period in the changes of the phases, however, and so depending on

[21] Pontecorvo was an Italian-born nuclear physicist who defected to the Soviet Union after having done atomic research in the United States.

the exact distance, they can all change back into ν_e as well, hence the name "neutrino oscillation".

For a toy model in which there are only two neutrino flavors, the mixing matrix is quite simple:

$$U = \begin{pmatrix} \cos\theta & \sin\theta \\ -\sin\theta & \cos\theta \end{pmatrix}. \tag{8.103}$$

The probability of a neutrino of flavor α changing into flavor β can be shown to be (in natural units):

$$\text{Prob.}(\alpha \to \beta) = \sin^2(2\theta)\sin^2\left(\frac{\Delta m^2 L}{4E}\right), \tag{8.104}$$

where L is the distance, and E the energy of the neutrinos. We call θ the *mixing angle*. The quantity Δm^2 is called the "mass difference". Neutrino oscillation physics is only sensitive to the squared mass differences $\Delta m_{ij}^2 := m_i^2 - m_j^2$, not the masses themselves. Thus we do not know the values of ν_1, ν_2, and ν_3 individually (they are ordered from the lightest to heaviest). If it turns out that $m_3^2 - m_2^2 > m_2^2 - m_1^2$, then we refer to it as the "normal mass ordering" or a normal hierarchy. In this scenario, the two lightest mass eigenstates have a mass difference of the order of 10 meV, while the third eigenstate has a mass about 50 meV higher. That is to say, the two lightest mass eigenstates have a relatively small mass difference. It is also possible that the order is reversed, such that $m_3^2 - m_2^2 < m_2^2 - m_1^2$; this is called the "inverted mass ordering" or inverted hierarchy, in which the heaviest and the second heaviest mass eigenstates have a relatively small mass difference. The correct ordering is, at the time of writing, unknown.

Neutrino physics plays an important role in astronomy. As mentioned in Section 6.1, unlike electromagnetic radiations that can be obscured by interstellar medium and dust, neutrinos can penetrate these without problems and therefore carry useful signatures that are otherwise not accessible to us. In addition, "high-energy and ultra-high-energy" neutrinos produced by extreme astrophysical events can probe fundamental physics from the TeV scale to the EeV scale and possibly beyond. This prospect has become even more important in recent years in our era of multi-messenger astronomy.

It should be mentioned that there are a lot of mysteries left in neutrino physics. For example, could a neutrino be its own antiparticle—a *Majorana fermion*, hypothesized by Ettore Majorana (1906–missing[22]) in 1937? To understand the Majorana neutrino, we need to first introduce the *Dirac equation*, named after the eccentric physicist Paul Dirac[23] (1902–1984). The Dirac equation is considered one

[22] Majorana disappeared in 1938 under unknown circumstances, but he seemed to have planned for his disappearance (he withdrew his savings from his bank, for example), which fueled even more mysteries as to why he chose to disappear. One possibility is that he withdrew to a monastery (Zichichi 2006).

[23] In an interview with Thomas Kuhn in 1963, Dirac said that "I don't suppose that applies so much to other physicists; I think it's a peculiarity of myself that I like to play about with equations, just looking for beautiful mathematical relations which maybe don't have any physical meaning at all. Sometimes they do."

of the greatest cornerstones in theoretical physics, as it predicted the existence of antimatter. Hidden in the equation, which was purportedly remarked by Dirac to be more intelligent than himself, is the possibility that there could exist in Nature as yet undiscovered Majorana fermions.

For particle physics, it is customary to use the $(+, -, -, -)$ spacetime signature, so we will do so here. Starting with the relativistic energy-momentum relation $E^2 - \vec{p}^{\,2} c^2 = m^2c^4$, or equivalently in the Einstein summation convention of 4-vector formulation $p^a p_a - m^2c^2 = 0$ (see Section 9.2, though beware of the opposite sign convention therein!), Dirac considered factoring this equation into two pieces:

$$p^a p_a - m^2c^2 = (\beta^b p_b + mc)(\gamma^d p_d - mc), \tag{8.105}$$

where β^b and γ^b are eight coefficients to be determined. Multiplying out the RHS and comparing with the LHS, we must have

$$\beta^b = \gamma^b, \quad p^a p_a = \gamma^b \gamma^c p_b p_c. \tag{8.106}$$

The second equation can be written out in full and compares with the LHS $(p^0)^2 - (p^1)^2 - (p^2)^2 - (p^3)^2$, and one finds that to get rid of the cross terms, we need to impose the following conditions:

$$\begin{cases} (\gamma^0)^2 = 1, \\ (\gamma^1)^2 = (\gamma^2)^2 = (\gamma^3)^2 = -1, \\ \gamma^a\gamma^b + \gamma^b\gamma^a = 0 \quad \text{for} \quad a \neq b. \end{cases} \tag{8.107}$$

Equivalently, the requirement is that

$$\{\gamma^a, \gamma^b\} = 2\eta^{ab}, \tag{8.108}$$

where $\{A, B\} := AB + BA$ is the anti-commutator and η_{ab} is the usual Minkowski metric diag$(1, -1, -1, -1)$ on $\mathbb{R}^{3,1}$. This requirement cannot be satisfied by numbers, but they *can* be satisfied by matrices, now called the *gamma matrices*. Dirac found the corresponding matrices in 1928. They are given explicitly by

$$\gamma^0 = \begin{pmatrix} 1 & 0 & 0 & 0 \\ 0 & 1 & 0 & 0 \\ 0 & 0 & -1 & 0 \\ 0 & 0 & 0 & -1 \end{pmatrix}, \quad \gamma^1 = \begin{pmatrix} 0 & 0 & 0 & 1 \\ 0 & 0 & 1 & 0 \\ 0 & -1 & 0 & 0 \\ -1 & 0 & 0 & 0 \end{pmatrix}, \tag{8.109}$$

and

$$\gamma^2 = \begin{pmatrix} 0 & 0 & 0 & -i \\ 0 & 0 & i & 0 \\ 0 & i & 0 & 0 \\ -i & 0 & 0 & 0 \end{pmatrix}, \quad \gamma^3 = \begin{pmatrix} 0 & 0 & 1 & 0 \\ 0 & 0 & 0 & -1 \\ -1 & 0 & 0 & 0 \\ 0 & 1 & 0 & 0 \end{pmatrix}. \tag{8.110}$$

Conventionally, we choose to describe the physics using the second factor of Equation (8.105): $\gamma^a p_a - mc = 0$. Upon the canonical substitution in quantum mechanics[24] $p_a \to i\hbar\partial_a$, we obtain the celebrated Dirac equation:

$$(i\hbar\gamma^a\partial_a - mc)\psi = 0.$$

It turns out that the defining relation of gamma matrices, namely Equation (8.108) actually defines an algebraic structure that was already discovered in 1878 by the mathematician William Kingdon Clifford (1845–1879), who was also probably the first to suggest that gravitation might be a manifestation of an underlying geometry.

Definition: The *Clifford Algebra* C_k is an associative algebra over $\mathbb{R}$ with multiplicative identity with k generators $e_1, \cdots, e_k$ satisfying $e_i^2 = -1$ for all $i = 1, \cdots, k$ and $e_i e_j = -e_j e_i$ for all $i \neq j$.

That is to say, we start with a multiplicative identity, and introduce a bunch of square roots of -1. This can therefore be viewed as a generalization of the *quarternions*:[25]

$$\mathbb{H} = \{a + bi + cj + dk \,|\, i^2 = j^2 = k^2 = ijk = -1\}, \tag{8.111}$$

which itself is a generalization of complex numbers. Indeed,

$$C_0 = \mathbb{R}, \quad C_1 = \mathbb{C}, \quad C_2 = \mathbb{H}. \tag{8.112}$$

We can generalize C_k to $C_{p,q}$, the algebra generated by the e_i's whose p of them are square roots of -1 and q of them are square roots of 1. The gamma matrices are then nothing but the generators of $C_{3,1}$.

The Dirac equation gives a good description of relativistic spin-1/2 particles like the electron and its antiparticle, the positron. The reason why Dirac's choice of gamma matrices are very useful for this purpose is that the entries of these matrices consist of both real numbers ± 1 and imaginary numbers $\pm i$, i.e., the corresponding quantum field ψ (the *Dirac spinor*) is complex, and thus is natural to describe charged particles. This is because in quantum field theory, if a given field ψ creates a particle A (and annihilates $\bar{A}$, the antiparticle), then its complex conjugate field ψ^* creates the antiparticle $\bar{A}$ (and annihilates A).

Of course, it is well known that the photon (spin-1) is its own antiparticle, and so are gravitons (spin-2). These are bosons. The question is whether there exists any *fermion* that is its own antiparticle. It turns out that Dirac's choice of gamma matrices are *not* the only set that satisfies the Dirac equation.

Dirac's choice of gamma matrices is called the *Dirac representation* of the Clifford algebra $C_{3,1}$. There are other representations. The choice of representation found by Majorana has the special property that all of the gamma matrices have imaginary entries; this leads to a Dirac equation that governs *real* fields, i.e., $\psi = \psi^*$, so that

[24] For spatial momenta, this reduces to the familiar looking $p_i = -i\hbar\partial_i$ in the $(-, +, +, +)$ convention, of usual quantum mechanics.

[25] Quarternions were first discovered by William Rowan Hamilton (1805–1865) in 1843. He got the idea in a flash of genius as he was on his way to the Royal Irish Academy on October 16, and carved the equation $i^2 = j^2 = k^2 = ijk = -1$ into the stone of Brougham Bridge as he walked passed it. There is now a plaque on the Bridge to commemorate this event.

the Majorana fermions are their own antiparticles. The Majorana representation is, explicitly,

$$\gamma^0 = \begin{pmatrix} 0 & 0 & 0 & -i \\ 0 & 0 & -i & 0 \\ 0 & i & 0 & 0 \\ i & 0 & 0 & 0 \end{pmatrix}, \quad \gamma^1 = \begin{pmatrix} 0 & 0 & i & 0 \\ 0 & 0 & 0 & i \\ i & 0 & 0 & 0 \\ 0 & i & 0 & 0 \end{pmatrix}, \tag{8.113}$$

and

$$\gamma^2 = \begin{pmatrix} i & 0 & 0 & 0 \\ 0 & i & 0 & 0 \\ 0 & 0 & -i & 0 \\ 0 & 0 & 0 & -i \end{pmatrix}, \quad \gamma^3 = \begin{pmatrix} 0 & 0 & 0 & -i \\ 0 & 0 & i & 0 \\ 0 & i & 0 & 0 \\ -i & 0 & 0 & 0 \end{pmatrix}. \tag{8.114}$$

There exists another representation of the Clifford algebra $C_{3,1}$ called the *Weyl representation* or *chiral representation*, which we mention here for completeness but will refrain from discussing in detail. The Weyl representation is related to the Dirac representation by the unitary transformation

$$\gamma^\mu_{\text{Weyl}} = U\gamma^\mu_{\text{Dirac}}U^{-1}, \quad U = \frac{1}{\sqrt{2}}\begin{pmatrix} 1 & -1 \\ 1 & 1 \end{pmatrix}. \tag{8.115}$$

(We remind the readers that here 1 in the matrix is actually a 2×2 identity matrix; so that U is a 4×4 matrix.) It turns out that the Weyl fermions are irreducible representations of the proper orthochronous Lorentz group, and is massless.

8.4.4 Quantum Cosmology

We have seen that quantum mechanics can be applied to various scenarios in astrophysics. What about the entire cosmos? Indeed, the general expectation is that in the very early Universe, during and right after the Big Bang, quantum effects should be very strong and we need a quantum gravity theory to properly understand the physics. However, so far there is no such theory available, although various approaches are being developed. It is also not clear if quantum gravity is a must.

For now, we note that we can *formally* write down a wavefunction for the Universe and its corresponding Schrödinger-like equation. This is known as the *Wheeler–DeWitt equation*,

$$\hat{H}(x)|\Psi\rangle = 0, \tag{8.116}$$

where $\hat{H}$ is the "Hamiltonian constraint"[26] in the (naively?) quantized version of general relativity. Ψ encodes all the information about the various fields as well as

[26] This comes from the naively quantized version of the classical splitting of the Einstein field equations into a manifestly "3+1" formulation, which is useful when we are interested in studying how a system evolves in time, given an initial condition. After the 3+1 split, there are evolution equations (which involve time derivatives) and the constraint equations. The constraint equations are further categorized into momentum constraint and the Hamiltonian constraint. The first Hamiltonian formulation of general relativity was in the 1952 study of Felix Pirani (1928–2015, spelled Pirant in some papers), Alfred Schild (1921–1977) and Ray Skinner (1927–1988) (Pirant et al. 1952), whose objective was to quantize gravity. The more famous (3 + 1)-split is the ADM formalism of 1959, by Richard Arnowitt (1928–2014), Stanley Deser (1931–2023) and Charles W. Misner (1932–2023) (Arnowitt et al. 1959).

geometry of the spacetime, so it is quite unlike the usual wavefunction in quantum mechanics. That is, schematically, the Euclidean wavefunction is

$$\Psi = \int \mathscr{D}[g]\mathscr{D}[\varphi]e^{-S[g_{\mu\nu}, \varphi]}, \tag{8.117}$$

for "path integrals" over the metric $g_{\mu\nu}$ and the field φ. Instead of a wavefunction of a particle defined on a fixed spacetime, now the geometry itself is only probabilistic (so that the classical spacetime is replaced by a superpositions of different possible geometries). We also see that the Wheeler–DeWitt equation has no time evolution (since the RHS is zero; and the Hamiltonian generates time evolution: $\hat{H} \leftrightarrow i\hbar\partial_t$). This is the famous "problem of time" first raised in the 1950s by Peter Bergmann (1915–2002) (Pitts 2022), a former assistant of Einstein—it seems that fundamentally there is no such thing as time in quantum cosmology, so how does time come to be? Is time an emergent quantity? These are, of course, unsolved problems.

A possibility is that the Universe was initially timeless, described by a four-dimensional Riemannian manifold, with four spatial dimensions. However, one of the dimensions became temporal. This is essentially the idea of the "no-boundary proposal" by Stephen Hawking and James Hartle. Unfortunately it is beyond the scope of this book to explore this interesting topic any further.

8.5 Exercises

(1) Verify that the Black–Scholes equation

$$\frac{\partial V}{\partial t} + rS\frac{\partial V}{\partial S} + \frac{1}{2}\sigma^2 S^2\frac{\partial^2 V}{\partial S^2} = rV \tag{8.118}$$

can be written in the form

$$\frac{\partial V}{\partial t} = -\frac{\sigma^2}{2}\left[S\frac{\partial}{\partial S} - \frac{1}{2}\left(1 - \frac{2r}{\sigma^2}\right)\right]^2 V + \frac{\sigma^2}{8}\left(1 + \frac{2r}{\sigma^2}\right)^2 V. \tag{8.119}$$

Hint: Note that this "completing the square" is not as obvious as completing the square for a quadratic equation. Namely, the following is *wrong*:

$$\frac{\partial V}{\partial t} = -\frac{\sigma^2 S^2}{2}\left(\frac{\partial^2 V}{\partial S^2} + \frac{2r}{\sigma^2 S}\frac{\partial V}{\partial S}\right) + rV \tag{8.120}$$

$$= -\frac{\sigma^2 S^2}{2}\left[\left(\frac{\partial}{\partial S} + \frac{r}{\sigma^2 S}\right)^2 V - \frac{r^2 V}{\sigma^4 S^2}\right] + rV. \tag{8.121}$$

The reason is that a differential operator is involved, thus the cross terms in the expansion of

$$\left(\frac{\partial}{\partial S} + \frac{r}{\sigma^2 S}\right)^2 V \tag{8.122}$$

leads to

$$\frac{\partial}{\partial S}(V)\frac{r}{\sigma^2 S} + V\frac{\partial}{\partial S}\left(\frac{r}{\sigma^2 S}\right) \neq \frac{2r}{\sigma^2 S}\frac{\partial V}{\partial S}. \tag{8.123}$$

(2) Students who have learned quantum field theory may recall that in the *Feynman expansion*, an integral over the space of some field Ω takes the form

$$\int_\Omega d\phi f(\phi)e^{iS(\phi)/\hbar} = \sum_{k=0}^{\infty} a_k g^k, \tag{8.124}$$

where g is a small coupling constant. The right-hand side is often pictorially represented by the Feynman diagrams. Such a series is often divergent and required "renormalization" and "regularization". Usually the first few terms are good enough to check with experiments. (Depending on the theory and the series involved, sometimes we will get an "asymptotic series" in which adding more terms would eventually make the number further *away* from the data. Having more terms does not always mean a better result when the series is not convergent!) How to make this process mathematical rigorous remains a work in progress. A recent approach is via the so-called *resurgence trans-series*. Find out more if you are interested.

(3) The commutator of the position and momentum operator is $[\hat{x}, \hat{p}] = i\hbar$. For any two matrices A and B, we have $\mathrm{Tr}([A, B]) = 0$. However, taking the trace of the right hand side gives $i\hbar\mathrm{Tr}(\mathbb{1})$, where $\mathbb{1}$ is the identity operator that has now been explicitly written. Since the matrix representation of the identity is just the identity matrix, its trace is the sum of the 1's on the diagonal. How can this give zero? Where is the mistake in this argument? (If you get stuck, see Gieres (2000).)

(4) Recall the interpolating function Equation (8.100) for the mass–radius relation for a white dwarf. Using the fact that $\rho \approx M/R^3$ and that $P \sim GM^2/R^4$ (why?), show that

$$R \approx \frac{K_1}{GM^{\frac{1}{3}}}\left(1 - \frac{G^2 M^{\frac{4}{3}}}{K_2^2}\right)^{\frac{1}{2}}. \tag{8.125}$$

Additional notes: A more accurate approximation would lead to

$$R \approx CM^{-\frac{1}{3}}\left[1 - \left(\frac{M}{M_{\mathrm{Ch}}}\right)^{\frac{4}{3}}\right]^{\frac{1}{2}}, \tag{8.126}$$

for some constant C, which shows that $R \to 0$ as $M \to M_{\mathrm{Ch}}$. Of course, realistically, the star would have collapse to form a black hole at some point, so an arbitrarily small white dwarf is not physical.

(5) Learn more about the various interpretations of quantum mechanics, such as the De Broglie–Bohm theory (Bohmian mechanics) and QBism. There

are also theories that are not just a different interpretation, but are genuine alternative theory of standard quantum mechanics, or contains additional structures/mechanisms on top of standard quantum mechanics. For example, the Diósi–Penrose model suggests that gravity can induce the objective collapse of the wavefunction (which occurs even without any observation or measurement).

References

Arnowitt, R., Deser, S., & Misner, C. W. 1959, PhRv, 116, 1322

Favella, L. F. 1967, PhT, 7, 77

Gieres, F. 2000, RPPh, 63, 1893

Handsteiner, J., Friedman, A. S., Rauch, D., et al. 2017, PhRvL, 11, 060401

Li, M.-H., Wu, C., Zhang, Y., et al. 2018, PhRvL, 121, 080404

Mermin, N. D. 2025, PhT, 78, 28

Physics Stack Exchange, https://physics.stackexchange.com/questions/60242/relation-between-black-scholes-equation-and-quantum-mechanics, (accessed 17 March 2025)

Olver, P. J. 2013, Introduction to Partial Differential Equations (Berlin: Springer Science & Business Media)

Ong, Y. C. 2016, Evolution of Black Holes in Anti-de Sitter Spacetime and the Firewall Controversy, Springer Theses (Berlin: Springer)

Pirant, F. A. E., Schild, A., & Skinner, R. 1952, PhRv, 87, 452

Pitts, J. B. 2022, SHPSA, 95, 1

Rauch, D., Handsteiner, J., Hochrainer, A., et al. 2018, PhRvL, 121, 080403

Tsvelik, A. M. 2003, Quantum Field Theory in Condensed Matter Physics (Cambridge: Cambridge Univ. Press)

Zee, A. 2003, Quantum Field Theory in a Nutshell (Princeton, NJ: Princeton Univ. Press)

Zichichi, A. 2006, CERNC, https://cerncourier.com/a/ettore-majorana-genius-and-mystery/

AAS | IOP Astronomy

Understanding the Universe
A conversational introduction to cosmology
Yen Chin Ong

Chapter 9

Staring into the Abyss: Black Holes

> *"After a little while I became possessed with the keenest curiosity about the whirl itself. I positively felt a wish to explore its depths, even at the sacrifice I was going to make; and my principal grief was that I should never be able to tell my old companions on shore about the mysteries I should see."*
> —Edgar Allan Poe, 'A Descent into the Maelstrom'.

The concept of black holes has undergone a drastic change over the past few decades. They were first treated as nothing but mathematical curiosity and unphysical solution of Einstein's theory of general relativity. As astronomical observations improved, however, the evidence for their actual existence has mounted and most astrophysicists now believe that black holes are not only physically real, but they are actually common in the cosmos. In this chapter we briefly introduce black hole solutions in general relativity. We then explore how particles—both massive and massless—move around a black hole. This is especially important to understand "black hole shadows" that were imaged by the Event Horizon Telescope. We should also take a closer look at the so-called "conserved energy" of a particle in black hole spacetime; it is not quite what it seems. Other astrophysically interesting structures in the environment of a black hole, such as astrophysical jet, are also briefly introduced. We end the chapter with some discussions on compact objects that are not black holes, including some highly hypothetical ones.

Black holes were once thought to be purely mathematical. Einstein himself, along with many of his contemporaries, never believed in the existence of black holes. He suspected that black holes are not physical, in the sense that a realistic stellar collapse would never result in a black hole. Einstein's suspicion was quite natural, given that the physics of stellar collapse was not well-understood back then.

doi:10.1088/2514-3433/ade963ch9

Around that time, the nuclear physicist J. Robert Oppenheimer (1904–1967) became interested in astrophysics, and published a series of three papers. His first paper, *On the Stability of Stellar Neutron Cores*, was in collaboration with Robert Serber (1909–1997). In this paper they discussed the properties of a neutron star. This was followed by *On Massive Neutron Cores*, coauthored by Oppenheimer and George Volkoff (1914–2000), in which they argued that a neutron star cannot be arbitrarily massive. Like the Chandrasekhar limit of white dwarfs, there exists a critical mass, now called the Tolman[1]–Oppenheimer–Volkoff limit, above which a neutron star must become unstable and continue to collapse further. Such a limit for any star supported by quantum degeneracy pressure was previously hypothesized to exist by Lev Landau (1908–1968) in 1932.

In the same year of Einstein's criticism, Oppenheimer and Hartland Snyder (1913–1962), who was then Oppenheimer's student, would go on to show how a spherically symmetric "star", which they assumed to be a homogeneous dust ball, can collapse into a black hole in their landmark paper *On Continued Gravitational Contraction*. Still, a realistic collapse will not be perfectly spherically symmetric. Would this prevent black hole formation? As it turned out black hole still formed in generic cases. The details become more complicated without spherical symmetry, and the collapse can only be studied numerically. Regardless, the qualitative features of Oppenheimer–Synder collapse continue to hold.

While initially skeptical, today astronomers have come to accept black holes as ubiquitous in the Universe, though some theorists still prefer to replace black holes by some other even more exotic objects. In this chapter we will study the black hole solutions in general relativity and some physics near black holes. After all, it is via the objects and light around a black hole that we can infer its existence. Typically, an astrophysical black hole is immersed in a very messy environment, with gas accreting and falling into the black hole, emitting high energy radiation of all kinds in the process. It is not my aim to discuss these complicated physics in detail; my objective is merely to expose students to these topics. We will end the chapter by discussing the aforementioned exotic compact objects, along with some other relatively more conventional objects and phenomena like magnetars.

9.1 Black Holes in General Relativity

John Michell (1724–1793), an English natural philosopher and clergyman with wide interests in the natural sciences, proposed a curious idea in 1783: a star that is so massive that even light cannot escape its gravitational field, and therefore the most massive stars in the universe would be invisible. Later, in 1796, this idea would occur independently to the mathematician Pierre-Simon Laplace (1749–1827), who published it in the earlier edition of his book *Exposition du système du Monde*. As we would see, this idea presupposes the corpuscular nature of light. Thus, after the

[1] Richard Tolman (1881–1948) laid the foundation by solving the Einstein field equations for spherically symmetric perfect fluid.

wave theory of light became more convincing (certainly Laplace was convinced), Laplace removed this discussion from the subsequent edition of his book.

The idea was quite simple indeed. If light is a corpuscle with some mass m, then by Newtonian mechanics, its kinetic energy is $mc^2/2$, where c is the speed of light. This is of course *not correct* as today we know that the formula $mv^2/2$ does not hold for $v = c$. But people didn't know this back then. The light corpuscle is subject to a gravitational potential given by $V = -GMm/r$. In order to escape the gravitational pull with escape velocity c, the particle must satisfy the equation

$$\frac{mc^2}{2} = \frac{GMm}{r}. \tag{9.1}$$

This leads to the expression

$$r = \frac{2GM}{c^2}, \tag{9.2}$$

the size of a "dark star". Remarkably in general relativity, this same expression remains valid—it is the *Schwarzschild radius* of a black hole. However, this is just a coincidence, since the derivation for the radius of the dark star is incorrect. In addition, r does not have the same meaning in general relativity, as we will soon discuss. Even if we take "dark star" as a plausible object, it is very different from a black hole. First of all, any star has a surface, even if it is not a well-defined hard surface like the Earth. On the other hand, a black hole's "surface"—the horizon—is a mathematical one, one does not feel anything special *across* the horizon[2], it's just empty spacetime. Furthermore, just like one can throw a ball into the air on Earth with speed less than that of escape velocity, light *can* leave the surface of a dark star momentarily before falling down onto the star. Therefore a dark star is only dark to a faraway observer; a nearby observer could intercept the light before it falls back onto the surface and thus *see* the "dark" star. However, light cannot leave the event horizon of the black hole at all.

What then, is a black hole[3]? Let us consider the simplest example: the Schwarzschild black hole. Its metric tensor is given by

$$g[\mathrm{Sch}] = -\left(1 - \frac{2M}{r}\right)dt^2 + \left(1 - \frac{2M}{r}\right)^{-1} dr^2 + r^2(d\theta^2 + \sin^2\theta d\phi^2), \tag{9.3}$$

[2] This doesn't mean that no effect is felt at all near a black hole. For example, an in-falling observer would experience a strong tidal force near a stellar mass black hole. The tidal force is smaller for larger black holes, thus one can fall into a supermassive black hole at the center of a galaxy without being torn apart by tidal forces. Black holes are gentle giant, they become increasingly dangerous when they are small!

[3] The history of the name "black hole" itself is somewhat convoluted. It is often said that John Wheeler (1911–2008) coined it, but Robert Dicke (1916–1997) had played a part in this too (Herdeiro & Lemos 2018). John Wheeler and many of his students contributed to the advancement of general relativity in the United States after World War II. Dicke himself is known for the *Brans–Dicke* theory, which is a generalization of general relativity that includes a new scalar field. He also founded the company Princeton Applied Research (PAR) to market the lock-in amplifier, a device that can extract a signal with a known carrier wave from an extremely noisy environment.

where M is the mass of the black hole. For simplicity we have set both c and G to unity. To obtain conventional units, just replace M by GM/c^2. This form of the metric, though referred to as the "Schwarzschild solution", was not the original form Schwarzschild had worked with. Rather, this choice of coordinates was due to Johannes Droste (1886–1963), a student of Hendrik Lorentz. The metric is asymptotically flat, as it reduces to flat Minkowski spacetime in spherical coordinates when $r \to \infty$. For every spherical surface at fixed r, its area is $4\pi r^2$, thus we refer to r as the areal radius. Note that r itself does not directly measure distance like in flat space. Instead, the distance from $r = a$ to $r = b > a$ (where both a and b are larger than the Schwarzschild radius) is given by the integral $\int_a^b (1 - 2M/r)^{-1/2} dr$. Thus, unlike r in the case of dark star, which is just the radial distance, the r in Schwarzschild radius is not a physical distance.

There are two peculiar values of r for which the metric becomes pathological: $r = 0$ where the term $2M/r$ blows up, and $r = 2M$ where the coefficient of dt^2 becomes zero, accompanied by the divergence of the coefficient of dr^2. We now know that $r = 0$ and $r = 2M$ correspond to, respectively, the singularity and the horizon.[4] In fact, Schwarzschild set out to solve the vacuum Einstein field equation (i.e., $T_{ab} = 0$, or equivalently $R_{ab} = 0$; see Exercise 9.5.(5)) for a spherically symmetric star (the Schwarzschild solution also describes the gravitational field outside a spherical star if the stellar radius R is larger than the Schwarzschild radius). In his choice of coordinates, it is considerably more difficult to notice that there is a "horizon" at all. The divergence at $r = 0$ is not surprising, since one could appreciate that as matter collapses into a black hole and concentrates into a point, density would diverge (but the singularity at $r = 0$ is actually quite different, more on this later). The nature of the horizon was much more puzzling, and remained so for a few decades. Einstein, Arthur Eddington, and even the great mathematician David Hilbert, all mistakenly thought that the horizon is a physical, impenetrable barrier. The issue was only cleared up much later in the 1960s, starting from the work of Martin Kruskal (1925–2006). We now understand that $r = 2M$ has no real singularity. The apparent divergence is due to the failure of the Schwarzschild–Droste's coordinate system. This is similar to the fact that despite latitude and longitude systems becoming degenerate at the North and South Poles, it does not mean there is anything special with these points on Earth. By changing to different coordinate systems that are better behaved across the horizon, there is no divergence at all in the metric coefficients. In hindsight, readers may find this rather perplexing: why decades of confusion, especially since many mathematicians like Hilbert are familiar with differential geometry? The reason may be that the different ways of thinking back that led to the philosophy that equations hold more physical significance than geometry. In fact, Einstein's own philosophy on general relativity is quite different from the modern version, since he never believed that the theory

[4] There are various notions of a horizon, which we will mention later in Section 10.4, so here I do not specify which horizon, since for stationary black holes most of the definitions coincide. If you insist on being specific, you can take it to be the event horizon; we will define this properly later.

had "geometrized" gravity. For a detailed discussion of Einstein's thought on these issues, see Norton (2024).

The singularity at $r = 0$ is, however, an actual singularity that cannot be removed by changing coordinates. How do we know that no coordinate exists that can remove the $r = 0$ singularity? To see this, we can check that the scalar quantity called the Kretschmann scalar, $K = R_{abcd}R^{abcd} = 48M^2/r^6$, is finite at the horizon, but is infinite at $r = 0$. Being a scalar quantity that has no lower or upper indices, K is invariant under coordinate transformation and so the divergence at $r = 0$ cannot be removed. Its nature, however, is rather subtle. It is *wrong* to consider $r = 0$ as the "center" of the black hole because the interior geometry is nontrivial. Just because the black hole horizon looks like a sphere from the outside, does not mean that the interior is anything like the inside of a beach ball. In fact, the Schwarzschild–Droste coordinate is valid inside the horizon (just not *on* the horizon), but since $r < 2M$ the coefficient of dt^2 is now positive, while the coefficient of dr^2 becomes negative. Therefore, instead of a $(-, +, +, +)$ signature of the metric, we now have $(+, -, +, +)$. In the causal structure of general relativity, the three spatial dimensions are the ones that have the same sign, which means that r now denotes a temporal dimension in the interior of a black hole. This does not mean something drastic like "space and time interchange their role" as sometimes claimed in popular science books—keep in mind that in general relativity coordinates are not "real", they are just useful labels. The correct interpretation is that while r has more or less the meaning of "radial direction" outside the horizon (though not physical distance), it no longer has this role inside. To better conform to our language preference, let us re-label $t \mapsto R$ and $r \mapsto T$. Then the interior Schwarzschild metric takes the form

$$g[\text{Sch-Int}] = -\left(\frac{2M}{T} - 1\right)^{-1} dT^2 + \left(\frac{2M}{T} - 1\right) dR^2 + T^2(d\theta^2 + \sin^2\theta d\phi^2). \quad (9.4)$$

Notably, the singularity is not a location in space, but an event in time. Note that the *coordinate* time T seems to run from initial time $T = 2M$ to $T = 0$, but the proper time of an observer at rest relative to the spatial coordinates satisfies

$$d\tau = \pm \frac{dT}{\sqrt{\frac{2M}{t} - 1}}, \quad (9.5)$$

so we can choose the appropriate sign such that the proper time increases even though the coordinate time decreases. Any observer who enters a black hole is doomed to arrive at $T = 0$, just like all of us is doomed to "arrive at" the next Monday. In fact, constant time (T=const.) slices is infinite in extent in the R-direction.[5] As one gets nearer to the eventual singularity at $T = 0$, one is stretched along the R direction and compressed along the other directions (since the S^2 part of the metric is shrinking in its areal radius). This can also be shown by checking the tidal

[5] To see this, simply note that for T = const.., we have $dT = 0$ in the metric, and from the remaining terms one recognizes the topology of $\mathbb{R} \times S^2$ as long as $T \neq 2M$, and R can range over the whole of $\mathbb{R}$. The spatial geometry is therefore an infinitely long cylinder.

forces correspond to the R, θ, ϕ directions. Another way to appreciate the interior Schwarzschild solution is to think of it as an anisotropic cosmological spacetime (in fact a special case of Kantowski–Sachs solutions) that has a future singularity.

It should be emphasized that the Schwarzschild solution is a *vacuum* solution of the Einstein field equation, so the mass M has nothing to do with matter; in fact there is no matter that plays the role of a "source" of the gravitational field. Of course, a realistic black hole is formed when a massive star collapses, and the mass of a star does originate from matter. Nevertheless, what general relativity is telling us is that *pure geometry* itself can give rise to what we measure as a mass.

If we add electrical charges into a black hole, the geometry changes into[6]

$$g[\mathrm{RN}] = -\left(1 - \frac{2M}{r} + \frac{Q^2}{r^2}\right)dt^2 + \left(1 - \frac{2M}{r} + \frac{Q^2}{r^2}\right)^{-1} dr^2 + r^2(d\theta^2 + \sin^2\theta d\phi^2). \tag{9.6}$$

This is known as the Reissner–Nordström solution. Now there are two horizons obtained by solving the equation $1 - 2M/r + Q^2/r^2 = 0$. This gives an outer horizon $r_+ = M + \sqrt{M^2 - Q^2}$ and an inner horizon $r_- = M - \sqrt{M^2 - Q^2}$. The case with $M = Q$ is called an extremal black hole, as $Q > M$ no longer corresponds to a black hole solution—it is a *naked singularity*, which we will discuss in Section 10.3.

As one crosses the outer horizon into the black hole, the spacetime signature flips into (+,−,+,+) just like in the Schwarzschild case. Thus r_- lies in the future of r_+ and one has no choice but to encounter it. However, the signature flips into (−,+,+,+) again as one passes r_-. Therefore unlike the case of Schwarzschild singularity, the Reissner–Nordström singularity $r = 0$ *is* a location in space. Technically, we say that a Schwarzschild singularity is a spacelike singularity, while the Reissner–Nordström one is a timelike singularity (because its worldline is in the timelike direction). Observers inside the charged black hole can safely avoid hitting the singularity. In fact, it is not easy to hit it because the singularity exhibits a *repulsive gravity*. An easy way to appreciate this is to consider the weak field approximation in general relativity (recall Section 5.1.2), which relates the Newtonian potential Φ and the metric function g_{tt}. For a Schwarzschild black hole we have

$$1 - \frac{2M}{r} = 1 + 2\Phi. \tag{9.7}$$

That is, $\Phi = -M/r$, which gives the Newtonian force $F = -\nabla\Phi = -M/r^2$. In the case of a Reissner–Nordström black hole, we would have instead,

$$1 - \frac{2M}{r} + \frac{Q^2}{r^2} = 1 + 2\Phi, \tag{9.8}$$

so that $\Phi = -M/r + Q^2/2r^2$. The potential is zero at

$$r = \frac{Q^2}{2M}. \tag{9.9}$$

[6] In the units $4\pi\varepsilon_0 = 1$.

Therefore the *force* is zero at[7]

$$r = \frac{Q^2}{M}. \tag{9.10}$$

Near the vicinity of the singularity smaller than this value of r, gravity becomes repulsive. Of course, it is debatable whether this regime qualifies as "weak field". A more rigorous approach can be made with the notion of "quasi-local energy", to show that the mass–energy of the singularity is negative (see, e.g., Maluf 2014; Szabados 2009), despite the overall mass of the spacetime being positive.

We will return to more discussions concerning the inner horizon in Section 10.3. For now, let us remark that in astrophysical contexts, one usually does not consider charged black holes because charged black holes in a realistic environment will quickly attract opposite charges from the surroundings and as a consequence the charges will be neutralized. However, this does not mean that black holes with *some* amount of electrical charges do not exist at all. In fact, it is hard to imagine that black holes are completely electrically neutral at all times.

Both the Schwarzschild and Reissner–Nordström black holes are non-rotating. This is not realistic especially if we consider black holes that are formed from gravitational collapse of massive stars. All stars have angular momentum, and although some amount of angular momentum can be lost due to shredding of mass prior to complete gravitational collapse, we expect that due to angular momentum conservation the resulting black holes should also be rotating. In fact, since the size of a black hole is much smaller than the original star, the black hole should be rotating at a much faster rate than the star, just like an ice skater rotates at a faster rate when she draws her arms closer to her body. The quest for a rotating black hole solution turned out to be extremely challenging, and was only achieved in 1963 by Roy Kerr (1934–present), almost 50 years after Schwarzschild's solution. In the so-called Boyer–Lindquist coordinates, the Kerr black hole is described by the metric of the form

$$\begin{aligned} g[\mathrm{Kerr}] = &-\left(1 - \frac{2Mr}{r^2 + a^2\cos^2\theta}\right)dt^2 - \frac{4Mra\sin^2\theta}{r^2 + a^2\cos^2\theta}dt d\phi + \left(\frac{r^2 + a^2\cos^2\theta}{r^2 - 2Mr + a^2}\right)dr^2 \\ &+ (r^2 + a^2\cos^2\theta)d\theta^2 + \left(r^2 + a^2 + \frac{2Mra^2\sin^2\theta}{r^2 + a^2\cos^2\theta}\right)\sin^2\theta d\phi^2. \end{aligned} \tag{9.11}$$

In addition to the mass M, there is an additional parameter a that is related to the angular momentum J by $J = aM(=aMc)$. Note that the Kerr solution is invariant under time translation $t \mapsto t$+const., just like the Schwarzschild and Reissner–Nordström solutions. However, unlike those two, it is *not* invariant under time reversal $t \mapsto -t$. Physically this can be easily appreciated: a rotating black hole has a clear sense of direction, say rotating clockwise as viewed from above its "north pole".

[7] Readers who are familiar with general relativity would know that the opening angle of light cone in the standard Schwarzschild–Droste coordinates becomes smaller as it approaches a Schwarzschild black hole, and "closes up" at the horizon. One can similarly compute the opening angle of a light cone in between the horizons of a Reissner–Nordström black hole. One finds that the angle is minimized at the hypersurface $r = Q^2/M$, which coincides with the onset of repulsive gravity.

Under time reversal the black hole would rotate in the counter-clockwise rotation. Technically we say that the Schwarzschild and Reissner–Nordström black holes are *static* solutions, whereas the Kerr black hole is only a *stationary* solution. If we add also electrical charges, the rotating solution is known as the Kerr–Newman black hole.

The Kerr solution too has two horizons given by $r_\pm = M \pm \sqrt{M^2 - a^2}$, and so also admits an extremal black hole solution with $a = M$. However, the Kerr black holes have some new properties that are absent in the Reissner–Nordström case. For example, the timelike singularity is no longer a point as in the Reissner–Nordström solution, but a ring. Passing through the ring one arrives at a region of spacetime (corresponds to $r < 0$ region—but remember that coordinates are just labels!) where there exist "closed timelike curves" (CTCs). Traveling along CTCs, one can arrive at one's own past—a time machine! These are mathematical properties of the solutions, but they are unlikely to be physical due to the inner horizon itself being unstable; see Section 10.3.

The most prominent difference, however, is the *ergoregion*, whose boundary the "ergosphere", is obtained by solving the equation $g_{tt} = 0$. Inside this region (which is still outside the outer horizon, so outside of the black hole), no object can go against the rotation of the black hole. No matter how fast a rocket fires, a spaceship cannot travel in the opposite direction but must co-rotate with the black hole. Only a hypothetical particle with faster-than-light velocity can counter-rotate the black hole inside the ergoregion. This is the consequence of the rotation of the black hole "dragging" spacetime around it. It is again worth emphasizing that even the Kerr solution is a vacuum solution, so we are talking about "rotation" of an empty spacetime, but nevertheless this has physical effects.

9.2 Physics Near Black Holes

Given that there is a black hole, how would objects behave around it? Or to put it in another way: how do we know that there is a black hole? One way to deduce this is by observing the motion of objects. For example, there is a supermassive black hole at the center of our own Milky Way Galaxy, called Sagittarius A*, which is estimated to be around 4.2 million solar masses (around 8.5×10^{36} kg). How do we know it is a black hole? If we want to be pedantic about it, we cannot be sure. Even the 2020 Nobel Prize in Physics awarded to Reinhard Genzel (1952–present) and Andrea Ghez (1965–present), was very careful in stating that the prize is for their discovery that Sagittarius A* is a supermassive *compact object*, not definitely a black hole. Still, once we know the mass (this can be deduced essentially from Kepler's third law, once we have gathered enough data to reconstruct the orbits of the stars around it;[8] see Figure 9.1. Of course, very near the black hole there will also be general relativistic corrections). The size of the object (the upper bound can be deduced from how close the orbits can get near the center of mass), we can deduce

[8] It is worth emphasizing that contrary to popular misunderstanding by the public, black holes do not actively "suck" stuff in like a galactic vacuum cleaner; it merely acts as a massive body, and general relativistic effects only become prominent close enough to the black hole.

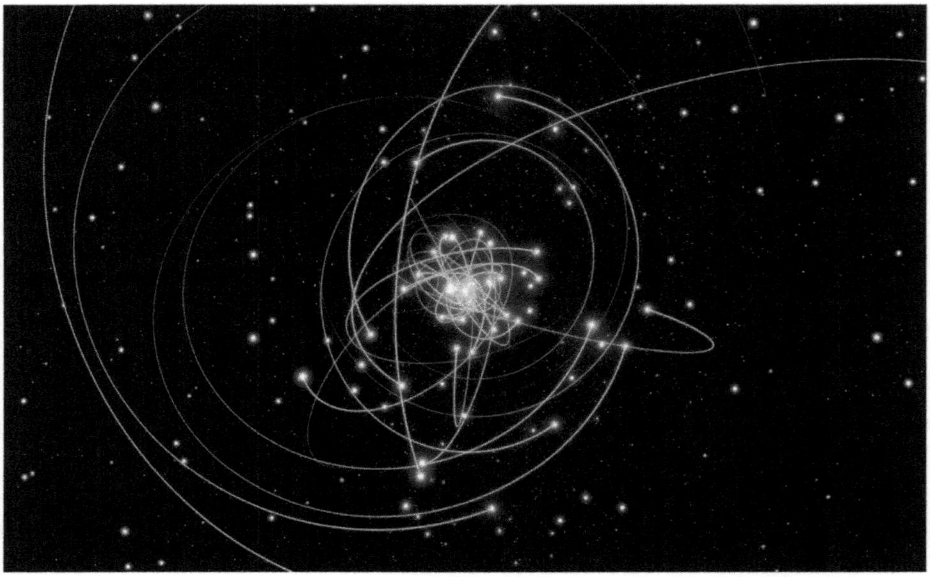

Figure 9.1. This simulation shows the orbits of stars very close to the supermassive black hole at the heart of the Milky Way, Sagittarius A*. *Credit:* ESO/L. Calçada/spaceengine.org. CC BY 4.0

indirectly that the object must be very compact and very massive. *As far as we know*, a black hole is the only explanation. There are other more exotic possibilities that we will mention briefly in Section 9.4, but these are speculations at best.

Other indirect observations can also be used to determine the size of a compact object. For example, the X-ray source Cygnus X-1, discovered in 1964, was later on determined to be a X-ray binary system about 7000 light-years from us: it consists of a compact object and a blue supergiant star (HDE 226868), the mass from the latter is being transferred and accreted into a disk around the compact object. The rapid variation in the intensity of the X-ray that occurs several times a second allowed astrophysicists to deduce that the X-ray generation occurs within a compact region no larger than about 10^5 km. By analyzing the motion of the compact object and the supergiant, their masses can be determined. The size and the mass taken together, allowed astrophysicists to then conclude that the compact object is, to our best understanding, most likely a black hole.

Understanding how stars and other objects orbit around black holes is therefore an important task. To prepare for the discussions ahead, it is best to first consider just flat Minkowski spacetime to introduce some techniques. For clarity, we will restore c explicitly until further notice. The Minkowski metric is of course

$$-c^2d\tau^2 = ds^2 = -c^2dt^2 + dx^2 + dy^2 + dz^2, \tag{9.12}$$

where τ denotes the proper time of a massive particle whose 4-velocity is

$$u^a := \frac{dx^a}{d\tau}. \tag{9.13}$$

We have

$$\eta_{ab}u^a u^b = \eta_{ab}\frac{dx^a}{d\tau}\frac{dx^b}{d\tau} = \frac{ds^2}{d\tau^2} = -c^2. \tag{9.14}$$

The 4-momentum is defined by

$$p^a := mu^a. \tag{9.15}$$

In particular, the spatial components p^i, $i = 1, 2, 3$, are just the usual 3-momentum $\vec{p} = \gamma m\vec{v}$; while $p^0 = E/c$; $E = \gamma mc^2$. We have $\eta_{ab}p^a p^b = -m^2c^2$. From this we can deduce the energy-mass relation

$$E^2 = m^2c^4 + p^2c^2 \tag{9.16}$$

of special relativity. Henceforth, we set $c = 1$ again.

We can obtain the energy-mass relation via the Lagrangian of the test particle:[9]

$$\mathsf{L} = -m\sqrt{\dot{t}^2 - \dot{x}^2 - \dot{y}^2 - \dot{z}^2} = -m, \tag{9.17}$$

where the dot denotes a derivative with respect to τ. We use L for the Lagrangian instead of L because later on L will be used to denote the conserved momentum. (Why is the Lagrangian essentially just the mass? See Exercise 9.5.(1).)

From a Lagrangian of this sort, one often defines a conjugate momentum via

$$p_a := \frac{\partial \mathsf{L}}{\partial \dot{x}^a}. \tag{9.18}$$

This is related to the 4-momentum by

$$p^a = \eta^{ab}p_b. \tag{9.19}$$

For example,

$$p_t = \frac{\partial \mathsf{L}}{\partial \dot{t}} = -\frac{m}{2}\frac{2\dot{t}}{\sqrt{\dot{t}^2 - \dot{x}^2 - \dot{y}^2 - \dot{z}^2}} = -m\dot{t} = -E. \tag{9.20}$$

Similarly one can compute p_x, p_y, p_z and check that

$$\sqrt{\left(\frac{E}{m}\right)^2 - \left(\frac{p_x}{m}\right)^2 - \left(\frac{p_y}{m}\right)^2 - \left(\frac{p_z}{m}\right)^2} = 1, \tag{9.21}$$

which then gives the energy-mass relation, Equation (9.16).

Now consider a test particle moving on the background of a Schwarzschild spacetime, then the Lagrangian Equation (9.17) is generalized into

$$\mathsf{L} = -m\left[\left(1 - \frac{2M}{r}\right)\dot{t}^2 - \frac{\dot{r}^2}{1 - \frac{2M}{r}} - r^2\dot{\theta}^2 - r^2\sin^2\theta\dot{\phi}^2\right]^{\frac{1}{2}}. \tag{9.22}$$

[9] To see that the square root term is just unity, observe that $d\tau = \sqrt{dt^2 - dx^2 - dy^2 - dz^2} = \sqrt{\dot{t}^2 - \dot{x}^2 - \dot{y}^2 - \dot{z}^2}\,d\tau$.

We note that the Euler–Lagrange equation for t is

$$\frac{d}{d\tau}\left(\frac{\partial L}{\partial \dot{t}}\right) - \underbrace{\frac{\partial L}{\partial t}}_{=0} = 0. \tag{9.23}$$

This means that $\partial L/\partial \dot{t}$ is conserved. A direct differentiation yields explicitly this conserved quantity:

$$\frac{\partial L}{\partial \dot{t}} = -m\dot{t}\left(1 - \frac{2M}{r}\right). \tag{9.24}$$

When $r \to \infty$, we know that the geometry is asymptotically Minkowski spacetime. But in Minkowski spacetime we know from Equation (9.20) that $\dot{t} = E/m$. Therefore we see that this conserved quantity is related to the energy of the particle. We therefore *define* E via

$$E := m\dot{t}\left(1 - \frac{2M}{r}\right). \tag{9.25}$$

However, note that what we refer to as the energy here is subtler than textbooks would have you believe. In special relativity, the energy of a particle is measured by an observer who is present at the site of the particle, whose 4-velocity is U^μ, such that $U^0 = 1$, i.e., the observer is locally at rest. (Because physically, in order to take measurement of the particle, you need to be *with* the particle. Mathematically, an inner product is only defined at the same spacetime point.) The energy is then the Minkowski product between the 4-velocity of the observer and the energy-momentum 4-vector of the particle, with only the zero-zero component:

$$-p_a U^a = -\eta_{00} p^0 U^0 = p^0. \tag{9.26}$$

In the Schwarzschild geometry, its equivalent is given by

$$\mathscr{E} := -g_{00} p^0 U^0 = \left(1 - \frac{2M}{r}\right) p^0 U^0. \tag{9.27}$$

However, in this case the observer's tangent vector is no longer just ∂_t and we have

$$U^0 = \left(1 - \frac{2M}{r}\right)^{-\frac{1}{2}}, \tag{9.28}$$

so that

$$p^0 = \frac{\mathscr{E}}{\sqrt{1 - \frac{2M}{r}}} = m\dot{t}. \tag{9.29}$$

In other words, $\mathscr{E}$ is no longer the first component of the energy-momentum 4-vector of the particle. Worse still, $\mathscr{E}$ is *different* from the E that is a conserved quantity defined in Equation (9.25)! They are related by $E = \mathscr{E}\sqrt{1 - \frac{2M}{r}}$. Neither p^0 nor $\mathscr{E}$ is conserved. If we *define* the energy as the first component of the energy-momentum 4-vector, or by $\mathscr{E}$, then energy is *not* conserved. Arguably $\mathscr{E}$ is the more natural choice (albeit not the one chosen in most textbooks!) since it naturally explains why photons undergo gravitational redshift as they "lose energy climbing out of the gravitational well". What happens to the energy of such a photon? It simply is not conserved, just like in the case of the cosmological redshift discussed in Section 5.3. If we choose to define E as the energy as is done in most textbooks, one has to explain the apparent loss of energy in the photon, which may call for the notion of gravitational energy, which many experts are not comfortable with. In Section 5.3, we also mentioned that energy is not conserved in cosmology because there is no time translation symmetry (and hence there is no conserved quantity associated to it, as per Noether's theorem). The difference here is that the Schwarzschild geometry is static and hence there is a time translation symmetry, and indeed there is a conserved quantity E. Whether we want to call this "energy" is, in some sense, a question of semantics and personal philosophy (whether you want to insist on energy being conserved). On a similar note, we used to say that "mass increases" as an object moves near the speed of light (this is still how some popular science books present it when introducing special relativity). Nowadays, however, we say that mass is an invariant, but the relativistic mass

$$m_{\text{rel}} = \frac{m}{\sqrt{1 - \frac{v^2}{c^2}}} = \gamma m, \tag{9.30}$$

where $\gamma = (1 - v^2/c^2)^{-1/2}$ is the Lorentz factor, is the quantity that increases with v. In the old nomenclature, we have equations like $p = m_{\text{rel}}v$ for momentum, which looks like how it was in Newtonian mechanics. With the new view that it is best to define mass m as the invariant mass and do away with the mention of relativistic mass, the equation instead becomes $p = \gamma mv$. The physics is the same, but how we talk about it differs.

Let us now look for another conserved quantity. Similarly, we note that $\partial L/\partial \phi = 0$, so the other conserved quantity is obtained from the equation

$$p_\phi = \frac{\partial \mathsf{L}}{\partial \dot{\phi}} = m(r^2 \sin^2 \theta \dot{\phi}). \tag{9.31}$$

This allows us to define an angular momentum by

$$L := p_\phi. \tag{9.32}$$

Note that this takes the same form as the angular momentum from classical mechanics: $\vec{L} = \vec{R} \times \vec{p}$, where $\vec{v} = \vec{r} \sin \theta \dot{\phi}$ and $\vec{R} = \vec{r} \sin \theta$. However, r in the black hole case is not the radial proper distance.

Similar to the relation $\sqrt{\dot{t}^2 - \dot{x}^2 - \dot{y}^2 - \dot{z}^2} = 1$ for the Minkowski case, we now have

$$1 = \left(1 - \frac{2M}{r}\right)\dot{t}^2 - \frac{\dot{r}^2}{1 - \frac{2M}{r}} - r^2\dot{\theta}^2 - r^2 \sin^2 \theta \dot{\phi}^2. \tag{9.33}$$

We want to consider orbits of test particles, and so by appealing to the spherical symmetry of the black hole, without loss of generality, we set $\theta = \pi/2$. This is often referred to as the "equatorial plane" (though of course its geometry is not flat!). With the two conserved quantities obtained above, we can re-write Equation (9.33) as

$$1 = \left(1 - \frac{2M}{r}\right)\dot{t}^2 - r^2\dot{\theta}^2 - r^2\sin^2\theta\dot{\phi}^2 \tag{9.34}$$

$$= \frac{E^2}{m^2\left(1 - \frac{2M}{r}\right)} - r^2\dot{\theta}^2 - \frac{L^2}{m^2r^2}. \tag{9.35}$$

Note that the conserved quantities allow us to re-write the derivative terms $\dot{t}$ and $\dot{\phi}$ as algebraic expressions. Therefore the whole equation now only involves the derivative of r. After some algebraic manipulations, we can obtain

$$m\left(\frac{dr}{d\tau}\right)^2 + \left(1 - \frac{2M}{r}\right)\left(m + \frac{L^2}{mr^2}\right) = \frac{E^2}{m}. \tag{9.36}$$

The first term on the LHS reminds us of kinetic energy[10] $mv^2/2$, and so we define the second term as an effective potential:

$$V_{\text{eff}}(r) = \left(1 - \frac{2M}{r}\right)\left(m + \frac{L^2}{mr^2}\right). \tag{9.37}$$

Thus on the LHS of Equation (9.36) we now have something that looks like "kinetic energy plus potential energy". Our Newtonian intuition would like this to be equal to the total energy. But why does E^2 appear on the RHS instead of E? In fact, there is no contradiction, we just have to be very careful in interpreting what E means. E is the relativistic "energy", which includes also the rest mass energy. Roughly, we can write $E \sim m + E_{\text{Newton}}$ so that $E^2 \sim m^2 + 2mE_{\text{Newton}}$, where the E^2_{Newton} term is negligible. Thus

$$\dot{r}^2 + \left(1 - \frac{2M}{r}\right)\left(\frac{L^2}{m^2r^2} + 1\right) \approx \frac{m^2 + 2mE_{\text{Newton}}}{m^2} = 1 + \frac{2E_{\text{Newton}}}{m}. \tag{9.38}$$

Multiplying both sides of the equation by $m/2$ we can solve for E_{Newton}:

$$E_{\text{Newton}} \approx \frac{1}{2}m\dot{r}^2 + \frac{L^2}{2mr^2} - \frac{GMm}{r} - \frac{GML^2}{mr^3}, \tag{9.39}$$

where we have restored G for clarity.

[10] To be pedantic, it's not quite the kinetic energy. In a central force problem, the Newtonian kinetic energy also has the angular part: $\frac{1}{2}m(\dot{r}^2 + r^2\dot{\phi}^2)$; note also that the Newtonian time derivative is of course not the same as the general relativistic one. However, just like in Newtonian physics, we can re-distribute the angular part into an effective potential term.

Comparing this result with the Newtonian central force case:

$$E_{\rm Newton} = \frac{1}{2}m\dot{r}^2 + V_{\rm eff}^{\rm Newton}, \tag{9.40}$$

where $V_{\rm eff}^{\rm Newton}$ is the Newtonian effective potential

$$V_{\rm eff}^{\rm Newton} = -\frac{GMm}{r} + \frac{L^2}{2mr^2}, \tag{9.41}$$

we see that general relativistic correction amounts to an additional term $-GML^2/mr^3$ in the effective potential. Since this term is a r^{-3} correction, it dominates over the Newtonian terms when r is sufficiently small. Furthermore, because it is *negative*, the potential tends towards negative infinity as $r \to 0$, unlike the Newtonian case, which approaches positive infinity instead. In the Newtonian case, if a particle has a nonzero angular momentum, it will always be deflected by the effective potential, and so no plunge orbit exists. This is no longer the case in general relativity.

Circular orbits can be obtained by solving $dV_{\rm eff}/dr = 0$, which yields

$$r = \frac{L}{2Mm^2}\left(L \pm \sqrt{L^2 - 12M^2m^2}\right). \tag{9.42}$$

The orbit further out is stable, whereas the one closer to the black hole is unstable. The latter corresponds to the local maximum of the effective potential. The *innermost (marginally) stable circular orbit* or ISCO is at $r = 6M$, which corresponds to the condition $L^2 = 12m^2M^2$. In the general relativistic case, due to the existence of an unstable circular orbit, a test particle can orbit the black hole several times near this orbit before being ejected out. This is the "zoom-whirl" orbit that is not possible in the Newtonian context.

We can also analyze the behavior of a massless test particle (photon) around a Schwarzschild black hole, but we will defer this to the next subsection. However, it is worth mentioning here another physical effect near a black hole, namely *gravitational time dilation.* At fixed r, a radially in-falling observer would find that

$$d\tau = \sqrt{1 - \frac{2M}{r}}\,dt. \tag{9.43}$$

Thus, when $r \to 2M$, even though the time interval measured by the in-falling observer $d\tau$ is finite, the time interval measured asymptotically far away diverges: $dt \to \infty$. This means that, as seen by an asymptotic observer, the clock (and hence the time) of the in-falling observer ticks slower and slower as it approaches the horizon. Eventually, the clock stops ticking at the horizon, and so the in-falling observer appears to be stuck on the horizon. This phenomenon has caused a great deal of confusion and misconceptions, but one should not confuse what is observed by a far away observer with what happens to the in-falling observer. In the proper time measured by the in-falling observer, the plunge into the black hole is definite and occurred in a finite proper time. In fact, the asymptotic observer cannot even see the in-falling observer too close to the horizon because of the redshift of the light: the image becomes redder and darker closer to the horizon. This is because the wavelength of the light λ is related to the wavelength measured at infinity λ_∞ by:

$$\lambda = \sqrt{1 - \frac{2M}{r}}\lambda_\infty, \tag{9.44}$$

so $\lambda_\infty \to \infty$ as $r \to 2M$. For Sagittarius A*, this effect has been successfully measured from the light of the star S2, one of the stars that orbits the black holes (its orbit has the closest approach to the black hole among other stars that are orbiting the black hole), see Figure 9.2.

The time dilation effect near black holes has been utilized by many science fiction stories. By having the characters spending some time near a black hole and then returning to Earth, they will find that years, even decades have passed on Earth even though they have only been orbiting the black hole for a few hours or days. To get a more drastic effect requires one to get closer to the black hole. If one wishes to save fuel of the spaceship, it would be best to enter a stable orbit around the black hole. However, since for Schwarzschild black hole the ISCO is at $r = 6M$, this is often not good enough. A better option is to utilize a rapidly rotating Kerr black hole, as rotation reduces the ISCO radius.

An astronaut that wishes to get close to a black hole must take note of its tidal force. This effect can already be appreciated in Newtonian gravity. First, consider an

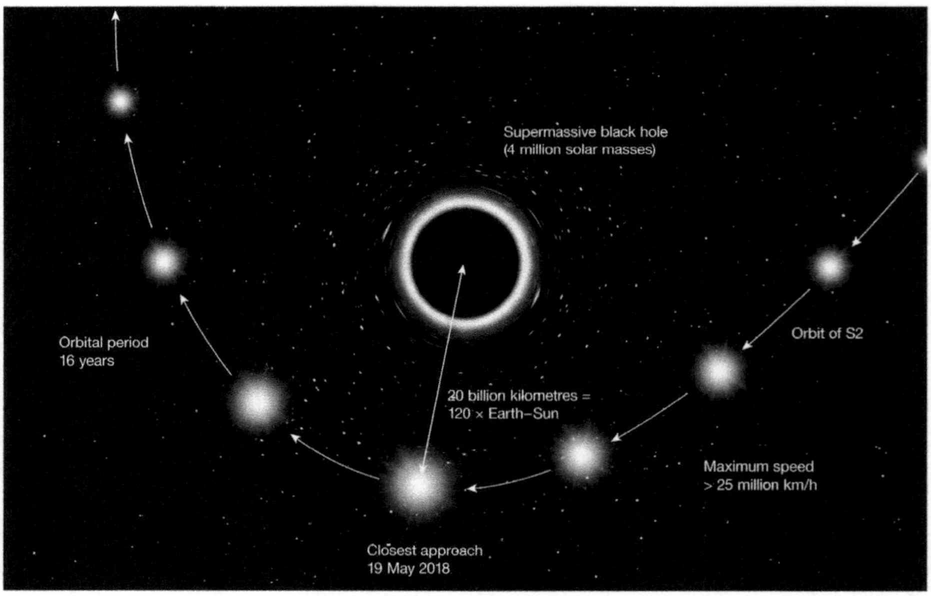

Figure 9.2. This artist's impression shows the path of the star S2 as it passes very close to the supermassive black hole at the center of the Milky Way. Close to the black hole horizon, gravitational redshift causes the color of the star to shift slightly to the red. In this graphic the color effect and size of the objects have been exaggerated for clarity. *Credit:* ESO/M. Kornmesser. CC BY 4.0

astronaut whose height is h floats feet down at a distance r from the center of the Earth. The gravitational acceleration is different at the head and the feet of the astronaut due to the height difference. Of course, if you try to substitute $r + h$ and r into the formula $a = GM/r^2$ directly, calculators will give virtually the same answer since $h \ll r$. The trick is to use the fact that

$$\frac{\delta a}{\delta r} \approx \frac{da}{dr} \Longrightarrow \delta a \approx da = -\frac{2GM}{r^3}dr, \tag{9.45}$$

where $dr \approx \delta r = h$. For a crude approximation, we can just take $h = 1$ m, and substituting in the mass of Earth $M \sim 10^{24}$ kg (we are only interested in the order of magnitude estimate). This yields the value of $|\delta a| = O(10^{-6})$ m s^{-2}, which is not something one would notice. For a black hole of about 10 solar masses, however, we would have $|\delta a| = O(10)$ m s^{-2}! A more careful analysis will tell us two important properties: (1) tidal force is weaker for a larger black hole; (2) as one falls into a black hole, one is stretched in the radial direction and squeezed from the sides.[11] Thus, it is safer to spend time near a large, supermassive black hole, rather than the smaller, stellar mass ones. Otherwise, death by spaghettification will happen even before one has the chance to pass through the horizon.

Rotating black holes can give rise to new phenomena that are absent in the non-rotating case. One example is the Lense–Thirring effect, which causes objects around a rotating black hole to be “dragged along” (hence also referred to as “frame dragging”), or more precisely, the orbit of a nearby test particle receives an additional precession induced by the black hole rotation. As mentioned before, once passed below the ergosphere, the dragging of spacetime becomes so intense that no object (not even light) can move against the black hole rotation. The presence of the ergoregion also permits the occurrence of another interesting physical phenomenon called the Penrose process, discussed in Box 7 below.

Box 7: Penrose Process and Superradiance

The Penrose process, named after Sir Roger Penrose (1931–present), is a mechanism that allows one to extract the rotational energy of the black hole. First, one drops an object into the ergosphere. As it gets close to the event horizon, the object drops a massive load, which due to the motion inside the ergosphere, will be shot forward and downward towards the horizon. By making use of the fact that particles in the ergoregion can have negative energy with respect to observers at infinity (since energy is normally associated with the Killing timelike vector $\partial/\partial t$, which becomes spacelike inside the ergosphere), one can devise trajectories such that the initial object would be sped up and leave the ergosphere with *more* energy than it originally possessed. In this process, the black hole would lose a bit of its angular momentum. Penrose imagined that an advanced civilization could build their cities around a rotating black hole, and

[11] Sometimes it is claimed that Stephen Hawking has coined the term “spaghettification” to describe this process, however the term is actually older. One can find it, for example, on p.143 in a book by Nigel Calder entitled *The Key to the Universe: A Report on the New Physics*, published in 1977 by Viking Press.

extract the rotational energy from the black hole in this way, while simultaneously dumping their rubbish into the black hole.

If instead of a particle one considers a wave, the wave can be amplified when scattered off the rotating black hole; this is called *superradiance*. Such a process can be used to constrain models of dark matter, whose existence should slow down black holes at a higher rate than expected from just having other processes like Hawking radiation (Cardoso 2018; Arvanitaki et al. 2025), though things can be more complicated if the black hole has a binary companion (Tong et al. 2022; Zhu et al. 2025). Superradiance is a more general phenomenon as it can occur even without an ergosphere or a black hole (Brito et al. 2020); Vicente et al. 2018). Actually the Penrose process can also happen without an actual geometric ergosphere; it suffices to have an *effective ergosphere*. This can happen, for example, in the Reissner–Nordtsröm background, with charged particles that can have negative energy (Denardo & Ruffini 1973). For some comparisons between the Penrose process and superradiance, see Vicente et al. (2018).

9.3 Black Hole Shadow, Accretion, and Jet

In 2017, history was made with humanity's first image of a black hole, taken by the "Event Horizon Telescope" (EHT). The EHT is not a single telescope, but rather a global network of radio telescopes, which act collectively as a giant telescope the size of our planet to be able to resolve a black hole that far away. The first imaged black hole was the supermassive black hole at the center of Messier 87 (a supergiant elliptical galaxy in the constellation Virgo), located about 55 million light-years from Earth. The black hole itself is called M87*. The galaxy itself has been imaged by many different telescopes, as it has been an object of interest for over a century. As far back as 1918 (just 3 years after the birth of Einstein's general relativity), astronomer Heber Curtis (1872–1942) noticed that the center of the galaxy seemed to exhibit "a curious straight ray" extending out of it. This straight line is actually an *astrophysical jet* produced by the disk of hot material that falls into the black hole, and we will look at another example later in this section. These jets traverse great distances in interstellar space and can heat up interstellar medium, which glows in infrared and radio waves (in the latter case the host galaxy is called a "radio-loud active galaxy" or simply "radio galaxy"). Figure 9.3 shows the M87 galaxy, the jet, and the black hole image of EHT. The black hole image itself looks like a deformed blob of doughnut, but the "hole" of the "doughnut" is actually larger than the size of the event horizon of the black hole. To understand why, we have to understand how the image of this so-called "black hole shadow" is formed.

For a massive particle, we have shown that its orbit satisfies the differential equation

$$\dot{r}^2 + \left(1 - \frac{2M}{r}\right)\left(\frac{L^2}{m^2 r^2} + 1\right) = \frac{E^2}{m^2}. \tag{9.46}$$

Define $\varepsilon := E/m$ and $\ell := L/m$, the "energy" and angular momentum per unit mass, respectively. The equation above is now equivalent to

Figure 9.3. The elliptical galaxy M87, taken by NASA's Spitzer Space Telescope, which specializes in infrared observation. Spitzer's infrared view shows a jet of material ejected from the central black hole M87* to the right of the galaxy. This jet, which extends up to some 5000 light-years (the material ejected has stretched some 1.5 kpc or 5 kly), is aimed almost directly toward Earth, so relativistic beaming effect makes it glows much brighter than the other jet that is emitted away from us, which is not visible in the image, though the shockwave it creates in the interstellar medium can still be seen. The central black hole image (the "hole" is technically called a "black hole shadow") is taken by the EHT collaboration. *Credit:* NASA/JPL-Caltech/IPAC/Event Horizon Telescope Collaboration

$$\dot{r}^2 + \left(1 - \frac{2M}{r}\right)\left(\frac{\ell^2}{r^2} + 1\right) = \varepsilon^2. \tag{9.47}$$

For a massless particle, even though $m = 0$, it still has energy and momentum, and the above equation still holds after a minor correction:

$$\dot{r}^2 + \left(1 - \frac{2M}{r}\right)\frac{\ell^2}{r^2} = \varepsilon^2. \tag{9.48}$$

Of course, the dot now denotes differentiation with respect to the affine parameter, not proper time, since the latter does not make sense for massless particles. The effective potential is now given by

$$V_{\text{eff}} = \left(1 - \frac{2M}{r}\right)\frac{\ell^2}{r^2}. \tag{9.49}$$

Remarkably it has a local maximum at $r = 3M$, which means that there is an unstable photon orbit there. Thus in general relativity, spacetime can be so curved that light can orbit around a black hole (although not for long).

In terms of the angular momentum, and on the equatorial plane, we note that

$$\frac{d\phi}{d\lambda} = \dot{\phi} = \frac{\ell}{r^2}, \tag{9.50}$$

so that the chain rule gives

$$\left(\frac{dr}{d\phi}\right)^2 = \left(\frac{dr}{d\lambda}\right)^2\left(\frac{d\lambda}{d\phi}\right)^2 = \left[\varepsilon^2 - \left(1 - \frac{2M}{r}\right)\frac{\ell^2}{r^2}\right]\frac{r^4}{\ell^2}. \tag{9.51}$$

Defining the *impact parameter* by $b := \ell/\varepsilon$, we now have

$$\left(\frac{dr}{d\phi}\right)^2 = \frac{r^4}{b^2}\left[1 - \left(1 - \frac{2M}{r}\right)\frac{b^2}{r^2}\right]. \tag{9.52}$$

When the derivative $dr/d\phi$ vanishes, we find that the radius of closest approach $r = r_{\text{min}}$ satisfies $b^2(r_{\text{min}} - 2M) = r_{\text{min}}^3$. As we have previously mentioned, the photon orbit (because of spherical symmetry, we actually have a *photon sphere*) corresponds to $r = r_{\text{min}} = 3M$. This in turn corresponds to the impact parameter of $b = \sqrt{27}\,M$. Physically this means that a light ray that starts far away with $b = \sqrt{27}\,M = 3\sqrt{3}\,M$, eventually enters the photon orbit; see Figure 9.4. If b is greater than this value, the light will only be bent or deflected by the black hole. Observers close to the horizon will only receive those rays of light with $b \leqslant 3\sqrt{3}\,M$ along their line of sight.

We can now ask what an asymptotic observer can see of a black hole (Figure 9.5). Of course, only light beams that don't end up in the black hole can be seen. For example, let us consider a light source very close to the horizon situated *behind* the black hole that we are looking at (say, a point on the accretion disk). The light that it directly emits towards the hole is absorbed, but even light rays that are emitted somewhat away from the black hole can be bent and absorbed. It turns out that the image taken by the Event Horizon Telescope[12] has a diameter of $2b$, not 2 times the Schwarzschild radius. If we perform a ray-tracing, we will see that the "black hole shadow" actually consists of the "image" (actually the lack therefore) of the entire black hole—front and back—but of course both appear black. The real observation has an extra complication—the black hole is rotating, which will distort the shadow

[12] Here we assume the Schwarzschild geometry for simplicity.

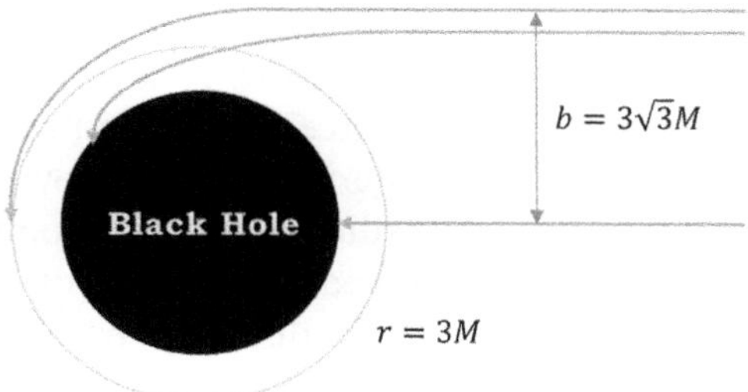

Figure 9.4. This is a schematic illustration of the light rays that travel towards a Schwarzschild black hole from the asymptotic region (not drawn to scale). A light ray that comes in with an impact parameter $b = 3\sqrt{3}M$ would enter the yellow circle, the photon sphere, located at $r = 3M$. All light rays with an impact parameter less than that will fall into the black hole. Light rays with a larger impact parameter will be deflected back out to space.

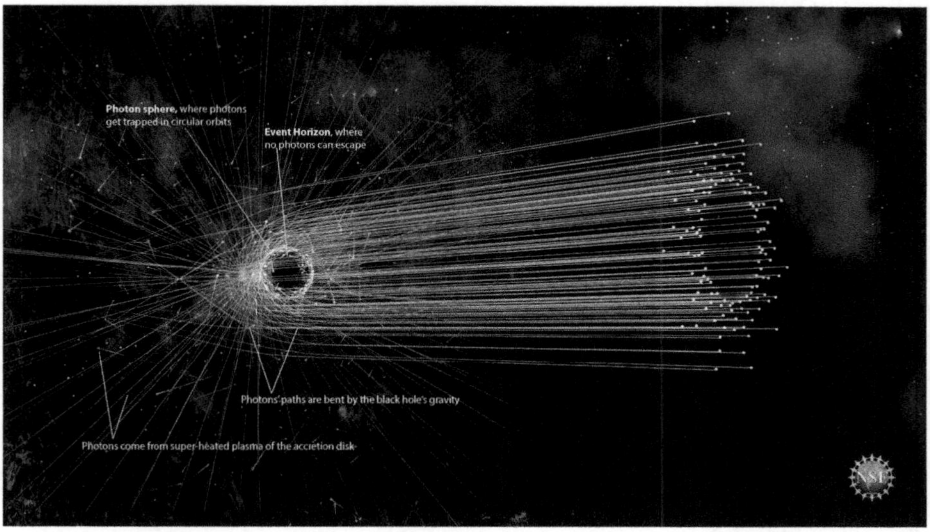

Figure 9.5. The paths of photons are bent by the gravitational field of the black hole at the center of M87* resulting in a shadow. In the language of general relativity, the photons merely follow the geodesics for massless particles in the curved spacetime around the black hole. Note that the shadow is larger than the size of the event horizon itself. This is essentially a three-dimensional view of Figure 9.4, but the source of the light rays are not from the right, but rather from the black hole environment. *Credit:* Nicolle R. Fuller/U.S. National Science Foundation.

image, but this can all be calculated, albeit tediously, using the Kerr metric in place of the Schwarzschild metric.

Such bending of light also occurs for neutron stars, so when we observe a neutron star, we don't just see 50% of its surface (namely the "front side") but more than that, since some light rays from the far side can also reach us. Depending on the orientation of the stars towards us, this means that sometimes we can see both the North and South poles of a neutron star. This phenomenon was observed in the X-ray band by the Neutron star Interior Composition Explorer (NICER) mission. The more compact the star is, the more we can see of its surface.

Note that if a black hole is sitting by itself in a relatively clean environment (an "isolated" black hole), we cannot see its shadow. A shadow is an absence of light, and so requires light around it to be perceived. A common light source is the *accretion disk*. As gas, dust and other material get pulled towards a black hole, they form a disk that swirls around the hole (due to angular momentum). Due to light-bending effect, the disk appears as if curved towards the observer from above and below even when the observer is viewing the disk edge-on. This is the famous black hole image shown in the movie *Interstellar*, though it is still not realistic enough (Doppler shift effect was not included, for example). While a new simulation was used for the movie (James et al. 2015), the first simulated image was done in 1979 by Jean-Pierre Luminet (1951–present) (Luminet 1979). The constant orbital radii were also shown therein. Even earlier, a number of such curves around an extremally rotating black hole were also shown in 1973 by James M. Bardeen (1939–2022) and his then PhD student (Cunningham & Bardeen 1973).

If the material gets too close, below the ISCO, then it will fall into the black hole. By observing the gap between the black hole (the size of the horizon has to be inferred by other means) and the inner boundary of the disk, we can in turn deduce how fast a black hole is rotating, as a rotating black hole has ISCO that is closer to the horizon. Still, generally speaking, it is very hard to accurately determine the spin of a black hole via astronomical observations. The physics of the accretion disk itself is rather rich. As the material swirls faster and faster closer to the black hole, friction heats up the accretion disk, so hot that they can emit brightly. However, sometimes *bursts* of X-ray emission are observed, and the exact mechanism behind such a flare-up behavior that can last a few hours (instead of a steady continuous emission) is yet unknown. One of the current models is that *above* the black hole, at a distance of $O(10)$ million kilometers, there is a structure known as the "corona", which is made up of highly energetic particles that generate X-ray light. The "lamppost" model proposes that the bursts of X-rays are produced by a relatively compact source, like a light bulb along the rotational axis of the black hole, that gets ejected away into space. The structure is highly dynamical and evolving.

Sometimes black holes are also accompanied by an astrophysical jet that stretches for many (thousands and even millions!) light-years, as depicted in Figure 9.6. These jets are most likely produced when ionized materials of the accretion disk interacted with strong magnetic fields that direct them away from the disk along the perpendicular directions. Though we know that the underlying mechanism involves the dynamical interactions between the rotation of the black hole and the magnetic field of the accretion disk (Blandford–Znajek process), the exact detail is yet to be figured out; general relativistic magnetohydrodynamics (GRMHD) is a challenging field. The fact that jets can stretch so far into interstellar space suggests that some black hole jets may be able to affect the distribution of matter and magnetism in the cosmic web. This in turn may mean that relatively small objects (small compared to a galaxy), if sufficiently "active", can nevertheless have bigger influence on the large-scale structure of the Universe than previously thought (Oei et al. 2024). Black holes may play more roles in cosmology than expected.

In astrophysics, a compact region at the center of any galaxy that emits a significant amount of energy across the electromagnetic spectrum at sufficiently

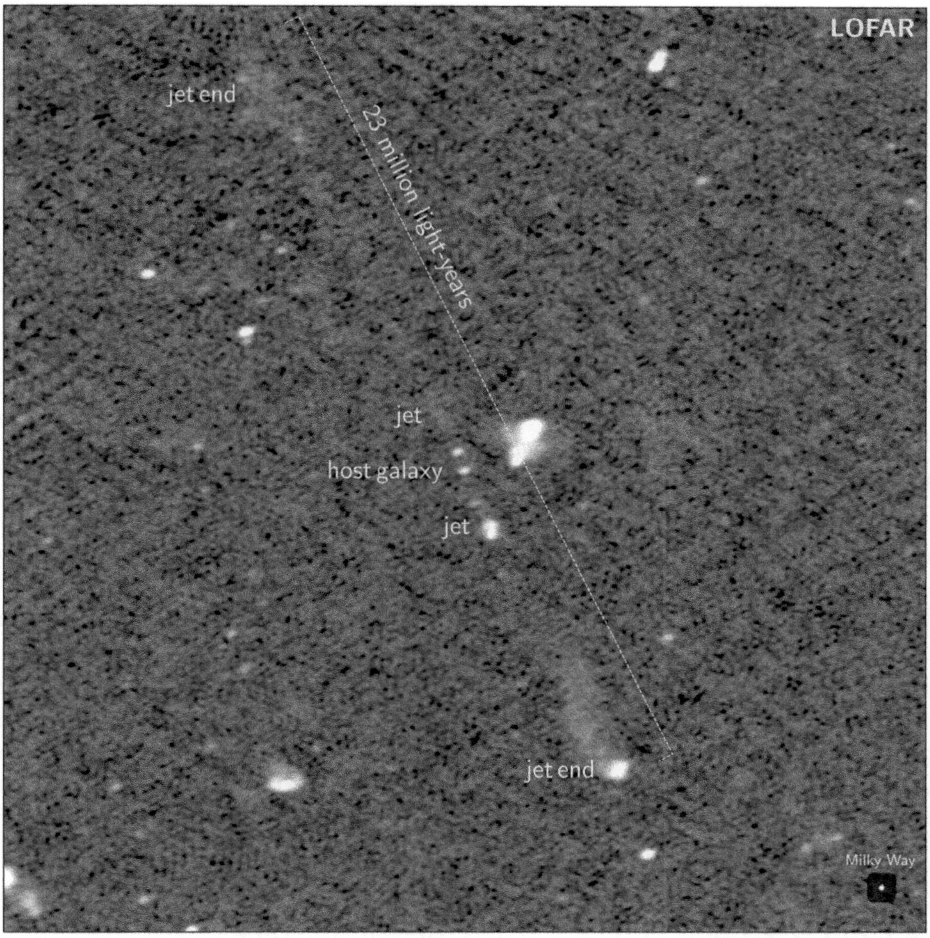

Figure 9.6. Radio image of a black hole jet system spanning an estimated 23 million light-years, nicknamed *Porphyrion* (after the king of the Greek "Gigantes", or giants, who waged war on the gods) and formally designated as J152932.16+601534.4, discovered by the team led by Martijn Oei (Leiden University/Caltech) (Oei et al. 2024). The host galaxy is classified as a radio galaxy (since the central black hole system emits strongly in the radio frequencies) located 7.5 billion light-years away from Earth. The image, taken with the LOFAR HBA at a central observing frequency of 144 MHz, has an effective resolution of $6.2''$ and covers $15' \times 15'$ angular size of the sky. The Milky Way (with assumed diameter 50 kpc) is shown for scale. Porphyrion has the longest jets of any radio galaxy identified thus far as per 2025, breaking the previous record holder *Alcyoneus*, discovered in 2022 by the same team, with lobed structures of 16 million light-years. *Credit:* This [Porphyrion LOFAR HBA 6"] image has been obtained by the author(s) from the Wikimedia website where it was made available by [Martijn Oei] under a CC BY-SA 4.0 licence. It is included within this book on that basis. It is attributed to [Martijn Oei].

large luminosity (far above the level emitted by ordinary stars) is referred to as an active galactic nucleus (AGN). The corresponding galaxy is then referred to as an "active galaxy". It is commonly thought that a galaxy becomes active when the supermassive black hole at its center is feeding on the accretion disk. Depending on the system, a jet may or may not be present. If a jet is oriented in such a way that it is pointing directly or almost directly at us, the AGN is called a *blazar*. High-energy gamma-ray photons have been linked to blazars. In September 2017, the IceCube Neutrino Observatory based in Antarctica detected a neutrino that was later traced to a blazar situated some 3.7 billion light-years away. This is a success of multimessenger astronomy. IceCube works by having an array of detectors buried into a kilometer cube of ice to increase the chance of the elusive neutrinos interacting with water molecules. As a neutrino interacts with protons or neutrons, a muon is created that moves in the same direction as the original neutrino, at a speed greater than the phase velocity of light in the ice. This creates a blue light in the ice (known as Cherenkov radiation[13]) that can be picked up by the detector. This allows the IceCube team to deduce which direction the neutrino comes from and put out an alert to the astrophysics community to follow up with other instruments. In this case, it was NASA's Fermi Gamma-ray Space Telescope that managed to link it with a blazar. In Section 11.2 we will see another success of multimessenger astronomy, with gravitational wave detectors joining in.

By the way, if you think jets that stretched for millions of light-years are shocking enough, there was a recent discovery of a supermassive black hole (about 20 million solar masses) that was ejected from its host galaxy. Strangely, it seems to be "dragging" behind a trail of stars, now 200,000 light-years long. These stars were not from the original galaxy that the black hole "took along", rather researchers theorized that they are *newly formed stars*. Thus black holes are not just destroyers but can also create stars. Apparently as the black hole moved through interstellar gas, it triggered new star formation, though the exact mechanism has not been clarified. Black holes, just like stars and planets, can become "rogue", being ejected from their own galaxy via gravitational interactions between other massive bodies (van Dokkum et al. 2023).

9.4 Compact Objects and Black Hole Mimickers

We shall now address the major question: is there really such a thing as a black hole? Do black holes actually exist in our Universe, or are they just fancy mathematical solutions of general relativity but nevertheless unphysical? As already mentioned, we don't really know for sure, but in the mainstream opinion, it is the *least* exotic option available. The other options are called "black hole mimickers" since they can mimic many properties of black holes. They are also called "exotic compact objects" (ECOs). Some authors include black holes of modified gravity solutions as part of the ECO family. In this case, black hole mimickers are just a subset of the ECO family.

It is ironic to think that black holes were at one point so exotic themselves that no self-respecting astronomer would seriously believe in them. They were treated as purely mathematical solutions to Einstein's theory, that surely cannot arise in the real Universe. Still, as more and more evidence became available we have now become so accustomed to black holes that they no longer feel exotic. So what are

[13] Named after the Soviet physicist Pavel Cherenkov (1904–1990). He shared the 1958 Nobel Prize in Physics with Ilya Frank and Igor Tamm "for the discovery and interpretation of the Cherenkov effect".

these ECOs? There is no point to list all possible candidates, but we will look at a few examples just to get the ideas. However, before that, it would be good to introduce some *known* compact objects. We have already introduced white dwarfs and neutron stars in Section 8.4.2, so we can continue from there.

Recall that neutron stars pack the mass somewhat larger than our Sun into the size of a city about 20 km across. Since stars have angular momentum, as it shrinks under gravitational collapse, it will spin even faster due to the conservation of angular momentum. PSR J0437-4715, located 509.8 light-years away, is a pulsar that has a radius of 11.4 km and a mass 1.4 times that of the Sun. It is classed as a *millisecond pulsar* because it rotates about its axis 173.7 times per second and thus completes a rotation every 5.76 milliseconds. Another pulsar, PSR J1748-2446ad, spins 716 times per second. Its tangential velocity reaches almost a quarter of the speed of light! A pulsar is just a neutron star that emits a radio beam like a spotlight, that sweeps past the Earth each time it rotates, and therefore "pulses" from our perspective. First discovered by Jocelyn Bell (1943–present) in 1967, there were initially wild speculations on whether such regular pulses may be signals from some extraterrestrial civilizations. Bell and her then PhD advisor Antony Hewish[14] (1924–2021) nicknamed the object "LGM-1", where LGM means "little green men". The detection of a second pulsar in a completely different part of the sky in the same year meant that this is highly unlikely.

Other neutron stars have extremely strong magnetic fields of at least 10^9 Tesla. In 2022, one such star designated J0243.6+6124 was found to possess a field strength of 1.6 billion Tesla by the Chinese Insight-HXMT observatory (Kong et al. 2022). These highly magnetic neutron stars are known as *magnetars*, and how they came to possess such a strong magnetic field is not fully understood. For comparison, the strongest magnet on Earth to date is "only" 42.02 Tesla, built by the High Magnetic Field Laboratory of Hefei Institutes of Physical Science, Chinese Academy of Sciences (CHMFL), in 2022. The CHMFL magnet is already 800,000 times stronger than Earth's magnetic field. The magnetic field of a magnetar is strong enough to kill any astronaut who ventures close enough (which can be as far as 1000 km away). This is achieved via distorting atomic structures and thereby destroying chemical bonds. Magnetic field is not the only thing around a magnetar that should worry a brave explorer since there are also plenty of radiation and high energy particles emitted from the star. Magnetars have also been associated with gamma-ray bursts. It has been suggested that magnetars may also be the source of *fast radio bursts* (FRBs), which are typically millisecond-long transient radio signals first identified in 2007, but whose underlying mechanism remains unknown.

Could there be other kinds of compact objects that we have not discovered? Could some of these more exotic types of compact objects be mistaken as black holes? Indeed, it has been suggested that what we thought to be black holes are actually just some kind of hitherto unknown stars. This is, in a sense, a natural guess.

[14] The fact that Hewish was awarded a Nobel Prize for Physics in 1974 for this discovery, but Bell, who did the hard work, didn't, is widely seen as a mistake.

We have seen that quantum mechanics can stabilize a star against total gravitational collapse in the case of white dwarfs and neutron stars. Is it possible that other, more compact, stars can exist? This is a very difficult question as it depends on what exactly the "equation of state" is that describes the interior matter of the stars. There is no shortage of possibilities. For example, a candidate is the so-called *quark star*. Recall that a neutron star mainly consists of neutrons because electrons and protons were fused during gravitational collapse. It is possible that if gravity is sufficiently strong, the neutrons would then break down into its constituent up and down quarks, hence becoming a quark star that is smaller and denser than a neutron star. At low densities, like in the states of everyday matter, quarks are confined by the strong force and cannot be separated from the nucleons. Similarly, consider a meson that is composed of a quark and its anti-quark. If we try to pull apart the quark-anti-quark pair, we would be injecting so much energy into the pair that we create an additional quark-anti-quark pair. Thus we would have two mesons, not a separated quark and anti-quark. However, at very high densities, the force between quarks may become sufficiently weak, and in that case we can treat them as free particles. In such an "asymptotically free" state, the quarks, along with the gluons, become a fluid known as the quark gluon plasma (QGP), in which the quarks are effectively non-interacting. A quark star, illustrated in Figure 9.7, is the hypothetical star in which this happens in its interior. The diameter of a quark star is about half that of a neutron star, and density about 2 orders of magnitude higher. So far there is no convincing observational evidence for the existence of quark stars, or any other exotic compact stars. What are these other exotic possibilities?

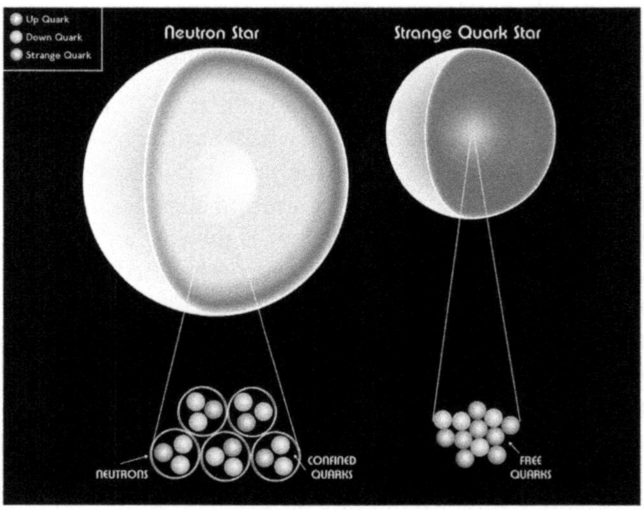

Figure 9.7. In a neutron star (left), the quarks that comprise the neutrons are confined inside the neutrons. In a hypothetical quark star (right), the quarks are effectively free, so they take up less space and the diameter of the star is smaller. *Credit:* NASA/CXC/M. Weiss.

Well, an even more exotic scenario is the *preon star*. Preons[15] are hypothetical point particles that are supposed to be sub-components of quarks and leptons. That is to say, in the preon models, quarks and leptons are not elementary particles. If this were the case, there could be preon stars that are only about 10 cm in diameter! Its density would be about 10^{24} kg m^{-3}.

Another candidate is a *boson star*. Elementary particles can be characterized as fermions or bosons depending on their quantum spin. Crucially, fermions cannot occupy the same state. This is known as *Pauli's exclusion principle*. An example of this can be seen in atomic physics: no two electrons in the same atom can have identical values for all of their quantum numbers. This principle is responsible for the relativistic motion of the electrons in a white dwarf even though the temperature is comparatively low, giving rise to the degeneracy pressure needed to counteract gravitational collapse. This will continue to hold even as the temperature drops to absolute zero as all the latent heat is gradually released (at which point the white dwarf dims and becomes a "black dwarf"; but it takes more than the current age of the Universe to reach such a state). Bosons are not subject to Pauli's exclusion principle, so one may wonder how it resists gravitational collapse. Indeed, not all bosons can do the job, only bosons with self-repulsive interaction can stabilize a star.[16] There are many different possible types of boson stars depending on the model. For example, if the boson has spin-1 (a massive vector field particle), it is referred to as *Proca star*. (A Proca field is like electromagnetism but with a "massive photon".) *Fermion-boson stars*, which possess both fermionic and bosonic constituents, have also been proposed. Another candidate, the *gravastar* proposed by Pawel O. Mazur and Emil Mottola in 2001 (but published only recently in 2023; Mazur & Mottola 2023), has a shell of exotic matter in place of the horizon, and its interior is that of a de Sitter spacetime. In other words, a gravastar is supported by an effective dark energy whose pressure p is the negative of its energy density ρ, namely $p = -\rho$ (within the shell itself, $p = \rho$).

There are also objects that for all practical purposes look like a black hole in classical observation but lack a well-defined horizon. For example, in string theory one can construct *fuzzballs* (Mathur 2005), which are essentially a ball of string. It has neither a horizon nor a singularity. This does not mean that there is no way to distinguish a black hole from a fuzzball by observations. For example, when two fuzzballs collide, how they "ring down" to a stable final fuzzball configuration is different from the case of black hole collisions.

For an ECO to be considered viable, it should not only be theoretically possible to construct, but one should be able to explain how it came to be dynamically. White dwarfs and neutron stars are well-understood as end states of compact stars. Quark stars can also be understood as such. However, it is harder to explain how, for example, a boson star could be formed in stellar collapse. In addition, it is important to study whether a proposed ECO is indeed stable. Some ECOs can give rise to a

[15] Not to be confused with *prion*, a misfolded protein that can induce misfolding in normal variants of the same protein, leading to cellular death. One example of a prion disease is the bovine spongiform encephalopathy (BSE), perhaps better known as the "mad cow disease".

[16] In model building, one usually uses a complex vector field to construct a boson star solution.

stable light right (photon orbit) that in turn leads to instability of the objects themselves. Intuitively we can appreciate how this instability arises: if there is a stable photon orbit, then light rays (and more generally, massless perturbations) can pile on and around it. The collective energy of these photons eventually become so great that it back-reacts on to the spacetime geometry, and so the original configuration cannot be stable. This being just a heuristic reasoning means whether this indeed happens for a specific model needs to be examined with care; some are indeed found to be stable.

Sometimes, hypothetical solutions in general relativity and modified gravity like wormholes are also considered part of the ECO family. We defer the discussions on wormholes to Section 13.6.

The issues of whether there are ECOs, and whether black holes are real, are not equivalent—it is possible that both black holes and some kind of ECOs exist. Personally I would be very surprised if black holes do not exist. The formation of classical black holes are well-understood at the theoretical level. Furthermore the curvature on (and near) a horizon can be very small for a large black hole, in the regime that general relativity is well understood. Given that general relativity is such a well-tested theory, there is no reason to think that it would fail to describe black hole physics as far as the horizon is concerned. The theory may cease to work near a singularity, but this is a different problem altogether. On the other hand, states of matter at high density are not well understood, so one should keep an open mind with regard to the possible existence of some kind of ECOs like quark stars. Readers interested in black hole mimickers can consult the articles collected in Bambi et al. (2025).

9.5 Exercises

(1) In Section 9.2 we mentioned that the Lagrangian of a free test particle is just $-m$. This may seem very strange because in classical mechanics the Lagrangian is $T - V$, and so in the absence of the potential it should be the kinetic energy. To see why there is no contradiction, consider its action

$$S = -\int m \quad ds, \tag{9.53}$$

show that Equation (9.53) is equivalent to

$$S = -\int m\sqrt{1 - \frac{v^2}{c^2}} \quad dt, \tag{9.54}$$

and so in the non-relativistic limit the integrand is just the kinetic energy minus the rest mass energy. The latter is the same for all paths in the variation of fixed endpoints, and thus it can be dropped from the action.

(2) Consider the $t =$ const., $\theta = \pi/2$ surface in the Schwarzschild geometry. This is a Riemannian manifold with metric tensor

$$dl^2 = \left(1 - \frac{2M}{r}\right)^{-1} dr^2 + r^2 d\phi^2. \tag{9.55}$$

Show that if we embed this surface into $\mathbb{R}^3$, whose metric tensor is given in the cylindrical coordinates

$$dl^2 = dz^2 + dr^2 + r^2 d\phi^2, \tag{9.56}$$

then the surface is described by the equation

$$z = 2\sqrt{2M(r - 2M)}. \tag{9.57}$$

Plot the surface by hand. (It is an important skill for students to develop mathematical intuition; don't overly rely on computers.)

Additional note: This is the so-called "Flamm's paraboloid" although it is not really a paraboloid. It is the diagram used to illustrate the curved space around black holes in most popular science textbooks. However, note that as far as this plot is concerned, the surface "ends" at $r = 2M$, and does not extend indefinitely "downward", which is sometimes wrongly illustrated.

(3) Even if the metric tensor has coefficients that explicitly depend on time, it does not necessarily mean that the geometry is time-dependent, as this could just be a coordinate effect. For example, consider the metric tensor

$$ds^2 = -dt^2 + \frac{4}{9}\left[\frac{9M}{2(r-t)}\right]^{\frac{2}{3}} dr^2 + \left[\frac{9M(r-t)^2}{2}\right]^{\frac{2}{3}} (d\theta^2 + \sin^2\theta d\phi^2). \tag{9.58}$$

Show that this is just the Schwarzschild black hole in disguise!

(4) Calculate the size of the shadow for a Reissner–Nordström black hole. How much does the size differ in percentage compared to a Schwarzschild black hole of the same mass, as a function of the charge-to-mass ratio $0 \leqslant Q/M \leqslant 1$?

(5) Consider the Einstein field equations without a cosmological constant. Show that $T_{ab} = 0$ is equivalent to $R_{ab} = 0$. We say that the vacuum case is equivalent to "Ricci-flat". Show that this is true in all spacetime dimensions above 3. If spacetime is two-dimensional, show that the Einstein tensor must vanish identically, and so if we define matter and gravity interaction via the Einstein field equations, then there cannot be any matter in two dimensions.

Note 1: Usually, instead of general relativity, one introduces a field ϕ called "dilaton" in two dimensions, and the gravity theory has ϕR in the action instead of just the Ricci scalar R. This is known as JT-gravity, short for Jackiw–Teitelboim gravity, named after Roman Jackiw and Claudio Teitelboim (who later changed his name to Claudio Bunster).

Note 2: In three dimensions, gravity is not dynamical. To see this, note that the Riemann tensor may be fully expressed via the Ricci tensor in three dimensions. Thus Ricci-flatness implies flatness. There can be no gravitational wave and hence no graviton (gravitons are the quanta of gravitational waves, just like photons are quanta of light). However, black holes can exist in three dimensions if we include a negative cosmological constant. This is the famous BTZ black hole found in 1992, named after Máximo Bañados,

Claudio Teitelboim, and Jorge Zanelli. Learn more about BTZ black holes if you are interested.

(6) Suppose we want to observe radio wave signals of 450 GHz from a black hole. In order to resolve a black hole located at about 10^{20} m away from us, find the aperture of the telescope and compare it to the size of Earth's diameter.

Hint: Use the Rayleigh criterion, named after Lord Rayleigh (John William Strutt; 1842–1919), which you can find in almost any introductory physics textbook.

References

Arvanitaki, A., Dimopoulos, S., & Galanis, M. 2025, PhRvD, 111, 055015

Bambi, C., Brustein, R., Cardoso, V., et al. 2025, Proceedings of the Black Hole Mimickers: From Theory to Observation workshop organized at the Princeton Center for Theoretical Sciences (PCTS) (Princeton, NJ: Univ. Princeton)

Brito, R., Cardoso, V., & Pani, P. 2020, Lecture Notes in Physics Superradiance—the 2020 Vol. 971 (Berlin: Springer)

Cardoso, V., et al. 2018, JCAP, 03, 043

Cunningham, C. T., & Bardeen, J. M. 1973, ApJ, 183, 237

Denardo, G., & Ruffini, R. 1973, PhLB, 45, 259

van Dokkum, P., Pasha, I., Buzzo, M. L., et al. 2023, ApJL, 946, L50

Herdeiro, C. A. R., & Lemos, J. P. S. 2018, GF, 41, 2

James, O., von Tunzelmann, E., Franklin, P., & Thorne, K. S. 2015, CQGra, 32, 065001

Kong, L.-D., Zhang, S., Zhang, S.-N., et al. 2022, ApJL, 933, L3

Luminet, J. P. 1979, A&A, 75, 228

Mathur, S. D. 2005, ForPh, 53, 793

Maluf, J. W. 2014, GReGr, 46, 1734

Mazur, P. O., & Mottola, E. 2023, Univ, 9, 88

Norton, J. D. 2024, PP, 2, 13

Oei, M. S. S. L., Hardcastle, M. J., Timmerman, R., et al. 2024, Natur, 633, 537

Szabados, L. B. 2009, LRR, 12, 4

Tong, X., Wang, Y., & Zhu, H.-Y. 2022, PhRvD, 106, 043002

Vicente, R., Cardoso, V., & Lopes, J. C. 2018, PhRvD, 97, 084032

Zhu, H.-Y., Tong, X., Manzoni, G., & Ma, Y. 2025, ApJ, 981, 165

Chapter 10

Beyond Black Hole Horizons

"Depth must be hidden. Where? On the surface." —Hugo Von Hofmannsthal, "Book of Friends".

This chapter has three objectives: (1) to explore what could be *inside* a black hole; (2) to introduce the thermodynamics of black holes, including Hawking radiation and the information paradox that it raises; and (3) to introduce various types of horizons in the contexts of both black holes and cosmology, such as the event horizon, trapped surface and the apparent horizon, the particle horizon, etc. Objective (1) may seem unnecessary since as external observers we can never see what's inside a black hole. However, it is important in the sense that most textbooks do not emphasize that a generic black hole has a *vast* volume far beyond what one naively expects from its horizon size. In addition, the nature of the so-called "central singularity" is a constant source of confusion in the literature and thus is worth some clarifications.

The word "beyond" can mean "on the other side of"; it can also mean "more than". The title uses both. In this chapter we will first discuss what the interior of a black hole looks like. Since all we can observe is the exterior of a black hole horizon, what lies on the other side of the horizon is opened to debate, especially if we allow for the possibility that general relativity may not be the correct theory, or at least that near the purported singularity of a black hole the theory needs to be modified by quantum gravity effects. It is a common misunderstanding that the volume of space inside a black hole with Schwarzschild radius r_S scales like r_S^3, as in Euclidean geometry. As we shall see, the curved geometry of black holes gives rise to a huge volume inside, in the standard, unmodified, general relativity. In addition, the interior geometry is not static; it gets larger in time. In fact, it is best to describe the interior as some kind of *cosmological* spacetime. Black hole interiors are like universes within the Universe.

doi:10.1088/2514-3433/ade963ch10

A black hole horizon is also more than just being a boundary of no return. It plays important roles in understanding the *thermodynamics* of black holes. In particular, the area of the horizon is the measure of the entropy of the black hole. This area law, instead of a more conventional volume law like the entropy of a box of gas, is a surprising feature of gravity, which was later generalized into the "holographic principle", with the "AdS/CFT" (anti-de Sitter/conformal field theory) correspondence being one special case but a major spin-off that developed into an entire field of its own, but ultimately beyond the scope of this book. Despite this, I think it is fair to say that, even to this day, we do not completely understand how the horizon area can encode all the information inside a black hole, and indeed there is a weaker alternative interpretation in which the horizon entropy does not encode all the information inside.

Black hole thermodynamics subsequently led to the discovery of Hawking radiation and thus the realization that black holes can become smaller in time by losing mass. This gave rise to the infamous information paradox and subsequently the firewall paradox. These paradoxes indicate that we still do not understand how quantum field theories and quantum information behave near black holes. There are many proposals that try to resolve the paradoxes, but everyone has their own opinion and no "resolution" is universally accepted.

In the final section we will introduce various types of horizons used in the context of black holes and cosmology, in addition to the event horizon. We will primarily focus on trapped surfaces, the apparent horizon, and the particle horizon. The concept of the Hubble radius is also introduced, though it does not necessarily coincide with a horizon.

10.1 What's Inside a Black Hole?

Black holes are exceptionally simple objects according to general relativity. It does not matter if you form a black hole of mass M with a star, or enough numbers of tables and chairs, or chocolates. Once the black hole is formed, it is only characterized by the mass M; all the other information about what formed the black hole is seemingly lost forever (this would lead to the famous information problem, to be discussed later in this chapter). At most, there are two more parameters: the charge Q and the angular momentum J, often characterized by the rotation parameter $a = J/M$. This is often referred to as the "no-hair theorem" of black holes, a "hair" being a physical parameter, so what is meant by "no-hair theorem" is that there is no other hair in addition to M, Q, J.[1] This is, however, just an exterior point of view, which is fair enough since we cannot see beyond the horizon to probe the interior spacetime. Nevertheless, the question "what is inside a black hole" is still a valid one.

[1] Note that the "theorem" is not so strict, additional hairs are possible if we work in other theories of gravity, or consider other matter fields beyond electromagnetism. One example is the Yang–Mills field. In some other scenarios, with suitable coupling constants, black holes can develop hair when its temperature is sufficiently low, much like a phase transition.

Let us start with the non-rotating and non-charged, idealized, Schwarzschild black hole. Recall that we have mentioned in the previous chapter that the $r = 0$ singularity is spacelike, and so it is an event that lies in the future of any in-falling observer. We also mentioned that the interior spacetime is not static. Instead, it is best to view the black hole interior as a *cosmological spacetime* that harbors a future singularity. This "universe" inside a black hole is not isotropic, since the radial direction gets stretched while the angular S^2 part is continuously squeezed until it reaches zero size in the limit when the singularity is "approached".[2] The main point is that although the horizon surface, as viewed from the outside, is spherical, this does not imply that the *volume* of the interior is that of a sphere, namely $4\pi r_h^3/3$, where r_h denotes the Schwarzschild radius. Although such a definition can be given and may be useful for certain purposes, it is not a natural one that reflects the *dynamical*, evolving, nature of the interior spacetime. For this, another notion of the volume is better suited. Known as the Christodoulou–Rovelli (CR) volume (Christodoulou & Rovelli 2015), it is the volume of the largest spacelike spherically symmetric surface bounded by the event horizon of the black hole. The CR volume is a geometrical property that is coordinate-independent, but best illustrated as below.

It is convenient to introduce the advanced time defined by

$$v := t + \int \frac{dr}{f(r)} = t + r + 2M \ln\left|\frac{r}{2M} - 1\right|; \quad f(r) := 1 - \frac{2M}{r}. \tag{10.1}$$

Then it can be shown that the standard Schwarzschild–Droste metric can be re-written as

$$ds^2 = -f(r)dv^2 + 2dvdr + r^2(d\theta^2 + \sin^2\theta d\phi^2). \tag{10.2}$$

The CR volume at time v^* is given by

$$\text{Vol.}(v^*) \sim \int^{v*} \int_{S^2} \max\left(r^2\sqrt{\frac{2M}{r} - 1}\right) \sin\theta \; d\theta d\phi dv. \tag{10.3}$$

The omitted lower limit in the integral with respect to v only contributes to a negligible finite value, while the integral is dominated by its upper limit v^*. The coefficient of the volume V can be maximized by maximizing the function

$$\mathscr{F}(r) := r^2\sqrt{\frac{2M}{r} - 1}\,. \tag{10.4}$$

A simple differentiation yields $r = 3M/2$, which corresponds to the maximal spacelike hypersurface. Thus, most of the volume comes from the contribution of this constant r slice, and it continues to grow in advanced time v because

$$\text{Vol.} \sim 3\sqrt{3}\pi M^2 v. \tag{10.5}$$

[2] See Page 5 of Chapter 9 for relevant discussions.

In other words, the CR volume grows asymptotically linearly in v. We emphasize that the $r = 3M/2$ slice is "in between" the horizon $r = 2M$ and the singularity $r = 0$ ($r = 3M/2$ is a spacelike hypersurface!). In fact, it is still quite far away (more accurately, "early away") from the eventual quantum effects near the future onset of the singularity, so what we have discussed are just standard general relativity in the regime that its results can be well-trusted. No conjectural new physics is required.

In their original publication, Christodoulou and Rovelli gave an estimate for the interior volume of the supermassive black hole at the center of our galaxy, Sagittarius A*. The result implies that it contains enough space to fit a *million* solar systems, despite its Schwarzschild radius being only a factor of 10 or so larger than the Earth-Moon distance. Taking into account the rotation of the black hole does not change this result by much (Bengtsson & Jakobsson 2015), even if we assume that the rotation rate of Sagittarius A* is about 90% of the extremal limit. The CR-volume for asymptotically flat black holes seems to be robust against rotational effects, as long as it stays below ~99% of the extremal limit.

Note that the above discussion assumes a static Schwarzschild black hole. Yet we find that the interior volume grows linearly with v. This means that although the "size" of the black hole appears unchanged, its volume is always becoming larger. Not knowing the age of a black hole, there is no way we can know how large a space it contains. As we shall soon discuss, upon considering quantum effects, black holes would undergo Hawking evaporation and lose mass over time. Thus, *as viewed from the outside*, a black hole will shrink. However, the interior volume does not decrease. Somewhat counter-intuitively it instead *increases*. To see this, we check that for a black hole whose mass is a function of time, $M = M(v)$, Equation (10.3) yields

$$\text{Vol.} \sim 3\sqrt{3}\pi \int^{v} M(v')^2 dv', \tag{10.6}$$

and so by the Fundamental Theorem of Calculus, $d\text{Vol.}/dt \sim 3\sqrt{3}\pi M(v)^2 \geqslant 0$. For more detailed discussions, see Ong (2015). This is a peculiar result, since it seems to suggest that at the end stage of Hawking evaporation, there is a huge "universe" inside the black hole, although from the outside the black hole may appear so small that it becomes "particle-like". If black holes completely evaporate at the end, this might even suggest that a new bubble universe could split off ours. However, these are only suggestive, because when a black hole becomes small enough and the curvature becomes huge, it is expected that new physics (quantum gravity) would kick in, which might change the eventual fate of the black hole.

Indeed, what happens inside a black hole also depends on one's attitude towards the singularity, and on the different models under consideration. There are some phenomenological models in which the singularity is removed, called "regular black holes", in which the interior could behave in different ways. For example, it could contain a de Sitter spacetime (a cosmological spacetime with a positive cosmological constant). We will have more to say about this later in this chapter when we discuss black hole singularities.

10.2 Thermodynamics of Black Holes: More than an Analogy?

In 1973, a remarkable paper by James Bardeen (1939–2022), Brandon Carter (1942–present), and Stephen Hawking (1942–2018) (Bardeen et al. 1973) revealed that black holes satisfy some properties that suggest they are thermodynamical systems. In some sense this is not that surprising, because as John Wheeler (1911–2008) once pondered, what happens if we drop a cup of hot tea into a black hole? The entropy contained inside the tea is absorbed into the black hole, whose mass increases slightly, but the overall entropy of the Universe would seemingly decrease. If black holes can act as an entropy eraser, this would spell trouble for the second law of thermodynamics, which says that entropy should be non-decreasing overall. Jacob Bekenstein (1947–2015) thought about this problem, and quite naturally proposed that the problem can be avoided if black holes have an entropy (Bekenstein 1973), such that when the black hole mass increases, its entropy also increases. Hawking's area theorem (Hawking 1971) shows that the surface area of a black hole is non-decreasing under any classical process, and so it can serve as a measure of the entropy. In the paper by Bardeen, Carter, and Hawking, which is titled *The Four Laws of Black Hole Mechanics*, however, they only claimed that the horizon area A and the surface gravity,[3] κ, of the black hole, "have a close analogy with entropy and temperature respectively". Furthermore, they wrote:

> "It can be seen that $\frac{\kappa}{8\pi}$ is analogous to temperature in the same way that A is analogous to entropy. It should however be emphasized that $\frac{\kappa}{8\pi}$ and A are distinct from the temperature and entropy of the black hole. In fact the effective temperature of a black hole is absolute zero."

This paper was written before the discovery of Hawking radiation, and so such a conclusion was only natural. We now know that with the Hawking temperature of a Schwarzschild black hole given by (Hawking 1975)

$$T = \frac{\hbar c^3}{8\pi k_B G M}, \tag{10.7}$$

where we have restored all physical constants for clarity. It has an associated Bekenstein–Hawking entropy[4]

[3] The physical intuition of surface gravity of the horizon is the tension of a string tied to a object such that it remains stationary at the horizon, with the other end situated at asymptotic infinity.

[4] In his 1973 paper, Bekenstein proposed the idea that black holes do indeed have entropy (instead of $A/4$, he had a different prefactor in place of $1/4$), it is not *just* an analogy. He defined also a temperature of a black hole via the standard thermodynamic relation $T^{-1} = (\partial S/\partial M)_{J,\,Q}$, but he still emphasized that this should not be taken as *the* temperature of the black hole, stating that "such an identification can easily lead to all sorts of paradoxes, and is thus not useful".

$$S = \frac{k_B c^3}{G\hbar}\frac{A}{4}. \tag{10.8}$$

Notice that the Hawking temperature is inversely proportional to the mass of the black hole, which means that as the black hole becomes smaller due to the mass loss, it becomes hotter, which allows it to emit even more particles.[5] In other words, its heat capacity is negative. The black hole therefore emits more and more particles and the temperature increases without bounds towards the end of the evaporation. What happens next is a big mystery, which we will discuss in the next section.

We can now state the laws of black hole thermodynamics as we understand them today:

(0) For stationary black holes, the surface gravity is constant over the horizon.
(1) For the Kerr–Newman family of black holes, the increment in the mass satisfies $dM = \frac{\kappa}{8\pi}dA + \Omega dJ + \Phi dQ$, where Ω denotes the horizon angular velocity of a rotating black hole, and Φ denotes the electrical potential at the horizon of a charged black hole.
(2) The horizon area is non-decreasing in time, i.e., $dA/dt \geqslant 0$.
(3) It is not possible to achieve $\kappa = 0$ by reducing its value by a finite sequence of operations.

These should be compared to the laws of thermodynamics for ordinary systems in Section 4.1.

For a Schwarzschild black hole, its first law is $dM = TdS$. Namely, by changing the mass of the black hole by dM, it increases its area and thus its entropy by dS, and these are related by $dM = TdS$. Interestingly, the first law of the Reissner–Nordström case can be obtained by noting that its metric function is

$$1 - \frac{2M}{r} + \frac{Q^2}{r^2} = 1 - \frac{2}{r}\left(M - \frac{Q^2}{2r}\right), \tag{10.9}$$

which, in comparison to the Schwarzschild case, we can define the effective mass as[6]

$$M_{\text{eff}} := M - \frac{Q^2}{2r}. \tag{10.10}$$

The term $Q^2/2r$ is related to the energy of the electric field (Hod 2018). Namely, given the field outside the horizon (recall that we are working in the unit $4\pi\varepsilon_0 = 1$)

[5] An astrophysical black hole is so massive that its temperature is extremely low; only massless particles like photons and gravitons are emitted. As the black hole becomes hotter, once the temperature scale reaches the mass scale of some heavier particle, that particular particle species can then be emitted.

[6] Recall from Section 8.1 that gravity is repulsive inside a Reissner–Nordström black hole in the region $r < Q^2/2M$. This corresponds nicely to $M_{\text{eff}} < 0$.

$$E = \frac{Q}{r^2}, \tag{10.11}$$

its energy $U(r)$ is

$$U(r) = \frac{\varepsilon_0}{2}\int_r^\infty E^2 \cdot 4\pi r'^2\, dr' = \frac{Q^2}{2r}; \quad 4\pi\varepsilon_0 = 1. \tag{10.12}$$

Now, we demand that $dM_{\text{eff}} = \tilde{T}dS$, where

$$\tilde{T} = \frac{1}{4\pi r_+} \tag{10.13}$$

is the effective Schwarzschild temperature (i.e., the Hawking temperature of a Schwarzschild black hole whose horizon radius is the same as the charged black hole). Substituting in Equation (10.10) and evaluating on the horizon yields

$$dM - \frac{QdQ}{r_+} + \frac{Q^2}{2r_+^2}dr_+ = \tilde{T}dS. \tag{10.14}$$

Using the fact that $dS = d(\pi r_+^2)$, this can be further simplified into the form

$$dM = \frac{Q}{r_+}dQ + \frac{1}{4\pi}\left(\frac{1}{r_+} - \frac{Q^2}{r_+^3}\right)dS. \tag{10.15}$$

We can check that

$$\frac{1}{r_+} - \frac{Q^2}{r_+^3} = \frac{r_+^2 - Q^2}{r_+^3} = \frac{r_+^2 - r_+ r_-}{r_+^3} = \frac{r_+ - r_-}{r_+^2}. \tag{10.16}$$

This gives the correct Hawking temperature for the Reissner–Nordström black hole, namely

$$T = \frac{1}{4\pi}\left(\frac{r_+ - r_-}{r_+^2}\right) = \frac{1}{4\pi}\left(\frac{1}{r_+} - \frac{Q^2}{r_+^3}\right), \tag{10.17}$$

as well as the first law

$$dM = \frac{Q}{r_+}dQ + TdS, \tag{10.18}$$

where $\Phi(r_+) = Q/r_+$ is the electric potential at the horizon. This type of argument does not work straightforwardly for the Kerr case, since the Kerr–Newman effective mass that takes into account the electric field energy turns out to be nontrivial (Hod 2020):

$$M_{\text{eff}} = M - \frac{Q^2}{4r_+}\left[1 + \frac{r_+^2 + a^2}{ar_+}\arctan\left(\frac{a}{r_+}\right)\right]. \tag{10.19}$$

Thus, for the Kerr case $Q = 0$ and one obtains $M = M_{\text{eff}}$, so the angular momentum term in the first law cannot be extracted in this way.

We note that classically, with $\hbar = 0$, the temperature is indeed absolute zero, and the entropy is infinite. Hawking radiation is a quantum phenomenon,[7] which is usually explained as due to particles being pair-produced near the horizon, and the one with positive energy being radiated off, while the negative energy particle falls into the black hole and decreases its mass. This itself is an extremely simplified heuristic picture[8], since in order to explain Hawking radiation properly, one has to understand quantum field theory and the fact that the definition of a vacuum state being observer-dependent via Bogoliubov transformation. We will not go into the details here.

A black hole radiates as a blackbody, and so its evaporation rate is given by the Stefan–Boltzmann law:

$$\frac{dM}{dt} = -CAT^4, \tag{10.21}$$

where C is some constant related to the particle species, known as the "graybody factor". This is not exactly true because the particles emitted by the Hawking process are quantum particles, so they also exhibit wave behavior. Note that even in the geometric optic limit the length scale is actually determined by the impact parameter of the photon sphere instead of just the Schwarzschild radius (for the same reason that the black hole shadow discussed in Section 9.3 is larger than the horizon area). Still, modulo these nitty-gritty details, the qualitative feature can be understood by Equation (10.21). Given that $T \sim 1/M$ and $A \sim M^2$, we see that $dM/dt \sim -M^{-2}$. This simple ODE can be straightforwardly integrated to obtain

---

[7] The first law $dM = TdS$ makes sense even classically because $\hbar$ in the numerator of T cancels with the $\hbar$ in the denominator of S. The quantity $\hbar$ provides the right dimension to define T and S separately (and give them finite nonzero values in general), but otherwise plays no role in the first law. We should emphasize, however, that although it is true that in this case the presence of $\hbar$ denotes a quantum phenomenon, it can be misleading to always rely on the argument that the presence of $\hbar$ automatically implies quantum physics. For example, the "classical" massive Klein–Gordon equation

$$\left(\Box + \frac{m^2c^2}{\hbar^2}\right)\phi(x) = 0. \tag{10.20}$$

Likewise, an *absence* of $\hbar$ does not mean that the underlying phenomena are classical. For example, *all* forces have dimension c^4/G, which is free of $\hbar$, but the Casimir force has a quantum origin.

[8] This already heuristic picture is sometimes made worse by misrepresenting the negative *energy* particle as being an *anti*-particle, which is an entirely different concept. If we go by the heuristic picture, we also need to explain why there is such a thing as a "negative energy particle" and why it must fall into the black hole. The important point is that asymptotic observers have a different notion of time compared to a putative observer inside the black hole, such that what appears to be a negative energy particle to the asymptotic observer actually has positive energy inside the black hole. (Recall that energy is the time-component of the 4-momentum vector.)

the result that a black hole whose initial mass is M will evaporate completely in a finite time that scales as $\sim M^3$.

One may wonder what happens to the second law of black hole thermodynamics when Hawking evaporation is considered. In this case the entropy of the shrinking black hole decreases, but one has to include the entropy of the Hawking radiation as well, so that overall the total entropy of the system is non-decreasing. This is known as the *generalized second law* (GSL). To appreciate how this works, consider a black hole of mass M. It has Bekenstein–Hawking entropy of $S = A/4 = 4\pi M^2$. Suppose it emits a photon with energy ε. Then the final system has entropy, which is the sum of the black hole entropy after emission, namely $4\pi(M - \varepsilon)^2$, and the entropy of the photon. The entropy of the photon can be crudely estimated by the relation $S = \varepsilon/T$ where T is the Hawking temperature $T = 1/8\pi M$. But ε itself is the characteristic energy set by the Hawking temperature $\varepsilon = nk_B T$, where k_B is the Boltzmann constant that we set to unity, and n some prefactor. Thus $S_{\text{photon}} = n = O(1)$. Consequently the system, after emission, has entropy given by

$$4\pi(M - \varepsilon)^2 + n = 4\pi(M^2 - 2M\varepsilon) + n + O(\varepsilon^2) \approx 4\pi M^2 - 8\pi M\varepsilon + n \quad (10.22)$$

$$= 4\pi M^2 + n - 8\pi M\left(\frac{n}{8\pi M}\right) \quad (10.23)$$

$$= 4\pi M^2, \quad (10.24)$$

where we have ignored the higher-order term $O(\varepsilon^2)$. Thus the total entropy of the system after emission seems to be the same as the entropy before emission. In fact, this is true even if we do not discard the higher-order terms, as a more rigorous analysis shown in Alonso-Serrano & Visser (2018), in which the authors stated: "Independent of the details of the microphysics, at the macroscopic level the Hawking radiation is essentially just (adiabatically) transferring the Bekenstein entropy from the black hole into the Clausius entropy of the radiation field; there are no significant qualifications or limitations to this result."

It should be noted that there are many ways to obtain the correct Hawking temperature in addition to Hawking's 1975 paper utilizing the Bogoliubov transformation. For example, in 2000 Maulik Parikh and Frank Wilczek[9] argued that Hawking radiation can be viewed as the result of quantum tunneling of particles from the interior across the horizon (Parikh & Wilczek 2000). Philip Candelas in 1980 (Candelas 1980), along with Dennis Sciama and David Deutsch in 1981 (Sciama et al. 1980), derived Hawking temperature by requiring that the quantum state of the black hole exterior should solve the semi-classical Einstein equations, which is possible only if the outgoing modes are in a thermal state at the correct temperature. In 2005, Sean Robinson and Wilczek (Robinson & Wilczek 2005) showed that a radiation flow (which gives rise to Hawking temperature) across the horizon is necessary to prevent anomalous breaking of the diffeomorphism

[9] The Nobel Prize in Physics 2004 was awarded jointly to David J. Gross, H. David Politzer, and Frank Wilczek "for the discovery of asymptotic freedom in the theory of the strong interaction".

symmetry. Advanced knowledge of quantum field theory is required to understand these, so we shall not elaborate further.

Path integral methods involving analytic continuation (Wick rotation; see Sections 8.1–8.2), pioneered by James Hartle and Stephen Hawking (Hartle & Hawking 1976), which avoids a conical singularity is worth explaining in some details, as it involves some nice geometrical ideas. Wick rotation provides a neat trick to calculate the temperature of a black hole. First, we consider any static black hole metric written in imaginary time $\tau = it$, which is therefore a Riemannian metric with $(+, +, +, +)$ signature:

$$ds^2 = f d\tau^2 + f^{-1} dr^2 + r^2(d\theta^2 + \sin^2\theta d\varphi^2). \tag{10.25}$$

The neighborhood near the horizon is trying to look like $\mathbb{R}^2 \times S^2$. This can be achieved if (r, τ) behave like polar coordinates on $\mathbb{R}^2$, i.e., if τ is periodic. This periodicity must behave appropriately so that conical singularity is avoided. This is ensured by imposing the condition that the infinitesimal ratio of the circumference (going around in the τ direction) to the radius (moving in the r direction) is 2π as we approach the origin of $\mathbb{R}^2$, which is at the horizon $r = r_h$. If the angle is less than 2π, there would be a angle deficit which leads to a conical singularity. This is like cutting out a sector of a disk, and gluing the resulting radial boundary lines, which gives a cone. Similarly, an angle excess is also problematic. The philosophy that every point of the geometry should look like any other, i.e., no special points like conical singularities, is ultimately why there is Hawking temperature.

For constant r and constant angular coordinates, we have $ds = f^{1/2} d\tau$. Thus the circumference C is given by

$$C = \int_0^\beta f^{\frac{1}{2}} d\tau = f^{\frac{1}{2}}\beta, \tag{10.26}$$

assuming the simple case that f is independent of τ[10]. Requiring regularity (i.e., avoiding conical singularity) means to impose:

$$\frac{\text{infinitesimal circumference}}{\text{infinitesimal radius}} = \lim_{r \to r_h} \frac{\beta}{f^{-\frac{1}{2}}} \frac{d(f^{\frac{1}{2}})}{dr} = \lim_{r \to r_h} \frac{\beta}{f^{-\frac{1}{2}}} \frac{\frac{1}{2} f^{-\frac{1}{2}} df}{dr} = 2\pi, \tag{10.27}$$

which then gives

[10] A keen-eyed reader may object that τ has physical dimension, but θ, measured in radians, does not. This is true; it is only an analogy. However, it is good to check that τ does have the same dimension as β for the circumference integral to make sense. To see that everything is consistent, we need to restore $\hbar$. Under Wick rotation, $\tau = it/\hbar$ (we are using the "original τ" here instead of the re-defined one discussed in Section 8.2), so its physical dimension is the same as that of $\beta = 1/k_B T \propto GM/\hbar c^3$ (recall that the Schwarzschild radius is $2GM/c^2$, so GM/c^2 has the dimension of length). The circumference $C = c\int_0^\beta f^{1/2} d\tau$ therefore has the physical dimension of [Length$\times\hbar^{-1}$]. The "actual circumference" is actually $\hbar C$. Thus, unlike the early practice in relativity that defines a positive definite metric by introducing $x_4 := ict$ in place of the temporal dimension (now so rarely used that the younger generations of physicists may not even know about), Wick rotation is inherently quantum mechanical. In the limit that $\hbar \to 0$, the whole edifice of periodic imaginary time will not work. Nuances like these are hard to notice if we set various physical constants to unity!

$$\left.\frac{df}{dr}\right|_{r=r_h} = \frac{4\pi}{\beta}. \tag{10.28}$$

The temperature is then simply the inverse of β (recall how this works from Section 8.2). For a Schwarzschild black hole, $f = 1 - 2M/r$ and we can check that this gives the correct Hawking temperature. Note that the black hole interior is absent in the Wick-rotated spacetime. One way to appreciate this is to note that the Wick-rotated metric (consider the Schwarzschild case for an example)

$$ds^2 = \left(1 - \frac{2M}{r}\right)d\tau^2 + \left(1 - \frac{2M}{r}\right)^{-1} dr^2 + r^2(d\theta^2 + \sin^2\theta d\phi^2) \tag{10.29}$$

has signature $(+, +, +, +)$ for $r > 2M$, but $(-, -, +, +)$ for $r < 2M$. This is no longer a Riemannian manifold, and worse there would be *two* temporal dimensions. Due to the signature difference, the Wick-rotated interior therefore cannot be connected with the Wick-rotated exterior.

For more complicated black hole solutions, Wick rotation can be quite subtle. For example, for a Reissner–Nordström black hole, although we can obtain the correct Hawking temperature by naively following the procedure described above, it is actually incorrect. To keep the electric potential 1-form real, the complexification of t means that the electric charge should also be complexified. So the metric function of the Euclidean Reissner–Nordström black hole is actually $f(r) = 1 - 2M/r - Q_E^2/r^2$, where Q_E is the "Euclidean electric charge". The Hawking temperature is obtained in the final step by Wick-rotating back to "Minkowskian charge". In the rotating case, the rotation parameter a also needs to be complexified. It should be emphasized that complexifying the metric is actually a rather tricky issue in general. Sidney Coleman once warned that (Coleman 1988)

> "[...] the Euclidean formulation of [quantum] gravity is not a subject with firm foundations and clear rules of procedure; indeed, it is more like a trackless swamp. I think I have threaded my way through it safely, but it is always possible that unknown to myself I am up to my neck in quicksand and sinking fast."

The same method can be used to derive the *Unruh temperature*, which is the temperature detected by a constant accelerating observer (often called the "Rindler observer") in Minkowski spacetime. Consider such an observer with proper time τ, whose uniform (proper) acceleration has magnitude α in the x-direction. The observer's spacetime trajectory is given by

$$\begin{cases} t(\tau) = \alpha^{-1}\sinh(\alpha\tau), \\ x(\tau) = \alpha^{-1}\cosh(\alpha\tau). \end{cases} \tag{10.30}$$

It is straightforward to verify, using the definition of 4-acceleration, that the magnitude of the acceleration is in fact α, and that the trajectory is a branch of the hyperbola $x^2(\tau) - t^2(\tau) = \alpha^{-2}$. If we introduce new variables $\rho := \alpha^{-1}$ and $\omega := \alpha\tau$, then the Minkowski metric becomes the Rindler metric:

$$ds^2 = -\rho^2 d\omega^2 + d\rho^2 + dy^2 + dz^2. \tag{10.31}$$

Now we perform a Wick rotation to imaginary time ω_E (or, equivalently, τ_E) and impose a periodic boundary condition on ω_E to avoid a conical singularity. Note that the (ω_E, ρ) part of the metric:

$$ds_E^2 = \rho^2 d\omega_E^2 + d\rho^2 \tag{10.32}$$

looks like the flat metric on $\mathbb{R}^2$ in polar coordinates (ρ, ϕ). Indeed, to avoid conical singularity means precisely that we want the geometry to be a flat space, and that there is *no special point*. This requirement on the geometry then implies that the period of ω_E must be 2π, and so that of τ_E must be $2\pi/\alpha$. The Unruh temperature, T_{U}, is therefore

$$T_{\mathrm{U}} = \frac{1}{\beta} = \frac{\alpha}{2\pi}, \tag{10.33}$$

which in SI units, reads

$$T_{\mathrm{U}} = \frac{\hbar\alpha}{2\pi c k_B}. \tag{10.34}$$

A temperature of 1 Kelvin would correspond to an acceleration of the order of $10^{20}\mathrm{ms}^{-2}$, which is why this effect is experimentally challenging to verify.

The Unruh effect, or Fulling–Davies–Unruh effect, is similar to the Hawking effect—both are due to the fact that the vacuum state in quantum field theory is observer dependent. In the case of the Unruh effect, what appears to be a particle-less vacuum to an inertial observer is seen to be teeming with radiation from the point of view of the Rindler observer. However, it should be emphasized that these two effects are still distinct. For the Schwarzschild black hole, it is true that the near-horizon geometry is a Rindler spacetime, as can be seen by writing $r = 2M + \varepsilon$ and expanding the Schwarzschild metric to first order in ε. This gives, for the (t, r)-part of the metric,

$$ds^2 = -\frac{\varepsilon}{2M}dt^2 + \frac{2M}{\varepsilon}dr^2. \tag{10.35}$$

By setting $\varepsilon = \rho^2/8M$, this becomes, with $dr = d\varepsilon$,

$$ds^2 = -\frac{\rho^2}{(4M)^2}dt^2 + d\rho^2. \tag{10.36}$$

This can be identified as the Rindler metric with proper acceleration $\alpha = 1/4M$, and so the Unruh temperature is $T = \alpha/2\pi = 1/8\pi M$, which agrees with the Hawking temperature. However, such a derivation can be misleading as it gives the impression that it is the near-horizon observer with acceleration $\alpha = 1/4M$ that detects the

Hawking temperature, whereas Hawking radiation is seen by inertial observers at infinity. The correspondence here works because the surface gravity of a black hole is $\kappa = 1/4M$, and the surface gravity can be understood as the tension of a rope with one end tied to the an object being held stationary at the horizon, and the other end at infinity (i.e., an acceleration as exerted at infinity needed to keep an object at the horizon). However, the surface gravity is not the same as the proper acceleration of an object held near the horizon. The magnitude of the proper acceleration of a local observer being held near the Schwarzschild horizon is

$$\alpha = \frac{1}{\sqrt{1 - \frac{2M}{r}}} \frac{M}{r^2}, \tag{10.37}$$

which *diverges* on the horizon. This is *not* the same as the finite value of the Rindler observer of the near-horizon metric. The redshift factor is important. The temperature seen by such a local observer also diverges at the horizon; it is called the *Tolman temperature*, which takes the form

$$T = \frac{1}{\sqrt{1 - \frac{2M}{r}}} T_\infty, \tag{10.38}$$

where T_∞ is the Hawking temperature seen at infinity. For an *extremal* Reissner–Nordström black hole, although $\kappa = 0$ it does not mean that local observer does not experience gravity, in fact α is finite and nonzero in that case. Of course, the near-horizon geometry for an extremal Reissner–Nordström black hole is no longer a Rindler spacetime, but rather $AdS_2 \times S^2$, where AdS_2 denotes a two-dimensional anti-de Sitter spacetime, so the analogy with Unruh radiation does not apply directly (though one can also generalize Unruh effect to arbitrary curved spacetime). In general, we have to be very careful when dealing with various vacuum states (the Hawking effect and the Unruh effect are in general, determined by different kind of vacuum states); analogies can be helpful but may also lead to confusions, especially since the Unruh effect itself is still not fully understood (Barbado et al. 2016). See also Earman (2011).

Why does the Wick rotation method work? The underlying reason goes beyond black holes and gravity, as we have discussed in detail in Section 8.2. As a reminder, recall the quote from Alexei Tsvelik in his textbook *Quantum Field Theory in Condensed Matter Physics* (Tsvelik 2003), which we also quoted in Section 8.2:

> "*The reason why the terms "quantum field theory" and "statistical mechanics" are used together so often is related to the essential equivalence between these two disciplines. Namely, a quantum field theory of a D-dimensional system can be formulated as a statistical mechanics theory of a (D+1)-dimensional system. This equivalence is a real godsend for anyone studying these subjects. Indeed, it allows one to get rid of noncommuting operators and to forget about time ordering, which seem to be characteristic properties of quantum mechanics. Instead one has a way of*

formulating the quantum field theory in terms of ordinary commuting functions, more or less conventional integrals, etc.".

A quantum field theory on the curved black hole spacetime (which governs the physics of Hawking radiation) can thus be described by statistical mechanics of a Wick-rotated (hence one spatial dimension higher) manifold. One underlying assumption was that thermal equilibrium is assumed between the static black hole and its Hawking radiation thermal bath. In a realistic situation, in asymptotically flat geometry, black holes evaporate away and never reach thermal equilibrium with their radiation and therefore strictly speaking, are *not* in thermal equilibrium. Nevertheless due to the slow evaporation timescale, the rationale here is that we can treat evaporating black hole as *quasi-static*, i.e., its temperature does not change too quickly, and can be treated to be in "equilibrium" if we only look at some sufficiently small time window.

The fact that Hawking temperature can be derived by various means is a double-edge sword. On one hand it shows that the phenomenon is robust, but on the other hand it raises the question concerning the true nature of Hawking radiation. What is it, really? In fact, if we pause and think a little deeper, we start to realize that Hawking temperature of a black hole is quite different from the temperature of an ordinary object. There are two major differences: (1) from where is the particle emitted? and (2) how often is the particle emitted?

The first question can be answered heuristically by recalling the Heisenberg uncertainty principle: $\Delta x \Delta p \sim \hbar$. The Hawking temperature is the characteristic energy scale (since $E = pc \sim k_B T$), so the peak of the variation in the momentum p is $\Delta p \sim T$. This gives $\Delta x \sim \hbar/T = 4\pi r_h$. Alternatively, the de Broglie wavelength of Hawking particle $\lambda = 2\pi\hbar/p = 8\pi^2 r_h$, which is in contrast with a typical black-body radiation that has wavelength much smaller than the size of the emitting body. Both of these arguments are only heuristic, so the prefactor should not be taken too seriously, but they suggest that Hawking particles can be emitted up to some distance $O(10)$ away from the horizon, a diffuse region known as the "quantum atmosphere" (Giddings 2016), not necessary "near"-horizon.

As for the second question, an important quantity is the *sparsity* (Visser et al. 1724), $\eta := \lambda^2/\sigma$, where σ is the geometric optics cross section. It turns out that Hawking emission is not continuous, but *sparse*, with $\eta \gg 1$, i.e., a particle is randomly emitted in a discrete manner, with "pauses" in between. Such emissions also contribute to the random walk of black hole, as the result of backreaction from Hawking emission. Over its long lifetime, a black hole can therefore drifts a tremendous distance away from its original position. For a black hole whose initial mass is M, its lifetime scales like M^3, and the position drift scales like M^2 (Page 1980; Nomura et al. 2013). On the contrary, if $\eta \ll 1$, then we have a typical blackbody like the Sun, which emits continuously.

A common point of view by experts in the field is that the Bekenstein–Hawking entropy measures *all* the underlying degrees of freedom of the black hole. This became the mainstream viewpoint mainly due to the rather successful program of "holography" or "gravity/gauge duality", in which a black hole in an asymptotically anti-de Sitter spacetime (which has a negative cosmological constant), can describe the physics of a quantum field theory (without gravity) that lives on the conformal boundary of the spacetime. Still, this "strong interpretation" of the Bekenstein–Hawking entropy, also known as the "central dogma", is not necessarily correct. An alternative viewpoint is that black holes could harbor way more states in the interior (especially given the large volume contained inside as discussed in the previous section), and that only the surface states are responsible for the thermodynamics and the Bekenstein–Hawking entropy. Indeed, the common view that black hole horizons have an underlying unknown microstates (whose degrees of freedom can explain the Bekenstein–Hawking entropy) is analogous to the microscopic explanation of the entropy of a vial of hot gas. Yet, while the temperature of the hot gas is the mean molecular kinetic energy of the gas particles, the same does not apply to black holes—the Hawking particles are the result of excitation in the quantum field around the black hole, not the mean motion of the black hole "molecules" that make up the horizon.

The Hawking area theorem of 1972, which shows the non-decreasing nature of the black hole horizon area under classical processes is a differential geometric proof, and therefore the non-decreasing property is absolute. This differs from the entropy of thermodynamical systems like that of gases, which can fluctuate and therefore *decrease* from time to time, although overall on average, it is non-decreasing. Historically, Max Planck had thought that the second law is absolute, and for this reason he initially objected the existence of atoms as proposed by Boltzmann (Kragh 2000). In this sense, black holes turn out to be more "Planckian" rather than "Boltzmannian". While this does not necessarily mean that there is no such thing as microstates of black holes, it does remind us that there are some aspects of black hole thermodynamics that remain mysterious, and just because we call it "thermodynamics" does not mean that it *is* thermodynamics in the usual sense of the word. As the philosopher of physics John Dougherty and Craig Callender pointed out (Dougherty & Callender 2016), there is a danger that the analogy between black holes and ordinary thermodynamics has been stretched too far. Indeed, even properties like "thermal equilibrium" in the zeroth law seems rather different for black holes. For black holes, the concept of stationary replaces that of thermal equilibrium, but it is conceptually difficult to define what it means for two black holes to be "stationary with" one another, while no confusion would arise for two objects to be in thermal equilibrium. See also discussions concerning other aspects in Wallace (2018), in which the author David Wallace also discussed the criticisms by Dougherty and Callender.

The issue goes further than black holes—for gravitational systems in general, the notion of entropy is a tricky one. The issue is too technical for us to get into details here. Still, it is worth mentioning that in a review paper on the second law of thermodynamics, mathematicians Elliot H. Lieb[11] and Jakob Yngvason also wrote (Lieb & Yngvason 1999):

[11] Elliott Lieb was awarded the Gauss Prize in 2022 "for deep mathematical contributions of exceptional breadth which have shaped the fields of quantum mechanics, statistical mechanics, computational chemistry, and quantum information theory".

> "Two suns cannot unite to form one bigger Sun with the same properties (the way two glasses of water can unite to become one large glass of water). A star with two solar masses is intrinsically different from a Sun of one solar mass. In principle, the two suns could be kept apart and regarded as one system, but then this would only be a 'constrained' equilibrium because of the gravitational attraction."

Having said that, they also stated that "it is possible to define an entropy for such systems by measuring its effect on some standard body". In any case, I think it is fair to say that the notion of entropy for a gravitating system, and that of the gravitational field itself, is still a subject that is far from being understood.

10.3 Hawking Radiation, Singularities, and Paradoxes

As mentioned in the last section, as a Schwarzschild black hole evaporates, its mass decreases, which causes its Hawking temperature to increase without bound. A realistic black hole would have also angular momentum and electrical charges, but the black holes always eventually shred these away at a rate faster than its mass, so that the end state is Schwarzschild. This is not to say that the evolution of the charge-to-mass ratio is simple; it is not—for a sufficient large black hole the charge-to-mass ratio would first *increase* due to Hawking radiation that emits mainly photons and gravitons (Hiscock & Weems 1990). The ratio can only decrease once charged particles are emitted, and this can only happen when the black hole is small enough and its temperature high enough to reach the mass scale of the lightest charged particles. In any case, a natural question to ask is: what happens during the very last stage of Hawking evaporation?

The mainstream opinion is that the black hole will completely evaporate away, and we are eventually left with nothing but empty spacetime (filled with Hawking particles). However, as the Hawking temperature becomes high enough, the energy scale becomes so high that new physics is expected to kick in. This should happen in the quantum gravity regime, when the black hole mass is about the Planck mass. It is also possible that new physics already occurs before that. In either case, the evaporation picture may be sufficiently affected, so we cannot be sure of what would happen to the black hole. It is possible that evaporation stops, in which case we have a black hole *remnant*. Depending on the models and quantum gravity phenomenology, the remnant can either be stable or meta-stable—the latter means that there is still diffuse emission during the remnant stage. Such a remnant can either evaporate completely in a finite time (at the order of at least M^4 for a black hole whose initial mass is M), or in an infinite amount of time. These remnants are potentially relevant to cosmology since they can serve as a dark matter candidate.

Black hole evaporation has led to many conceptual paradoxes. These paradoxes indicate that our understanding of Hawking radiation is still rudimentary. The information paradox, for example, concerns the apparent loss of information that falls into a black hole. The unitarity property of quantum mechanics essentially means that quantum information is conserved. In a classical black hole, we can argue that the information is just stored inside the black hole interior in some way, but if the black hole completely evaporates away, then the information would disappear, in stark contradiction to unitarity. Assuming the central dogma, the same problem would arise even if there is a Planck size remnant, since eventually the black hole area becomes too small to encode all the quantum information, and this occurs way before the Planck scale. One way out of this is to let the information comes out via quantum entanglement between the apparently thermal Hawking radiation. The information would then be recoverable in theory[12]—but not in practice (much like recovering information from the ashes after burning a book)—by collecting all the Hawking particles and running them through a quantum supercomputer to decode the information. Daniel Harlow and Patrick Hayden had argued that such a decoding takes an extremely long time that scales like e^S, where S is the Bekenstein–Hawking entropy of the initial black hole (Harlow & Hayden 2013). This is of course far longer than M^3, the lifetime of the black hole.

Technically speaking, if we start with a pure state and collapse a region to form a black hole, we have a mixed state outside the horizon. The Bekenstein–Hawking entropy has been interpreted as the entanglement entropy between the exterior and the interior regions. The entanglement entropy among the radiation would first increase, and then starts to decrease with the onset of the *Page time* (Page 1993, 1993, 2013), which is the time the entropy of the black hole is half of the initial black hole before evaporation. Only after the Page time can the Hawking particles start to "contain information". Eventually the late-time Hawking entropy starts to purify the earlier emitted radiation, driving the entanglement entropy of the radiation to zero as the black hole completely evaporates. We would have a pure state again. This is assuming the validity of the central dogma.

Unfortunately, allowing information to come out along with Hawking radiation in such a manner led to another paradox, called the firewall paradox or the AMPS paradox (AMPS are the first letters of the surnames of the authors: Ahmed Almheiri, Donald Marolf, Joseph Polchinski, and James Sully), raised in 2012 (Almheiri et al. 2013). According to the AMPS argument, after Page time, the black hole horizon ceases to be smooth ("uneventful") and would instead be replaced by a high energy region, dubbed a "firewall". Any observer who falls through a sufficiently massive[13] black hole in classical general relativity would just pass right through the horizon (which is a mathematical surface, not a physical one), but once the firewall sets in, an in-falling observer would now be incinerated at the firewall. The onset of the firewall is due to the interior and exterior spacetimes across the horizon no longer being entangled after Page time. Hawking particles are formed by pair-production from the vacuum "near" the horizon. The usual argument is that the pair-produced particles are maximally entangled. However, the Hawking particles emitted after Page time need to purify the earlier Hawking radiation, and therefore they cannot be maximally

[12] Is something that is only verifiable in principle but not in practice really verifiable? This is a philosophical question. There's a whole discipline dedicating to various versions of "verificationism", first conceived by the logical positivists of the famous Vienna circle in the 1920s.

[13] This is so that the tidal force is negligible.

entangled with their own partner that falls into the black hole. This is prevented by the monogamy theorem of quantum information, which basically says that a system cannot be maximally entangled with two other systems at the same time. Since the outgoing and ingoing Hawking pairs are no longer entangled, the quantum field configuration across the horizon need not necessarily be continuous. The Hamiltonian (energy) of the field contains terms like $\partial^2\phi$, which involves the *derivative* of the field. If ϕ is not continuous, its derivative would diverge, hence the firewall. Many arguments for and against the firewall has appeared in the literature, ranging from doubting the necessity of maximally entangled Hawking pairs *ab initio*, to questioning whether *locality* (see Section 8.3), as is usually assumed in quantum field theory, really holds. One could also ask whether a well-defined firewall replacing the horizon can really occur, in view of our discussion in the last section that Hawking pairs are produced in a diffused quantum atmosphere and not necessarily just right outside the horizon.

Firewalls, if they exist, are even worse than initially thought, as there is a nonzero probability for them to arise arbitrarily far away from the black hole. Since the divergent energy density is much like a singularity, such a firewall is dubbed a "naked firewall" (Chen et al. 2016). (In fact, Leonard Susskind has proposed another possibility: as the interior and exterior regions become disentangled, the singularity itself moves "outward" and coincides with the horizon (Susskind 2025), which effectively means that firewall is a naked singularity.) This would be a good time to discuss the idea of "cosmic censorship" as it is also relevant for our discussions.

Classically, a Schwarzschild black hole possesses a spacelike singularity, which is really a future event—an end of time. The charged Reissner–Nordström black hole and the rotating Kerr black hole have a timelike singularity—point-like for the former and a ring for the latter (some people dubbed it a "ringularity"). Timelike singularities are particularly bad because it is a *place*, and as such is present at any instant of time once they are formed. This spells trouble for the predictability of general relativity. In physics, we are often interested in calculating what will happen in the future of a given system once an initial condition is given. For example, if we start with two black holes of masses M and m, respectively, orbiting each other at some distance d apart at time $t = 0$, how would they spiral into each other? If a singularity is present, we have to know how they influence spacetime outside of it, but this is not known since we cannot understand what singularities are in general relativity, with their divergence curvature.[14] In technical mathematical language, we

[14] Mathematically, there is another notion of singularity that is not necessarily tied with curvature divergence, namely geodesic incompleteness. This is the kind of singularity that usually appears in the context of singularity theorems. Physically interesting singularities are, however, those with curvature blow-up. After all, if geodesic is incomplete, it is only natural to ask *why* this is physically so.

[15] Einstein and Rosen wrote in 1935 that: "For a singularity brings so much arbitrariness into the theory that it actually nullifies its laws" (Einstein & Rosen 1935). In 1956, Einstein further elaborated that "the introduction of singularities is equivalent to postulating boundary conditions (which are arbitrary from the point of view of the field equations) on 'surfaces" which closely surround the singularities." In the same work, he also remarked that singularities could just be a sign that a better theory beyond general relativity is required: "It may be plausible that the theory is for this reason inadequate for very high density of matter. It may well be the case that for a unified theory there would arise no singularity" (Gutfreund & Renn 2018). These are pretty much still the same view physicists have today.

do not know what kind of boundary condition to impose on the singularity in order to render the Cauchy problem well-posed. This is the real problem of singularity: it may affects things *away from itself* in unknown and uncontrolled manners[15]; it is not that singularities are inherently *bad*, just like the function $1/x$ is not "bad" at $x = 0$, as we can (and we do) just exclude that point from the domain of the definition. We now also see why spacelike singularities are not usually problematic: future singularities are just the end of time, if there is no spacetime after that, it cannot affect anything. Likewise, a spacelike singularity at the beginning of time, such as at the Big Bang, is only problematic if we want to understand the origin of the Universe. If we are only concerned about physics after the Big Bang has occurred, then we can take initial conditions starting at those times, when physics is already well-understood. Having said this, there is of course the worry that under perturbations, a spacelike singularity can become at least partially timelike.

The issues concerning the loss of predictability can be avoided if singularities are always hidden behind a black hole horizon, since anything behind the horizon cannot affect the exterior region. This is the so-called "cosmic censorship conjecture", or more specifically, the weak cosmic censorship conjecture. Naked singularities that are not shielded by a horizon are known to form under some specific conditions, but the conjecture requires that under *generic* conditions naked singularities never form. Of course the problem now becomes: what do we mean by generic? This is a tough mathematical problem. There is also the strong cosmic censorship conjecture, which essentially means that *no* observer can ever see a naked singularity, including observers who dare to venture inside a black hole. In the case of a Reissner–Nordström or a Kerr black hole, there is an inner horizon r_-. We also call the inner horizon the "Cauchy horizon", meaning that once we get past it, the Cauchy problem is no longer well-posed. The reason is due to the singularity, which is "naked" with respect to anyone in the region $0 < r < r_-$. Since any observer should be able to do sensible physics, we should not discriminate against such observers, hence the strong cosmic censorship proposal.

In fact the situation here is quite analogous to a negative mass Schwarzschild spacetime, which is obtained by simply setting the mass M of a normal Schwarzschild black hole spacetime to be negative. We can write the metric as

$$ds^2 = -\left(1 + \frac{2|M|}{r}\right)dt^2 + \left(1 + \frac{2|M|}{r}\right)^{-1} dr^2 + r^2(d\theta^2 + \sin^2\theta d\varphi^2). \quad (10.39)$$

This spacetime is not physical, but it serves as an interesting comparison. There is no event horizon in this spacetime, and $r = 0$ is a *timelike* singularity unlike the positive mass Schwarzschild case. Consider any non-singular point p in this spacetime. Its causal past defined by its past light cone would intersect with the timelike singularity, which means that information from the singularity would affect p. In other words, we must impose some boundary conditions on the singularity in order to determine what happens at any point in the spacetime. As mentioned, this we cannot do, because we do not know the physics of singularities. Therefore, determinism is completely spoiled if such a singularity is formed. This is exactly the same problem faced by the interior observer in the case of Reissner–Nordström and Kerr black holes.

Thus, essentially, the strong cosmic censorship conjecture requires that naked singularities should not be present to an interior observer. Thankfully this can be achieved physically due to the instability of the inner horizon. Classically the instability is due to the blueshifting of lights from the exterior spacetime, which back-reacts on the geometry. This can be avoided in some non-general relativity black holes specifically constructed such that the inner horizon has zero surface gravity (Carballo-Rubio et al. 2022). These black holes are said to possess an "inner extremal horizon". However, even then, they still suffer from semi-classical instability (McMaken 2023), and so are likely to be unstable. Once instability is taken into account, the inner horizon would likely be destroyed and then turn into a new singularity, some portion of which could be null. Null singularities are also known to exist, for example, when a dilaton charged black hole (the "GMGHS" solution) becomes extremal. Such black holes are a solution to a low energy limit of string theory, and they have only a single horizon. But in the extremal limit the horizon itself becomes a null singularity, with a divergent curvature.

It should be emphasized that a truly divergent curvature invariant is not the only problem—it is equally bad if the curvature is unbounded though still finite. The modern point of view is that general relativity should be viewed as an *effective field theory*, i.e., there is a regime for which the theory is valid. It is generally accepted that general relativity is valid for an energy scale that is below the Planck scale. Specifically, a black hole with a Planck mass has a Planckian curvature, and can no longer be described by general relativity. As mentioned before, other new physics may enter the picture way before the Planck scale is reached, so general relativity may have to be modified to include higher curvature terms and possibly other terms before that. Effective theories allow us to deal with low energy physics with effective descriptions without worrying about all the "actual underlying physics", much like we can do thermodynamics or hydrodynamics without worrying about what the molecules are doing. This conforms to what the philosopher David Hume said in his *Treatise* (Book I, Part II, Sec.4), that "the notion of any correction beyond what we have instruments and art to make is a mere fiction of the mind, and useless as well as incomprehensible".

It is generally believed that singularities should ultimately be cured by quantum gravity, much like how the "ultraviolet catastrophe" is avoided in quantum mechanics (though historically speaking, there was never such a catastrophe, see Section 7.4). Since a working full theory of quantum gravity is still not available, this belief has led to a variety of ad hoc proposals as to how the singularity can be avoided or smoothed out. A black hole with no singularity is known as a "regular black hole". However, "ultraviolet catastrophe" is a prediction of classical physics that is *not* observed in reality, and so the absence of such a phenomenon *necessitates* its removal by a better theory. In the case of black hole singularities, however, various studies have shown that cosmic censorship does indeed usually hold, i.e., Nature seems to try very hard to prevent naked singularities. If there is no singularity to worry about, why such a persistent effort from Nature? In other words, there is nothing to *necessitate* the removal of spacetime singularities other than our expectation of how quantum gravity should behave. It is of course entirely possible that singularities are indeed smoothed out, but not entirely removed.

Thus, returning to the end state of Hawking evaporation, we notice that if black holes completely evaporate, or if the remnants are Planckian, then they would lead to problems with predictability, just like naked singularities would, because the curvature at the horizon becomes unbounded. Therefore, if a singularity is undesirable, it is *not* enough to “cure” the black hole singularities, as ultimately all black holes become effectively singular at the end. This can be avoided if new physics kicks in to prevent further evaporation (or allow the black hole to undergo a phase transition into another state) way before the curvature becomes Planckian, and indeed there are proposals for such a macroscopic remnant. Another possible way to obtain a remnant is if the black hole approaches an extremal state with zero temperature as it evaporates. This is usually achieved by modifying the Schwarzschild solution so that it has an inner horizon, as in the case of most regular black holes. Putting aside the instability of the inner horizon, such near-extremal black holes can be macroscopic in size if the extremal state is approached fast enough and therefore seemingly able to avoid the aforementioned problem. However, numerous recent studies have shown that extremal and near-extremal black holes are highly quantum objects (see, e.g., Horowitz et al. 2023; Turiaci & ICTSN 2023; Emparan 2025), and therefore again new physics may arise and we cannot be confident about the end state. In fact, there is evidence that suggests such black holes are effectively singular, with the singularity being close to or coinciding with the horizon, which means that the same problem arises.

If the central dogma does not hold, then black holes can have vastly more states than its Bekenstein–Hawking entropy measures. This permits Hawking radiation to be information free, with all the information being hidden inside the ever-shrinking black hole (though its volume is ever-increasing!). If the remnant is semi-stable, then eventually all the information is emitted via a diffuse emission of particles during the remnant stage. The large number of states together with the small mass implies that the remnant must decay into a huge number of soft (extremely low energy) particles (Kazemian et al. 2023). This explains why we do not see remnants being copiously produced in high energy collisions, or randomly fluctuate out from the vacuum (Rovelli & Vidotto 2025). Such a criticism for remnants is based on the assumption that, although the rate of producing a single remnant “species”[16] is suppressed and small, given that there are infinitely many “species” the rate would still be large if not infinite. There are many problems with this criticism. First of all, often one has to be careful when “infinity” gets into an argument; it is not at all obvious that there are indeed infinitely many states, especially when quantum mechanics is taken into account. Next, when calculating production rate in quantum field theory, which is

[16] “Species” refers to the internal state configuration. Once a black hole is formed, we can only characterize them by their mass M, charge Q, and angular momentum J. However, there are potentially infinitely many ways to form a black hole with a given set of $\{M, Q, J\}$. Each of these way is a different “species”.

an effective theory, one needs to introduce a cutoff scale $\lambda_c \gg \ell_P$, above which the remnant is treated as a point particle. However, due to the highly curved spacetime of a remnant, it is nontrivial to define such an effective field theory. Indeed, due to gravitational redshift/blueshift, a longer wavelength far away from the remnant, is short close to it, and long again in the interior. This means that a fixed cutoff scale does not seem to make much sense (Hossenfelder & Smolin 2010). This is not to say that we cannot have an effective field theory on curved spacetime at all, but rather that it would be too premature to rule out remnants based on an argument that is itself rather non-rigorous. But most crucially, it is important to distinguish between a newly formed Planck-sized black hole from a Planckian remnant even if they have the same mass and looks the same from the outside, as their interior volume is entirely different. A remnant would have a vastly larger volume. In fact, the most plausible way to form a semi-stable remnant "artificially" would be the time-reversed process of its evaporation—by collecting a large number of ultra-*low* energy particles focusing into a Planckian region, *not* via *high* energy experiments.

We should also mention some other proposals in which the interior solution is replaced by other cosmological spacetimes, such as the de Sitter spacetime. This is similar to the gravastar proposal mentioned in Section 9.4 but in this case the horizon is still intact (not replaced by material shell). Such an idea was proposed already in 1989 by V.P. Frolov, M.A. Markov, and V.F. Mukhanov (Frolov et al. 1989); see also Dymnikova (2019). Together with my collaborators, I have also presented a possibility in which a 3-form field can regularize a black hole interior and replace it with $dS_2 \times S^2$ geometry (Bouhmadi-López et al. 2021). The S^2 part of the geometry is fixed, so as the de Sitter part expands, the interior Universe becomes increasingly effectively two-dimensional.

10.4 Different Types of Horizons

We have avoided the technicalities of defining a horizon thus far to get into the interesting physics first. However, it is still important to know the difference between an event horizon and, say, an apparent horizon, a concept that one frequently encounters in the literature, even in theoretical cosmology. Therefore, this section is aimed to give a quick introduction to various horizons, and even so the list is non-exhaustive (for example, we will not cover the Killing horizon).

An event horizon is what we usually think of when we talk about a black hole. Naively we imagine that anything inside the black hole, not even light, can escape. The definition for an event horizon is thus the entire spacetime region *excluding* the *causal past* of the *future null infinity*. Roughly speaking, the future null infinity is essentially the idealized "location" where any light emitting in spacetime will end up in the far future. The causal past of any point p in spacetime is the set for which signals can reach p. Therefore the causal past of the future null infinity refers to all the points in spacetime whose signals can escape to infinity unhindered. The black hole region is thus the complement of this set. The event horizon is then defined as the boundary of the black hole region (see Box 8). The problem of such a definition is that it is *teleological* in the sense that the determination of the event horizon requires the knowledge of the entire

Box 8: Mathematical Definition of a Black Hole and Event Horizon
For readers who are mathematically inclined, here is, in mathematically precise, set-theoretical language, what we said in words earlier. Given a point p in a spacetime $\mathscr{M}$, we define the *causal past* of p, denoted by $J^{-}(p)$, as the set

$$J^{-}(p) := \{q \in \mathscr{M} \mid \exists \text{ a past-directed causal curve from } p \text{ to } q.\} \tag{10.40}$$

The causal past can be defined for a set S, simply as

$$J^{-}(S) = \bigcup_{p \in S} J^{-}(p). \tag{10.41}$$

The causal future of a point and a set can be defined in the same way, denoted by $J^{+}(p)$ and $J^{+}(S)$, respectively.

Light rays travel on null geodesics all the way to the "future null infinity", which is denoted by $\mathscr{I}^{+}$. A *black hole*, BH, is the region from which light cannot escape, so

$$\text{BH} := \mathscr{M} \backslash J^{-}(\mathscr{I}^{+}). \tag{10.42}$$

The *event horizon*, EH, is its boundary

$$\text{EH} := \partial(\mathscr{M} \backslash J^{-}(\mathscr{I}^{+})). \tag{10.43}$$

Note that defined in this way, EH is three-dimensional (extending in time). In practice, we also refer to the EH at a particular instance of time as an "event horizon", in which case it is a two-dimensional entity.

future null infinity; it is *not* locally associated with the notion of a strong gravitational field.

Imagine an evil advanced extraterrestrial empire builds a large sphere surrounding Earth at some large distance R away and from each point of the sphere a strong laser is fired towards us. At some radius $r < R$, the energy contained in the laser would be focused into a sufficiently small region to form a black hole. This is a "collapsing null shell". The geometry can be described by the *Vaidya metric*:

$$ds^2 = -\left[1 - \frac{2m(v)}{r}\right]dt^2 + 2dvdr + r^2(d\theta^2 + \sin^2\theta d\phi^2), \tag{10.44}$$

where v is the advanced time in the ingoing Eddington–Finkelstein coordinates, and the mass function is

$$m(v) = \begin{cases} 0, & \text{for} \quad v < 0, \\ M = m(v_0), & \text{for} \quad v > v_0, \end{cases} \tag{10.45}$$

where $dm/dv > 0$ for $0 \leqslant v \leqslant v_0$. This is a dynamical version of the Schwarzschild metric. For $v < 0$, the geometry is *flat*, and for $0 \leqslant v \leqslant v_0$ the black hole forms and its mass grows as a function of v until the last ray of the laser has fallen into the black hole and the final mass is $M = m(v_0)$. However, there exists a period during $v < 0$, *before* the first ray of the laser was emitted and the spacetime is locally *Minkowski*, during which there is already an event horizon. This is because nothing can travel faster than the speed of light, so we can no longer "escape to infinity" if we leave Earth too close to the time when a null shell is collapsing towards us—i.e., at some point we cannot overrun the formation of the black hole. To make things even more confusing, sometimes in the literature "event horizon" only refers to part of the mathematical event horizon *after* the black hole formation. This is not the standard meaning of "event horizon".

The fact that the event horizon is teleological also allows it to respond in advance to what will happen in the future. For example, consider a black hole that has already formed. Suppose at some time $t = t_1$ (say, in the Schwarzschild–Droste coordinates) a bunch of stuff falls into the black hole. The event horizon would not "suddenly" become larger at $t = t_1$, instead it would start to increase already before that, despite nothing crossing the horizon. The growth is continuous instead of sudden, but this seems rather strange—how does the horizon know about the incoming stuff? There is no great mystery here, however, as the event horizon is *defined* in such a way that it knows about everything in spacetime. The event horizon is a global, not local, construct.

For the Vaidya solution, $r = 2m(v)$ is actually an *apparent horizon*, which we have more to say on later. The *event horizon* is generated by radially outgoing null geodesics,[17] which obey (solving $ds^2 = 0$ for the radial part)

$$\dot{r} := \frac{dr}{dv} = \frac{1}{2}\left[1 - \frac{2m(v)}{r}\right], \tag{10.46}$$

and are on the boundary of such null geodesics that propagate towards future null infinity. Assuming that the event horizon grows at a steady rate, it can be shown that (Nielsen 2010)

$$r = 2m(v)[1 + 4\dot{m} + 32\dot{m}^2 + O(\dot{m}^3)]. \tag{10.47}$$

Since $\dot{m} > 0$, we see that the location of the event horizon is larger than $r = 2m(v)$.

In general relativity, the apparent horizon of a black hole always lies inside the event horizon provided that the null Ricci condition (equivalently, the null energy condition)—$R_{ab}l^a l^b \geqslant 0$ for all null vector l—is satisfied. In other theories of gravity, this need not be the case. For example, the apparent horizon of a black hole can lie outside the event horizon during spherical collapse in Brans–Dicke theory, although once the black hole settled down into a static configuration it is indistinguishable from the one in general relativity. When the null energy condition is not satisfied in semi-classical general relativity, for example, when quantum effects are involved during Hawking evaporation, apparent horizon can also be outside the event horizon.

To understand apparent horizons, we first need to introduce the concept of a *trapped surface*. Consider the Minkowski spacetime and consider a sphere specified by t and r being constant. The surface has two null normal vector fields. The first one is the future *outward* null vector

[17] In (1+1)-dimensions, any null curve is a null geodesic. In (3 + 1)-dimension this is true when the spacetime is spherically symmetric.

$$N_1 := \frac{\partial}{\partial t} + \frac{\partial}{\partial r}. \tag{10.48}$$

If we draw a light cone centered at the origin in a (1 + 1)-dimensional spacetime diagram, this describes the ray that emanates towards the right. Similarly, the left ray has direction specified by the future *inward* null vector, and is given by

$$N_2 := \frac{\partial}{\partial t} - \frac{\partial}{\partial r}. \tag{10.49}$$

The names "outward" and "inward" don't really make sense in (1 + 1)-dimensions. The name only becomes obvious if we draw more spatial dimensions. With (2 + 1)-dimensions, we can imagine "rotating" the (1 + 1)-dimensional light cone along a circle, then the "left" rays point towards the "inside" of the circle, forming a cone. We check that the Lorentzian inner product between N_1 and N_2 is

$$g(N_1, N_2) \equiv \langle N_1, N_2 \rangle = \left\langle \frac{\partial}{\partial t} + \frac{\partial}{\partial r}, \frac{\partial}{\partial t} - \frac{\partial}{\partial r} \right\rangle \tag{10.50}$$

$$= \left\langle \frac{\partial}{\partial t}, \frac{\partial}{\partial t} \right\rangle - \left\langle \frac{\partial}{\partial r}, \frac{\partial}{\partial r} \right\rangle = -1 - 1 = -2. \tag{10.51}$$

Note that the mixed terms of the inner product canceled. To normalize the inner product to -1 instead of -2, the null vector N_2 is often defined with an additional factor $1/2$.

Why did we bother with introducing the null vector fields? The idea is that we want to capture the behavior of light near a black hole. Inside a black hole we would expect that even if we shine a torchlight, the beam cannot get out of the black hole. Thus, even the light-fronts consisting of the "outgoing" null vectors cannot expand outward. This can be made rigorous as follows.

Technically, for any closed embedded spacelike 2-surface Σ inside a spacetime manifold M, there are four linearly independent vectors $\{e_1, e_2, e_3, e_4\}$. The vectors e_1 and e_2 denote the orthonormal basis of the tangent bundle $T\Sigma$. Suppose Σ is a boundary component of a spacelike hypersurface, e_3 denotes the outward spacelike unit normal of Σ with respect to the hypersurface, such that $\langle e_3, e_4 \rangle = 0$, where e_4 is the future timelike unit normal of the hypersurface. We then define (generalizing the Minkowski space results above) $N_1 := e_3 + e_4$ and $N_2 := (e_4 - e_3)/2$. The *expansion* of Σ along N_i is

$$\theta_i := \sum_{A=1}^{2} \langle \nabla_{e_A} N_i, e_A \rangle = -\langle H, N_i \rangle, \tag{10.52}$$

where $H := \sum_{A=1}^{2} (\nabla_{e_A} e_A)^{\perp}$ is called the *mean curvature vector*. It can be expanded in the basis of N_1 and N_2 as

$$H = \theta_1 N_2 + \theta_2 N_1, \tag{10.53}$$

and so its norm is $\langle H, H\rangle = -2\theta_1\theta_2$. The geometric picture can be further understood by looking into the area variation of Σ, but we will not get into that. Now, if Σ is a closed embedded spacelike 2-surface, and N_1, N_2 are null normal vector fields of Σ such that $\langle N_1, N_2\rangle = -1$, then Σ is called a trapped surface if *both* θ_1 and θ_2 are negative.

The aforementioned construction can be somewhat confusing for readers who are not familiar with Lorentzian geometry or advanced general relativity. Thankfully, in practice, if the spacetime is spherical symmetric, we can always write the metric in the form

$$ds^2 = g_{AB}dx^A dx^B + r^2 d\Omega^2, \tag{10.54}$$

where r is the areal radius. For the Schwarzschild solution, for example, $\{x^A\}_{A=0,\,1} = \{t, r\}$. The mean curvature vector H can be readily computed using the expression

$$H = -\frac{2}{r}\nabla r; \quad \nabla r = \frac{\partial r}{\partial x^A} g^{AB} \frac{\partial}{\partial x^B}. \tag{10.55}$$

For the Schwarzschild black hole, this gives

$$H = -\frac{2}{r}\left(1 - \frac{2M}{r}\right)\frac{\partial}{\partial r}. \tag{10.56}$$

Its Lorentzian norm is

$$\langle H, H\rangle = \frac{4}{r^2}\left(1 - \frac{2M}{r}\right)^2\left(1 - \frac{2M}{r}\right)^{-1} = \frac{4}{r^2}\left(1 - \frac{2M}{r}\right). \tag{10.57}$$

The sign of this norm becomes negative when Σ is trapped. Thus all constant r surfaces in the region $r < 2M$ are trapped surfaces.

Finally, we can introduce the apparent horizon: a closed embedded spacelike 2-surface is an apparent horizon or a *marginally outer trapped surface* (MOTS) if the "outward" null expansion is identically zero (the "inward" one is negative). Using this definition, we can compute the expansions θ_i for the Vaidya geometry and show that its apparent horizon is indeed at $r = 2m(v)$. This is left as an exercise for readers who have the necessary background in advanced general relativity.

Based on the behaviors of the local expansion, various other types of horizons can be defined, which include trapping horizon (Hayward 1994), isolated horizon (Ashtekar et al. 1999), dynamical horizon (Ashtekar & Krishnan 2003), and the slowly evolving horizon (Booth & Fairhurst 2004).

Aside from black holes, we also encounter various horizons in cosmological settings. There is also the concept of "event horizon" in cosmology: it is the surface that encloses the set of points from which signals sent at a given moment of time will never be received by an observer in the future. In other words, the region of space from which a photon that is emitted "today" will still be able to reach us at some point in the future is inside of the cosmological event horizon. (Note that we are inside the horizon, so outside/inside is opposite to the black hole case.)

Let us consider the FLRW metric but written in *conformal time* defined by

$$\eta := \int \frac{dt}{a(t)}, \tag{10.58}$$

in which the behavior of light rays is easier to analyze. The FLRW metric reads

$$ds^2 = a^2(\eta)[-d\eta^2 + d\chi^2 + \text{sinn}^2\chi(d\theta^2 + \sin^2\theta d\phi^2)], \tag{10.59}$$

where

$$\text{sinn}\chi := \begin{cases} \sin\chi, & k = +1; \\ \chi, & k = 0; \\ \sinh\chi, & k = -1. \end{cases} \tag{10.60}$$

Radial light rays satisfy $ds^2 = 0$, which gives $\chi(\eta) = \pm\eta + \text{const}$. Spacetime points beyond the event horizon have "comoving distance" χ that satisfies (subscript E for event horizon)

$$\chi > \chi_E(\eta) := \int_\eta^{\eta_{\max}} d\eta' = \eta_{\max} - \eta. \tag{10.61}$$

In terms of the physical size, the event horizon is given by

$$d_E(t) = a(t)\int_t^{t_{\max}} \frac{dt'}{a(t')}. \tag{10.62}$$

The upper limit of the integral $t_{\max}$ is simply the age of the Universe, and is infinite if the Universe does not contract to a point in the future. Not all cosmological models have an event horizon. The simplest example to illustrate this is the $k = 0$ model filled with dust, i.e., the Einstein–de Sitter model mentioned in Chapter 5, whose scale factor is given by $a(t) = Ct^{2/3}$, where C is a constant. We check that the integral

$$\int_t^{t_{\max}} \frac{dt'}{a(t')} \propto \int_t^{t_{\max}} t'^{-\frac{2}{3}} dt' = 3(t_{\max}^{\frac{1}{3}} - t^{\frac{1}{3}}) \tag{10.63}$$

is divergent when $t_{\max} \to \infty$ since the flat Universe expands forever. Thus, there are no spacetime points that can be *beyond* $d_E = \infty$. In other words, there is no boundary beyond which light or information cannot reach an observer due to the expansion of the Universe. On the other hand, there is an event horizon in the contracting phase of $k = 1$ FLRW cosmology.

For another example, consider the de Sitter space with flat slicing ("flat de Sitter"):

$$ds^2 = -dt^2 + a^2(t)[dx^2 + dy^2 + dz^2], \quad a(t) = e^{Ht}, \tag{10.64}$$

where H is the Hubble constant. In this case we observe that the integral

$$d_E(t) = e^{Ht}\int_t^{t_{\max}} e^{-Ht'}dt' = e^{Ht}\left[\frac{1}{H}(e^{-Ht} - e^{-Ht_{\max}})\right], \tag{10.65}$$

converges to $1/H$ in the limit $t_{\max} \to \infty$, so there exists an event horizon, which is independent of time (since H=const. in this case). The scale $1/H$, or c/H if we restore the speed of light explicitly, is also called the "Hubble radius". In general, the Hubble parameter is not a constant, and the Hubble radius is simply *defined* by $1/H(t)$. This does not have to coincide with the event horizon in a general cosmological spacetime.

The size of the Hubble radius, as it evolves, plays a prominent role in cosmology. For example, black holes can form in the very early Universe (these are called primordial black holes) via various possible channels. One of these is simply that an over-dense region larger than the Hubble radius at that time, when collapsed into a region smaller than the Hubble radius, formed a black hole. See Box 9 for more details.

Box 9: Superhorizon Mode and Primordial Black Holes

A wavenumber k in a Fourier mode $\sim e^{ikx}$ is associated with a wavelength by $\lambda \propto 1/k$. In cosmology, if we consider the wavenumber k to be the comoving one with the expansion, then it is related to the physical wavenumber by $k_{\rm phys} = k/a$. If a mode $k_{\rm phys}$ is smaller than the Hubble parameter, then its associated wavelength is larger than the Hubble radius. This is called a "superhorizon" mode. (Likewise, if $k_{\rm phys} > H$, it is called "subhorizon"). Recall that the Hubble radius is not necessarily a horizon, but the names "superhorizon" and "subhorizon" have become standard. A superhorizon mode is often said to be "not evolving" or "frozen". This is referring to the fact that if the wavelength is larger than the Hubble radius, its frequency would be extremely small. Since frequency is inversely proportional to a timescale, the timescale of its evolution would be very long compared to the age of the Universe at the time, hence effectively frozen. As the Hubble radius grows, superhorizon modes would eventually "re-enter" the "horizon" (or rather the horizon "catches up" to them), and start to evolve. If a primordial fluctuation is sufficiently dense, as it re-enters the horizon, it would collapse into a *primordial black hole*.

Primordial black holes can also be formed from *quantum* fluctuation of the vacuum, in addition to the classical collapse process discussed above. Given a background temperature T of the Universe, the rate of black hole formation due to vacuum fluctuation is proportional to $e^{-M^2/m_{\rm Pl}^2}$, where $m_{\rm P}$ is the Planck mass (Gross et al. 1982; Kapusta 1984). (In terms of the Bekenstein–Hawking entropy S of the black hole, we note that the rate is exponential in $-S$.) Since $M \propto 1/T$, this process is only important in the very early Universe when the temperature is extremely high.

Primordial black holes have received more attentions recently, not just because they may constitute some part of the dark matter, but may explain why black holes detected via gravitational waves are larger than the typical stellar mass black holes. The size of primordial black holes can be quite varied and still not well constrained. If one allows non-standard physics like modified gravity or modification to quantum mechanics, then the constraints can be relaxed even further. For example, if primordial black holes are too small, we would expect that they would have completed evaporated by Hawking radiation. However, if Hawking evaporation stops due to new physics at a sufficiently high energy scale, this would change the picture (Arbey 2025).

If primordial black holes exist, they may also explain why most large galaxies today have supermassive black holes at their center. Perhaps primordial black holes served as the seed for galaxy formation (Cappelluti et al. 2022).

Apparent horizon is also a helpful concept in cosmology, and unlike the event horizon, FLRW universes have an apparent horizon. To see this, we write the FLRW metric in the spherically symmetric form

$$ds^2 = g_{AB}dx^A dx^B + R^2(d\theta^2 + \sin^2\theta d\phi^2), \tag{10.66}$$

where

$$g_{AB}dx^A dx^B = -dt^2 + \frac{a^2(t)}{1 - kr^2}dr^2, \tag{10.67}$$

and $R = a(t)r$ is the areal radius. From the discussion above, the mean curvature vector is

$$H = -\frac{2}{R}\nabla R; \quad \nabla R = \frac{\partial R}{\partial x^A}g^{AB}\frac{\partial}{\partial x^B}. \tag{10.68}$$

From this we can find the apparent horizon, which is

$$d_A = R = \frac{1}{\sqrt{H^2 + \frac{k}{a^2}}}. \tag{10.69}$$

When $k = 0$, we get $R = 1/H$, which is the same as the Hubble radius.

In cosmology, there is another horizon of interest, the *particle horizon.* It is defined in the same way as the event horizon, except that one integrates from the birth of the Universe if there is a Big Bang ($t = 0$) to the current time. This marks the edge of the observable Universe, i.e., the maximum distance that a ray of light has traveled from the Big Bang to the present era. For example, the Einstein–de Sitter model has a particle horizon at

$$d_p(t) = t^{\frac{2}{3}}\int_0^t (t')^{-\frac{2}{3}}dt' = t^{\frac{2}{3}} \cdot (3t^{\frac{1}{3}}) = 3t. \tag{10.70}$$

At the current epoch, $d_p(t_0) = 3t_0 = 3/H_0$, which is larger than the apparent horizon and the Hubble radius. For the flat de Sitter case, we have

$$d_p(t) = e^{Ht}\int_0^t e^{-Ht'}dt' = e^{Ht}\left[\frac{1}{H}(1 - e^{-Ht})\right]. \tag{10.71}$$

Therefore, at late time, as $t \to \infty$, the particle horizon behaves like

$$d_p(t) \sim \frac{e^{Ht}}{H}. \tag{10.72}$$

This means that the causally connected region grows exponentially, whereas the event horizon that we computed before stays fixed at $1/H$. The fact that the particle horizon is larger than the event horizon may be confusing to some, but recall that the particle horizon marks the furthest distance from which light *has already* reached us, i.e., what we can currently see. On the other hand, the event horizon marks the furthest distance from which light emitted *now* could ever reach us in the *future*. For example, assuming that we live in a flat de Sitter space, the event horizon can be computed to be roughly 17.5 Gly. This does not mean that we cannot see anything that is currently located at 20 Gly from us. The observable universe is about 46.5 Gly in radius. What it means is that if a galaxy G is currently located at 20 Gly from us, we will never be able to see the light that it emits *now*. The light we see today was emitted from the past.

We should emphasize that the apparent horizon is observer- and coordinate-dependent, but its application in cosmology is not so much of a problem because there is *already* a preferred frame in cosmology anyway. More specifically, the FLRW metric describes a homogeneous and isotropic universe *only for observers like us* who follow the "Hubble flow". The apparent horizon depends on the same FLRW cosmic time t. Furthermore, in $k = 0$ FLRW geometry, the radius of the apparent horizon is just $R_H = 1/H$, i.e., the Hubble radius, which is a physically relevant scale that can be computed from observation (of the Hubble constant). Some detailed clarifications can be found in Melia (2018), in which the name "gravitational horizon" is used to emphasize its physical nature (as opposed to a purely mathematical and arbitrary nature, as one may fear given the coordinate-dependent nature of the apparent horizon). Similar to black hole thermodynamics, we can also discuss the thermodynamics that are associated with cosmological apparent horizons.

10.5 Exercises

(1) In Section 10.2, we derived the Hawking temperature for a static spherically symmetric black hole, whose Wick-rotated metric tensor has the form

$$ds^2 = f d\tau^2 + f^{-1} dr^2 + r^2(d\theta^2 + \sin^2\theta d\varphi^2). \tag{10.73}$$

Use the same method to show that for a more general form

$$ds^2 = f d\tau^2 + g^{-1} dr^2 + r^2(d\theta^2 + \sin^2\theta d\varphi^2), \tag{10.74}$$

where f is not necessarily the same as g, we have

$$T = \frac{1}{4\pi}\sqrt{f'(r_h)g'(r_h)}\,. \tag{10.75}$$

(2) Consider the *parabolic LTB spacetime* described by the metric tensor

$$ds^2 = -dt^2 + [R'(r)]^2 dr^2 + R^2(d\theta^2 + \sin^2\theta d\phi^2) \tag{10.76}$$

$$= h_{ab}dx^a dx^b + \tilde{r}^2(d\theta^2 + \sin^2\theta d\phi^2); \quad \tilde{r} \equiv R, \tag{10.77}$$

where the radial function is

$$R(r) = \frac{1}{2}[9F(r)]^{\frac{1}{3}}[t + \beta(r)]^{\frac{2}{3}} \tag{10.78}$$

for some functions $F(r)$ and $\beta(r)$. We followed the notation of Wang et al. (2000). Show that its apparent horizon is given by

$$R_{\rm AH} = \frac{3}{2}[t + \beta(r_{\rm AH})]. \tag{10.79}$$

(3) In Section 10.4, we introduced the null energy condition. This is the weakest energy condition among all the classical energy conditions in general relativity. An energy condition is required to impose some constraints on the energy-momentum tensor T_{ab}, so that the solutions of Einstein field equations are physical. For students who are familiar with general relativity,

(a) Show that the Einstein field equations

$$R_{ab} - \frac{1}{2}g_{ab}R = \kappa T_{ab} \tag{10.80}$$

can be re-written in the so-called "trace-reverse" form:

$$R_{ab} = \kappa\left(T_{ab} - \frac{1}{2}g_{ab}T\right), \tag{10.81}$$

where $T \equiv g^{ab}T_{ab}$.

(b) Note that the terms in the bracket on the right-hand side is precisely the combination that goes into the *strong energy condition*, which is the statement that for every timelike vector field X, we have

$$\left(T_{ab} - \frac{1}{2}g_{ab}T\right)X^aX^b \geqslant 0. \tag{10.82}$$

Argue that the strong energy condition implies the null energy condition.

(c) Consider a real scalar field ϕ with energy-momentum tensor given by

$$T_{ab} = \nabla_a \phi \, \nabla_b \phi - \frac{1}{2}g_{ab}[\nabla_c \, \nabla^c\phi + V(\phi)]. \tag{10.83}$$

Show that the null energy condition holds regardless of what the potential V is.

References

Alonso-Serrano, A., & Visser, M. 2018, PhLB, 776, 10

Almheiri, A., Marolf, D., Polchinski, J., & Sully, J. 2013, JHEP, 2013, 62

Arbey, A. 2024, [arXiv:2405.08624 [gr-qc]] 2024arXiv240508624A

Ashtekar, A., Beetle, C., & Fairhurst, S. 1999, CQGra, 16, L1

Ashtekar, A., & Krishnan, B. 2003, PhRvD, 68, 104030

Bardeen, J. M., Carter, B., & Hawking, S. W. 1973, CMaPh, 31, 161

Barbado, L. C., Barceló, C., Garay, L. J., & Jannes, G. 2016, JHEP, 2016, 161

Bekenstein, J. D. 1973, PhRvD, 7, 2333

Bengtsson, I., & Jakobsson, E. 2015, MPLA, 30, 1550103

Bouhmadi-López, M., Chen, C.-Y., Chew, X. Y., Ong, Y. C., & Yeom, D.-h. 2021, EPJC, 81, 278

Booth, I., & Fairhurst, S. 2004, PhRvL, 92, 011102
Candelas, P. 1980, PhRvD, 21, 2185
Carballo-Rubio, R., Di Filippo, F., Liberati, S., Pacilio, C., & Visser, M. 2022, JHEP, 09, 118
Cappelluti, N., Hasinger, G., & Natarajan, P. 2022, ApJ, 926, 205
Chen, P., Ong, Y. C., Page, D. N., Sasaki, M., & Yeom, D.-h. 2016, PhRvL, 116, 161304
Christodoulou, M., & Rovelli, C. 2015, PhRvD, 91, 064046
Coleman, S. 1988, NuPhB, 310, 643
Dougherty, J., & Callender, C. 2016, Preprint
Dymnikova, I. 2019, Univ, 5, 111
Earman, J. 2011, SHPSB, 42, 81
Einstein, A., & Rosen, N. 1935, PhRv, 48, 73
Emparan, R. 2025, JHEP, 2025, 122
Frolov, V. P., Markov, M. A., & Mukhanov, V. F. 1989, PhLB, 216, 272
Giddings, S. B. 2016, PhLB, 754, 39
Gutfreund, H., & Renn, J. 2018, The Formative Years of Relativity: The History and Meaning of Einsteinas Princeton Lectures 326 (Princeton, NJ: Princeton Univ. Press)
Gross, D. J., Perry, M. J., & Yaffe, L. G. 1982, PhRvD, 25, 330
Hawking, S. W. 1971, PhRvL, 26, 1344
Hawking, S. W. 1975, CMaPh, 43, 199
Hod, S. 2018, EPJC, 78, 1013
Hod, S. 2020, EPJC, 80, 982
Hartle, J. B., & Hawking, S. W. 1976, PhRvD, 13, 2188
Hiscock, W. A., & Weems, L. D. 1990, PhRvD, 41, 1142
Harlow, D., & Hayden, P. 2013, JHEP, 2013, 85
Horowitz, G. T., Kolanowski, M., Remmen, G. N., & Santos, J. E. 2023, PhRvL, 131, 091402
Hossenfelder, S., & Smolin, L. 2010, PhRvD, 81, 064009
Hayward, S. A. 1994, PhRvD, 49, 6467
Kragh, H. 2000, PhyW, 13, 31
Kazemian, S., Pascual, M., Rovelli, C., & Vidotto, F. 2023, CQGra, 40, 087001
Kapusta, J. I. 1984, PhRvD, 30, 831
Lieb, E. H., & Yngvason, J. 1999, PhR, 310, 1
McMaken, T. 2023, PhRvD, 107, 125023
Melia, F. 2018, AmJPh, 86, 585
Nomura, Y., Varela, J., & Weinberg, S. J. 2013, PhRvD, 87, 084050
Nielsen, A. B. 2010, CQGra, 27, 245016
Ong, Y. C. 2015, GReGr, 47, 88
Parikh, M. K., & Wilczek, F. 2000, PhRvL, 85, 5042
Page, D. N. 1980, PhRvL, 44, 301
Page, D. N. 1993, PhRvL, 71, 1291
Page, D. N. 1993, PhRvL, 71, 3743
Page, D. N. 2013, JCAP, 09, 028
Robinson, S. P., & Wilczek, F. 2005, PhRvL, 95, 011303
Rovelli, C., & Vidotto, F. 2024, [arXiv:2407.09584 [gr-qc]] 2024arXiv240709584R
Sciama, D. W., Candelas, P., & Deutsch, D. 1980, AdPhy, 30, 327
Susskind, L. 2012, [arXiv:1208.3445 [hep-th]] 2012arXiv1208.3445S

Tsvelik, A. M. 2003, Quantum Field Theory in Condensed Matter Physics (Cambridge: Cambridge Univ. Press)

Turiaci, G. J. 2023, [arXiv:2307.10423 [hep-th]] 2023arXiv230710423T

Visser, M., Gray, F., Schuster, S., & Van-Brunt, A. 1724, Part of Proceedings, 14th Marcel Grossmann Meeting on Recent Developments in Theoretical and Experimental General Relativity, Astrophysics, and Relativistic Field Theories (MG14) (Singapore: World Scientific)

Wallace, D. 2018, SHPSB, 64, 52

Wang, B., Abdalla, E., & Osada, T. 2000, PhRvL, 85, 5507

AAS | IOP Astronomy

Chapter 11

Making Sense of Spacetime

> *"To see a world in a grain of sand, and a heaven in a wild flower, hold infinity in the palm of your hand, and eternity in an hour."* —William Blake, 'Auguries of Innocence'.

We learned from general relativity that gravity is the manifestation of spacetime curvature. Indeed in general relativity, spacetime is mathematically modeled as a four-dimensional manifold. However, this does not tell us what space and time "really" is at the fundamental level, or whether they are even fundamental! We shall review some debates concerning the nature of space and time—whether space and time can exist without matter, and why there are only three spatial dimensions. We then discuss gravitational waves, the so-called "ripples in the fabric of spacetime". As the latest window to the Universe, gravitational wave research will be a very active field in the foreseeable future and thus any serious student in astrophysics and cosmology should learn a thing or two about this subject. Finally we end with some musings on the nature of spacetime, and why it may be a mistake to treat gravity as just another field.

In ancient Egypt, infinity and eternity were personified into a deity named "Heh" (sometimes translated as "Huh"). Similarly, the word "chronology" came from Greek's personification of time, Cronos. As we have seen in preceding chapters, humanity's fascination with time (which is tied to fate) and space has been around since antiquity. Nowadays, we have learned that the Universe in general relativity is a spacetime manifold. But what *is* spacetime, really? Can time and space exist independently of matter? The latter question is not as obvious as it first seems: if there is nothing at all, who or what is there to perceive or interact with spacetime? How then can we know if spacetime with nothing in it can even exist? Even prior to general relativity, philosophers have debated on the nature of space and time.

doi:10.1088/2514-3433/ade963ch11

We have seen in Chapter 3 that some thinkers once identified space with God ("divinized" space), or at least considered God to be "co-extensive" with the infinitude of space. As for time, the theologian and philosopher Augustine of Hippo (354–430; perhaps better known as St. Augustine) wrote in his *Confessions*, Chapter XIV:[1]

> "For what is time? Who can readily and briefly explain this? Who can even in thought comprehend it, so as to utter a word about it? But what in discourse do we mention more familiarly and knowingly, than time? And, we understand, when we speak of it; we understand also, when we hear it spoken of by another. What then is time? If no one asks me, I know: if I wish to explain it to one that asketh, I know not: yet I say boldly that I know, that if nothing passed away, time past were not; and if nothing were coming, a time to come were not; and if nothing were, time present were not. Those two times then, past and to come, how are they, seeing the past now is not, and that to come is not yet? But the present, should it always be present, and never pass into time past, verily it should not be time, but eternity. If time present (if it is to be time) only cometh into existence, because it passeth into time past, how can we say that either this is, whose cause of being is, that it shall not be; so, namely, that we cannot truly say that time is, but because it is tending not to be?"

We now understand that according to general relativity, space and time can be so curved that they form black holes, and recent discovery of gravitational waves show that spacetime can "vibrate". Yet, spacetime is not a "material substance" like the æther of the old days, so what does it really mean for empty spacetime to vibrate and propagate? We shall discuss some properties of spacetime and the basics of gravitational waves in this chapter, but unfortunately many of the big questions concerning the ultimate nature of spacetime, such as whether it is fundamental or emergent, would remain unsolved.

11.1 Properties of Spacetime

We are now used to thinking about gravity as just a manifestation of spacetime curvature. Sometimes when we are so used to a concept, we start to take it for granted and lose the appreciation for how remarkable the concept is. Consider a Kerr black hole, for example, which is a vacuum solution in general relativity. Being a vacuum means that there is no matter. Yet, there is *mass* and *angular momentum*. When we say that a black hole is rotating, it is not like the surface of a planet that is rotating, but rather *space* itself. What does that mean, really? How can an empty space expand or rotate or do whatever it is a certain solution of the Einstein field equations dictates it to do? I don't think anyone has any satisfactory answer to these questions. In fact, what space *is*, or rather, how should we think of space and time,

[1] Available online at https://ccel.org/ccel/augustine/confess.xii.xiv.html.

has always been a difficult philosophical problem. Nowadays we are also used to thinking about empty space because of our deeply-ingrained education in mathematics, namely we think of the Cartesian three-dimensional space $\mathbb{R}^3$. This was also essentially how Newton thought about space. The philosopher Samuel Clarke (1675–1729) agreed that space necessarily exists even without matter, because, as written in his *A Discourse Concerning the Unchangeable Obligations of Natural Religion, and the Truth and Certainty of the Christian Revelation,*[2]

> "To suppose any part of space removed, is to suppose it removed from and out of itself: and to suppose the whole to be taken away, is supposing it to be taken away from itself, that is, to be taken away while it still remains: which is a contradiction in terms. There is no obscurity in this argument but what arises to those who think immense space to be absolutely nothing: which notion is itself likewise an express contradiction; for nothing is that which has no properties or modes whatsoever; that is to say, it is that of which nothing can truly be affirmed, and of which every thing can truly be denied; which is not the case of immensity or space."

Clarke would enter into a letter exchange with Leibniz debating the absolute nature of space. This episode in the history of science is now known as the "Leibniz–Clarke correspondence". The French philosopher and bishop Nicole Oresme (1325–1382) also argued for absolute space; note that this was centuries before Newton's time. In fact, similar ideas can be traced back to John Philoponus (490–570) of Alexandria, Egypt.

However, there was another opposing point of view by Leibniz and others, that space by itself has no objective meaning, only a relational meaning. More precisely, space consists solely in the relations among bodies. That is to say, we know that object A and object B are two different objects even though they are identical, if they are separated in space. This is a view similar to that of "time is what keeps everything from happening at once."[3] This relational view may seem odd to us, but we can consider another scenario: if nothing ever happens, is there time? You may argue that this is absurd. Surely "time passes" even when there is no one there to measure it ("does a tree really fall in the forest if no one hears it?"), but if physics is to be a science, there has to be some kind of measurement. In this view, it is perhaps easier to understand why the relational view of spacetime has its appeal. After all, if there is nothing, how do we know there is space? We can, of course, conceive of a *mathematical* empty space $\mathbb{R}^3$, but this is not physics. The moment that you have an observer in your space, it ceases to be empty. The psychoanalyst Carl Jung agreed with this view, as he wrote in a letter to André Barbault in 1954 (Adler et al. 1976),

---

[2] Available online at https://www.ccel.org/ccel/clarke_s/being.iii.iv.html.

[3] This quote is often attributed to Einstein or Wheeler, but seems to first appear in March 1919, in a story titled *The Girl in the Golden Atom* by Ray Cummings in the magazine "All-Story Weekly".

> "Time in itself consists of nothing. It is only a *modus cogitandi* that is used to express and formulate the flux of things and events, just as space is nothing but a way of describing the existence of a body. When nothing occurs in time and when there is no body in space, there is neither time nor space. Time is always and exclusively 'qualified' by events as space is by the extension of bodies."

Descartes also argued that there can be no empty space by itself, since empty space is nothingness, and we cannot think of nothingness. (To which Gassendi would object that one can clearly conceive of space without any object. It's somewhat ironic that today we think of the Cartesian space $\mathbb{R}^3$ as empty space, but the father of Cartesian geometry thought otherwise!) Thus from the relational view, empty space (nothingness) is nonsensical without any matter, while matter itself of course requires space to occupy. One is reminded of the *Heart Sutra* of Buddhism, in which a famous passage reads:[4]

> "Form is not distinct from emptiness, emptiness is not distinct from form;
> Form is emptiness, emptiness is form."

Beyond the nature of space, one also finds a relational view in an interpretation of quantum mechanics called relational quantum mechanics (RQM) (Rovelli 1996, 2022), proposed by Carlo Rovelli, in which a quantum state is relational, i.e., quantum states do not describe an absolute property of a system but rather a relationship between systems. After all, when an electron is not interacting with anything, it is as good as it being non-existent[5].

Of course, when discussing the nature of spacetime, we should not forget the view of the philosopher Immanuel Kant, although it is somewhat misguided. He believed in the existence of "synthetic a priori", which are truths about the world that one can deduce even without having any experience about it. Truths about space and time, as well as geometry, he argued, were such examples. In a sense, we have innate intuitions about how the world works. By geometry he meant Euclidean geometry, but he did not anticipate other Non-Euclidean geometries.[6] Likewise, he didn't know about future developments of general relativity that gave rise to an entirely different understanding of space and time; intuition can only go so far.

---

[4] Of course the meaning of this passage goes beyond the debate on the nature of space.

[5] The relational viewpoint of the world is an old philosophy. For example, in Buddhism, "emptiness" is called *śūnyatā*. In addition one finds the concept of *Pratītyasamutpāda* or "dependent arising" (缘起 in Mandarin), the idea that everything arises only in dependence upon other things. Every phenomenon and object is the results of a web of relations intertwined, without which their essence is emptiness (性空). For example, think about *who you are*. The answer is always in relation to someone and some place. But who are you, really, without those webs of relationships that define you?

[6] Recent investigations have provided evidence that the actual geometry of spatial *representation* in the brain is non-Euclidean. More specifically, neurons in the so-called "CA1 region" of rat hippocampus that mediate spatial perception represent space using a dynamic hyperbolic geometry. Thus, even if our "intuition" perceives the world as locally Euclidean, the perception space itself is not (Zhang et al. 2023).

There are other problems concerning space and time. Let us discuss the obvious one: why are there three dimensions of space and one of time? Why not other numbers? Such a question is quite natural to ask given our mathematical education background. Otherwise, three-dimensionality of space seems like something that is so obvious that no one is likely to ask. Thus it might surprise some readers that scholars were already thinking about the number of spatial dimensions in the Middle Ages. Recall the condemnations of 1277 we mentioned at the beginning of Chapter 2. One of the articles discussed therein was whether God can create more than three dimensions!

The fact that we have only one well-defined arrow of time is tied to the second law of thermodynamics. If there are two temporal dimensions, say t_1 and t_2, then the linearly independent vectors in these two directions can span a *temporal plane*, just like x and y directions of space span a two-dimensional plane. It is not clear to me—perhaps due to the lack of imagination on my part—how we can have a well-defined arrow of time that in turn gives rise to some kind of "order" in such a hypothetical universe. Thus requiring that there can only be one temporal dimension seems to be preferable if not necessary. The number of the dimensionality of space is harder to explain. One possibility is the stability of orbits. In fact, in a system governed by a central force, bound orbits could occur only if the number of spatial dimensions is 3. If planetary orbits are not bounded, we wouldn't be here to ask the question.[7] This is what we today call the "weak anthropic principle", but such a reasoning is quite an old one, before the term "anthropic principle" became popular. For example, Gerald J. Whitrow has made such an argument in his 1955 paper "Why Physical Space Has Three Dimensions" (Whitrow 1955), in which he wrote:

> "[...] physical conditions on the Earth have been such that the evolution of Man has been possible. Is this possibility in any way dependent on the three-dimensional nature of space?"

Withrow then appealed to the known mathematical results (Bertrand's theorem) which guarantees orbital stability for central force law only if the force has the form $f = -k/r^n$, with $n \leqslant 2$. The case $n = 2$ corresponds to three spatial dimensions. Lower dimensions would likely not allow organisms that have enough complexity to evolve. For example, in his famous book *A Brief History of Time*, Hawking argued that in two spatial dimensions, an animal cannot have a digestive system analogous to ours, because if you draw a tube from the mouth to the anus of such a creature, the poor creature would no longer be continuous and instead split into two halves![8] Of course other forms of digestion can be feasible, and perhaps it is our lack of imagination that is the problem here. At the very least, it has been demonstrated that certain families of planar graphs share properties which are observed in real-life

[7] Such an argument is not necessarily convincing since we know that orbits can be unstable in reality. As long as the instability *timescale* is sufficiently long, this would not affect the development of life and civilizations.

[8] This is the only picture I can vividly remember from the book, which I read in high school.

biological neural networks, and so some complexities are allowed in two spatial dimensions (Scargill 2020).

Anyway, let's say that three spatial dimensions is the only viable option.[9] There are in fact other unique properties that three dimensions enjoy.[10] For example, we have previously discussed in Section 8.4.2 that the degeneracy pressure due to quantum mechanics is what kept white dwarf stars stable against gravitational collapse. However, this cannot happen if the spatial dimension is larger than 3 (Chavanis 2007). Of course, it is theoretically possible that there are indeed more than three spatial dimensions, but that either they are compactified[11] and too small for us to perceive, or that we are stuck on a three-dimensional world that is floating inside an even higher dimensional space (so-called "brane world scenario"). So far there is no experimental or observational evidence for either of these possibilities.

With hindsight, the idea of extra dimensions has been quite useful. Not only that it allows us to formulate the theory of general relativity, which treats spacetime as a whole four-dimensional entity, but more generally it allows us to define abstract spaces. A space of n dimensions, be it phase space in physics, or other kind of spaces in other fields, can be defined as long as we have n-independent variables. They need not correspond to actual spatial or temporal dimensions. In this we see the power of abstraction in mathematics. Still, as pointed out by Whitrow (1955), the study of dimensions above 3 was historically expressively rejected by geometers. Since geometry is linked to algebra (e.g., the quadratic equation is related to "completing a square"), even higher-order equations beyond the cubic were likewise avoided for being "unreal" until the 16th and 17th centuries. Withrow specifically cited Michael Stifel who wrote in 1553, "going beyond the cube just as if there were more than three dimensions [...] which is against nature". Withrow also mentioned that even Galileo failed to recognize the possibility of going beyond three dimensions. There are lessons we can learn from this episode of the history of science: it is true that we should try to focus on theories and models that are as "physical" as possible, but at

[9] It is interesting to note that the psychoanalyst Carl Jung had discussed a great deal about the archetype of the *quaternity*, and he pointed out that a quaternity often has a "3+1" structure, meaning that one of the four often has either an exceptional position or has a nature that is different from the rest. One of his examples being that three of the symbols of the Evangelists are animals while the fourth is an angel. An older example would be the four sons of Horus from Egyptian mythology. Indeed, in actual physics, the temporal dimension is different from the three spatial ones due to the fact that we can only move in a single direction in time. Our consciousness is tied to the arrow of time; we can't "remember" the future.

[10] It is intriguing to note that, Geroch's theorem states that a non-compact four-dimensional spacetime admits a spin structure if and only if it is parallelizable (i.e., there is a global smooth vector field that at every point of spacetime, the value of the field provides a basis of the tangent space there). Steenrod's theorem, on the other hand, states that all orientable 3-manifolds are parallelizable. Thus, all four-dimensional spacetimes with orientable spatial section are parallelizable. This is important as spin structure is crucial for physics (spinors are used to describe the intrinsic angular momentum, or "spin", of the electron and other subatomic particles). It is interesting that the only requirement for three-dimensional spatial sections to admit a spin structure is that it is orientable, as many manifolds are *not* parallelizable. For example, among the n-spheres, R. Bott and J. Milnor proved in 1958 that only S^0, S^1, S^3, S^7 are parallelizable. This is in turn related to the fact that the only normed division algebras are $\mathbb{R}$, $\mathbb{C}$, $\mathbb{Q}$, $\mathbb{O}$.

[11] As in string theory, as well as the Kaluza–Klein theory, an early failed attempt to unify general relativity and electromagnetism by introducing an additional 5th, compactified spatial dimension.

the same time, we should allow ideas that seem too far-fetched to be pursued. Especially when mathematics is concerned, ideas that do not seem physically relevant can turn out to be useful for applications in the future. The mathematics of higher dimensions—specifically sphere packing[12] in 24 dimensions—turned out to be useful in the design of error-correction codes for the Voyager spacecraft! Operating in the abstract four-dimensional space, the "Binary Golay code" developed in 1949 can correct up to 7 errors within a 24-bit data block. The engineering applications then feed itself back to pure mathematics. In 1964, while studying the Binary Golay code, John Leech (1926–1992) noticed that there are still "holes" in the packing that are large enough to fit more spheres of the same size. This led to the doubling of the packing density, in what we would today call the *Leech lattice*. Roughly speaking, spheres can only be packed the "tightest" possible in 24 dimensions and 8 dimensions (in a configuration known as the E_8 lattice). Both the Leech lattice and the E_8 lattice have mysterious links with many areas of mathematics that are still not fully understood. Both of them have also been applied in theoretical physics (especially string theory).

So far we have only discussed classical spacetimes. But what happens at the small scale, when quantum gravity is supposed to reign? Well then it depends on the specific theories or models of quantum gravity. There is no consensus on this issue. Often we rely on some guiding principles to construct a theory, but which guiding principle is correct is hard to say in the absence of experimental data and observation. For example, at the fundamental level spacetime may be discrete instead of continuous, or spacetime may fluctuate wildly at the Planck scale (a scenario referred to as "spacetime foam") with mini wormholes forming and disappearing. Neither discreteness nor "foaminess" is observable at the scale of everyday life, which appears anything but continuous and smooth. This is much like those "pointillism" paintings, which if looked at from far away, the "points" are not visible (Figure 11.1). The pixels on your computer screen is another analogy; you don't notice them unless up-close!

It is also possible that the very symmetry of spacetime changes when the energy scale is sufficiently high, so that in place of Lorentz invariance we have something else. For example, in Hořava–Lifshitz gravity Lorentz invariance is broken via a Lifshitz-type anisotropic scaling, whereas in the so-called "Rainbow gravity" paradigm, the metric tensor that describes the geometry of spacetime depends on the energy of the test particle used to probe the structure of that spacetime manifold, thus light of different frequency is affected by gravity in a different manner. There are many theories that break Lorentz invariance in one way or another, but they are subject to tight observational constraints. One popular way to constrain Lorentz invariance is via ultra-high-energy cosmic ray (UHECR), those cosmic radiation with energy greater than 1 EeV (Risse 2023; Boncioli et al. 2015; Bi et al. 2009; Anchordoqui & Soriano 2018), or GRB photons (Xu & Ma 2016; Song & Ma 2025).

[12] Sphere packing has a long history. Johannes Kepler conjectured in 1611 that the densest way to pack equal-sized spheres in three dimensions is equivalent to the familiar way we see in fruit stores, a pyramidal stack. This was only proved in 1998 by Thomas Hales (1958–present); the proof is some 250 pages (Hales 2005)!

Figure 11.1. *A Sunday Afternoon on the Island of La Grande Jatte* painted from 1884 to 1886 by Georges Seurat (1859–1891), is one of the most famous examples of "pointillism", a technique of painting in which images are formed by tiny distinct dots of various colors. The dots are not visible when viewed from afar. Some physicists argue that spacetimes too may be discrete if we zoom in close enough. The term "pointillism" was coined by art critics as a pejorative term to ridicule the approach, much like the term "impressionism" was used in the beginning of that art movement (similarly mannerism, Fauvism, baroque, and Gothic were all derogatory or pejorative terms in the beginning). Curiously, even in physics, "Big Bang" was coined by its fierce critic Fred Hoyle, who instead believed in steady-state cosmology. Many have interpreted this choice of name as somewhat pejorative (whether Hoyle meant it to be so is unclear. See Kragh (2013) for more discussions). Till his death, Hoyle never accepted the Big Bang theory. *Credit:* Helen Birch Bartlett Memorial Collection Source: https://www.artic.edu/artworks/27992/a-sunday-on-la-grande-jatte-1884

A new promising tool to test Lorentz violation (LV) is via gravitational wave, as LV can cause a variety of signatures, including birefringence of gravitational waves (Kostelecky & Mewes 2016). Now would be a good time to briefly discuss gravitational waves, which are ripples of spacetime itself.[13] Again, if you think about it, it is quite remarkable for empty spacetime devoid of anything to vibrate and ripple, as if it is a material!

[13] By the way, do not write "gravity waves" in place of gravitational waves. The former already has its own definition in fluid dynamics! Gravity waves are waves in a fluid medium or at the interface between two media when the force of gravity or buoyancy tries to restore equilibrium.

11.2 Gravitational Waves

Although gravitational wave is a prediction of general relativity, it is perhaps not so surprising that it exists. Gravity and electromagnetism share many similarities (Gomes & Rovelli 2024). For example, they both give rise to an inverse-square law, and more importantly they are both *field theories*. Given that electromagnetism admits electromagnetic waves, one should expect that a gravitational field should admit gravitational waves. Of course, given that there are also *differences* between the two theories, the properties of the waves should be quite different. Incidentally, Maxwell, in his 1865 *The Dynamical Theory of the Electromagnetic Field*, also mentioned his failed attempt to understand gravity:

> "The assumption, therefore, that gravitation arises from the action of the surrounding medium in the way pointed out, leads to the conclusion that every part of this medium possesses, when undisturbed, an enormous intrinsic energy, and that the presence of dense bodies influences the medium so as to diminish this energy wherever there is a resultant attraction. [...] As I am unable to understand in what way a medium can possess such properties, I cannot go any further in this direction in searching for the cause of gravitation. "

The phrase "surrounding medium" we can today replace by "field", but in Maxwell's understanding it would be something akin to the æther. Even in electromagnetism, Maxwell believed that an æther is required to explain electrical force between two charged particles (so the æther was not only proposed to be necessary for light to propagate in, but also for other purposes), with its constituents acting as dielectrics. However, unlike electromagnetism in which there are both attractive and repulsive forces, gravity only attracts, and this led to various difficulties for Maxwell to understand how gravity can work as a "field theory". We shall not get into the details here. In any case, his eventual conclusion, as quoted above, was that the only possibility this can work is if the "field" gives rise to an enormous background energy,[14] and that the presence of massive bodies somehow *reduce* the background energy.

In 1893, Heaviside tried to study gravity by writing down a system of equations that the gravitational field might satisfy, in analogy with electromagnetism. Thus, instead of just one gravitational field, he needs two "components": the gravitoelectric and the gravitomagnetic fields (in analogy to the electric and magnetic fields) (Heaviside 1893, 1893). In hindsight, this theory cannot work because it is a vector field theory, but gravity as described by general relativity is a tensor field theory. (Today, one still speaks of gravitoelectric and gravitomagnetic fields, by splitting the gravitational field into an "electric" and "magnetic" components, but these are different from pre-general relativity construction of Heaviside.)

[14] This sounds strangely similar to the huge vacuum energy that would permeate space if the cosmological constant is interpreted as such.

As we mentioned in Section 5.1, William Clifford (1845–1879) proposed in 1857 that the curvature of space can continuously propagate from one point to another "after the manner of a wave". Today we know that gravitational waves involve the geometry of spacetime, not just space. One often says that gravitational waves are vibrations in spacetime that propagate at the speed of light. To describe gravitational waves mathematically, however, requires a wave equation. We first recall the Einstein field equations

$$R_{ab} - \frac{1}{2}Rg_{ab} = \kappa T_{ab}, \tag{11.1}$$

where $\kappa \equiv 8\pi G/c^4$. We consider a perturbation of the flat metric

$$g_{ab} = \eta_{ab} + h_{ab}, \tag{11.2}$$

where each of the components of the matrix (h_{ab}) has magnitude much smaller than 1, and define

$$\bar{h}_{ab} := h_{ab} - \frac{1}{2}\eta_{ab}h_{cd}\eta^{cd} = h_{ab} - \frac{1}{2}\eta_{ab}h. \tag{11.3}$$

(Here η_{ab} is the Minkowski metric, and $h := h_{ab}\eta^{ab} = h_a^{\ a}$ is the trace of the metric perturbation.) The box operator or the d'Alembertian is defined as $\Box := \eta^{ab}\partial_a\partial_b$. We then have, in the *linearized theory* (namely we neglect all higher-order perturbation terms),

$$\Box\bar{h}_{ab} = -2\kappa T_{ab}. \tag{11.4}$$

Note the similarity between this equation and the Poisson's equation in Newtonian gravity:

$$\nabla^2\Phi = 4\pi G\rho. \tag{11.5}$$

Outside a gravitating source, i.e., in vacuum $T_{ab} = 0$, one obtains a wave equation $\Box\bar{h}_{ab} = 0$, or more explicitly,

$$\left(-\frac{\partial^2}{\partial t^2} + \nabla^2\right)\bar{h}_{ab} = 0. \tag{11.6}$$

It should be emphasized that much like electromagnetism, general relativity has a "gauge freedom" related to the choice of coordinate systems. Usually in the context of gravitational waves, one chooses to work in the so-called "transverse-traceless" (TT) gauge in which $h_{0a} = 0$ and $h_a^{\ a} = 0$. In this gauge, $h_{ab} = \bar{h}_{ab}$. This wave equation admits a "plane wave" solution of the form $h_{ab} = A_{ab}\exp(ik_c x^c)$, where the components of both A_{ab} and the wave vector k^a are assumed to be constant. We can check that the wave equation implies $\eta^{ab}k_a k_b h^{cd} = 0$. For this to be true generically the wave vector must satisfy $k_a k^a = 0$, i.e., it must be null. Thus we see that the gravitational wave travels at the speed of light. In addition, in the TT gauge, the Riemann curvature tensor satisfies

$$R_{i0j0}^{\mathrm{TT}} = -\frac{1}{2}\partial_t^2 h_{ij}^{\mathrm{TT}}. \tag{11.7}$$

Thus we see that curvature perturbation propagates in time. This is the "ripples of spacetime" often referred to in popular science descriptions.

The wave equation discussed above has an additional gauge condition: $\partial_a h^{ab} = 0$, known as the "de Donder gauge" (sometimes "Lorentz gauge", "Hilbert gauge", or "harmonic gauge"). The issues with gauge freedom are quite important as choosing a suitable gauge simplifies our calculations. In principle we can work in any coordinate system, but in general it is very difficult to separate real physics from "coordinate effects". This had in fact, historically speaking, caused a great deal of confusion regarding whether gravitational waves are even real[15]. To appreciate why coordinate systems can be misleading, consider for example the metric tensor (Hill & Nurowski 2025)

$$\begin{aligned} ds^2 &= -\,dt^2 + dx^2 + dy^2 + dz^2 \\ &= -\cos(t-x)[2+\cos(t-x)]dt^2 \\ &\quad + 2\cos(t-x)[1+\cos(t-x)]dtdx - \cos^2(t-x)dx^2. \end{aligned} \tag{11.8}$$

It appears "wave-like", and satisfies the vacuum Einstein equation. However, it also satisfies $R_{abcd} \equiv 0$, so what is going on? Well, this is because the metric just describes the Minkowski spacetime in an unusual coordinate system (obtained by $t \mapsto t + \sin(t-x)$). It is the coordinate system that is "oscillating", not the spacetime itself. Arthur Eddington once computed that some fake gravitational waves that are not physical but merely coordinate effects, can travel at any speed, and famously referred to those "waves" as traveling "at the speed of thought". Good gauge conditions remove the waving of the coordinates. In addition to the gauge confusions, the existence of gravitational waves also requires proving that the waves not only exists in the linearized theory, but the full theory. That is, the wave solution should solve the fully nonlinear Einstein field equations (otherwise it could just be an artifact in the linearized theory). This took quite some time to be sorted out by the community of physicists and mathematicians (Hill & Nurowski 2025).

It can be shown that the components of A_{ab} in the TT gauge can be written as the matrix

$$(A^{\mathrm{TT}})_{ab} = \begin{pmatrix} 0 & 0 & 0 & 0 \\ 0 & A_{xx} & A_{xy} & 0 \\ 0 & A_{xy} & -A_{xx} & 0 \\ 0 & 0 & 0 & 0 \end{pmatrix}. \tag{11.9}$$

The nonzero components are related to the two polarizations of the gravitational waves, and we often write

[15] Einstein's opinion on the existence of gravitational waves changed at least twice; for the interesting history, see Weinstein (2025).

$$A^{ab} = h_+ e_+^{ab} + h_\times e_\times^{ab}, \tag{11.10}$$

where $+$ and x are called the plus and cross polarization states, respectively.

In electromagnetism, the electromagnetic waves at lowest order can only be emitted by a dipole source, whereas in general relativity, the lowest allowed multipole is the quadrupole. The gravitational wave h_{ab} can be solved formally by integrating over the source (i.e., the energy-momentum tensor), just like in electromagnetism the vector potential A^a can be obtained by integrating over the current J^a. Thus we can write

$$h_{ab}(t, \vec{x}) = \frac{4G}{c^4} \int d^3x' \, \frac{T_{ab}(\vec{x}', t - |\vec{x} - \vec{x}'|/c)}{|\vec{x} - \vec{x}'|}. \tag{11.11}$$

We have restored both G and c for clarity. In the weak field regime where the source is at a distance $r = |\vec{x} - \vec{x}'|$ away, we can simplify the expression into

$$h_{ij}^{\mathrm{TT}} = \frac{2G}{c^4 r} \ddot{Q}_{ij}^{\mathrm{TT}}(t - r/c), \tag{11.12}$$

where

$$Q_{ij} := I_{ij} - \frac{1}{3}\delta_{ij}\delta_{mn} I^{mn} \tag{11.13}$$

is the trace-free quadrupole moment, defined in terms of

$$I_{ij} := \int d^3x \quad \rho(t, \vec{x}) x^i x^j. \tag{11.14}$$

Note the use of the *retarded time* $t - r/c$ simply reflects that gravitational waves travel at a finite speed (of light).

The power emitted by gravitational waves (i.e., its luminosity L_{GW}) can be obtained from Q_{ij} as

$$L_{\mathrm{GW}} = \frac{dE_{\mathrm{GW}}}{dt} = \frac{G}{5c^5} \langle \dddot{Q}_{ij} \dddot{Q}^{ij} \rangle, \tag{11.15}$$

where $\langle \cdot \rangle$ indicates an averaging over wavelengths has been performed, but we need not go into the details here. This is the famous *quadrupole formula* of Einstein, developed in 1918. His first investigation into this issue was published in 1916, though the quadrupole formula therein was incorrect. Nevertheless, Einstein essentially noticed that the tiny numerical value of the combination of physical constants, G/c^5, means that gravitational energy emission is practically unnoticeable. Nevertheless, in the same paper, he went on to comment that (Einstein 1916):

> "Due to the electron movement occurring within atoms, atoms must emit not only electromagnetic but also gravitational energy, albeit by a tiny amount. Since this should not actually occur in Nature, it appears that

quantum theory will have to modify not only Maxwell's electrodynamics but also the new theory of gravitation."

This is a rather strange comment because Einstein realized how small the effect of the gravitational wave is, and so it is not so obvious that one should worry about the stability of the atoms (a calculation will need to establish the instability timescale); here we of course have in mind the obsolete idea that atoms are "classical", like a tiny solar system. The lifetime of a circular binary system consisting of masses m_1 and m_2 is (Peters 1964)

$$\tau = \frac{5}{256}\frac{c^5 a_0^4}{G^3 m_1 m_2 (m_1 + m_2)}, \tag{11.16}$$

where a_0 is the initial semimajor axis. (A generic orbit is elliptical, but with the gravitational wave emission, the orbit tends to circularize.) For a hydrogen atom, if we consider the typical scale of radius of 10^{-10} m, electron mass about $9.11 \times 10^{-31} \approx 10^{-30}$ kg and the mass of a proton $1.67 \times 10^{-27} \approx 10^{-27}$ kg, then the lifetime works out to be

$$\tau \approx 10^{115} \quad \text{seconds}, \tag{11.17}$$

which is indeed longer than the age of the Universe, which is of the order of 10^{17} seconds. Thus Einstein need not have worried.

As for electromagnetism, as we discussed in Section 7.4, the classical picture of an electron orbiting the atom would indeed cause the orbit to decay, and the characteristic timescale can be shown to be about 0.1 nanosecond. The solution was that in the quantum theory, we should *not* think of electrons as actually "orbiting" in a classical sense, so that this problem never arises. Thus there is no need to modify Maxwell's equations.[16]

Another useful result is the power emitted, which for two bodies of masses m_1 and m_2 that orbit each other about their common center of mass, is given by

$$\langle P \rangle = -\frac{32}{5}\frac{G^4}{c^5}\frac{m_1^2 m_2^2 (m_1 + m_2)}{a^5}, \tag{11.18}$$

where a denotes their orbital separation. Usually in the gravitational wave physics literature, the *chirp mass* $\mathcal{M}_c$ is used. It is defined as

$$\mathcal{M}_c = \frac{(m_1 m_2)^{3/5}}{(m_1 + m_2)^{1/5}}. \tag{11.19}$$

[16] It is nevertheless interesting to note that Einstein not only already considered the need to modify his own general theory of relativity just a year after its publication (this is not really surprising as his eventual aim was to get to a unified field theory), but also the possibility that Maxwell's theory may need to be modified, by *quantum mechanics*. It is often said that Einstein didn't like quantum mechanics, but this does not mean that he rejected quantum mechanics.

Note the power of the speed of light c^5 in the denominator of the equations for $\langle P \rangle$. This makes the effect rather small and difficult to detect, unless the masses are huge, or if the separation is small. Joseph H. Taylor Jr. (1941–present) and Russell A. Hulse (1950–present) were awarded the Nobel Prize in Physics in 1993 for the *indirect* confirmation of gravitational waves. They studied a binary pulsar, first discovered in July 1974, that are gradually inspiraling towards each other as their mutual orbits shrink due to gravitational wave emission (which carries away energy). The observed cumulative period shift of the Hulse–Taylor binary matched the prediction of general relativity very well.

The *direct* detection of gravitational waves, however, had to wait till September 14 of 2015, when the two detectors of the Laser Interferometer Gravitational-Wave Observatory (LIGO) simultaneously observed a signal from some 1.3 billion years ago that resulted from the collision of two black holes with masses of about 36 and 29 solar masses, respectively, resulting in a final black hole of about 62 solar masses. Thus about three solar masses of energy was radiated in gravitational waves. Three solar masses of pure "nothingness" rippling through spacetime! This event is designated GW150914. Kip Thorne (1940–present), Rainer Weiss (1932–present), and Barry C. Barish (1936–present) received the 2017 Nobel Prize in Physics for this remarkable achievement of LIGO. The story of LIGO is a testament of the virtue of being patient and never giving up. It is also a story of why a continual governmental support for fundamental science research is important. It took almost half a century of efforts[17] to finally detect the feeble signal, once thought to be impossible to detect by Einstein. Such detections are now commonplace and don't make the news anymore.

Modern gravitational wave detectors like LIGO are laser interferometers set up in an L-shape, with each arm a few kilometers long. (The idea utilized by inteferometers is common place in physics experiments. See Box 10 for further discussions.)

Box 10: Experiments That Use Interferometer and Its Idea

The interferometer of LIGO is much larger than the Michelson–Morley interferometer that was designed to detect the æther, however its basic idea is the same. One starts with firing a coherent beam of light (nowadays, laser) through a beam-splitter in the middle of the apparatus, which splits the original beam into two, sending them in two different, perpendicular, directions. The two beams both hit a mirror and are reflected back, and then combined, at a detector. The set-up is such that the two waves of light are perfectly canceled. In the case of LIGO, gravitational waves would stretch and squeeze the apparatus ever so slightly and the cancellation wouldn't be perfect, allowing us to detect gravitational waves. For the æther, it is assumed that Earth and the æther are in relative motion, which means that with respect to us, there is an

[17] LIGO began operating in 2001 with multiple upgrades since then, but the initial research and planning started back in the 1970s. It is worth mentioning that Joseph Weber (1999–2000) designed the resonance bar detector (which consisted of large cylinders made of aluminum, about 2 meters in length and 1 meter in diameter) and claimed that he had detected gravitational waves back in 1969 via the vibration of the cylinders. However, others could not reproduce his results. The general consensus is that he became too convinced and started to see patterns in the noise that weren't there.

"æther wind", the direction and speed of which should differ through the seasons as the Earth orbits the Sun. The presence of the æther should then affect the speed of the two beams of light ever so slightly to produce an interference pattern.

There is another type of interferometer called the Mach–Zehnder interferometer, in which each of the light paths is traversed only once (no reflection). After being split by the beam-splitter, the two perpendicular beams are each reflected by a mirror so that both meet at another beam splitter. The combined light source is then split into two other beams, and two detectors will be present at the end of the two paths. One can use such a set-up as the modern version of the double-slit experiment in quantum mechanics. When a photon encounters the first beam splitter, it would be in a superposition between two possible paths. The second beam splitter is set up in a way that causes these paths to interfere constructively at one of the detectors so that it would detect the photon with probability 1; and the other 0 due to destructive interference. In other words, one of the detectors will never detect anything if the photon behaves as a wave. We have mentioned in Section 7.4 that if we know which path the photon takes, there would be no interference pattern (i.e., the photon behaves like a particle instead of a wave). In the interferometer set-up, we can simply block off one of the paths after the photon went through the first beam splitter. This is also equivalent to having no second beam splitter. Many other quantum experiments are variations of the Mach–Zehnder interferometer. A famous example is Wheeler's "Delayed choice experiment". Suppose we don't have the second beam splitter. Let a photon pass through the first beam splitter. Without the second beam splitter, the path taken to either detector is known, so we expect the photon to behave like a particle, i.e., it should have taken either one path or the other, but not both. However, *after* it has passed the first beam splitter and *before* it hits the detector, we can then insert the second beam splitter. The final result is an interference pattern, as if the photon suddenly realizes that it needs to go into a superposition.

The idea of detecting the æther wind is also used in some modern experiments to search for dark matter, although the detector is not an interferometer. The DAMA/LIBRA experiment in Italy utilizes a matrix of NaI(Tl) scintillation detectors to detect dark matter particles and has been operating for over 20 years. The idea is that dark matter particles also interact with ordinary atoms in the detectors and emit flashes of light, albeit very rarely (they are the WIMPS—"weakly interacting massive particles".) Suppose that the Milky Way Galaxy has a dark matter halo, as the Earth orbits the Sun, which itself orbits the galaxy, the amount of dark matter that Earth passes through would be different in different seasons. When the Earth and the Sun move in the same relative direction, the detectors should detect a higher number of interactions, and conversely, six months later, when the Earth and the Sun move in the opposite direction, there should be fewer detections. In other words, there should be an annual modulation of the signals, like a sine wave. This was indeed what the team reported. However, other experiments could not replicate the results, and the community remains skeptical of the claim.

A similar annual modulation signal was also claimed to have detected the æther. Not everyone immediately stopped working on the æther after the publication of special relativity. In particular, Dayton Miller summarized the search for the æther in his 1933 paper *The Ether-Drift Experiment and the Determination of the Absolute Motion of the Earth.* He repeated the Michaelson–Morley experiment at the Mount Wilson laboratory with improved technology, and claimed that he had observed the annual modulation signal. However, modern standard statistical analysis (Roberts 2025) (using concepts and techniques that were not available to Miller) indicated that the "detection" is consistent within the margin of errors of having no æther.

Gravitational waves give rise to a *strain* $h = \Delta L/L$, and astrophysical signal gives rise to h that is extremely small, at the order of 10^{-21} or 10^{-22}. Even with $L \sim O(1)$ km of a typical ground-based detector arm, this means that one would have to measure a change of $\Delta L \sim 10^{-19}$ m, which is much smaller than the diameter of a hydrogen atom $\sim 10^{-10}$ m! This is an over-simplification however, since in the LIGO set-up, the laser beam is reflected over hundreds of time in a Fabry–Pérot cavity (the photons bounce up and down between the mirrors in the cavity for about a millisecond), so that its effective length is about 1200 km instead of 4 km. Still, one still has to be able to measure $\Delta L \sim 10^{-16}$ m, about a tenth the width of a proton. (This is *still* an over-simplification, experimentalists are extremely skilful in how to devise methods to improve sensitivities, so the actual LIGO's sensitivity turns out to be about two orders of magnitude better than this.) Because the signals from gravitational wave events are so weak, this requires that on the experimental side, environmental noises are reduced as much as possible. Interestingly, this has made it possible to detect the effect of *quantum fluctuation* from the laser on the LIGO's mirrors. The mirrors weight 40 kg, and the quantum fluctuation can move it by 10^{-20} m, *and this was successfully measured* (Chu 2020)! The small signal-to-noise ratio also means that detection requires the theorists to first develop the waveforms from theoretical predictions, which were then used to "sieve out" the weak signal. The successful detection of gravitational waves is a triumph of human engineering. As Harry Collins (1943–present), a sociologist of science, wrote in his book *Gravity's Shadow: The Search for Gravitational Waves* (Collins 2004):

> "Gravitational wave detection is about seeing the biggest things that ever happen—the collisions, explosions, and quakings of stars and black holes —by measuring the smallest changes that have ever been measured: the effect of gravitational waves on experimental apparatus mounted on the surface of the earth."

Every time we have a new window to observe the Universe, there are new discoveries to be made. Among other promises, gravitational wave observation can join forces with other more established channels like gamma-ray and X-ray observations, to detect collisions of compact stars and black holes. The power of multi-messenger astronomy was successfully demonstrated in another landmark event GW170817, observed by the LIGO and Virgo detectors on 17 August 2017. It originated from the shell elliptical galaxy NGC 4993, located about 140 million light-years away from Earth. The use of the Virgo detector, located near Pisa in Italy, allows a better localization of the direction of the source (the first joint detection of LIGO and Virgo was the event GW170814, shown in Figure 11.2), thus allowing other astronomical telescopes to follow up and pin-point the possible source. As luck would have it, this time the stellar collision involved two neutron stars and not black holes, thus there was also an abundance of electromagnetic signals, which were subsequently observed by over 70 observatories on 7 continents (including the Antarctic Survey Telescopes) as well as in space (such as the Fermi and INTEGRAL telescopes, both specialize in gamma-rays

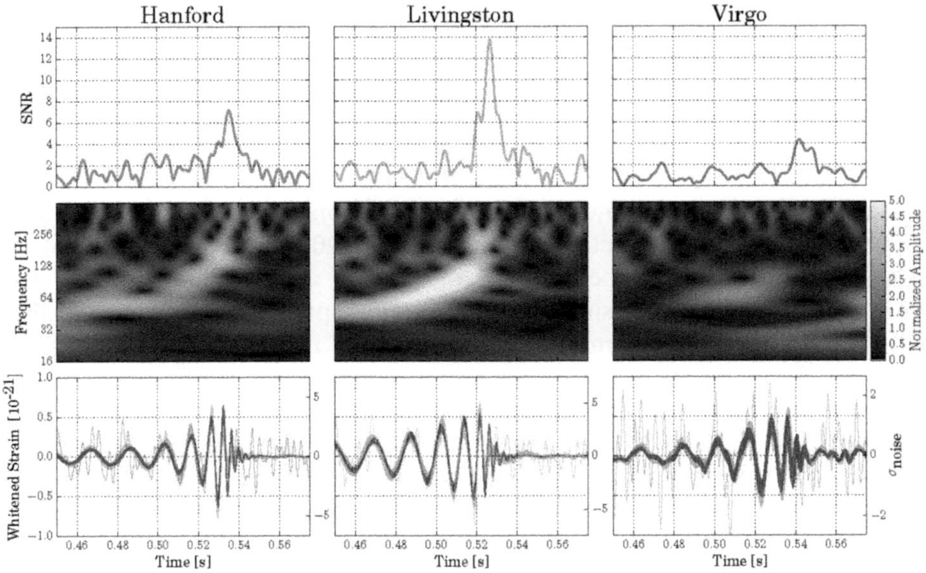

Figure 11.2. This graphic shows the data from each of the three observatories: LIGO Hanford, LIGO Livingston, and Virgo, for the signal from GW170814, caused by the merger of two black holes (which are about 30.5 and 25.3 solar masses, respectively, producing a black hole of about 53.2 solar masses), and the first event seen simultaneously with the Virgo detector and the two LIGO detectors. The top graph shows the signal-to-noise ratio. The middle graph shows the "chirp" or frequency shift across time. The bottom graph shows the wave form. *Credit:* LIGO/Caltech/MIT/LSC.

observation; INTEGRAL, short for the INTErnational Gamma-Ray Astrophysics Laboratory, has since retired). This event provides strong evidence that at least some short gamma-ray bursts (GRBs) are caused by neutron star collisions. There is also evidence that supports the astrophysical theory that the merger of neutron stars is the primary source of the so-called "*r* process" (rapid neutron capture nucleosynthesis) that produces heavy elements like gold and platinum (instead of supernovae). Perhaps with more future events of neutron star mergers, we will be able to figure out what the interior of neutron stars looks like. There are many theoretical and phenomenological models for the interior "equation of states", but so far there is no consensus.

As mentioned in the previous section, gravitational wave signals can be used to test Lorentz invariance. Similarly, it can be used to test alternative theories in which the speed of gravitational waves is different from the speed of light. GW170817 has constrained the difference between these two speeds to between 3×10^{-15} and 7×10^{-16} times the speed of light (Virgo Collaboration 2017). Gravitational wave physics is expected to make more discoveries in the coming decades, with new detectors on Earth and in space already planned. In addition to the aforementioned LIGO detectors, LIGO-India is targeted to be operational by 2030.

At the time of writing, the ESA/NASA joint space-based laser interferometer called LISA (Laser Interferometer Space Antenna) has been approved and is planned to be launched in 2035. Other proposed space-based detectors are the Chinese

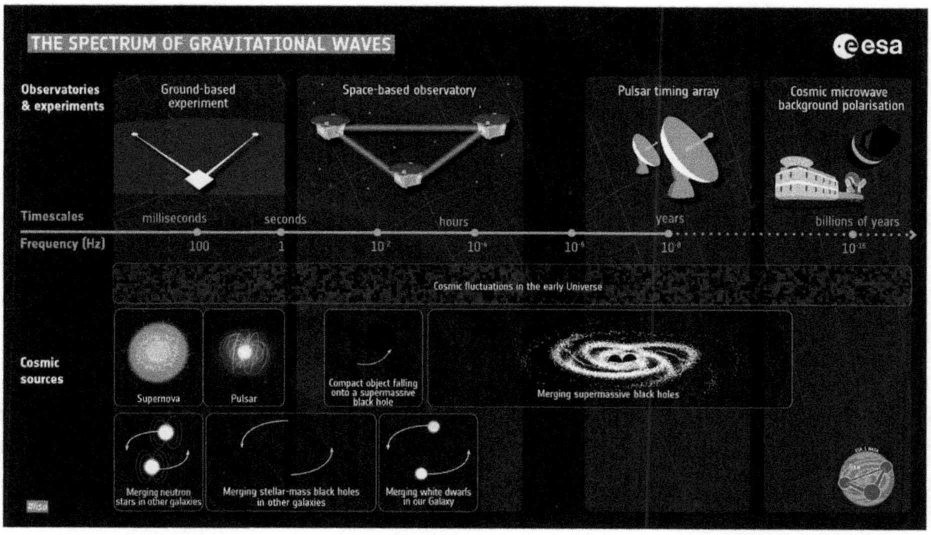

Figure 11.3. The spectrum of gravitational waves and their sources, as well as the corresponding detectors that can probe those frequencies. *Credit:* ESA.

TianQin[18] (天琴) and *Taiji* (太极). Taiji is similar to LISA in design (but with longer arm separation), both of which consist of three spacecraft that are linked up with laser, and the entire configuration will be orbiting the Sun. TianQin, however, will have the three spacecraft be positioned in an equilateral triangular formation around the Earth and thus orbits our own planet. Japan also has a planned program for a space-based detector called DECi-hertz Interferometer Gravitational wave Observatory (DECIGO). Space-based laser interferometers would be sensitive to gravitational waves in the frequency range 10^{-4} to 10^{-1} Hz, instead of the 10 to 10^3 Hz band of the ground-based detectors. The space-based detectors would therefore allow us to probe different kind of physics from different systems, such as the so-called "extreme mass ratio inspiral" (EMRI)—in which a relatively light object spirals towards a massive one, with their mass ratio being about at least 10^5. The smaller object can essentially be a probe that "maps" the spacetime geometry of the massive body in unprecedented details. Future ground-based detectors have also been planned, such as the European *Einstein Telescope* whose arm length would be 10 km long. Like the Japanese Kamioka Gravitational Wave Detector (KAGRA),[19] it will be located underground to reduce "gravity gradient noise" caused by nearby moving objects. Different astrophysical and cosmological phenomena produce gravitational waves at different frequencies, and they would require different detectors to observe. This is summarized in Figure 11.3.

---

[18] It means "heavenly harp".

[19] KAGRA also utilizes cryogenic mirrors to further reduce the thermal noise. "KAGRA" comes from the combination of **Ka**mioka and **gra**vity. It is also meant to remind one of the Japanese traditional sacred Shinto dance called "Kagura" (神楽), literally "entertainment for the gods".

Hopefully future gravitational wave signals can help us to answer a few fundamental questions, such as whether black hole solutions are indeed as predicted by general relativity (or some other modified gravity), whether there is any signature of horizon structures that could indicate quantum gravity effects, and how gravity really behaves in a highly dynamical strong gravity regime. Perhaps it can even reveal the presence of extra compactified dimensions and other exotic physics. Only time will tell.

11.3 What Is Spacetime?

Since spacetime curvature is tied to gravity in general relativity, asking "what is spacetime" is also to some extend, asking "what is gravity". To the best of our understanding, there are four fundamental forces in our Universe—the weak and strong nuclear forces, electromagnetic force, and gravity. Historically gravity is the first force to be studied scientifically. Ironically it is, in some sense, the least understood force of all. In popular science books it is often written that gravity is the weakest force because the ratio of the magnitude of gravitational force

$$F_{\mathrm{G}} = \frac{GMm}{r^2}, \tag{11.20}$$

to that of the electrostatic force

$$F_{\mathrm{EM}} = \frac{1}{4\pi\varepsilon_0}\frac{Qq}{r^2}, \tag{11.21}$$

between two electrons is so minute:[20] $F_{\mathrm{G}}/F_{\mathrm{EM}} = \mathcal{O}(10^{-42})$. Such a statement, however, is not quite a precise characterization of the weakness of gravity, since a hypothetical particle with Planck mass $m_{\mathrm{P}} := \sqrt{\hbar c/G}$ and Planck charge $q_{\mathrm{P}} := \sqrt{4\pi\varepsilon_0 \hbar c}$ would have $F_{\mathrm{G}}/F_{\mathrm{EM}} = 1$. The real mystery is why the charge of the electron is so much higher than its mass? Or rather, why the aforementioned hypothetical particle does not exist? Of course, charge and mass have different physical dimensions so they cannot be directly compared. The precise statement is that the *dimensionless ratio* $m_e/m_{\mathrm{P}} = \mathcal{O}(10^{-22}) \ll q_e/q_{\mathrm{P}} = \mathcal{O}(10^{-1})$. This observation —along with the fact that gravity is also significantly weaker when compared to the strong and weak nuclear forces—is known as the *hierarchy problem* (Randall & Sundrum 1999a, 1999b), which remains unsolved. There are of course many attempted explanations. For example, in string theory, it was proposed that gravity is weak because our universe is a brane embedded in a five-dimensional bulk spacetime, and gravity, unlike other forces that are confined to propagate on our brane, can propagate through the higher dimensional ambient space, thus appears to us to be much weaker than it actually is. Note that we are speaking of gravity as a

[20] The number 42 reminds me of the answer to "The Ultimate Question of Life, the Universe, and Everything" in Douglas Adams' *The Hitchhiker's Guide to the Galaxy*.

force, which from the perspective of general relativity, is instead described in terms of geometry. So, what *is* gravity?

Most general relativists would prefer gravity as geometry. A particle physicist, on the other hand, is used to thinking in terms of forces. In quantum field theory, it is the field that is fundamental, as particles are just excitations of the field. Each field has an associated "spin". For example, a scalar field has 0 spin, whereas a vector field has spin equals to 1. An example of a spin-1 particle is the photon, which mediates electromagnetic interactions. This has led to the interpretation that the metric is just a spin-2 field, whose perturbation gives rise to particles that mediate gravitational interaction, the "gravitons". For the spin-2 tensor theory to be mathematically consistent, if it is not a "free theory" (i.e., no matter coupling), then it must require infinitely many correction terms, by including "gravitational energy-momentum" into the field equations (essentially, the field is coupled to its own energy-momentum, a process which iterates indefinitely). This is a very difficult program, no less since the concept of gravitational energy is a pseudo-tensor.[21] Still, if everything is done correctly and carefully, one should be able to obtain general relativity (perhaps even including higher-order terms) in all its glory. More specifically, as we have seen, in linearized gravity, the metric tensor has the component $g_{ab} = \eta_{ab} + h_{ab}$, where η_{ab} is the background Minkowski metric, and h_{ab} is the perturbation. Roughly speaking, the quantization of h_{ab} gives rise to the graviton. The graviton self-interaction then back-reacts on the geometry and eventually makes it curved. Presumably if the history of physics were to be different, with quantum field theory developed way before general relativity, perhaps someone will work out a theory of gravity following this path, and may even work out general relativity as an "emergent theory" in the sense that curvatures like the Ricci tensor arise from nonlinear interaction among the gravitons (this has been attempted, see Straumann (2025) for a review). This already sounds a lot more complicated than just having a theory based on a bona fide curved spacetime from the beginning.

In addition, treating gravity as just another field is unsatisfying in my opinion, the reason is nicely explained already by Buchdahl (1981):

> "I have spoken of 'energy', for instance. The energy of what? Of the field perhaps? Field?—We have no field in the sense in which one has a Maxwell field. Whenever I have used the term 'field' [in the context of the metric] I have done so as a matter of mere verbal convenience. [...] The classical fields—the electrostatic field, for example—in the first instance had, so to speak, a subjunctive existence. Let *P* be a particle satisfying all the criteria of being free except in as far as it carries an electric charge. Then if *P* were placed at some point it would be subject to a force depending on the value of the field intensity there. In particular, when the latter is zero there is no force. [...] the 'true' fields subjunctively quantify

[21] This does not mean that the concept is not useful. In the gravitational wave community, one often does talk about the energy carried by gravitational waves, and the energy-momentum pseudo-tensor of the gravitational waves.

> the extent to which a given particle P is not free, granted that P would be free if any charges it may carry were neutralized [...] this field [the metric], unlike the 'true' fields, cannot be absent, cannot be zero."

While the other three forces have been successful quantized, the quantization of gravity has been met with serious difficulties. While this does not necessarily mean that gravity cannot be quantized,[22] it *might* suggest that gravity is quite different from the other three forces.

It should be emphasized that physics can never tell us what Nature truly *is*, all it can do is to provide a good theory that describes the phenomena we observe, and if the theory is good enough, to provide predictions that can later on be tested and verified. An apple falling is an observation, but whether gravity is fundamentally a force, or spacetime geometry, we cannot know "for sure". In the so-called "teleparallel equivalent of general relativity", the spacetime geometry is *flat*; there are however nonzero torsion tensors.[23] This allows a "torsion scalar" to be constructed, which differs from the original scalar curvature of general relativity by a total divergence term. This means that at the level of the action, the torsion scalar can replace the curvature scalar and the theory is equivalent to general relativity despite its geometry being entirely different (its description of gravity is also different since test mass does not follow the geodesics of the new torsion-full connection). What gravity "really is" is therefore not a scientific question, at best we can only ask whether a specific description of gravity is a good one. So far general relativity has passed all the tests that we can throw at it, and so we often talk of gravity *as* spacetime curvature, as such a description is very useful. This might well change in the future if and when we have a new, better, theory of gravity. When there are multiple theories which are equivalent, one may wish to choose the one that is "more natural" on some philosophical grounds.

From our discussions of black holes, we have also seen that certain spacetime geometries can exhibit thermodynamical properties like temperature. Consider a Schwarzschild black hole that undergoes Hawking evaporation and thereby slowly loses its mass. The Schwarzschild black hole solution is a vacuum solution, and so the "mass" is the property of spacetime itself—a manifestation of spacetime curvature. As the black hole evaporates, its mass decreases, and *matter* is produced. Thus Hawking evaporation showcases the remarkable property that as general

---

[22] Some physicists have considered such a possibility. For example, if gravity is not fundamental but only emergent, it would not make sense to quantize the theory, just like it doesn't make sense to quantize hydrodynamics. There are also other points of view that entertain the possibility that gravity is fundamentally classical (Oppenheim 2023, 2023).

[23] Because general relativity is a theory that utilizes curvature, this has caused some physicists to think that torsion can be viewed as "just another tensor field" in the spacetime. This is rather unnatural, especially from the point of view of initial value problem (Nester & Wang 2012). A geometric quantity should be treated with respect—not every quantity with tensor indices is "just a field". Treating even the metric tensor as just a field naturally gives rise to various modifications of gravity that involve more than one metric, whether dynamical or not.

relativity tells us, spacetime geometry and matter affect each other. The famous Wheeler's slogan that "matter tells spacetime how to curve, spacetime tells matter how to move" is not radical enough; sometimes spacetime can *create* matter (from the quantum field that permeates it). So perhaps, matter and spacetime is more intimately related than we thought? This is a possibility that many have considered, even before the conception of general relativity. For example, the mathematician William Clifford once wrote in *The Common Sense of the Exact Sciences* in 1885 the idea that matter is just structure *of* space instead of merely being *in* space:

> "We may conceive our space to have everywhere a nearly uniform curvature, but that slight variations of the curvature may occur from point to point, and themselves vary with the time. These variations of the curvature with the time may produce effects which we not unnaturally attribute to physical causes independent of the geometry of our space. We might even go so far as to assign to this variation of the curvature of space 'what really happens in that phenomenon which we term the motion of matter.' "

Like Riemann, he had the idea that space can have non-Euclidean geometries, and even foresaw that curvature can propagate. Neither Clifford nor Riemann had anticipated that *spacetime* as a whole, not just space, is crucial to understanding gravity. Perhaps if there is indeed a theory that can unify gravity with all other forces and matter, we can finally understand what spacetime really is. If matter too is just a manifestation of spacetime geometry, suitably generalized, it would be the ultimate realization of Johannes Kepler's philosophy that "*At ubi materia, ibi Geometria*" (where there is matter, there is geometry).

As we already mentioned, there is another school of thought that says we should not attempt to quantize gravity because gravity (and spacetime) is emergent and "not fundamental", in the sense that hydrodynamics is emergent and quantizing hydrodynamics would not make much sense. How might gravity be emergent? In recent years it has been suggested that quantum entanglement gives rise to connectedness of spacetime, thus if region A and region B are not entangled at the quantum level, then they are not connected at the macroscopic level (i.e., geometrically or even topologically). A specific example is the so-called "ER=EPR" proposal (Maldacena & Susskind 2013), in which two entangled particles are connected by a wormhole (Einstein–Rosen bridge; hence ER. "EPR" stands for the famous Einstein–Podolsky–Rosen (EPR) thought experiment in quantum mechanics). This had led to the idea that perhaps fundamentally quantum entanglement is more fundamental, that somehow they gave rise to spacetime. However, in quantum mechanics a wavefunction is a function of space and time, so how can one think of quantum entanglement before first having spacetime? This seems like a chicken-and-egg conundrum! We can of course refer to quantum states that live in an abstract Hilbert space, but are we to think of said Hilbert space in the Platonic sense, living "somewhere" outside the physical spacetime? If so, how does an abstract mathematical quantity give rise to *physical* spacetime[24]? Perhaps it is more plausible that spacetime and quantum entanglement are intertwined in some way, or perhaps they are even "equivalent" (note the *equal sign* in "ER=EPR"), but then neither one is *more* fundamental than the other.

11.4 Exploration

Instead of exercises, this chapter probably calls for more advanced activities. Given that we have discussed the number of spacetime dimensions, the readers may want to look into the effects of changing the number of dimensions in various contexts. For example, as already mentioned, in spacetime dimensions above 4, quantum mechanics cannot balance gravitational collapse and therefore compact stars like white dwarfs cannot exist (Chavanis 2007). On the other hand, in addition to black holes, we can have other "black objects" such as a black ring (Emparan & Reall 2006) in higher dimensions, so black hole physics is richer.

There is also a difference in odd and even dimensions when some physics are concerned. For example, consider the Huygen's principle for waves, which is only valid if the number of spatial dimensions is odd. Essentially, a wave in odd spatial dimensions has a characteristic speed. For light this would be the speed of light, so it travels on a light cone. However, in even spatial dimensions, the wave travels also at all speeds within the light cone. This means that instead of a flash of light, we will continue to observe an "afterglow" long after the light has been turned off. The afterglow lasts indefinitely, though its intensity would decrease (Balazs 1955). As another example, readers who are well-versed in quantum field theory can consider a uniformly accelerated monopole detector in the Minkowski vacuum. The response of the detector is also different in odd and even dimensions (Sriramkumar 2002, 2003).

References

Adler, G., Jung, C. G., & Aniela, J. 1976, Letters of C. G. Jung Vol. 2 (London: Routledge) 176

Anchordoqui, L. A., & Soriano, J. F. 2018, PhRvD, 97, 043010

Boncioli, D., di Matteo, A., Salamida, F., et al. 2015, 34th Int. Cosmic Ray Conf. (ICRC 2015) (Trieste: PoS) 521

Bi, X.-J., Cao, Z., Li, Y., & Yuan, Q. 2009, PhRvD, 79, 083015

Buchdahl, H. A. 1981, Seventeen Simple Lectures on General Relativity Theory (New York: Wiley)

Balazs, N. L. 1955, PPSA, 68, 521

Chavanis, P. H. 2007, PhRvD, 76, 023004

Chu, J. 2020, MIT News, https://news.mit.edu/2020/quantum-fluctuations-jiggle-objects-0701

Collins, H. 2004, Gravity's Shadow: The Search for Gravitational Waves (Chicago, IL: Univ. Chicago Press)

Einstein, A. 1916, SPAW, 1916, 688

Emparan, R., & Reall, H. S. 2006, CQGra, 23, R169

[24] Recall from Chapter 1 the old problem regarding how the spiritual soul can control a physical body in mind-body dualism!

Gomes, H., & Rovelli, C. 2024, PP, 2, 2
Hales, T. C. 2005, AnMat, 162, 1065
Heaviside, O. 1893, Electr, 31, 281
Heaviside, O. 1893, Electr, 31, 359
Hill, C. D., & Nurowski, P. 2016, [arXiv:1608.08673 [physics.hist-ph]] 2016arXiv160808673D
Kostelecky, A., & Mewes, M. 2016, PhLB, 757, 510
Kragh, H. 2013, [arXiv:1301.0219 [physics.hist-ph]] 2013arXiv1301.0219K/abstract
LIGO Scientific Collaboration, Virgo Collaboration, Fermi Gamma-Ray Burst Monitor, INTEGRAL 2017, ApJL, 848, L13
Maldacena, J., & Susskind, L. 2013, ForPh, 61, 781
Nester, J. M., & Wang, C.-H. 2012, IJMP, 07, 158
Oppenheim, J. 2023, PhRvX, 13, 041040
Oppenheim, J. 2023, IJMPD, 32, 2342024
Peters, P. C. 1964, PhRv, 136, B1224
Rovelli, C. 1996, IJTP, 35, 1637
Rovelli, C. 2022, The Oxford Handbook of the History of Quantum Interpretations, ed. O. Freire, Jr. G. Bacciagaluppi, O. Darrigol, et al. (Oxford: Oxford Univ. Press)
Risse, M. 2023, Proc. of the Ninth Meeting on CPT and Lorentz Symmetry (Singapore: World Scientific) 44
Roberts, T. J. 2006, [arXiv:physics/0608238 [physics.class-ph]] 2006physics...8238R
Randall, L., & Sundrum, R. 1999a, PhRvL, 83, 3370
Randall, L., & Sundrum, R. 1999b, PhRvL, 83, 4690
Scargill, J. H. C. 2020, PhRvR, 2, 013217
Song, H., & Ma, B.-Q. 2025, PDU, 47, 101808
Straumann, N. 2000, [arXiv:astro-ph/0006423] 2000astro.ph..6423S
Sriramkumar, L. 2002, MPLA, 17, 1059
Sriramkumar, L. 2003, GReGr, 35, 1699
Whitrow, G. J. 1955, BJPS, 6, 21
Weinstein, G. 2016, [arXiv:1602.04040 [physics.hist-ph]] 2016arXiv160204040W
Xu, H., & Ma, B.-Q. 2016, APh, 82, 72
Zhang, H., Rich, P. D., Lee, A. K., & Sharpe, T. O. 2023, NN, 26, 131

AAS | IOP Astronomy

Understanding the Universe
A conversational introduction to cosmology
Yen Chin Ong

Chapter 12

Modifying Physics: Case Studies

> *"Pull one hair and the whole body shifts."* (牵一发而动全身) —Chinese idiom.

This chapter is a relatively technical one. It is also very speculative and could be wrong. The aim is to expose students to some examples of topics that are at the forefront of research. Specifically, we will look at how modifications to the Bekenstein–Hawking entropy and the generalized version of the Heisenberg uncertainty principle can lead to many subtleties. Just like modified gravity, any change to most laws of physics, however small, needs to be handled with care, otherwise unwanted problems can quickly arise. The path to progress requires the courage to forge new paths, but anyone doing so also needs to examine closely one's path lest one steps on a venomous snake.

As we are approaching the end of this book, we will now look at some attempts to modify physical laws (though we have encountered some examples very briefly in previous chapters, such as the various nonlinear electrodynamics theories, including Mie's attempt at generalizing electromagnetism in Section 7.3; and modified gravity theories in Section 6.3). The content in this chapter is therefore rather speculative (and heavily biased since they are topics of my own research interests), but the reason for including it is to show how some of the current theoretical physics research is done, which would be especially useful for beginning graduate students. They will see that ideas are slowly developed and debated, instead of appearing fully formed. There will be mistakes and disagreements, as some concepts are sometimes not even well-defined. Science in the making is often messy, and that's part of the fun of doing science. In his "four

doi:10.1088/2514-3433/ade963ch12

golden lessons",[1] particle physicist Steven Weinberg wrote: "My advice is to go for the messes—that's where the action is." Indeed, we have seen throughout the book that modifying known physics has always been one of the approaches physicists employed when facing anomalies that the existing theories or models could not explain. We will consider two different types of modification: (1) generalizing the Bekenstein–Hawking entropy of black holes, and (2) generalizing the Heisenberg uncertainty principle to include the effects of gravity. The main lesson one should draw from this chapter is that it is not easy to modify any part of physics without at the same time affecting it in some unexpected and unwanted ways.

12.1 From Modified Gravity to Modified Entropies

We have previously discussed modified gravity in Section 6.3, which aims to explain dark energy and/or dark matter as just a manifestation of gravity. We have also discussed black holes in Chapter 9 and have seen that they have properties that look very similar to thermodynamics, and in fact generally believed to be bona fide thermodynamical objects. Still, there are some differences between black holes and ordinary thermodynamic systems as also discussed in the aforementioned section. Still, instead of claiming that perhaps black holes are so special that they are not really thermodynamical *despite* the resemblances, it makes more sense that thermodynamics should be suitably generalized to include black holes *because of* the resemblances. For example, in ordinary systems, entropy is extensive as a function of the energy. In particular, the entropy of a box of gas can be written as the function of three extensive properties—internal energy E, volume V, and number of particles N: $S(E, V, N)$. Then for any scaling factor λ, we have $S(\lambda E, \lambda V, \lambda N) = \lambda S(E, V, N)$. The Bekenstein–Hawking entropy is, however, $S \propto A \propto M^2$, in which A is the horizon area and M is the mass of the black hole, identified as its internal energy. Thus, S is not extensive in this case. This opens the door to consider other non-extensive statistics that can be applied to black holes, which result in generalization of the Bekenstein–Hawking entropy. It should be emphasized that there are ongoing debates as to whether this is well-motivated.

[1] Weinberg also suggested in this article that one should learn the history of science, or at least the history of the branch of science that we specialized in, because

> "The least important reason for this is that the history may actually be of some use to you in your own scientific work. [...] More importantly, the history of science can make your work seem more worthwhile to you. As a scientist, you're probably not going to get rich. Your friends and relatives probably won't understand what you're doing. And if you work in a field like elementary particle physics, you won't even have the satisfaction of doing something that is immediately useful. But you can get great satisfaction by recognizing that your work in science is a part of history." (Weinberg 2003)

This was also one of my views expressed in Chapter 1.

However, my opinion is that even if they are not viable, we can learn a lot from studying what happens when such a generalization is made. This is the same philosophy we can apply to modified gravity; from figuring out how difficult it is to consistently modify general relativity we may come to understand the theory better (i.e., if the theory is so rigid, why?)

There are many ways to generalize the Bekenstein–Hawking entropy, for example Barrow entropy (Barrow 2020), Tsallis(–Cirto) entropy (Tsallis 1988; Tsallis & Cirto 2013), Kaniadakis entropy (Kaniadakis 2002, 2005; Drepanou et al. 2022), and Sharma–Mittal entropy (Jahromi et al. 2018). We will discuss Barrow entropy in the next section. Here we can write down the Tsallis modification as an example (Tsallis & Cirto 2013):

$$S = \frac{A_0}{4}\left(\frac{A}{A_0}\right)^{\delta} = \frac{A_0^{1-\delta}}{4}A^{\delta}, \tag{12.1}$$

where A_0 is a constant and δ a parameter of the theory. Note that this is itself inspired by the original Tsallis entropy formulation. The original Tsallis entropy is a non-extensive extension of the Boltzmann–Gibbs entropy in the following sense: it is given by

$$S_q(p_i) := \frac{k}{q-1}\left(1 - \sum_i p_i^q\right) = k\sum_i p_i \ln_q\left(\frac{1}{p_i}\right), \qquad q \neq 1. \tag{12.2}$$

where

$$\ln_q x := \frac{x^{1-q} - 1}{1 - q}, \tag{12.3}$$

and $q \in \mathbb{R}$ with the constraint $\sum_i p_i = 1$. The limit $q \to 1$ recovers the Boltzmann–Gibbs entropy form

$$\lim_{q\to 1} = -k\sum_i p_i \ln p_i, \tag{12.4}$$

and we see that in this limit k can be identified as the Boltzmann constant k_B. It is often assumed that $k \equiv k_B$ even when the limit is not taken. This entropy function is not additive since

$$\frac{S_q(A+B)}{k} = \frac{S_q(A)}{k} + \frac{S_q(B)}{k} + (1-q)\frac{S_q(A)}{k}\frac{S_q(B)}{k}. \tag{12.5}$$

For the continuous probability distribution, we have

$$S_q(p) := \frac{k}{q-1}\left(1 - \int p(x)^q dx\right). \tag{12.6}$$

This generalization of the Boltzmann–Gibbs statistics was introduced by Constantino Tsallis in 1988, but it is identical in form to the *Havrda–Charvát structural* α*-entropy* introduced in 1967 within information theory (Havrda & Charvát 1967), by showing that if some axioms are satisfied, then there is a unique extension of said form. The Tsallis entropy has been applied to understand various physical systems, from galaxies to spin-glass relaxation. Tsallis and Cirto further generalized the idea to include another parameter δ, so that we have $S_{q,\delta}$, defined by

$$S_{q,\delta} := k\sum_i p_i \left[\ln_q\left(\frac{1}{p_i}\right)\right]^\delta. \tag{12.7}$$

For black holes, $S_\delta \equiv S(1, \delta)$ is used. Note that the entropy is also non-additive as it satisfies

$$\frac{S_\delta(A+B)}{k} = \left[\left(\frac{S_\delta(A)}{k}\right)^{\frac{1}{\delta}} + \left(\frac{S_\delta(B)}{k}\right)^{\frac{1}{\delta}}\right]^\delta. \tag{12.8}$$

Note that the Boltzmann–Gibbs entropy is extensive and additive, but the Tsallis one is not. However, the Bekenstein–Hawking entropy is already not extensive before the Tsallis extension.

Another example is the Rényi entropy (Rényi 1961). For black holes, this takes the form (Biró & Czinner 2013; Czinner & Iguchi 2016)

$$S_R := \frac{1}{\alpha}\ln\left[1 + \alpha\left(\frac{A}{4G}\right)\right], \quad \alpha > 0, \tag{12.9}$$

for which the limit $\alpha \to 0$ recovers Bekenstein–Hawking entropy.

12.2 Barrow Entropy and Cosmological Implications

This section is mainly based on my papers (Di Gennaro & Ong 2022; Xia & Ong 2024). In 2020, John Barrow proposed that due to quantum gravity corrections, black holes might exist with extremely wrinkled surfaces such that the event horizon becomes “fractalized”.[2] Fractals exhibit a property known as *self-similarity*, such that no matter how much we zoom in, the structure remains the same. An example is the “Koch snowflake”:[3] start with a equilateral triangle, then divide each side into three equal parts, using the middle part as the base of a new equilateral triangle, thus adding a new triangle to every side of the original triangle. The process is then repeated ad infinitum. For more discussions and detailed calculations regarding the Koch snowflake and fractal dimension, see Box 11 below.

For the black hole case, Barrow considered the surface to be a “sphereflake”, a three-dimensional analogy of the Koch snowflake using layers of spheres. Care has

[2] Actually Rafael Sorkin already considered possible fractal structures in 1996 (Sorkin 1996), but he did not consider the entropy modification.

[3] See Demichel (2025) for the interesting history of the Koch snowflake.

Box 11: Koch Snowflake and the Fractal Dimension

Let us take the Koch snowflake as an example. We denote the number of sides at each iteration by N_n, and L_n the length of each side (sometimes called the *scaling factor*). Let the area be denoted by A_n and its perimeter by P_n. The original triangle is the 0th iteration, whose each side length is scaled to be 1. It can be shown that:

$$N_n = 3 \cdot 4^n; \qquad L_n = \left(\frac{1}{3}\right)^n; \qquad P_n = N_n L_n = 3\left(\frac{4}{3}\right)^n, \tag{12.10}$$

The fractal dimension, d_f, which can be generalized into a more measure-theoretic notion called the *Hausdorff dimension* (introduced by Felix Hausdorff (1868–1942)) for more complicated fractals, can be calculated as

$$d_f := -\lim_{n\to\infty} \frac{\ln N_n}{\ln L_n} = \frac{\ln 4}{\ln 3} \approx 1.2619, \tag{12.11}$$

which is indeed in between the dimension of a regular curve ($d = 1$) and the dimension of the embedding plane ($d = 2$).

The area satisfies the recurrent relation:

$$A_n = A_{n-1} + \frac{1}{4} N_n L_n^2 A_0 \tag{12.12}$$

$$= A_{n-1} + \frac{1}{3}\left(\frac{4}{9}\right)^{n-1} A_0. \tag{12.13}$$

We can iterate the recurrent relation to write:

$$A_n = A_{n-1} + \frac{1}{3}\left(\frac{4}{9}\right)^{n-1} A_0 \tag{12.14}$$

$$= \left[A_{n-2} + \frac{1}{3}\left(\frac{4}{9}\right)^{n-2} A_0\right] + \frac{1}{3}\left(\frac{4}{9}\right)^{n-1} A_0 \tag{12.15}$$

$$= \cdots = A_0 + \frac{A_0}{3}\sum_{k=0}^{n-1}\left(\frac{4}{9}\right)^k. \tag{12.16}$$

Recognizing that the sum is a geometric series, we have

$$\sum_{k=0}^{n-1}\left(\frac{4}{9}\right)^k = \frac{1-\left(\frac{4}{9}\right)^n}{1-\frac{4}{9}} = \frac{9}{5}\left[1-\left(\frac{4}{9}\right)^n\right]. \tag{12.17}$$

This finally yields

$$A_n = \frac{1}{5}\left[8 - 3\left(\frac{4}{9}\right)^n\right] A_0 \to A_\infty = \frac{8}{5} A_0, \tag{12.18}$$

in the limit $n \to \infty$. The area is finite while the perimeter is divergent. Such a property is not unique to fractals. For example, *Gabriel's horn* is a surface with a finite volume but infinite surface area.

to be taken during the construction to ensure that the area converges (see Exercise 12.5.1), otherwise the entropy becomes divergent and thus non-physical. However, in addition to the normal "area", one can also define area based on fractal dimension. For example, a "Hilbert curve" starts out as a one-dimensional curve that in the limit fractalized to cover the entire unit square (it is an example of a *space-filling curve*), so its fractal dimension is 2. A generic fractal curve would have a fractal dimension in between 1 and 2. Similarly, a fractal surface would have a fractal dimension in between 2 and 3.

Barrow proposed that the Bekenstein–Hawking entropy of the black hole can be generalized into

$$S = \left(\frac{A}{A_0}\right)^{1+\frac{\Delta}{2}}, \tag{12.19}$$

where $0 \leqslant \Delta \leqslant 1$ is the parameter that governs how "fractalized" the surface has become and we shall refer to it as "Barrow entropy index" (BEI). Some authors have subsequently considered the possibility that Δ may be negative, based on the fact that some "porous" fractals have dimensions *less* than its original shape before being fractalized. For example, the Sierpiński triangle, a fractal created by repeatedly removing smaller triangles from the interior of an equilateral triangle, has a fractal dimension of about 1.585, which is less than 2. It is not clear whether we should have $S = \frac{k_B}{4}\left(\frac{A}{\ell_P^2}\right)^{1+\Delta/2}$ or $S = k_B\left(\frac{A}{4\ell_P^2}\right)^{1+\Delta/2}$. In Barrow's work, he used $S \propto A/4 \approx A$ from the beginning (before the correction) and thus this issue was not discussed. Regardless, the constant factor $1/4$, whether it is raised to the power of $1 + \Delta/2$ or not, does not affect the results much except possibly when dealing with microscopic black holes, since the area would be very large anyway. Thus we will simply use the form of Equation (12.19), and for definiteness, we will set $A_0 = 4\ell_P^2$, where ℓ_P is the Planck length $\ell_P = \sqrt{G\hbar/c^3}$. Henceforth in this section, we shall employ the Planck units in which $\hbar = c = k_B = G = 1$.

This form of correction is very similar to the Tsallis form introduced in the previous section, but Δ now has natural lower and upper bounds coming from the spatial dimensions. (We focus on the original proposed bound $0 \leqslant \Delta \leqslant 1$.) We immediately notice a peculiarity if we expand the Barrow entropy assuming Δ is small. To the lowest order in Δ we have

$$S = \frac{A}{4} + \frac{A}{8}\ln\left(\frac{A}{4}\right)\Delta + O(\Delta^2). \tag{12.20}$$

We note the presence of the area as a coefficient of the logarithmic correction. This is unlike the result of many quantum gravity approaches, which gives $\ln(A)$ as the first order correction term. For large black holes, $\ln(A)$ is always smaller than $A/4$, so the Bekenstein–Hawking entropy term always dominates. Here, however, for any fixed $\Delta \ll 1$, the putative "correction term" $A \ln A$ would instead become larger than the

$A/4$ term if A is sufficiently *large*. This is rather strange because we expect that a large black hole should be more classical (i.e., receive less quantum gravitational correction), and so its entropy should be well-described by the original Bekenstein–Hawking expression. One possible resolution of this apparent paradox is to allow the BEI to be a function of A. For example, if Δ scales inversely proportional to $A\ln(A)$, we can keep the sub-leading term small at $O(1)$. Such a running[4] BEI would guarantee that the quantum gravity correction is small when the black hole is large (i.e., when the energy scale is small, since the Hawking temperature is inversely proportional to the mass). Another option is to consider Δ that runs like $1/A$, then the Barrow entropy would just be the same as the standard logarithmic correction (at least up to the first order in Δ; they still differ after that), but in this work we take the former view, with the hope of uncovering new physics. Note that both choices $\Delta \sim 1/(A\ln A)$ and $\Delta \sim 1/A$ are distinct from the standard quantum gravitational correction, which can be seen by considering small BEI expansion. The choice $\Delta \sim 1/(A\ln A)$ yields higher-order correction terms, which are inverse powers of the area, i.e., of the same form as the standard quantum gravity version, whereas the choice $\Delta \sim 1/A$, while providing a logarithmic term, also gives rise to a higher-order correction of the form $(\ln A)^k/A^{k-1}$. In principle this is possible, and would provide a larger correction to small quantum black holes, but we will not discuss it here.

How large can Δ be? We can use a simple argument to put an upper bound on Δ. We know from observations that black holes do not fragment, i.e., a black hole cannot split into two. In general relativity this is prohibited by deep mathematical theorems, but also by the fact that entropy (which is identified with $A/4$), cannot decrease. It is easy to show that the splitting process would in fact decrease the Bekenstein–Hawking entropy (the original black hole area is larger than the sum of the area of the two black holes after splitting), and hence is prohibited by the second law. However, once the entropy function has been modified under Barrow correction, we can show that fragmentation is entropically preferred if Δ is sufficiently large.[5] Thus, we can use the fact that no black hole has ever been seen to fragment to put an upper bound on Δ. The lightest black holes observed thus far is about $O(1)$ solar mass. For a solar mass black hole, $M = M_\odot \sim 10^{38}$, so $\Delta \sim 10^{-80}$ if $A \sim 1/A\ln A$—an extremely small number. But how large can BEI be for smaller black holes? Clearly infinitely many functions would satisfy $\Delta \sim 1/(A\ln A)$, and there is no additional equation that governs the dynamic of Δ. In other words, there is a huge amount of freedom for how the function $\Delta(A)$, or equivalently $\Delta(M)$ as a function of black hole mass, might actually behave. This seems to open up the possibility that Δ can be quite large, i.e., close to unity, until a

[4] Here we borrow the term from particle physics or quantum field theory: a *running* coupling constant is a parameter that describes how the strength of interactions between particles changes with energy scale.

[5] This also hints at there must be some modifications on the gravity theory as well, otherwise Hawking's area theorem that prevents fragmentation would still apply, as it is a geometric proof.

certain critical M is reached, whereby its value abruptly drops. However, we shall argue that this is not feasible.

To that end, we start with the claim that perhaps there exists a mass scale $M^* \gg 1$ such that the usual area law breaks down for black hole whose mass is smaller than M^*. This is unlikely, but the point we wish to make is that even if we consider such a scenario, the value of Δ remains small. If $M^* = M_\odot$, by definition black holes with mass $M \geqslant M^*$ satisfies $\Delta \sim 1/(A \ln A)$, whereas it is possible for a black hole whose mass is smaller than M^* to admit a much larger value for Δ. We can construct such a Δ smoothly via bump functions with arbitrarily small transition width ε, so that $\Delta \gg 1/(A \ln A)$ for $M < M^* - \varepsilon$. We want to investigate how large Δ can be under such a scenario. (Note that our subsequent argument does not depend on the choice of either $\Delta \sim 1/(A \ln A)$ or $\Delta \sim 1/A$).

The claim is that a black hole with mass M^* cannot be stable, as it will fragment into two black holes (for simplicity, we assume that these have equal masses $M^*/2$). To see this, we consider the initial entropy

$$S_B(M^*) = (4\pi)^{1+\frac{\Delta_i}{2}}(M^*)^{2+\Delta_i}, \tag{12.21}$$

and compare this with the entropy of the final configurations of two black holes, whose BEI is denoted by Δ_f:

$$2S_B(M^*/2) = 2 \times (4\pi)^{1+\frac{\Delta_f}{2}}\left(\frac{M^*}{2}\right)^{2+\Delta_f}. \tag{12.22}$$

We can neglect the initial BEI $\Delta_i \sim 10^{-80}$ for $M^* = M_\odot$. Fragmentation can happen if the final configuration is entropically favored, i.e., if

$$4\pi(M^*)^2 < 2\pi^{1+\frac{\Delta_f}{2}}(M^*)^{2+\Delta_f}. \tag{12.23}$$

A straightforward calculation shows that with $M^* = M_\odot$, fragmentation happens if $\Delta_f \gtrsim 0.00787$. Therefore, the fact that we observed astrophysical black holes with $M \sim M_\odot$ implies that $\Delta \lesssim 0.00787$. If we ever observe an even smaller black hole in the future, the upper bound for Δ can be increased. But even if $M^* = 10^2$, just 100 times greater than Planck mass, we have $\Delta \lesssim 0.13388$. The implication is that introducing the cutoff mass M^* does not allow Δ to be larger than $O(0.1)$. Other effects may come into play for even smaller M^*, so it is not useful to consider those black holes. For example, if spacetime is non-commutative, the non-commutative scale can be much larger than the Planck scale.[6] If we stay in the regime $M^* \gg 1$, say $M^*_{\min} = 10^6$ for definiteness, then $\Delta \lesssim 0.048176$. The bound $\Delta \lesssim O(10^{-3})$, rather surprisingly, falls in the same range as the bound obtained from various cosmological considerations assuming fixed BEI (Anagnostopoulos et al. 2020; Barrow

[6] According to Mendes (2019), the upper bound is some 10 orders of magnitude larger than Planck length—by considering gamma-ray burst data (since non-commutativity can influence the group velocity of massless particle wave packets).

et al. 2021; Leon et al. 2021; Luciano & Saridakis 2022). Thus, even allowing BEI to run does not allow larger values of Δ simply by thermodynamic stability.

Even if Δ is small except possibly at the Planck scale, it could have nontrivial significance in cosmology. The idea is to apply the Barrow correction to the apparent horizon in FLRW cosmology. The expression for the apparent horizon is (recall Section 10.4)

$$r_{\mathrm{AH}} = \left(H^2 + \frac{k}{a^2}\right)^{-\frac{1}{2}}. \tag{12.24}$$

(This r is not the r in the standard FLRW metric, but rather the areal radius, i.e., the FLRW's r multiplied by $a(t)$, which we call R in Section 10.4.) In the following we will assume a spatially flat universe so that $k = 0$.

If the BEI runs, we can compute the change in entropy as

$$\begin{aligned} dS &= d\left(\frac{4\pi r^2}{A_0}\right)^{1+\frac{\Delta}{2}} \\ &= \left(\frac{4\pi r^2}{A_0}\right)^{1+\frac{\Delta}{2}}\left[\frac{(2+\Delta)\dot{r}}{r} + \ln\left(\frac{4\pi r^2}{A_0}\right)\frac{\dot{\Delta}}{2}\right]dt. \end{aligned} \tag{12.25}$$

Using the first law of thermodynamics and the continuity equation, we can further show that (Sheykhi 2021)

$$\begin{aligned} dE &= -\frac{1}{2\pi r}\left(\frac{4\pi r^2}{A_0}\right)^{1+\frac{\Delta}{2}}\left[\frac{(2+\Delta)\dot{r}}{r} + \ln\left(\frac{4\pi r^2}{A_0}\right)\frac{\dot{\Delta}}{2}\right]dt \\ &= -4\pi r^3 H(\rho + p)dt, \end{aligned} \tag{12.26}$$

and consequently

$$\frac{(4\pi)^{\frac{\Delta}{2}}}{2\pi A_0^{1+\frac{\Delta}{2}}}\left[(2+\Delta)r^{\Delta-3} + \ln\left(\frac{4\pi r^2}{A_0}\right)\frac{r^{\Delta-2}}{2}\frac{d\Delta}{dr}\right]dr = -\frac{\dot{\rho}}{3}dt. \tag{12.27}$$

In standard cosmology, H decreases with time, so $r_{\mathrm{AH}} = H^{-1}$ is an increasing function of time. Let us write $r \equiv r_{\mathrm{AH}}$ for simplicity. We want Δ to at least decrease as $\Delta \sim \mathrm{const.}/r^2 \ln(r^2)$ at large r. For definiteness we can take a function that is always smaller than the asymptotic behavior $\mathrm{const.}/r^2 \ln(r^2)$, so a natural choice is $\Delta(r) = e^{-r}$.

The differential form of the modified Friedmann equation is given by

$$(2-\Delta)r^{\Delta-3}dr + \frac{2-\Delta}{2+\Delta}\ln\left(\frac{4\pi r^2}{A_0}\right)\frac{r^{\Delta-2}}{2}\frac{d\Delta}{dr}dr = -\frac{8\pi}{3}\underbrace{\left[\frac{2-\Delta}{2+\Delta}\frac{A_0^{1+\frac{\Delta}{2}}}{4(4\pi)^{\frac{\Delta}{2}}}\right]}_{=:G_{\mathrm{eff}}}d\rho, \tag{12.28}$$

where G_{eff} is an effective gravitational constant.

At sufficiently late time, let us choose some sufficiently large r^* such that $1/r^2\ln(r^2) \ll 1$ is as small as desired. We then have $\Delta \sim 0$ (more precisely, we assume that Δ is small enough so that it is slowly varying compared to r, so we can take some small value of fixed Δ when integrating with respect to r). Then, with $A_0 = 4$ in Planck units, the first term gives (with $R > r^*$)

$$2\int_{\tilde{r}}^{R} r^{-3}dr = 2\left[-\frac{1}{2r^2}\right]_{r*}^{R} = \frac{1}{r^{*2}} - \frac{1}{R^2}. \tag{12.29}$$

The right-hand side of Equation (12.28) can be integrated to give the matter density ($G = 1$):

$$-\frac{8\pi}{3}[\rho]_{\rho*}^{\rho(R)}. \tag{12.30}$$

Thus, in the absence of the second term in Equation (12.28), this will just give the usual Friedmann equation (as we take $R \to \infty$; note $\rho(R) \to 0$ in this limit because matter density decreases as the universe expands):

$$H^2 = \frac{8\pi\rho}{3} \tag{12.31}$$

evaluated at the time that corresponds to r^*.

The second term in Equation (12.28) can thus be interpreted as the differential form of the effective dark energy (thus in this model, dark energy is obtained from modification of thermodynamics, instead of modified gravity), so that Equation (12.28) yields the form[7] $H^2 - \Lambda/3 = 8\pi\rho/3$. We now have

$$\frac{d\Lambda_{\text{eff}}}{3} = \frac{2-\Delta}{2+\Delta}\ln\left(\frac{4\pi r^2}{A_0}\right)\frac{r^{\Delta-2}}{2}\frac{d\Delta}{dr}dr. \tag{12.32}$$

Substituting in the ansatz $1/r^2\ln(r^2)$ for Δ, we have

$$\frac{d\Lambda_{\text{eff}}}{3} = -\frac{2-\Delta}{2+\Delta}\ln\left(\frac{4\pi r^2}{A_0}\right)\frac{r^{\Delta-2}}{2}\left[\frac{2}{r^3\ln(r^2)} + \frac{2}{r^3(\ln(r^2))^2}\right]dr. \tag{12.33}$$

At late time, the BEI $\Delta \sim 0$, we have an upper bound:

$$-\frac{\Lambda_{\text{eff}}}{3}\Bigg|_{\Lambda(r*)}^{\Lambda(R)} \sim \frac{1}{2}\int_{r*}^{R}\frac{\ln(\pi r^2)}{r^2}\left[\frac{2}{r^3\ln(r^2)} + \frac{2}{r^3(\ln(r^2))^2}\right]dr \tag{12.34}$$

$$< \frac{1}{2}\int_{r*}^{R}\frac{\ln(\pi r^2)}{r^2}\frac{4}{r^3\ln(r^2)}dr \tag{12.35}$$

[7] Note that $\frac{1}{3}\int_{\Lambda_{\text{eff}}(r*)}^{\Lambda_{\text{eff}}(R)} d\Lambda_{\text{eff}} = \frac{1}{3}[\Lambda_{\text{eff}}(R) - \Lambda_{\text{eff}}(r^*)]$. This means that the observed value of the dark energy would be $\Lambda_{\text{eff}}(r^*) - \Lambda_{\text{eff}}(\infty)$ in the limit $R \to \infty$. Without loss of generality, we can take $\Lambda_{\text{eff}}(\infty) \equiv 0$ so that $\Lambda_{\text{eff}}(r^*)$ is the observed value of the effective cosmological constant.

$$= 2\int_{r*}^{R} \frac{1}{r^5} \frac{1}{\ln(r^2)} [\ln \pi + \ln(r^2)] dr \tag{12.36}$$

$$< 2\int_{r*}^{R} \left(\frac{1}{r^5} + \frac{1}{r^5}\right) dr = 4\int_{r*}^{R} \frac{1}{r^5} dr \tag{12.37}$$

$$= \left[-\frac{1}{r^4}\right]_{r*}^{R} = \frac{1}{r^{*4}} - \frac{1}{R^4}. \tag{12.38}$$

Thus in the limit $R \to \infty$, we have the inequality

$$\frac{\Lambda_{\text{eff}}(r^*)}{3} < \frac{1}{r^{*4}}. \tag{12.39}$$

Since r^* is chosen to be very large, Λ_{eff} must be very small, and furthermore it is getting smaller at later time. This is good, since it conforms to the observational fact that the cosmological constant is small. In fact, the integral Equation (12.34) yields explicitly

$$\Lambda_{\text{eff}}(r^*) \sim \frac{3}{4}\left\{\frac{1}{r^{*4}} \frac{\ln(\pi r^{*2})}{\ln(r^{*2})} - 2[1 - \ln(\pi)]\,\text{Ei}[-4\ln(r^*)]\right\}, \tag{12.40}$$

where Ei is known as the exponential integral function. One can verify numerically that this expression is positive and decreases very rapidly towards 0. There are of course other ways to obtain an asymptotically vanishing effective cosmological constant scenario, see Wetterich (1995, 2010, 2017), Overduin et al. (2007), Mashhoon & Wesson (2004), and Babourova & Frolov (2022) for some approaches in the literature.

So far we have only shown that a running BEI can conform to late time cosmology and explained why the cosmological "constant" is so small at late time. But do we get something new in the early time, right after the Big Bang? Of course, the effective dark energy term depends on the profile of Δ. Let us consider as before, $\Delta(r) = e^{-r}$, then the effective dark energy at the neighborhood of the Big Bang is, with $\Delta \sim 1$ and $0 < \varepsilon \ll 1$,

$$\left.\frac{\Lambda_{\text{eff}}}{3}\right|_{\Lambda_{\text{eff}}(\varepsilon)}^{\Lambda_{\text{eff}}(2\varepsilon)} \sim -\frac{1}{3}\int_{\varepsilon}^{2\varepsilon} \ln(\pi r^2) \frac{e^{-r}}{2r} dr. \tag{12.41}$$

The integration interval is kept small near the Big Bang to ensure that $\Delta = e^{-r}$ is nearly constant. This allows us to approximate the integral as

$$\Lambda_{\text{eff}}\Big|_{\Lambda_{\text{eff}}(\varepsilon)}^{\Lambda_{\text{eff}}(2\varepsilon)} \sim -\int_{\varepsilon}^{2\varepsilon} \ln(\pi r^2) \frac{e^{-r}}{2r} dr \sim -\varepsilon \ln(\pi\varepsilon^2) \frac{e^{-\varepsilon}}{2\varepsilon}. \tag{12.42}$$

That is,

$$\frac{\Lambda_{\text{eff}}(2\varepsilon) - \Lambda_{\text{eff}}(\varepsilon)}{\varepsilon} \sim -\ln(\pi\varepsilon^2) \frac{e^{-\varepsilon}}{2\varepsilon}. \tag{12.43}$$

In the limit $\varepsilon \to 0$, this gives $d\Lambda_{\text{eff}}/d\varepsilon \to \infty$. It follows that[8] $\Lambda_{\text{eff}} \to -\infty$ at the Big Bang.[9]

We note that as long as Δ is a strictly decreasing function, the integral

$$\Lambda_{\text{eff}}(\varepsilon) \sim -\int_{\varepsilon}^{2\varepsilon} \ln(\pi r^2) \frac{1}{2r}\frac{d\Delta}{dr} dr \tag{12.44}$$

is always negative because $d\Delta/dr < 0$ *and* $\ln(\pi r^2) < 0$ for small r. The choice $\Delta(r) = e^{-r}$ is thus not so special for the following important implication: the effective dark energy Λ_{eff} started out *negative* (and formally divergent), though it is small and positive at late time. This is an example of what is known as the "sign-switching" dark energy scenario, which has been argued to be able to alleviate some cosmological tensions, though this is still debatable.

12.3 How Generalized Entropy Affects Gravity

Parts of this section are reprinted from (Lü et al. 2025), with permission from Elsevier. Despite the apparent interesting implications for cosmology and black hole physics, it is important to further investigate whether it is consistent to modify thermodynamics while still keeping the black hole solution from general relativity. This is because there appears to be a strong connection between gravity and thermodynamics. In fact, various approaches have tried to derive general relativity from thermodynamic considerations. One of the more established approaches is a method pioneered by Ted Jacobson (Jacobson 1995), which essentially makes use of the Clausius relation $\delta Q = TdS$. The derivation here assumes more advanced knowledge in general relativity than is typically covered in the first course of the subject (and certainly beyond what is covered in this book). However the basic idea can be understood even if one glossed over some details. The following discussions are based on my papers Di Gennaro et al. (2022) and Lü et al. (2025).

We first consider the heat flow across the horizon given by the matter flux

$$\delta Q = -\kappa \int_{\mathscr{H}} \lambda T_{ab} k^a k^b d\lambda dA, \tag{12.45}$$

where A is the area of the horizon $\mathscr{H}$ with k^a denoting the tangent vector of the horizon generators,[10] κ is the surface gravity, and λ is an appropriate affine parameter. The variation of the area can be described by the Raychaudhuri equation, named after the Indian physicist Amal Kumar Raychaudhuri (1923–2005). It governs the

---

[8] The observed value of the effective cosmological constant over the interval $(\varepsilon, 2\varepsilon)$ is $\Lambda_{\text{eff}}(\varepsilon) - \Lambda_{\text{eff}}(2\varepsilon) < 0$. Since they are both comparable in magnitude (negative and with the same divergence properties), we can just refer to $\Lambda(\varepsilon)$ as the effective cosmological constant at the Big Bang.

[9] Note that a function $f(\varepsilon) = \ln(\pi\varepsilon^2)e^{-\varepsilon} < 0$ for small ε satisfies $f(2\varepsilon) - f(\varepsilon) \to \infty$ and $f'(\varepsilon) \to \infty$. Another example that can illustrate the same phenomenon, albeit with a different divergence, is $f(x) = -1/x$. This function satisfies $f(2\varepsilon) - f(\varepsilon) \to \infty$ and $f'(\varepsilon) = 1/\varepsilon^2 \to \infty$.

[10] The horizon generator is a null geodesic. They are neither ingoing nor outgoing. Every point of an event horizon $\mathscr{H}$ belongs to a null geodesic that lies within $\mathscr{H}$. A generator cannot have a future endpoint, in the sense that it cannot leave $\mathscr{H}$.

evolution of a congruence of geodesics (it is purely geometric and does not rely on the field equations). This gives

$$\delta A = \int_{\mathscr{H}} \theta d\lambda dA = -\int_{\mathscr{H}} \lambda R_{ab} k^a k^b d\lambda dA, \tag{12.46}$$

where θ is the “expansion” of the horizon generators.

Now we apply $\delta Q = TdS$, where $T = \kappa/2\pi$ is proportional to the surface gravity, after which we obtain

$$-\kappa \int_{\mathscr{H}} \lambda T_{ab} k^a k^b d\lambda dA = -\frac{\kappa}{2\pi} \int_{\mathscr{H}} \lambda R_{ab} k^a k^b d\lambda dS. \tag{12.47}$$

Then, if $S = A/4G$ holds,[11] this becomes (we shall keep G explicit because later on we will have a modification of G)

$$-\kappa \int_{\mathscr{H}} \lambda T_{ab} k^a k^b d\lambda dA = -\frac{\kappa}{2\pi} \cdot \frac{1}{4G} \int_{\mathscr{H}} \lambda R_{ab} k^a k^b d\lambda dA. \tag{12.48}$$

This can only be generically valid if $T_{ab} k^a k^b = (1/8\pi G) R_{ab} k^a k^b$ for all null vector k^a, therefore we must have

$$(8\pi G) T_{ab} = R_{ab} + f g_{ab} \tag{12.49}$$

for some function f. Finally, taking the covariant derivative and applying the contracted Bianchi identity (which is purely geometric)

$$\nabla^a R_{ab} = \frac{1}{2} \nabla_b R, \tag{12.50}$$

fixes f to be

$$f = \Lambda - \frac{R}{2}, \tag{12.51}$$

where Λ is an integration constant, which would serve as the cosmological constant once we substitute f back into Equation (12.49) and obtain the Einstein field equations.

What happens when one replaces the Bekenstein–Hawking entropy $S = A/4G$ with any generalization thereof? We can repeat the calculation using the Tsallis entropy as an explicit example. The left hand side of Equation (12.47) remains *unchanged* because the Tsallis modification only modifies the entropy expression; it does not modify the horizon area. Assuming that the parameter δ is constant, we would obtain

$$\begin{aligned} -\kappa \int_{\mathscr{H}} \lambda T_{ab} k^a k^b d\lambda dA &= -\frac{\kappa}{2\pi} \frac{A_0^{1-\delta}}{4G} \int_{\mathscr{H}} \lambda R_{ab} k^a k^b d\lambda dA^\delta \\ &= -\frac{\kappa}{2\pi} \frac{A_0^{1-\delta}}{4G} \int_{\mathscr{H}} \lambda R_{ab} k^a k^b \delta A^{\delta-1} d\lambda dA. \end{aligned} \tag{12.52}$$

[11] It is true that we are assuming the existence of a horizon before we know for sure that it can exist (since we don’t yet have a field equation.) Personally I don’t think this is a big problem. This is more like a consistency requirement in the sense that if there is a horizon, we would need to satisfy the premise.

That is to say,

$$-\kappa \int_{\mathscr{H}} \left[T_{ab} - \frac{1}{8\pi G} A_0^{1-\delta} \delta A^{\delta-1} R_{ab} \right] \lambda k^a k^b d\lambda dA = 0. \tag{12.53}$$

This implies

$$8\pi G_{\text{eff}} T_{ab} = R_{ab} + f g_{ab}, \tag{12.54}$$

where the effective gravitational constant is

$$G_{\text{eff}} = \frac{G}{\delta} \left(\frac{A}{A_0} \right)^{1-\delta}. \tag{12.55}$$

To obtain the modified version of the Einstein field equations, we need to take a covariant derivative and apply the contracted Bianchi identity. This results in

$$8\pi \ \nabla^a \ (G_{\text{eff}} T_{ab}) = \frac{1}{2} \ \nabla_b R + \nabla_b f. \tag{12.56}$$

At this stage, we have to assume some kind of "conservation law" for the energy-momentum tensor, or more accurately, the combination $G_{\text{eff}} T_{ab}$. A natural choice seems to be $\nabla^a (G_{\text{eff}} T_{ab}) = 0$. This finally implies the field equations

$$R_{ab} - \frac{1}{2} g_{ab} R + g_{ab} \Lambda = 8\pi G_{\text{eff}} T_{ab}. \tag{12.57}$$

We note that this is the same as the field equations in the standard general relativity, but with the gravitational constant G replaced by G_{eff}. The tricky aspect is that this new G_{eff} depends on the area of the horizon, and it is not clear what this means. In the original work of Jacobson, he actually employed a *local Rindler horizon* (only dA appears in the derivation), instead of a black hole horizon or cosmological apparent horizon. Of course the derivation also works with black hole horizons, but the crucial difference is that in the former, this holds *locally* even if there is no global event horizon. Indeed, this is consistent with the field equation of general relativity being a set of partial differential equations, which is by nature, local. In our case, once thermodynamics has been modified, however, G_{eff} depends on the *area* of the horizon (A appears in the result, not just dA). In Lü et al. (2025), possible interpretations and implications have been discussed in detail, here it suffices to mention two very different choices on how to deal with G_{eff} that is area dependent.

The first approach is to claim that this is not desirable, and take this result as a hint that the parameter δ *must not* be a constant, but must be itself a function of A. This would be consistent with the argument for Barrow's Δ being energy dependent in the previous section. If so, then for any generalized entropy, its associated parameter must be a function chosen in such a way so that the area dependence in G_{eff} can be canceled. A more radical approach is to accept that G_{eff} can indeed be area dependent. Often in research, when we obtain surprising results, we try to find supporting evidence in the literature. In this case, surprisingly there is at least one. Although the calculation above uses Tsallis entropy as an explicit example, the area dependence is generic: if we have an entropy correction to the horizon area law of the form $S = F(A)/4G$ for some differentiable function F, then $G_{\text{eff}} = G/F'(A)$. Specifically, if we were to consider the standard quantum gravity logarithmic correction $S = A/4G + \text{const.}\ln(A)$, then we would obtain

$$G_{\text{eff}} = \frac{G}{1 + \frac{C_1}{A}}, \tag{12.58}$$

where C_1 is a constant. This is not too different from the running gravitational constant scenario one obtains from renormalization group argument in asymptotically safe gravity (ASG; see, for example, Niedermaier 2007; Pawlowski & Stock 2018; Eichhorn & Held 2023; Platania 2023, 2025):

$$G(k) = \frac{G(k_0)}{1 + C_2 k^2}, \tag{12.59}$$

where k_0 is a reference energy scale and C_2 another constant. (Another lesson for students: familiarize yourself with the literature and read widely, so that you can recognize similar expressions when they occur!) The quantity $G(k_0)$ is the measured gravitational constant at our relatively low energy scale. In the literature k is often taken to be inversely proportional to a distance scale. Interestingly, it has been argued that Chen et al. (2022) (see also Chen et al. 2024) k is horizon area dependent and is explicitly given by $k = \text{const.}/\sqrt{A}$. This in turn implies $G(k)$ would recover the form in Equation (12.58). The Jacobson's method thus provides a link between the standard quantum gravitational correction of Bekenstein–Hawking entropy and the ASG approach.

The Barrow case is more subtle. In order to write down a Schwarzschild-like spherically symmetric spacetime metric, we have to treat the Barrow correction as an effective one that is not visible at the macroscopic scale. In other words, we keep the smooth geometry and treat the modification as only a modified entropy function, although fundamentally it arises from the microscopic fractal correction. This allows the same argument as before to follow through. In an earlier work (Di Gennaro et al. 2022), we have argued that for the Barrow case both the side of Equation (12.47) should be modified from dA to $dA^{1+\frac{\Delta}{2}}$, but only *re-interpreted* in the subsequent step. That is, one starts with

$$\int_{\mathscr{H}} \lambda T_{ab} k^a k^b d\lambda dA^{1+\frac{\Delta}{2}} = \frac{\eta}{2\pi} \int_{\mathscr{H}} \lambda R_{ab} k^a k^b d\lambda dA^{1+\frac{\Delta}{2}}, \tag{12.60}$$

from which we get

$$\int_{\mathscr{H}} \lambda \left(T_{ab} - \frac{\eta}{2\pi} R_{ab} \right) \left(1 + \frac{\Delta}{2} \right) A^{\frac{\Delta}{2}} k^a k^b d\lambda dA = 0, \tag{12.61}$$

and consequently

$$\left(T_{ab} - \frac{1}{8\pi G} R_{ab} \right) \left(1 + \frac{\Delta}{2} \right) A^{\frac{\Delta}{2}} = f g_{ab}. \tag{12.62}$$

If at this step one re-interprets A as the area of the black hole (instead of $A^{1+\frac{\Delta}{2}}$), as was done in Di Gennaro et al. (2022), then the resulting field equations has no G_{eff}, but instead (in the case of stationary spacetimes) the cosmological constant is rescaled

$$\Lambda \mapsto \frac{\Lambda}{\left(1 + \frac{\Delta}{2}\right)A^{\frac{\Delta}{2}}}. \tag{12.63}$$

The result is interesting, as it may suggest that if a similar result holds also in non-stationary cosmological spacetime, it may explain why the observed cosmological constant is so small, as the "bare" cosmological constant is divided by a power in the apparent horizon area (which is large). However, I am now skeptical that this is the correct approach, as before we can make the "re-interpretation", the very first step of Equation (12.60) is somewhat problematic. If at this step the geometry is still considered as a fractalized area, it is hard to make sense of what the tangent vector means. (Also note that the G_{eff} obtained via the Jacobson's method in this section is different from the one we defined in the previous subsection via the modified Friedmann equation.)

We end this section with another cautionary tale. It is often tempting to work with a metric tensor since it allows us to compute a lot of quantities. One way to obtain a metric is to derive it from the first law of thermodynamics. When entropy is modified, one may be tempted to use directly the first law $dM = TdS$, assuming M is still the same as the black hole mass in general relativity. This may not be consistent in general as the thermodynamic energy need not coincide with the mass (in fact in many cases, such a choice would lead to a gross violation of the Bekenstein bound; see below). Putting this issue aside, some works in the literature then considered integrating the first law to obtain the "modified Hawking temperature", and using the Wick rotation trick (see Section 10.2) in reverse, finds the modified metric. Again, this may not be consistent because the Wick rotation method was developed assuming the Boltzmann–Gibbs distribution, so whether the modified temperature still relates to the regularization to avoid conical singularity in the Wick-rotated geometry needs to be investigated when the statistical distribution is modified. Even if this works, the metric tensor obtained usually amounts to a re-definition of the Schwarzschild mass. The Barrow case, for example, would lead to the metric (in Planck units)

$$ds^2 = -\left(1 - \frac{(\Delta + 2)M^{\Delta+1}(4\pi)^{\Delta/2}}{r}\right)dt^2 + \left(1 - \frac{(\Delta + 2)M^{\Delta+1}(4\pi)^{\Delta/2}}{r}\right)^{-1}dr^2 + r^2(d\theta^2 + \sin^2\theta d\phi^2). \tag{12.64}$$

If we know the mass M, we can use astrophysical data to constrain Δ. However, this would entail a God's eye view. In practice, we can only measure the effective mass $\tilde{M} = (1 + \Delta/2)M^{\Delta+1}(4\pi)^{\Delta/2}$ as a whole. That is, what we call the "mass" would actually be $\tilde{M}$ and not M. Ontologically we cannot constrain α in such a manner. It is therefore important to not only figure out what is the consistent approach but also the physical observables once we consider modification to known physics.

Finally, a quick comment about the Bekenstein bound (Bekenstein 1981), which is given by the inequality

$$S \leqslant \frac{2\pi k_B}{\hbar c} RE = 2\pi RE \tag{12.65}$$

in our units. Essentially, the idea is that the entropy of a system of size R and energy E cannot be arbitrarily large. Instead, it has to be bounded by the product RE up to a constant prefactor. Note that the Schwarzschild black hole saturates the inequality Equation (12.65) if S is the Bekenstein–Hawking entropy, $E = M$, and R is the Schwarzschild radius. This can be interpreted as the statement that if a bounded system has too much entropy (given a fixed energy), then it will collapse to form a black hole. The bound is generally believed to be quite fundamental, since even non-gravitational quantum field theory satisfies it (Casini 2008); of course one has to interpret what is meant by R and E carefully in different systems. If we replace the Bekenstein–Hawking entropy with some other entropies—say, the Tsallis(–Cirto) entropy—in a naive manner, then the Bekenstein bound is drastically violated. To see this, recall that the Tsallis entropy for a black hole is

$$S_T = \frac{A_0}{4G}\left(\frac{A}{A_0}\right)^{\delta} = \frac{A_0}{4G}\left(\frac{4\pi R^2}{A_0}\right)^{\delta}. \tag{12.66}$$

The RHS of the Bekenstein bound is

$$2\pi RE = 2\pi RM = \frac{\pi R^2}{G}. \tag{12.67}$$

We can show the bound is only satisfied if

$$S_T^{1-\frac{1}{\delta}} \leqslant \left(\frac{A_0}{4G}\right)^{1-\frac{1}{\delta}} = \text{const}. \tag{12.68}$$

Evidently, since S_T is a function of A, we see that the bound fails for *large* black holes. This was first pointed out in Abreu & Neto (2022) (the example we use here is similar to their computation for the Barrow case). See also Shokri (2025).

However, if we use the Jacobson's approach, then

$$G_{\text{eff}} = \frac{G}{\delta}\left(\frac{A}{A_0}\right)^{1-\delta}, \tag{12.69}$$

and it can be shown that the thermodynamic energy is not M, but rather (Lü et al. 2025)

$$E = \frac{M}{2\delta - 1}. \tag{12.70}$$

Thus,

$$S_T = \frac{A_0}{4G}\left(\frac{4\pi R^2}{A_0}\right)^{\delta} = \frac{A_0}{4G}\left(\frac{4\pi(2G_{\text{eff}}M)^2}{A_0}\right)^{\delta}. \tag{12.71}$$

On the other hand, Equation (12.69) yields a relationship between G_{eff} and G:

$$G = \frac{M^{2(\delta-1)}(16\pi)^{\delta-1}A_0^{1-\delta}\delta}{G_{\text{eff}}^{1-2\delta}}. \tag{12.72}$$

We can substitute this into Equation (12.71) and obtain

$$S = 4\pi\delta^{-1}G_{\text{eff}}M^2. \tag{12.73}$$

On the other hand, the RHS of the Bekenstein bound is

$$2\pi RE = 2\pi(2G_{\text{eff}}M)\frac{M}{2\delta - 1} = \frac{4\pi}{2\delta - 1}G_{\text{eff}}M^2. \tag{12.74}$$

Thus the Bekenstein bound holds if and only if

$$\frac{1}{\delta} \leqslant \frac{1}{2\delta - 1} \Leftrightarrow \delta \geqslant 2\delta - 1, \tag{12.75}$$

and thus for all $\delta \leqslant 1$. Otherwise, we can consider the "weak form" of the Bekenstein bound by replacing the constant 2π with some constant C. Then the weak Bekenstein bound

$$S \leqslant CRE \tag{12.76}$$

is satisfied for S_T if $C = 2\pi(2\delta - 1)/\delta$. Personally, I don't really consider the weak form as a true violation of the Bekenstein bound, since its essential physics that the entropy should be bounded above by the product RE remains intact. Thus we see that the Jacobson approach is preferable as it prevents a gross violation of the Bekenstein bound. See Lü & Ong (2025) for more general discussions.

12.4 Generalized and Extended Uncertainty Principles

In the last section of this chapter, we shall turn to modifications of the uncertainty principle in quantum mechanics. There are two main types of modifications (there are sub-varieties for each, which we will not get into here): the *generalized* uncertainty principle (GUP), which effectively incorporates the effect of gravity into quantum uncertainties, and the *extended* uncertainty principle (EUP), which takes into account the effect of local curvature scale. Though gravity is also a manifestation of spacetime curvature, these two corrections turn out to be quite different. This section is mainly based on my works (Ong 2018, 2018, 2023, 2020). These references should be organised chronologically.

Let us again restore for now, $\hbar$, c and G for clarity. The most basic form of GUP is given by[12]

$$\Delta x \Delta p \geqslant \frac{1}{2}\left(\hbar + \alpha \ell_{\text{P}}^2 \frac{\Delta p^2}{\hbar}\right), \tag{12.77}$$

[12] The notation Δx^n is a short hand of $(\Delta x)^n$. Similarly for Δp^n.

where α is a dimensionless parameter, which is typically taken to be of order unity (so that the correction only becomes important at the Planck scale). We will first discuss the case in which $\alpha > 0$. From Equation (12.113) we can obtain

$$\frac{\hbar}{\alpha \ell_P^2}\left[\Delta x\left(1 - \sqrt{1 - \frac{\alpha \ell_P^2}{\Delta x^2}}\right)\right] \leqslant \Delta p \leqslant \frac{\hbar}{\alpha \ell_P^2}\left[\Delta x\left(1 + \sqrt{1 - \frac{\alpha \ell_P^2}{\Delta x^2}}\right)\right]. \quad (12.78)$$

The presence of the square root in this inequality makes it clear that GUP imposes a minimum uncertainty in the position given by

$$\Delta x_{\min} = \sqrt{\alpha}\,\ell_P. \quad (12.79)$$

The fact that GUP incorporates gravitational "uncertainty " can be seen by taking the limit $\hbar \to 0$, which leads to $\Delta x \sim G\Delta p/2c^3$. This is the statement that in the classical limit, but when gravitational effect is dominant, a particle of mass m cannot be localized to within an uncertainty Δx. In the language of Heisenberg's microscope thought experiment, if we focus a beam of light with a small enough wavelength (equivalently, high enough energy) in order to localize the particle, an event horizon is formed and so we cannot locate the particle to within the horizon scale characterized by $GM/c^2 \sim 2GM/c^2$. (The discussion here is heuristic so we don't really care about the exact prefactor.) Indeed the change in the momentum of the particle is γmv, which for v close to the speed of light but still bounded away from it, is $O(1)mc$ (even at $v = 0.9c$, the quantity γmv is still only approximately $2mc$). Therefore $\Delta x \sim G\Delta p/2c^3$ does indeed lead to $\Delta x \sim GM/c^2$. In other words, GUP includes both quantum uncertainty and gravitational (classical) uncertainty due to horizon formation. One may question whether these two very different kinds of "uncertainties" can be considered together on an equal footing, but let us accept this and continue.

Since the Hawking radiation is a quantum effect, it should be affected by GUP modification. For an asymptotically flat Schwarzschild black hole this modification was obtained heuristically in Adler et al. (2001) by considering the characteristic energy of the Hawking particle as $E = pc$ and by identifying Δx with the horizon scale $\Delta x = 2GM/c^2$. This gives (upon multiplying with a normalization constant prefactor $1/2\pi$ in order to obtain the correct temperature when $\alpha \to 0$):

$$T_{\mathrm{GUP}} = \frac{Mc^2}{\pi\alpha}\left(1 - \sqrt{1 - \frac{\alpha\hbar c}{4GM^2}}\right). \quad (12.80)$$

By the first law of black hole thermodynamics $dS = dM/T$ (in the standard approach that assumes the thermodynamic energy is unmodified), we can integrate to find the expression of the entropy (see the Exercise section at the end of the chapter), whose series expansion is (from now onwards we again set $c = \hbar = G = 1$ for simplicity)

$$S = 4\pi M^2 - \frac{1}{2}\pi\alpha \ln M + \sum_{n=1}^{\infty} c_n (4\pi M^2)^{-n} + \text{const.}, \quad (12.81)$$

where the c_n's are some coefficients.

The temperature in Equation (12.80) has the property that it is only defined (i.e., $T_{\rm GUP} \in \mathbb{R}$) for black holes whose mass is larger than the minimum mass

$$M_{\rm min} = \frac{\sqrt{\alpha} m_{\rm P}}{2}. \tag{12.82}$$

This suggests that as a black hole evaporates to this minimum mass, the evaporation should stop. However, curiously the temperature is not zero when this happens, though it is thermodynamically inert since its specific heat can be shown to be zero. How does one interpret such a nonzero temperature but a stable object? Could this be a sign that modifying the Hawking temperature without modifying the Schwarzschild metric may be problematic? Furthermore, if we look at the temperature of a Reissner–Nordström black hole in standard general relativity, there is also a square root involved, namely a factor of $\sqrt{M^2 - Q^2}$. In this case the temperature tends to zero as $M \to Q$, so the $M = Q$ state is an extremal black hole. Still, even then, there are considerable interests in the possibility of a naked singularity formation. That is, whether some kind of perturbation can bring Q to exceed M. Crucially, note that $\sqrt{M^2 - Q^2} \notin \mathbb{R}$ does not imply there is no solution such that $Q > M$, only that such a solution would no longer be a black hole. Therefore, in the GUP case discussed above, it is actually not necessary that the $T \neq 0$ remnant "ceases radiating" because the temperature becomes imaginary. Rather, precisely because $T \neq 0$, there is a risk that it may drive the evolution further beyond a black hole stage into another horizonless configuration, if not a naked singularity. Whether this is the case would require a thorough investigation.

Another issue is that the heuristic derivation above cannot be justified in general. After all, while the method works for the Schwarzschild case, it does not lead to the correct known temperature for the Kerr or Reissner–Nordström black hole (Good & Ong 2025), so how could we be confident that it works when there is a GUP correction? For other criticisms for heuristic treatments of GUP, see Ong (2023, 2020); Bosso et al. (2023). In addition, in this approach, we also assumed that the Schwarzschild solution is unchanged. It is not clear if this is the correct choice. One finds in the literature numerous other approaches that did try to also change the metric, but it is not so clear which approach is the correct one. Let us take a look at some examples.

We can assume, as above, that the horizon is unchanged. However, this does not mean that the *metric* is unchanged. This is because we can have a metric of the form

$$ds^2 = -f(r)dt^2 + f(r)^{-1}dr^2 + r^2 d\Omega^2, \tag{12.83}$$

but with

$$f(r) = \left(1 - \frac{2M}{r}\right) g(r). \tag{12.84}$$

Then the horizon remains at the Schwarzschild radius if $g(r)$ has no zero for $r > r_h$. The Hawking temperature is, with $r_h = 2M$ denoting the horizon, given by

$$T_{\mathrm{GUP}} = \frac{f'|_{r=r_h}}{4\pi} = \frac{1}{4\pi}\left[\frac{2M}{r_h^2}g(r_h)\right] = \frac{1}{8\pi M}\cdot g(r_h). \tag{12.85}$$

On the other hand, from $1/T = dS/dM$, one could obtain from the series expansion of the entropy (Equation (12.81)), that

$$T_{\mathrm{GUP}} = \left(8\pi M - \frac{\pi\alpha}{2M} + \cdots\right)^{-1}. \tag{12.86}$$

This approach was first discussed in Contreras et al. (2016), in which the authors proceeded to match the first two terms of Equation (12.86) with Equation (12.85), obtaining[13]

$$T = \frac{1}{8\pi M\left[1 - \frac{\alpha}{(2(2M))^2}\right]} = \frac{1}{8\pi M}\cdot g(r_h). \tag{12.87}$$

Consequently they obtained

$$g(r) = \left(1 - \frac{\alpha}{4r^2}\right)^{-1}. \tag{12.88}$$

However, this form of $g(r)$ gives rise to a new curvature singularity when the minimum mass $M = \sqrt{\alpha}/2$ is reached. This is, however, an artifact caused by using the series expansion outside its domain of validity (the *radius of convergence* of a power series). Indeed, if one works with the *full* expression of the modified Hawking temperature, i.e., Equation (12.80), instead of the first few terms of the series expansion, we would obtain

$$T_{\mathrm{GUP}} = \frac{1}{8\pi M}\left[\frac{8M^2}{\alpha}\left(1 - \sqrt{1 - \frac{\alpha}{4M^2}}\right)\right], \tag{12.89}$$

and so we can identify

$$g(r_h) = \frac{2}{\alpha}(2M)^2\left(1 - \sqrt{1 - \frac{\alpha}{(2M)^2}}\right). \tag{12.90}$$

If we follow the same approach and naively take[14]

$$g(r) := \frac{2r^2}{\alpha}\left(1 - \sqrt{1 - \frac{\alpha}{r^2}}\right), \tag{12.91}$$

---

[13] Note that our α is proportional to the α^2 in Contreras et al. (2016).

[14] Note that even if we work with the full form of GUP, this function is still not unique. The reason is that a function f cannot be determined by its value at a single point. For example, $g(r) = \frac{2r^2}{\alpha}\left(1 - \sqrt{1 - \frac{\alpha}{r^2}}\right)\tilde{g}(r)$ would also satisfy Equation (12.90) for any smooth function $\tilde{g}$ satisfying $\tilde{g}(r_h) = 1$ and $\lim_{r\to\infty}\tilde{g}(r) = 1$ (to maintain asymptotic flatness). Equation (12.91) is an obvious "minimal" choice based on Equation (12.90), but one will need more justification for such a choice.

we will find that there is no curvature singularity anywhere aside from $r = 0$, which is irrelevant since a GUP black hole cannot reach zero mass. Often physicists like to work with series expansion because it is easy, but we have to be careful when doing so may lead to incorrect conclusions. For other examples of this type, see Ong (2023).

Another approach to obtain a GUP-corrected metric is to consider the inequality Equation (12.78), and obtain the bound

$$\Delta p \geqslant \frac{1}{\alpha}\Delta x\left(1 - \sqrt{1 - \frac{\alpha}{\Delta x^2}}\right) = \frac{1}{2\Delta x}\left[1 + \frac{1}{4}\frac{\alpha}{\Delta x^2} + \frac{1}{8}\frac{\alpha^2}{\Delta x^4} + \cdots\right]. \quad (12.92)$$

The original Heisenberg uncertainty principle gives the characteristic energy for a photon in flat spacetime as $E\Delta x \sim 1/2$, so the GUP-corrected energy is presumably given by

$$E_{\mathrm{GUP}} \sim E\left[1 + \frac{1}{4}\frac{\alpha}{\Delta x^2} + \cdots\right]. \quad (12.93)$$

Neglecting the higher-order terms, and considering $\Delta x \sim r_h$ as usual, the method of Anacleto et al. (2020, 2021) considers the same modification for the mass[15] leading to

$$M_{\mathrm{GUP}} := M\left[1 + \frac{1}{4}\frac{\alpha}{(2M)^2}\right]. \quad (12.94)$$

The effective metric is then taken to be

$$ds^2 = -\left(1 - \frac{2M_{\mathrm{GUP}}}{r}\right)dt^2 + \left(1 - \frac{2M_{\mathrm{GUP}}}{r}\right)^{-1}dr^2 + r^2 d\Omega^2. \quad (12.95)$$

There are other methods to construct modified metrics from GUP. The problem is that they are mostly heuristic, and as such cannot be fully trusted. (Based on the Jacobson's approach that obtains G_{eff} as discussed in the previous section, I offered another possible approach in Ong 2025.) Worse, as we have seen numerous times in this chapter, one modification at one spot may lead to inconsistency in some other aspects; physics is inter-related after all. For example, if we fix the horizon position to still be at $r = 2M$, the Hawking temperature obtained via the metric is often *not* the same as, or even proportional to, the characteristic energy scale of the Hawking photon deduced from GUP. This could either mean that there are inconsistencies (perhaps the assumption that $r = 2M$ may be to blame), or that the standard identification of Hawking temperature $T = F'(r)\big|_{r_h}/4\pi$ for a static black hole with metric function $F(r)$ no longer holds under GUP correction.

Ultimately the position occupied by GUP metrics is a rather awkward one: we want to consider corrections from quantum gravity on the spacetime metric, but on the other hand, the common belief is that at the Planck scale the notion of metrics

[15] Anacleto et al. (2020) actually considered also the possibility of including the linear term in Δp on the RHS in GUP, while Anacleto et al. (2021) considered an even more general case.

would eventually break down (even spacetime itself may fluctuate wildly at that scale, leading to the so-called "spacetime foam"). Thus, even if we have an effective GUP metric, we probably cannot trust its conclusion very close to the Planck scale. But how close is close? This we do not know. Remarkably, Bernhard Riemann, the founder of Riemannian geometry himself, has thought about the limitation of a metric in his famous inaugural lecture *On the Hypothesis Which Lie at the Foundation of Geometry* of 1854:

> "Now it seems that the empirical notions on which the metrical determinations of space are founded, the notion of a solid body and of a ray of light, cease to be valid for the infinitely small. We are therefore quite at liberty to suppose that the metric relations of space in the infinitely small do not conform to the hypotheses of geometry; and we ought in fact to suppose it, if we can thereby obtain a simpler explanation of phenomena."

Before moving on to EUP, we shall mention the possibility that the GUP parameter α may be negative. One possible motivation is this: if we consider GUP as a higher-order correction to the metric function, and so $1 - 2M/r$ becomes, at the first higher-order expansion,

$$F(r) = 1 - \frac{2M}{r} + \varepsilon\frac{M^2}{r^2}, \tag{12.96}$$

then by identifying the Hawking temperature T_{GUP} with $F'/4\pi$ evaluated at the horizon (Scardigli & Casadio 2015), one can argue that the GUP parameter should be proportional to $-\varepsilon^2$ and is therefore negative. It has also been shown that quite a few quantum gravity models are consistent with $\alpha < 0$; see Ong (2018). For a negative GUP parameter, the Hawking temperature can be written in the form

$$T[\alpha < 0] = -\frac{M}{4|\alpha|\pi}\left(1 - \sqrt{1 + \frac{|\alpha|}{M^2}}\right). \tag{12.97}$$

In this case there is no stable remnant since there is no minimum mass. The mass of the black hole can go to zero when temperature reaches $T_{\max} = 1/\sqrt{4\pi|\alpha|}$. Recall that for the case $\alpha > 0$, we have a finite nonzero temperature when the minimum mass is reached; now with $\alpha < 0$, the situation seems even more peculiar—how can there be a finite nonzero temperature when the mass is already zero (no black hole left)? However, such a "zero mass" state is never reached because it takes an infinite time for a black hole to lose its mass. This can be seen from the Stefan–Boltzmann equation

$$\frac{dM}{dt} \sim -AT^4, \tag{12.98}$$

where we take $A \sim M^2$ as the emission area. With GUP correction this gives

$$\frac{dM}{dt} \sim -M^2 \cdot \left(\frac{M}{4\alpha\pi}\right)^4\left(1 - \sqrt{1 - \frac{\alpha}{M^2}}\right)^4. \tag{12.99}$$

For the case of negative α, the mass evolution follows

$$\frac{dM}{dt} \sim -\frac{M^6}{(4|\alpha|\pi)^4}\left(1 - \sqrt{1 + \frac{|\alpha|}{M^2}}\right)^4. \tag{12.100}$$

The numerically solved plots can be seen in Figure 12.1. When M has become sufficiently small, we can use the series expansion of the right-hand side:

$$\frac{dM}{dt} \sim -\frac{M^2}{(4\pi)^4\alpha^2}. \tag{12.101}$$

This has solution of the form

$$M = M_0\left(\frac{256\pi^4\alpha^2}{256\pi^4\alpha^2 + M_0 t}\right), \tag{12.102}$$

where M_0 is the "initial" (small) mass. Thus we see that $M \to 0$ as $t \to \infty$, as claimed. Therefore there is still an *effective* remnant, whose lifetime is infinite. Since the zero mass state is never reached, the paradox of such a state having nonzero

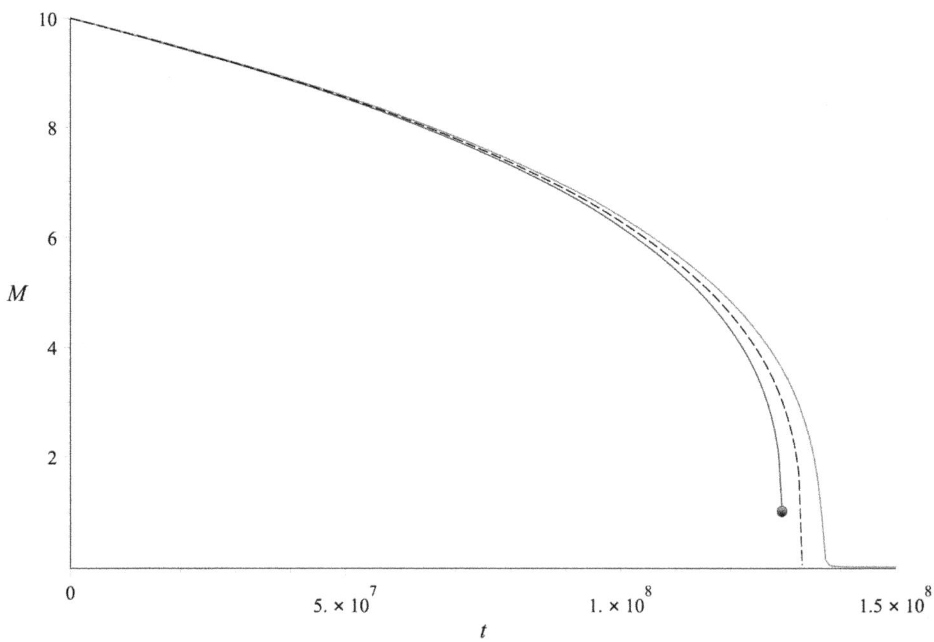

Figure 12.1. Mass evolution of Schwarzschild black holes with the same initial mass. The middle black dashed curve has no GUP correction. Beside it are those with a positive GUP correction (blue, left curve), and a negative GUP correction (red, right curve). The positive GUP correction leads to a remnant in a finite time (it stops evaporating at some $M \neq 0$ near if not at the Planck scale, indicated by the blue dot), whose temperature remains finite and thus could indicate an instability. On the other hand, a negative GUP parameter leads to an infinite lifetime, as the black hole slowly evaporates indefinitely. The asymptotic state, which is unattainable, has a finite temperature but zero mass.

temperature never arises. (Of course, one may question if such a state can form spontaneously from quantum fluctuation, and how to make sense of such an event.)

The negative GUP parameter is interesting because it seemingly avoids a problem that arises from GUP with $\alpha > 0$. Recall from Section 8.4.2 that the Chandrasekhar mass—the lowest mass of a white dwarf star—can be obtained from a back-of-the-envelope calculation utilizing the Heisenberg uncertainty principle. If we were to use GUP with $\alpha > 0$ instead, the Chandrasekhar mass would seemingly disappear. Let us see why this is so.

Consider the GUP estimate (with $\alpha > 0$)

$$\Delta x \Delta p \sim 1 + \alpha \Delta p^2, \tag{12.103}$$

which allows us to solve for Δp:

$$\Delta p \sim \frac{\Delta x}{2\alpha}\left[1 \pm \sqrt{1 - \frac{4\alpha}{\Delta x^2}}\right]. \tag{12.104}$$

The sign in front of the square root can be chosen definitively, but let us keep both signs for now for clarity. Now we simply repeat the calculations following Section 8.4.2, which yields, for the non-relativistic case (the conclusion for the relativistic case is similar; see Ong 2018):

$$M^{\frac{5}{3}} \sim \frac{1}{8Gm_e^{\frac{4}{3}}\alpha^2} R^3 \left(1 \pm \sqrt{1 - \frac{4\alpha M^{\frac{2}{3}}}{m_e^{\frac{2}{3}} R^2}}\right)^2. \tag{12.105}$$

We are only interested in the mathematical properties of the equation (i.e., in the qualitative picture), so let us now set m_e to be unity as well. Of course we are working in the units $G = \hbar = c = 1$, so we are using the Planck units in which $m_P = 1$. This means that the electron mass is actually some small number ($\sim 4.1854 \times 10^{-23}$), *not* unity. However, to keep the equation clean and not having to deal with very large (or very small) factors throughout, we are setting $m_e = 1$ purely for convenience! After all we are only interested in the qualitative picture.

If α is sufficiently small, we can expand to first order in α and obtain

$$M^{\frac{5}{3}} \sim \frac{R^3}{8\alpha^2}\left[1 \pm \left(1 - \frac{2\alpha M^{\frac{2}{3}}}{R^2}\right)\right]^2. \tag{12.106}$$

From this step it becomes clear that we should choose the minus sign in front of the square root, so that in the small α limit

$$M^{\frac{5}{3}} \sim \frac{R^3}{8\alpha^2}\left(\frac{2\alpha M^{\frac{2}{3}}}{R^2}\right)^2 = \frac{M^{\frac{4}{3}}}{2R}, \tag{12.107}$$

that is, $R \sim M^{-\frac{1}{3}}$, as obtained using the usual Heisenberg's uncertainty principle. From this we obtain the standard non-relativistic white dwarf's property: its radius goes to zero when M tends to infinity. Taking relativistic correction into account gives a finite nonzero mass as the minimum mass of a white dwarf—the Chandrasekhar limit.

Now, Equation (12.106) can be solved analytically to obtain the stellar radius R as a function of the mass M:

$$R(M) := \frac{1}{6}\left[\frac{1}{M^{\frac{1}{3}}} + \frac{1}{M}f(\alpha, M) + \frac{M^{\frac{1}{3}}(1 + 24\alpha M^{\frac{4}{3}})}{f(\alpha, M)}\right], \tag{12.108}$$

where

$$f(\alpha, M) := \left[M^2\left(216\alpha^2 M^{\frac{8}{3}} + 24\sqrt{3}\,\alpha M^2\sqrt{\alpha(27\alpha M^{\frac{4}{3}} + 1)} + 36\alpha M^{\frac{4}{3}} + 1\right)\right]^{\frac{1}{3}}. \tag{12.109}$$

Again, it is helpful to check that as $\alpha \to 0$, we do obtain the correct behavior $R \to 1/(2M^{\frac{1}{3}})$. With GUP in place, however, we note that for large R, the leading term goes like $M^{\frac{5}{9}}$, that is, $\lim_{M\to\infty} R = \infty$. In other words, we should not expect that there is a Chandrasekhar limit. This can be confirmed by considering the relativistic correction. Although this is a heuristic argument, it agrees with other results obtained via a more rigorous analysis of the GUP-corrected Lane–Emden equation (Rashidi 2016).

One might mistakenly think that this is somewhat obvious. The degenerative pressure is the result of Δp, which in the case of Heisenberg's uncertainty principle, becomes larger when Δx becomes smaller. If the star is too big, the motion of the constituent particles in the stellar core need not have so large a momentum, and so fail to support a pressure strong enough to counter-balance gravity. (To be more precise one should have worked with a three-dimensional version of the uncertainty principle.) However, Δp in the GUP case is linearly proportional to Δx when the latter is large. This seems to imply that pressure is large even when the star is large. However this is *not* the correct reasoning, because by choosing the negative sign branch in Equation (12.106), we have imposed the requirement that $\Delta p \sim 1/\Delta x$ when Δx is large, instead of $\Delta p \sim \Delta x$. Thus, the reason of the failure of the Chandrasekhar limit is not as trivial as it appears.

GUP with a negative α can avoid this problem and help to restore the Chandrasekhar limit (which is important because we do not see arbitrarily large white dwarfs in the actual Universe). However, it is not necessary. It can be shown that white dwarf stars that are too large in the $\alpha > 0$ case, at least in some range of α that is still not ruled out by experiments, are simply unstable (Mathew & Nandy 2021). Nevertheless, we see that the Chandrasekhar limit now requires a stability analysis and not straightforwardly obtained as in the $\alpha = 0$ case (though one should of course still show, even in that case, that the star is stable). The lesson, as in the other scenarios we have seen in this chapter, is worth emphasizing: a small modification to any physics is likely to affect others, and so modification or generalization is not as easy as it sounds. Not only that we have to make sure the results are sensible and do not violate known observations, but also that the mathematical structures remain problem free and consistent throughout.

Lastly, we briefly discuss the case of EUP, which takes the form

$$\Delta x \Delta p \geqslant \frac{\hbar}{2}\left[1 + \frac{\beta(\Delta x)^2}{L^2}\right], \tag{12.110}$$

where $1/L^2$ is the local sectional curvature of the geometry. For example, in a de Sitter (dS) spacetime, the cosmological constant is related to the curvature length scale L by $\Lambda = 3/L^2$. The parameter β is often taken to be of order unity, and just like α in the case of GUP, it is unclear what the sign of β should be. If we use the heuristic derivation of the Hawking temperature as in the original GUP approach, then applying the EUP to an asymptotically anti-de Sitter (AdS) black hole, we can match the Hawking temperature T with Δp if $\beta > 0$. This suggests that β is positive in AdS but negative in dS, i.e., it has the opposite sign of the underlying sectional curvature.

EUP has a similar form to the uncertainty principle on a circle, which takes the form

$$\Delta x \Delta p \geqslant \frac{\hbar}{2}[1 - C(\Delta x)^2], \tag{12.111}$$

where C is a constant, argued to be $3/\pi^2$ (Judge 1963, 1964; see also Brown 2018). This result is developed by D. Judge, who was actually discussing an equivalent problem: the uncertainty principle between angular momentum L_z and angle φ. This suggests that the underlying *topology* of the manifold might also be relevant for the form of the uncertainty principle. In particular for S^1, which is closed, the sign of the EUP parameter would be negative. However, in Ong (2020), I found that if we use a heuristic derivation that is based on the Schwinger effect (pair production of electrons by a strong electric field) instead of Hawking temperature, then in order to match known results from quantum field theory, β should be negative in AdS spacetime, which is a completely different conclusion. Which heuristic derivation is correct? I don't know. The lesson is that heuristic argument has its limit, different heuristics may not even agree on the sign of some parameters. Unfortunately, for some theoretical physics topics, it is difficult to make progress rigorously, and so one relies on insights from certain heuristic calculations. Sometimes this can be misleading.

12.5 Exercises

(1) Consider the sphereflake fractal. Suppose we start with a sphere with radius r_0. At each step of the iteration we add m spheres of radius $r_{n+1} = Cr_n$ for some constant $C < 1$ to each sphere of radius r_n; $n \geqslant 0$. For simplicity, we ignore the possible intersection of the spheres. If a black hole horizon has sphereflake geometry, we would want the horizon to have a finite area (so that the entropy remains finite). Find the relation between the proportionality constant α and m such that the area is finite.

(2) Find out more about the fractal dimension d_f introduced in Equation (12.11) and its various generalizations, e.g., the "box dimension" (or the *Minkowski–Bouligand dimension*) of any bounded set in $\mathbb{R}^n$. One starts by covering the figure with an evenly spaced grid, then use a box-counting algorithm to shrink the boxes so we have a finer and finer covering to approximate the actual shape. If each box is length ε and there are $N(\varepsilon)$ boxes, the box dimension is defined by

$$d_{\text{box}} := \lim_{\varepsilon \to 0} \frac{\ln N(\varepsilon)}{\ln(1/\varepsilon)}. \tag{12.112}$$

How is the *Hausdorff dimension* defined? This may be challenging to understand if you are not familiar with measure theory, or at least some real analysis.

(3) The generalized uncertainty principle of the form

$$\Delta x \Delta p \geqslant \frac{1}{2}\left(\hbar + \alpha \ell_{\mathrm{P}}^2 \frac{\Delta p^2}{\hbar}\right) \tag{12.113}$$

may be only the first order correction of quantum gravity. Due to the series $e^x = 1 + x + x^2/2! + \cdots$, one possible form of GUP that reduces to this in the small α limit is the exponential GUP:

$$\Delta x \Delta p \geqslant \frac{\hbar}{2} \exp\left(\frac{\alpha \ell_{\mathrm{P}}^2 \Delta p^2}{\hbar^2}\right). \tag{12.114}$$

Following the approach in deriving Equation (12.80), find the modified Hawking temperature under the exponential GUP. Assuming $\alpha > 0$, does such a GUP still lead to a black hole remnant?

(4) Using Equation (12.80) and assuming that $\alpha > 0$, show that the first law of black hole thermodynamics implies that the entropy is given by (in Planck units):

$$S_1 = 2\pi M^2 + 2\pi M\sqrt{M^2 - \frac{\alpha}{4}} - \frac{\alpha\pi}{2}\ln\left(M + \sqrt{M^2 - \frac{\alpha}{4}}\right) + C_1(\alpha). \tag{12.115}$$

For large M, show that this reduces to the form in Equation (12.81). An alternative form is

$$S_2 = \frac{\pi\alpha}{2 - \sqrt{4 - \frac{\alpha}{M^2}}} - \frac{\alpha\pi}{2}\tanh^{-1}\left(\frac{1}{2}\sqrt{4 - \frac{\alpha}{M^2}}\right) + C_2(\alpha). \tag{12.116}$$

Here C_1 and C_2 are constants that only depend on α. Show that these two expressions are equivalent. What are C_1 and C_2 explicitly? Conclude that neither S_1 nor S_2 make sense if $\alpha < 0$. What is the correct expression in this case?

References

Abreu, E. M. C., & Neto, J. A. 2022, PhLB, 835, 137565
Adler, R. J., Chen, P., & Santiago, D. I. 2001, GReGr, 33, 2101
Anacleto, M. A., Brito, F. A., Campos, J. A. V., & Passos, E. 2020, PhLB, 810, 135830
Anacleto, M. A., Campos, J. A. V., Brito, F. A., & Passos, E. 2021, AnPhy, 434, 168662
Anagnostopoulos, F. K., Basilakos, S., & Saridakis, E. N. 2020, EPJC, 80, 826
Babourova, O., & Frolov, B. 2022, Univ, 6, 230
Barrow, J. D., Basilakos, S., & Saridakis, E. N. 2021, PhLB, 815, 136134
Barrow, J. D. 2020, PhLB, 808, 135643
Bekenstein, J. D. 1981, PhRvD, 23, 287
Biró, T. S., & Czinner, V. G. 2013, PhLB, 726, 861
Bosso, P., Luciano, G. G., Petruzziello, L., & Wagner, F. 2023, CQGra, 40, 195014
Brown, A. R. 2018, PhRvD, 98, 036008
Casini, H. 2008, CQGra, 25, 205021
Chen, C.-M., Chen, Y., Ishibashi, A., Ohta, N., & Yamaguchi, D. 2022, PhRvD, 105, 106026
Chen, C.-M., Chen, Y., Ishibashi, A., & Ohta, N. 2024, ChJPh, 92, 766
Contreras, E., Villalba, F. D., & Bargueño, P. 2016, EL, 114, 50009
Czinner, V. G., & Iguchi, H. 2016, PhLB, 752, 306
Demichel, Y. 2023, [arXiv:2308.15093 [math.HO]] 2023arXiv230815093D
Di Gennaro, S., & Ong, Y. C. 2022, Univ, 8, 541
Di Gennaro, S., Xu, H., & Ong, Y. C. 2022, EPJC, 82, 1066
Drepanou, N., Lymperis, A., Saridakis, E. N., & Yesmakhanova, K. 2022, EPJC, 82, 449
Eichhorn, A., & Held, A. ed. C. Bambi 2023, Regular Black Holes, Springer Series in Astrophysics and Cosmology (Singapore: Springer)
Good, M. R. R., & Ong, Y. C. 2025, AnPhy, 474, 169919
Havrda, J., & Charvát, F. 1967, Kyber, 3, 30
Jacobson, T. 1995, PhRvL, 75, 1260
Jahromi, A. S., Moosavi, S. A., Moradpour, H., et al. 2018, PhLB, 780, 21
Judge, D. 1963, PhL, 5, 189
Judge, D. 1964, NCim, 31, 332
Kaniadakis, G. 2002, PhRvE, 66, 056125
Kaniadakis, G. 2005, PhRvE, 72, 036108
Lü, H., Di Gennaro, S., & Ong, Y. C. 2025, AnPhy, 474, 169914
Lü, H., & Ong, Y. C. 2025, PhLB, 868, 139691
Leon, G., Magaña, J., Hernández-Almada, A., et al. 2021, JCAP, 12, 032
Luciano, G. G., & Saridakis, E. N. 2022, EPJC, 82, 558
Mashhoon, B., & Wesson, P. S. 2004, CQGra, 21, 3611
Mathew, A., & Nandy, M. K. 2021, RSOS, 8, 210301
Mendes, R. V. 2019, PhRvD, 99, 123006
Niedermaier, M. 2007, CQGra, 24, R171
Ong, Y. C. 2018, JCAP, 09, 015
Ong, Y. C. 2018, JHEP, 10, 195
Ong, Y. C. 2023, EPJC, 83, 209
Ong, Y. C. 2020, EPJC, 80, 777
Ong, Y. C. 2025, PhLB, 870, 139936
Overduin, J. M., Wesson, P. S., & Mashhoon, B. 2007, A&A, 473, 727

Pawlowski, J. M., & Stock, D. 2018, PhRvD, 98, 106008
Platania, A. 2023, Handbook of Quantum Gravity, ed. C. Bambi, L. Modesto, & I. Shapiro (Singapore: Springer)
Platania, A. 2025, GReGr, 57, 58
Rényi, A. 1961, Proc. 4th Berkeley Symp. on Mathematics, Statistics and Probability, (Budapest: Hungarian Academy of Sciences) 547
Rashidi, R. 2016, AnPhy, 374, 434
Scardigli, F., & Casadio, R. 2015, EPJC, 75, 425
Sheykhi, A. 2021, PhRvD, 103, 123503
Shokri, M. 2025, PhLB, 860, 139193
Sorkin, R. D. 1996, Proc. 1st Australasian Conf. on General Relativity and Gravitation, ed. D. Wiltshire (Adelaide: Univ. Adelaide) 163
Tsallis, C. 1988, JSP, 52, 479
Tsallis, C., & Cirto, L. J. L. 2013, EPJC, 73, 2487
Weinberg, S. 2003, Natur, 426, 389
Wetterich, C. 1995, A&A, 301, 321
Wetterich, C. 2010, PhRvD, 81, 103507
Wetterich, C. 2017, PhLB, 73, 6
Xia, J., & Ong, Y. C. 2024, Univ, 10, 177

Chapter 13

The Big Questions

> *"And therefore as a stranger give it welcome. There are more things in heaven and earth, Horatio, than are dreamt of in your philosophy."* — William Shakespeare, "Hamlet".

In this chapter, we explore some "big questions", such as whether the multiverse exists; what is cosmic inflation and whether it really happened; how the Universe might end; and whether there are extraterrestrial civilizations elsewhere, or are we the only intelligent species capable of figuring out how the Universe works. Who knows, maybe we are just living in some kind of simulation? It is very difficult to answer questions like these and many others. Nevertheless, big questions stimulate our curiosities and attract future talents into physics and astronomy. Therefore, let us indulge ourselves with some fun speculations, while maintaining a good dose of our scientific skepticism.

We are now at the final chapter before the conclusion, and I would like to discuss some "big questions". There are of course many such questions, and everyone has their own list of what questions are considered big. This list is therefore neither objective nor exhaustive. Humans as a species has always been interested in questions like "what is the meaning of life", which many philosophers have attempted to provide some answers to. Figure 13.1 depicts the famous painting *Where Do We Come From? What Are We? Where Are We Going?* by Paul Gauguin (1848–1903). These are certainly big questions.

As far as physics is concerned, however, the ultimate origin of us and everything else in the Universe came from the Big Bang, and our eventual fate depends on the mass and energy content of the Universe. These are cosmological questions. "What are we?" can also be given a cosmological explanation. Elements that made us were produced in and by the stars during various processes, including supernovae explosions. For reasons that we still don't understand, some arrangement of matter gave rise to

doi:10.1088/2514-3433/ade963ch13

Figure 13.1. *Where Do We Come From? What Are We? Where Are We Going?* is the largest painting by Paul Gauguin. He wrote to de Monfried in 1898, explaining the painting: (Paulson 1921) "To the right at the lower end, a sleeping child and three crouching women. Two figures dressed in purple confide their thoughts to one another. An enormous crouching figure, out of all proportion and intentionally so, raises its arms and stares in astonishment upon these two, who dare to think of their destiny. A figure in the center is picking fruit. Two cats near a child. A white goat. An idol, its arms mysteriously raised in a sort of rhythm, seems to indicate the Beyond. Then lastly, an old woman nearing death appears to accept everything, to resign herself to her thoughts. She completes the story! At her feet a strange white bird, holding a lizard in its claws, represents the futility of words.... So I have finished a philosophical work on a theme comparable to that of the Gospel." *Credit:* Tompkins Collection—Arthur Gordon Tompkins Fund (MFA Boston).

consciousness (how consciousness arose is another big question), which allows us to contemplate the Universe[1]. As the writer Alan Wilson Watts (1915–1973) wrote:

> "Through our eyes, the universe is perceiving itself. Through our ears, the universe is listening to its harmonies. We are the witnesses through which the universe becomes conscious of its glory, of its magnificence."

The great science communicator Carl Sagan (1934–1996) also expressed a similar sentiment:

> "The cosmos is within us. We are made of star-stuff. We are a way for the universe to know itself."

Many big questions do not have clear answers. However, they are crucial to motivate us to explore, to seek answers. Science needs big questions to keep our curiosities and dreams intact. We may be only capable of solving smaller problems in our everyday research, but it is good to keep something grander in view, questions that we strive to answer someday. Big questions are also the sort of questions that tend to attract young students into Science. This is why good science communication

[1] Brain injury induced phenomena like "blindsight", in which the patient can see and react to visual stimuli but not consciously aware of their ability, suggests that consciousness may not be necessary for survival of any animal species. If so, why did we evolve to become conscious? On the other hand, could what we think as awareness be an illusion?

to the public is extremely important. It's not just that an educated public can in turn support government policies that are conducive of scientific research, but also kids and talented young students can be inspired to pursue a career in the sciences.

13.1 Why Is There Something Rather Than Nothing?

Why do we, and indeed, the entire Universe, exist at all? Is it not simpler to have absolutely nothing than our complicated world? This may sound like a strange question, but what we are actually asking is whether there is any law of physics that *necessitates* the birth of our Universe. More poetically, did God have any choice to not create the Universe?[2] Perhaps the state of nothingness is somehow unstable, so it has to "decay" into *something*. Clearly such a question is almost impossible to answer to everyone's satisfaction. The philosopher Martin Heidegger (1889–1976) called it "the fundamental question of metaphysics".

The reason why this question is so difficult is we have to first agree on what do we mean by "nothing". This is quite different from asking whether there is *something* else before the Big Bang, in which case there are many possibilities. For example, there may not be a Big Bang singularity, but rather a "bounce". In such a scenario, there was a contracting phase of the cosmos that shrunk to a sufficiently small region before quantum gravity effects caused it to bounce instead of collapsing fully into a Big Crunch singularity. It is also possible that, as Gibbons and Hawking proposed, asking what was there before the Big Bang does not make sense, since time itself begins at the Big Bang, much like there is no "northward" at the North Pole[3]. Indeed, in the Gibbons–Hawking no-boundary proposal, "before" there was time, the Universe could be described as a manifold of $(+, +, +, +)$ signatures. In popular science parlance, time was imaginary (Wick-rotated). However, this was just a mathematical description for a quantum tunneling process from what is technically called an *instanton* solution with topology of a four-dimensional sphere S^4 to the de Sitter spacetime dS_4, for example. But one could argue that even an instanton is not "nothing".[4] Carrying the question further, even if we agree on what

---

[2] "Who knows whence this creation had its origin?
He, whether He fashioned it or whether He did not,
He, who surveys it all from the highest heaven,
He knows—or maybe even He does not know."

—Rig Veda, Hymn 129

[3] In Book 11, Chapter 12 of the *Confessions of St Augustine*, St. Augustine wrote:

> "Behold, I answer to him who asks, 'What was God doing before He made heaven and earth?" I answer not, as a certain person is reported to have done facetiously (avoiding the pressure of the question), "He was preparing hell," says he, "for those who pry into mysteries.' "

Note that contrary to popular belief, St. Augustine did not give such an answer to the question, but merely stated that someone else did! (*The Confessions* is available at https://www.newadvent.org/fathers/110111.htm.)

[4] For some technical attempts to understand nothing as the absence of spacetime, see Brown & Dahlen (2012).

"nothing" is, do we then have to explain *why* there was nothing? Where did nothing come from? What was the "first cause"?

A more well-defined question, as far as science is concerned, is why there is any *matter* at all. This is the problem of matter and antimatter asymmetry that we mentioned in Section 5.4. To recap: quantum field theory and particle physics tell us that at the beginning when the Universe is extremely hot, there is enough energy to pair-produce particles from the vacuum. For example, if the available energy is more than twice the electron mass, then a pair of electron and its antiparticle—the positron—can be created from the vacuum. However, if for each particle there is an antiparticle partner, then they will all annihilate back to pure energy (photons) via collisions. The fact that we are here today means that there is a tiny asymmetry between the number of particles and anti-particles, though explaining *why* this is the case has been difficult. An intriguing possibility involves the topology of the Universe. If the spatial section has a nontrivial topology, a particle that circumnavigates the cosmos can return to its initial position as an antiparticle. This is similar to a left-handed glove that goes around a non-orientable surface of a Möbius strip and comes back being right-handed. Such a universe is called an "Alice universe", named after Alice in Lewis Carroll's children's book *Through the Looking-Glass.* In other words, particle/antiparticle distinction cannot be defined consistently at every point of space. Thus, if we imagine that when all the particles and anti-particles were produced they have 1: 1 ratio, this would not be the case after some time just by the fact that some of them have circumnavigated the Universe. Such a journey quickly became impossible after the Universe has expanded large enough, and the ratio would then be fixed. A "freeze-out[5] by topology" if you will. However, for this idea to be taken seriously one has to devise a concrete model that can get the right number that matches the observation.

As for the original question concerning true "nothingness", it is likely that this question would be forever outside the reach of science. Philosopher Sidney Morgenbesser (1921–2004) once said that "if there were nothing, you'd still be complaining!", knowing full well that no one would be around to complain if there is nothing! This then ties in with the "anthropic principle", the weakest version of which simply says that the Universe is as it is because if otherwise we would not be here to ask why the Universe is like this. We began this section asking if there is any law of physics that necessitates the existence of the Universe, but can there be any law of physics without the Universe being already in existence in some form? The laws may not be unique of course, not to mention the values of the constants of nature, in which case some universes may harbor life and some would not, but perhaps it does not make sense to contemplate the existence of any law prior to the Universe (or multiverse?) came into being?

We end our discussion with a quote from William Clifford in the article "The First and the Last Catastrophe", published in *Popular Science Monthly*, Volume 7 in July 1875:

---

[5] In Big Bang cosmology, we say that a process "freezes out" if it is no longer reversible. The relative abundance of particles becomes "frozen" at that point. This is usually caused by the temperature being too low to enable the reverse process. See Section 5.4.

"If we consider the case of the whole universe we should be able, supposing we had paper and ink enough, to write down an equation which would enable us to make out the history of the world forward, as far forward as we liked to go; but if we attempted to calculate the history of the world backward, we should come to a point where the equation would begin to talk nonsense—we should come to a state of things which could not have been produced from any previous state of things by any known natural laws."

13.2 Universe or Multiverse?

In the previous section, we contemplated the possibility that the laws of physics, as well as physical constants, could be different. Our universe may just happen to pick a set of these. Recall from Section 2.2 that Kepler once tried to explain why there were only six planets in the solar system (before Uranus and Neptune were discovered), believing that there exists a law of physics that dictates said number. We now know that there is no such law, only initial conditions. Right now we don't know whether the laws governing our Universe *have* to be so, uniquely, or indeed there are other possible ones. If so, could there exist out there other universes separated from our own, that have these different laws? The collection of all these universes is called a multiverse (this is not the only kind; see below).

Well, of course we can always contemplate such a possibility, the question for physics is whether such a scenario is ever testable. Imagine our universe as a bubble floating in some higher dimensional ambient space, with other bubble universes around. If those universes are not in causal contact with ours, there is no way we can detect them. But there could be interactions. For example, Justin Khoury, Burt Ovrut, Paul Steinhardt and Neil Turok proposed the "Ekpyrotic universe" scenario in 2001 (Khoury et al. 2001). The name is based on the "ekpyrosis" idea of ancient Greece, in which the cosmos ends with fire, follows up by a cycle of rebirth. The Ekpyrotic universe model envisions that our Universe can collide with another universe, separated by some distance in the extra dimension. The collision gives rise to a Big Bang, and the two universes bounce away from each other, possibly colliding again in the future and the cycle repeats. It is very hard to test such a theory. There is also the possibility that two universes can collide and merge. In fact, as discussed in Section 10.1, black hole interiors can be viewed as a cosmological spacetime, and we know that black holes do collide, so it is quite natural to think about what a black hole merger looks like from the perspective of an interior observer. Interestingly, there were claims that some models of colliding universes can fit current observation data (Osborne et al. 2025; Wainwright et al. 2014; Johnson et al. 2016).

In Section 10.1 we also mentioned that the interior volume of a black hole continues to grow even though its area shrinks due to Hawking radiation. This suggests that at the end, either the huge volume is still connected to a Planck size remnant outside, or it completely "snaps" off as the black hole completely evaporates from the exterior point of view. The former describes a baby universe that is still connected to the mother universe, whereas in the latter, the baby universe is completely split off, becoming a bubble universe. Other bubble universes can also form in the inflationary multiverse picture, which we shall discuss in the next subsection.

Of course, just because there can be other physical laws does not imply that these universes are actually realized. It is often said that string theory allows 10^{500} different universes (technically, "vacua" of the theory). The number can be higher; one paper claimed the number 10^{272000} for F-theory flux vacua (Taylor & Wang 2015). It would have been better if we have a theory of quantum gravity that gives a unique solution that actually describes the actual Universe (i.e., predicts the right number of forces, elementary particles and their masses, the values of the coupling constants, etc.), but alas we have a lot more, way more possibilities. What's even worse is that this so-called "string landscape", vast though it is, may not contain the actual observed Universe as a solution! Our accelerated expansion cosmology seems to be best described by an effective low-energy physical theory whose UV-completion is not compatible with quantum gravity (i.e., de Sitter spacetime is not a solution of quantum gravity). Such theories are said to lie in the "swampland". Efforts have been made to devise clever ways to incorporate de Sitter solutions or de Sitter-like solutions into the landscape. Popular in recent years are a set of conjectured criteria for theories to belong to the quantum gravity landscape, known collectively as the "Swampland conjectures". As summarized in Cassé & Silk (2022),

> "For example, the no-global symmetry conjecture avoids overclosure of the Universe by black hole remnants of the Planck mass, the *distance conjecture* holds that scalar fields cannot exhibit field excursions much larger than the Planck scale, while the *weak gravity conjecture* states that the lightest particles of a theory cannot carry a mass larger than their charge in Planck units. The transplanckian Swampland conjecture forbids trans-planckian quantum fluctuations from becoming classical."

At least to me, these seem way too many conjectures to worry about. I wonder if this approach to quantum gravity is on the right track.

There is another type of multiverse which has to do with the interpretation of quantum mechanics. The usual interpretation taught in textbooks (which mostly even avoid mentioning that it is an interpretation), is the "Copenhagen interpretation", for which the quantum wavefunction "collapses" when a measurement is made, where the probability is given by the Born rule: Prob. $=|\Psi|^2$, named after Max Born (1882–1970). The "collapse" of the wavefunction is neither built into any mathematics of Hilbert space underlying quantum mechanics, nor is it dictated by the Schrödinger equation. To many philosophers of physics, the collapse is thus an ad hoc addition to the theory. There are many alternative interpretations, and a better-known one to most physics enthusiasts is the many-world interpretation (MWI), in which all possible outcomes are realized in some branches of the "parallel universes". As an interpretation, its prediction is just the same as the Copenhagen interpretation, and hence we cannot conclude whether this "multiverse" of parallel universes actually exists "somewhere". For all practical purposes we are stuck to only observing our own branch.

13.3 Was There An Inflation?

Some cosmologists may not like that I have included inflation in such a speculative chapter. However, since there is still no concrete proof for inflation, I think it is quite fair. In any case, the fact that I include the topic means that I do think that inflation is an important and interesting big question in need of more concrete answers.

Historically, the idea of cosmic inflation, first proposed by Alan Guth in 1980, in which the very early Universe went through a short but extremely pronounced accelerated expansion, was motivated by some puzzles that concern the initial conditions of the Universe. We will first discuss some of these issues, but we will also explain why they are no longer the best motivations today.

We begin with the *horizon problem*. Consider the initial conditions at early time t^*, say, during the radiation-dominated era, in which $\rho \propto a(t)^{-4}$ and $a(t) \propto \sqrt{t}$. Whether one region is causally connected to another is determined by the *particle horizon* with radius $2t^*$ (see Section 3.4 for the definition of a particle horizon). No region larger than twice this distance should be in causal contact, and therefore should not have the same temperature (i.e., should not appear homogeneous). In FLRW cosmology, the temperature scales as the inverse of the scale factor: $T(t) \propto 1/a(t)$, so that the size of homogeneous region is of radius

$$2t^* \times T(t^*)/T(t_0). \tag{13.1}$$

(We use subscript 0 to denote the current values.) The current temperature of the CMB is roughly $T_0 \sim 2.7$ K, back then in the units of GeV, we have $T(t^*) \sim 10^{15}$, and using FLRW model we can compute t^*. In the radiation-dominated era, $a(t) \propto t^{1/2}$, so with $T(t) \propto 1/a(t)$, we get $t \propto T^{-1}$. More precisely it can be shown that (Mukhanov 2012)

$$t \approx 1.39\kappa^{-\frac{1}{2}}T_{\text{MeV}}^{-1}, \tag{13.2}$$

where

$$\kappa := \frac{\pi^2}{30}\left(g_b + \frac{7}{8}g_f\right), \tag{13.3}$$

with g_b and g_f referring to the numbers of internal degrees of freedom of all relativistic bosons and fermions, respectively. With the known numbers substituted into Equation (13.1), one finds that the current size of the homogeneity region should be about the order of 1 m, instead of the CMB homogeneity size of $O(10^{26})$ m, which is the present horizon scale! Why should regions that are out of causal contact have the same temperature (with variation of only 1 part in 100,000)?

The *flatness problem* concerns why the spatial geometry of the Universe seems to be so flat, i.e., $k \approx 0$. We first present the standard argument of many textbooks (not wrong but misleading!). From the Friedmann equation for the FLRW geometry, one can obtain the equation for the sectional curvature k:

$$\frac{k}{a(t)^2} = \left(\frac{\rho}{\rho_c} - 1\right)\frac{\dot{a}(t)^2}{a(t)^2}, \tag{13.4}$$

where $\rho_c(t) = 8\pi G/3H(t)^2$ is the critical density. The density parameter is $\Omega(t) = \rho/\rho_c$. Then during the radiation-dominated era with $a(t) \propto t^{1/2}$, we have

$$\frac{k}{a(t)^2} = \frac{\Omega - 1}{4t^2}. \tag{13.5}$$

The ratio of this to the current value of

$$\frac{k}{a(t_0)^2} = (\Omega_0 - 1)H_0^2, \tag{13.6}$$

leads to, using the fact that $a(t) \propto 1/T(t)$,

$$\Omega - 1 = 4H_0^2 t^2\left(\frac{T^2}{T_0^2}\right)(\Omega_0 - 1). \tag{13.7}$$

The near-flatness now means that k is very small, which means that $|\Omega_0 - 1|$ is very small. So let us be generous and use $|\Omega_0 - 1| \sim O(1)$. Again, plugging in the known values we can estimate that at time t^* during the radiation-dominated era, we get a ridiculously small number $|\Omega(t^*) - 1| \sim O(10^{-53})$. This is because if Ω is not exactly unity, then $|\Omega - 1|$ will always grow as the Universe expands. This can be seen from the relation

$$\frac{k}{a(t)^2 H(t)^2} = \Omega - 1, \tag{13.8}$$

given that in the standard FLRW cosmology, $H(t) := \dot{a}(t)/a(t)$ is a decreasing function of time. Furthermore in the standard non-inflationary FLRW model, the combination $a(t)H(t)$ is also a decreasing function. This is because the expansion decelerated:

$$\frac{d}{dt}(\dot{a}) = \frac{d}{dt}(aH) < 0. \tag{13.9}$$

Therefore it seems that the fact that the Universe is very flat now requires that it is even flatter the further we go back in cosmic time. This seems to suggest a fine-tuning of the value of the sectional curvature k. *However*, note that this derivation is simply the property of any FLRW universe, as long as we assume the metric to hold! In other words, the so-called fine-tuning in the above derivation is *not* a fine-tuning of k to be near zero in early time, it is a necessary generic behavior as long as we assume the FLRW geometry to hold. Peter Coles and George Ellis emphasized that "*there is no flatness problem* in a purely classical cosmological model" [emphasis in the original] (Coles & Ellis 2010). For more discussions on this and related issues, see Helbig (2021); Singal (2024); Carroll & Tam (2025); Helbig (2012); Holman (2018). In order to argue that there is a flatness problem, one has to show that the "fine-tuning" of the sectional curvature remains even if we start with a geometry that is *not* FLRW in the beginning, which somehow evolves into near-FLRW later on.

In any case, it is said that inflation solves the horizon problem and the flatness problem because (1) it allowed regions that were otherwise far apart to be way closer back then, and (2) it "smoothed out" all local inhomogeneities. To see how (2) happened, it is important to realize that inflation means accelerated expansion, and so $\ddot{a}(t) > 0$. That is to say,

$$\frac{d}{dt}(\dot{a}) = \frac{d}{dt}(aH) > 0. \tag{13.10}$$

Therefore from Equation (13.8), we see that the denominator increases with time, meaning that $\Omega(t) - 1$ is becoming smaller as the Universe expands. This is precisely the opposite behavior from the one that leads to the flatness problem. An explicit simple model is to consider $a(t) \sim e^{Ht}$ during inflation. Then $|\Omega - 1| \sim e^{-2Ht}$, and this exponential decay quickly drives Ω towards unity. We may consider H as a constant and write

$$a(t) \sim \exp\left(H \int dt\right) = e^{-N(t)}. \tag{13.11}$$

This yields $dN = -d \ln a = -Hdt$. That is, upon integrating from some initial time t_i to final time t_f,

$$N(t_i) - N(t_f) = \ln \frac{a(t_i)}{a(t_f)} = H(t_f - t_i), \tag{13.12}$$

where $a(t_f)$ and $a(t_i)$ are the scale factor at the beginning and the end of inflation, respectively. Inflation ends when exponential expansion ceases, i.e., when $N(t_f) = 0$. The initial value $N(t_i)$ has to be determined from observational requirement. It has been estimated that the amount of inflation needed to resolve both the horizon and flatness problem is $N(t_i) \sim 60$. We refer to this as the number of "e-folds". As a consequence of this the horizon size shrinks by e^{-60} due to inflation, so that two points that are in causal contact during inflation need not necessarily continue to be so after inflation ends. The horizon of course grows again during the usual radiation-dominated era after the end of inflation. This means that a density perturbation can start with being inside the horizon to being outside, and then re-enters the horizon again later. This horizon exit and re-entry allows what we observe to be in our horizon now to be in contact during or before the inflationary epoch, thus solving the horizon problem.

It is instructive to consider one simple model of inflationary cosmology, in which inflation was caused by a scalar field ϕ, the "inflaton", whose equation of motion satisfies

$$\ddot{\phi} + 3H\dot{\phi} = -\frac{\partial H}{\partial \phi}. \tag{13.13}$$

The second derivative term and the derivative of the potential look familiar, it is just the equivalent of $F = ma$. The term $3H\dot{\phi}$ can thus be seen as a viscosity or "Hubble drag" term, so that overall we have an equation that is analogous to $F = mg - bv$, for a body falling in the presence of a drag term $-bv$.

Treated as a perfect fluid, the energy density and pressure of the inflaton are

$$\rho = \frac{1}{2}\dot{\phi}^2 + V(\phi), \tag{13.14}$$

and

$$p = \frac{1}{2}\dot{\phi}^2 - V(\phi), \tag{13.15}$$

respectively. From the acceleration equation,

$$\frac{\ddot{a}}{a} = -\frac{4\pi G}{3}(\rho + 3p), \tag{13.16}$$

we see that in order to have $\ddot{a} > 0$, we must have $\rho + 3p < 0$. From the above equations for ρ and p, this leads to the requirement that

$$\dot{\phi}^2 < V. \tag{13.17}$$

It turns out that typically we require a stronger relation $\dot{\phi}^2 \ll V$, known as the "slow-roll" condition. Another condition that is related to the duration of the inflation (and thus the e-folding number) is that $\ddot{\phi} \ll H\dot{\phi}, \quad V'$. This gives $3H\dot{\phi} + V' \approx 0$. We have used a prime to denote the derivative with respect to the field ϕ. To see why this condition is related to the duration, simply notice that the analogy $F = mg - bv$ also implies that there is a terminal velocity $v = mg/b$ for which the force is zero. Here it is similar, but with $\ddot{\phi} = 0$ instead when inflation ends. The "terminal velocity" is $\dot{\phi} = -V'/3H$.

The Hubble parameter satisfies the Friedmann equation

$$H^2 = \frac{8\pi G}{3}\left(\frac{1}{2}\dot{\phi}^2 + V\right). \tag{13.18}$$

The ratio of the two terms in the bracket is

$$\frac{\frac{1}{2}\dot{\phi}^2}{V} \approx \frac{\frac{1}{2}\left(-\frac{V'}{3H}\right)^2}{V} \approx \frac{1}{48\pi G}\frac{V'^2}{V^2}. \tag{13.19}$$

Thus we expect that the ratio $(V'/V)^2$ to play an important role in the dynamics. One often denotes

$$\varepsilon := \frac{1}{16\pi G}\frac{V'^2}{V^2}, \tag{13.20}$$

as well as

$$\eta := \frac{1}{8\pi G}\frac{V''}{V}. \tag{13.21}$$

These dimensionless quantities are the so-called *slow-roll parameters*; they need to satisfy $\varepsilon \ll 1$ and $|\eta| \ll 1$. We can classify inflationary models according to the magnitude and sign of these parameters. In particular, if $0 < \eta \leqslant \varepsilon$, we have a "large field inflation", whereas if $\eta < 0 < \varepsilon$ we have a "small field inflation". A "hybrid inflation" satisfies $0 < \varepsilon < \eta$. At the end of the inflation, the energy of the inflaton has to be converted into radiation so that the usual radiation-dominated era starts. How to make inflation ends after the required amount of e-folds, without messing with the radiation-dominated era, is known as the *graceful exit problem*. How this can be achieved depends on how ϕ couples to the standard model particles, which is highly model-dependent. There are also models that utilize more than one inflaton field, but we will not get into the details here.

These are the usual stories told about inflation. However, it should be emphasized that whether the horizon and flatness problems are indeed problematic has been called into question (Helbig 2021; Singal 2024). It is also possible that we can explain why the Universe has to be born *exactly* flat mathematically, perhaps by relating it to the gravitational entropy being low in the past (see, e.g., McInnes 2007, 2025). Indeed, there are possible solutions of the flatness and monopole problems[6] that do not rely on inflation. It has also been argued that trying to avoid fine-tuning of the initial conditions (that result in the horizon and flatness problems) by appealing to inflation alone cannot work, because inflation itself requires special initial conditions to start, for example the "slow-roll" condition for the simplest model, as we have seen above. This means that at least some models of inflation are also fine-tuned to some degrees. In fact, from the view of the arrow of time (see Section 4.3), this is only natural—entropy increases in time, so since inflation occurs before a radiation-dominated era, the overall entropy has to be *lower* back then. Therefore, inflation itself cannot serve to answer the question of why entropy was so low in the past (Page 1983, 1984). What the arrow of time problem is telling us is that the pre-inflationary Universe cannot be arbitrarily generic.

These do not mean that inflation is a bad idea, but it is good for other reasons, as stated by Thanu Padmanabhan (1957–2021) in his book *Gravitation* (Padmanabhan 2010), p.631:

> "Originally inflationary scenarios were suggested as a pseudo-solution to certain pseudo-problems; these are only of historical interest today and the only reason to take the possibility of an inflationary phase in the early universe seriously is because it provides a mechanism for generating the initial perturbations."

Likewise, in the book *The Primordial Density Perturbation* by David Lyth and Andrew Liddle (Lyth & Liddle 2012), the authors wrote on p.307 that

> "In the modern view, by far the most important function of inflation is to generate the primordial curvature perturbation [$\cdots$] It may generate other

[6] The monopole problem concerns the absence of the magnetic monopoles. Their presence in the very early Universe is expected from theoretical prediction of the Grand Unified Theories. The inflationary explanation is that their number density has been considerably diluted by inflation.

primordial perturbations too, including the isocurvature and tensor perturbations [⋯] However, the historical motivation for inflation was rather different, and arose largely on more philosophical grounds concerning the question of whether the initial conditions required for the unperturbed Big Bang seem likely or not."

In addition, it is thought that inflation "magnified" the quantum fluctuation in the very early Universe, which then led to density perturbations that later grew via gravitational instability into large cosmological structures like galaxy clusters.

We should also briefly discuss some problems of inflation. The potential $V(\phi)$ of the inflaton can have many local minima, each of which is a vacuum. The Universe before inflation was in a false vacuum state, which then underwent a quantum tunneling into a true vacuum. However, it is quite possible that not everywhere made the transition at one go, some regions may tunnel into a true vacuum earlier than the others. These localized "bubbles" of new vacua each corresponds to an individual universe, while the whole multiverse is in a state of *eternal inflation*.[7] This was pointed out by Paul Steinhardt in 1983. Although individual bubbles have exited the inflationary phase, the inflationary phase lasts forever throughout most of the multiverse, with new bubble universes being continuously created. Each universe may have its own laws of physics, as mentioned in Section 13.2. What is the probability that we find ourselves in a universe that is habitable? Such a question is notoriously difficult to answer since there is potentially an infinite number of possible universes, and making sense of the correct statistical distribution requires us to define a useful "measure" (in the technical sense of measure theory) of the infinity. This *measure problem* remains controversial and unresolved. Paul Steinhardt subsequently gave up on the idea of inflation.

Another question regarding inflation is whether the Universe can be eternally inflating without a past Big Bang singularity. It has become a "folklore" that all inflationary spacetimes are past-incomplete (i.e., singularity cannot be avoided; i.e., the Universe had a beginning at some finite time in the past) as the result of the Borde–Guth–Vilenkin ("BGV") theorem, named after Arvind Borde, Alan Guth and Alexander Vilenkin (Borde et al. 2003). This is not quite true, as there are cases in which an inflationary universe can be past-eternal, though they are arguably "fine-tuned" with very special initial conditions (Geshnizjani et al. 2023).

13.4 How Might the Universe End?

The fate of the Universe largely depends on the true nature of the dark energy. For example, there would be a Big Rip singularity if dark energy is a phantom field, as discussed in Section 6.2; see also Nojiri et al. (2005). If dark energy is a quintessence, it is possible that the current accelerating phase is only temporary, and if furthermore the true cosmological constant is negative, then it will dominate at

[7] In the eternal inflationary multiverse, the bubble universes are not located in a higher dimensional bulk. They are localized regions in the multiverse that have stopped inflating. Meanwhile space between them are still undergoing inflation.

late time. This will eventually cause the Universe to re-collapse (this will also happen if the effective cosmological constant due to the quintessence is negative at late time). The Universe would then either end at a Big Crunch singularity, or a Big Bounce if a certain quantum gravitational effects kicks in, leading to the birth of the next Universe. Of course, if dark energy is simply a positive cosmological constant, the Universe will go on expanding forever and becomes asymptotically de Sitter. These possibilities are depicted in Figure 13.2.

Recall from Section 6.2 that in the presence of a phantom dark energy, there will be a Big Rip singularity in a *finite* time, and all structures from galaxies to planets

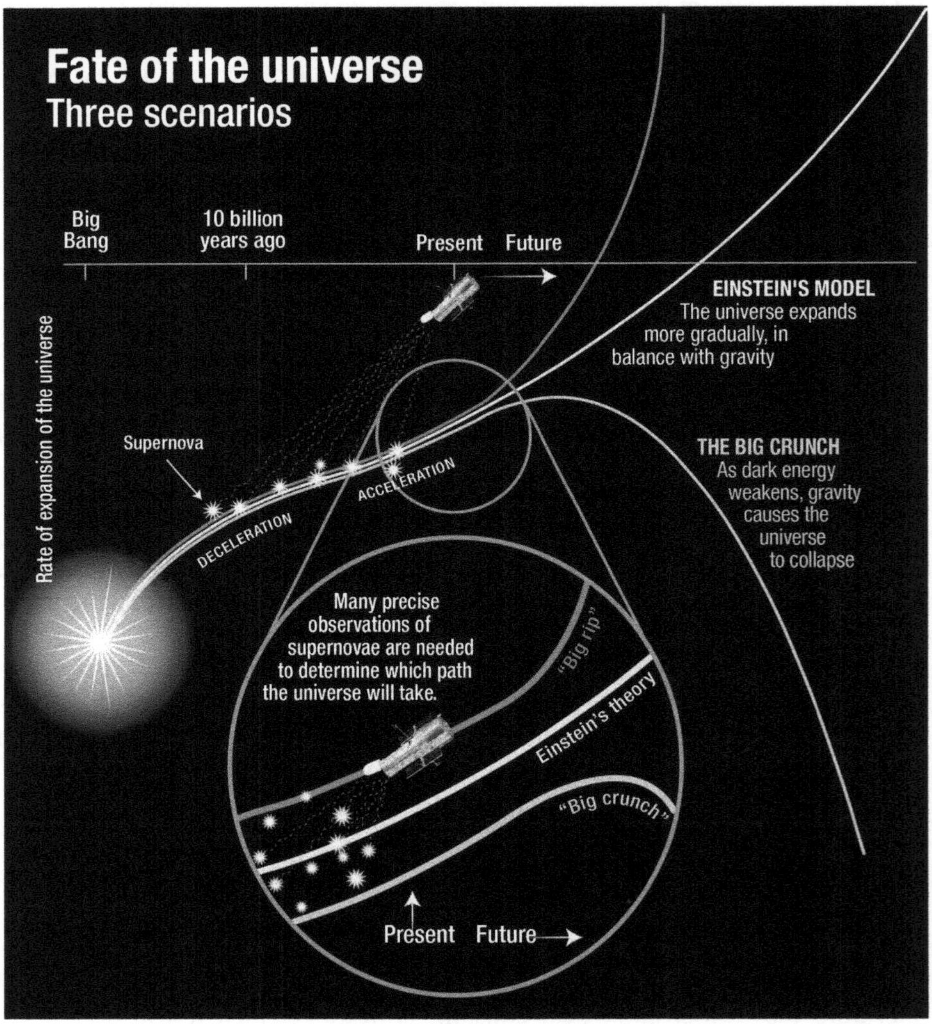

Figure 13.2. We deduced that the expansion of the Universe is accelerating by observing supernovae from the past. The fate of the Universe, however, is still unclear. It depends on the nature of dark energy. In particular, we still cannot rule out the possibility that the Universe might re-collapse (the "Big Crunch" scenario), or the possibility that the expansion becomes so drastic that it "tears" itself apart (the "Big Rip" scenario). *Credit:* NASA and A. Feild (STScI).

and eventually down to molecules and atoms are ripped apart by the expansion of space. There are variations to this scenario, such as the "Little Rip" (Frampton et al. 2011), characterized by the equation of state w_{DE} approaching -1, while the scale factor a, the Hubble parameter H and its derivative $\dot{H} \to \infty$ as $t \to \infty$. This should be contrasted from the Big Rip scenario in which $a \to \infty$ in finite time. Yet another scenario called the "Little Sibling of Big Rip", is characterized by a, $H \to \infty$ as $t \to \infty$, but $\dot{H}$ approaches a finite value (Bouhmadi–Lopez et al. 2015).

In the standard ΛCDM picture, however, things are nowhere near as dramatic. The Universe is asymptotically de Sitter, and so continues to expand indefinitely. Galaxy clusters become further apart due to the expansion of space, stars eventually burn out, and even black holes evaporate away (if there is no stable remnant). The future Universe would then essentially be empty space filled with massless particles (mainly photons; but also some massive particles like electrons). The Universe reaches the maximal entropy state, a state with no thermodynamic free energy available. This is the "heat death" of the Universe, with nothing interesting happening anymore, except some fluctuations away from equilibrium that happens from time to time, as proposed by Boltzmann long ago (see Section 4.3). Heat death does *not* mean that the Universe will become hot. "Heat" is a reference to the temperature gradient that is necessary for work to be done in the thermodynamical sense. Heat death therefore means no more work can be done. An alternative phrase for heat death is the "Big Freeze" or the "Big Chill".

Sir Roger Penrose proposed another possibility, which is known by the name "cyclic conformal cosmology" or simply CCC. The idea is to identify the late time Universe, which has only massless particles (if all massive particles eventually also decay) with the early time of the next universe. In fact, our Universe could be just one of these many "aeons" of universes. The trick here is that without a mass scale, there is no length scale as well. Hence the geometry is only a *conformal* one. Without a length scale, we cannot say whether the size is large or small. Hence it makes sense to identify what we thought of as a large late time Universe, with the early time of the next aeon (Penrose argued that during the Big Bang, particles are essentially effectively massless.). Note that unlike other cyclic cosmological models in which the previous universe has to shrink and then re-expand, there is *no* contracting phase in CCC.

A much scarier scenario is the following: we already mentioned in the previous section that the inflaton field tunnels from a false vacuum to a true vacuum. Such a phenomenon can happen also for other fields, if it is not already in the true vacuum. The Higgs field, for example, given the measured Higgs mass (~125 GeV) at the Large Hadron Collider, is only metastable at best.[8] There is a danger that the Universe might one day catastrophically tunnel into the true stable vacuum state, destroying everything we know in the process. Typically, a metastable physical system has a stability timescale, and it is generally believed that it will take a time much longer than the age of the Universe for this tunneling to happen. Still, this is

[8] Technically, the instability bound is analyzed by examining the renormalization running (the renormalization group flow or "RG"-flow) of the Higgs self-coupling "constant", to see if it becomes negative at a higher energy scale.

somewhat unsatisfactory since one should ideally be able to explain *why* the Universe is in a metastable state, instead of a true vacuum. Related to the measure problem of eternal inflation, one should also expect that most universes in the inflationary multiverse has made the transition to a true vacuum of the Higgs field. Why do we find ourselves in one of the infinitesimal fractions that is in the metastable state? Perhaps the anthropic principle can simply say that we wouldn't be here if the vacuum had been different. But it is not clear at all if other, more stable, true vacua, can give rise to intelligent observers. If so, then the question becomes why we find ourselves to be one of the rarest species of all.

Also, could some super-advanced extraterrestrial civilization achieve a high enough energy scale in their high energy physics experiments that can, just maybe, trigger the vacuum transition much earlier than its natural timescale, and therefore destroy our Universe? Maybe. But personally it's not a scenario that is serious enough to keep me awake at night...

13.5 Are We Alone? Where Is Everybody Else?

Do advanced extraterrestrial civilizations exist somewhere in our Universe, if not our Milky Way Galaxy? Given the tremendous size of our galaxy, not to mention the Universe as a whole, it is almost unfathomable that we are the *only* technological civilization, yet we have not seen any sort of evidence thus far that could indicate the existence of other civilizations. This is known as the *Fermi paradox* (though Enrico Fermi was not the first to raise this question)—given that the chance of intelligent life forms arising in the cosmos is nonzero, and given the age of the Universe, it seems inevitable that civilizations far more advanced than us, those with millions of years head start than us, would be all over our galaxy. So, where are they? Note that the question is not just about the lack of *active communication*,[9] i.e., signals we receive from other planets, but the complete lack of observational evidence. If interstellar civilizations exist, we would expect that they would leave some evidence behind. For example, if they build a *Dyson sphere* around their star to gather energy, or otherwise try to astro-engineer stars and other planets in some unexpected ways, there should be at least some measurable bizarre signatures that we can observe. Indeed there are some stars that exhibit peculiar properties, but there are none thus far that cannot be explained in more mundane manners. Recently there have even been studies of gravitational wave signatures that should be produced by rapidly accelerated spacecraft that are sufficiently massive. The lack of detection of gravitational waves of this sort has put a bound on the existence of such extraterrestrial vehicles. For example, the current LIGO observatory is sensitive to such craft of 1 Jupiter mass accelerating to a fraction of the speed of light (e.g., 30%) from 10 to 100 kpc or a Moon mass from 1 to 10 pc (Sellers et al. 2025).

[9] The lack of active communication can be explained in many ways, such as the "zoo hypothesis", the idea that the human species is not yet ready for contact and is therefore put under observation with a strict no-contact policy; c.f. the "Prime Directive" in the fictional *Star Trek*, which prohibits the interference with the natural development of (unprepared) civilizations.

In addition, even if there is only *one* advanced extraterrestrial civilization in our galaxy that can manufacture self-replicating probes (SRPs)—tiny machines that consume interstellar resources to build more copies of themselves—to explore the galaxy, there would have been enough time for the entire galaxy to be filled with such probes. The fact that we don't see them is strong evidence *against* the existence of such a civilization at least in our galaxy. I don't find the explanation that *all* advanced civilizations have, for one reason or another, decided not to broadcast their existence and even stop exploring to be convincing. As far as SRPs are concerned, there could be the worry that if the programming in such a probe went wrong (perhaps cosmic rays hitting their "circuits" can, overtime, create enough "mutation"), the probes may consume everything in their paths, including their own civilization. Therefore perhaps there is a good reason not to launch SRPs into space. Yet, can we rule out just *one* civilization that decided that the gain outweighed the risk, rule out *all* possible accidents that led to such a probe being released, rule out the possibility of rogue private groups who decided to launch such probes despite the majority of their planet denizens objecting to the idea? It is difficult to rule out so many possibilities, especially when considering that we as humans cannot hope to understand extraterrestrial psychology, much less exo-politics and philosophy! And it only takes *one* civilization to launch SRPs to populate the entire galaxy by now. There have been suggestions that perhaps over time some probes mutated into "predators" that treat the original probes as consumable resources, thus keeping the entire population number in check. However, careful analysis utilizing the Lotka–Voltera equations (the same set of equations used to model population dynamics in biological systems; see Section 1.5) revealed that in such a case, the predatory probe would simply take over as the dominant probe "species" that continues to grow exponentially in number (Chen et al. 2022), so the original puzzle remains.

Personally I think there is another argument against the existence of super-advanced civilization in our galaxy, or at least any such civilization who is keen to show its existence to others. The usual argument for the lack of any signal from the SETI (Search for Extraterrestrial Intelligence) program that monitors electromagnetic radiation (especially radio waves) for signs of transmissions from civilizations on other planets, is that we have only been listening for a relatively short amount of time in a very tiny region of space. Maybe we haven't been looking at the right place. Yet, if a super-advanced civilization wants to be noticed, there are plenty of ways to achieve this. Not every place in the galaxy is equal; some locations are more interesting than others. This is the idea of *Schelling point*—not every location or time is on an equal footing. Schelling asked in 1960: "if you are to meet a stranger in New York City, but you cannot communicate with the person, then when and where will you choose to meet?" The answer then was "noon at Grand Central Terminal" (Schelling 1981). For any galaxy, the obvious place would be the supermassive black hole at the center (the time is less relevant here).

I would think, for example, that Sagittarius A*, the supermassive black hole at the center of our galaxy would be of interest to any civilization for scientific reason. Thus, knowing that most sufficiently advanced species would one day observe Sagittarius A* in detail, any advanced civilization looking for contact could put an artificial star around the orbit of the black hole that clearly emits artificial pulses of signals, or with signals embedded into the spectrum of the light emitted. So far, we have not seen anything like that around Sagittarius A*. However, this does not rule out advanced civilizations who just want to study the black hole without broadcasting their presence. If any artificial satellites are orbiting the black hole, provided that they are massive enough (say, a Jupiter mass), they would emit a gravitational wave signature that could be picked up by future observatories like LISA (Abramowicz et al. 2020).

Therefore, we should take seriously the possibility that, however much we want to believe otherwise, there is no super-advanced civilization around in our galaxy. This may seem unlikely to most physicists, who are used to physical laws being universal, but the emergence of life to its development into a technological civilization is a multi-step process, each of which may be more unlikely than we thought. It is true that building blocks of life such as amino acids and the nucleobases[10] are probably quite common in space, as they have been confirmed to be present on the asteroid Bennu after analysis was performed on the sample collected by NASA's OSIRIS-REx[11] (delivered back to Earth in 2023). Other organic molecules like isopropanol, ethanol, urea, and even the "buckyball" buckminsterfullerene have also been found in interstellar medium. The idea that life originated from organic molecules arriving from space, rather than being produced directly on Earth, is known as "pseudo-panspermia", as opposed to the more radical idea that primitive life forms arrived fully formed from space, known as "panspermia". At the time of writing (April 2025), the news announced the discovery of dimethyl sulfide (DMS) and dimethyl disulfide (DMDS) on the exoplanet K2-18b, located 124 light-years away from Earth (very close by astronomical standard!). These molecules are produced on Earth by marine microorganisms like phytoplankton, and thus could be a bio-signature (or chemical processes that we aren't familiar with on Earth).

However, although life on Earth started rather soon after the formation of the planet, as far as we know, the first multi-cellular life form developed only some 600 million years ago. The first known single-celled organisms appeared on Earth about 3.5 *billion* years ago, roughly only a billion years after Earth formed. Perhaps multi-cellular life is rare[12], and rarer still the fraction that went on to develop technology. With only one example, namely Earth itself, we often imagine that life on other planets

[10] In DNA, we have four nucleobases: adenine (A), cytosine (C), guanine (G), and thymine (T), whereas in RNA uracil (U) replaces thymine. In a base pair, A is always paired with T (or U), and G with C.

[11] OSIRIS-REx stands for "Origins, Spectral Interpretation, Resource Identification and Security—Regolith Explorer". Its mission has since been extended and renamed OSIRIS-APEX (Apophis Explorer), as it heads towards the asteroid Apophis, which still poses a risk to Earth.

[12] Even before multi-cellular life, the emergence of *eukaryotes* represent a major step forward in the evolution of life (as opposed to *prokaryotes*, which include bacteria and archaea, whose simple cell lacks a nucleus and other membrane-bound organelles). Most eukaryotic cells contain mitochondria, which are membrane-bound organelles that are responsible for energy production (in the form of adenosine triphosphate or ATP). Evolutionary biologists believe that mitochondria evolved from separate organisms that ended up living inside the ancestors of eukaryotic cells in a symbiotic relationship ("endosymbiosis"). This was a chance encounter. Would something like this happen elsewhere, and if not, would energy production in cells ever become efficient enough to support complex life forms?

would be similar, that science would develop in more or less the same way. However, we know creatures like whales, dolphins, and octopuses are intelligent, but they did not develop technology (yet), perhaps due to life underwater having many different challenges than life on land. The existence of moons may also be crucial to provide enough tides to create an environment for terrestrial life to gradually evolve (since organisms that can adapt to temporarily live out of the water will not perish when the tide recedes. Some of them may then continue to evolve to become a complete land-dweller). And what if we have more than one moon, how would the history of astronomy change, and how would that influence the subsequent development of philosophy and science? While the notion of the habitable zone may be less restrictive than we think (for example, exomoons can have oceans heated by tidal friction from the gravity of the planet, thus sustaining life), especially since there is the possibility of life as we *do not* know it. However, the probability for actually achieving a technological stage advanced enough to explore the entire galaxy may turn out to be too small. Readers who are interested in the Fermi paradox can consult the comprehensive book by Webb (2015), which discusses 75 possible solutions!

This section is incomplete without a mention of UFOs (Unidentified Flying Objects), or now UAP (Unidentified Anomalous Phenomena). The U.S. government currently has a specialized team under the All-domain Anomaly Resolution Office (AARO) investigating these. There are many reports of unusual objects in our sky and even out/in of bodies of water (the latter category used to be referred to as USO —Unidentified Submarine/Submerged Objects). The fact that such sightings exist is not up for debate;[13] some rather credible cases involve also very unusual physical and photographic evidence. It is indeed possible that there are some phenomena as yet unknown to our current knowledge of science. However, this does not necessarily mean that UAP are extraterrestrial spacecraft (they are after all, *unidentified*). From time to time, there are reports of "occupants" or creatures associated with the UAPs. See Box 12 for discussions.

Box 12: Close Encounters of the Third Kind

In 1977, Steven Spielberg directed a science fiction movie *Close Encounters of the Third Kind.* This phrase was actually coined by UFO researcher and astronomer J. Allen Hynek (1910–1986), who was the scientific advisor to the U.S. Air Force "Project Blue Book" during 1952–1969 that investigated UFO reports. He devised a "close encounter classification" in which close encounters of the first kind (CE1) is the visual sightings of

[13] In 1953, a scientific committee later known as the "Robertson panel" reviewed the reports concerning UFOs and concluded that UFOs were not a *direct* threat to national security, but standard military communications or intelligences may be overwhelmed by public interests and reports about such phenomena. It was therefore recommended that the public should be discouraged from paying too much attention to UFOs. Note that the conclusion was never that there is nothing new to be learned from UFO sightings, but rather that they are not direct security threats. The Robertson who headed this committee was none other than the "R" in the FLRW metric. He also contributed to a deeper understanding of the Heisenberg uncertainty principle by relating it to the commutativity of observables in quantum mechanics.

an unidentified flying object within about 150 m distance, and which details of the object can be discerned to some degree (not just a blob of light). Close encounters of the second kind (CE2) refer to physical effects on the environment when the object is sighted. This can include interference to equipment and vehicles, physiological effects like paralysis or heat, and trace evidence like unusual impressions left in the ground, unusual magnetic fields or radioactive signatures above the normal level, etc. Close encounters of the third kind (CE3) involve sightings of entities associated to the UFO, which can be either biological humanoids or robots of some kind, i.e., what a layperson would call "alien sightings". Starting out as a skeptic who actively tried to debunk the UFO phenomena, Hynek eventually changed his mind when some cases became too difficult to explain via "conventional means".

Another astronomer Jacques Fabrice Vallée (1939–present), who started out working together with Hynek on the UFO phenomena, extended Hynek's system to also include close encounters of the fourth kind (CE4), in which the human witness is reportedly "abducted" or invited into the UFO. Vallée noticed that close encounters resemble many old stories throughout human cultures, such as those about mythical beings like fairies, and suggested that they have the same source. Even if there is an external unknown phenomenon at work, perhaps much of the experiences are psychological in nature (this was also Carl Jung's opinion). This could explain why there are so many different types of entities reported by seemingly sensible and reliable witnesses. In short, even if there is a true mystery, it does *not* mean that these are necessarily extraterrestrial crafts and occupants. Though started out believing in the extraterrestrial hypothesis (ETH) for UFOs, Vallée came to the conclusion that the available data do not support ETH.

In 1953, Hynek wrote in the Journal of the Optical Society of America, in an article titled *Unusual Aerial Phenomena* that (Hynek 1953):

> "I could not help, as an astronomer, recalling another wave of stories—stories of stones that fell from heaven. Because of poor reporting and poor imagery, scientific progress in meteorites had been held back for a good century. What a difference in imagery there is between "a stone falling from the sky" and "the interception by the earth of a particle pursuing an orbit around the Sun." The use of improper and inaccurate description of what actually happened kept meteorites in the category of old wives' tales and out of the niche that celestial mechanics had made ready for them a century before! In 1801, Thomas Jefferson said that he would sooner believe that two Yankee professors had lied than that stones had fallen from heaven. And the French Academy of Sciences branded stories of meteorite falls as fanciful and absurd and dismissed a bona fide meteorite whose fall had been sworn to—as an ordinary stone that had been struck by lightning. Perhaps the moral of this is: Beware the ready explanation!"

Let us therefore keep an open mind, but not blindly believe in every extraordinary claim. The right scientific attitude is to approach the phenomena and investigate them like any other more established natural phenomena. Of course, this requires governmental supports. Unfortunately, due to how academia works, no scientist can hope to base his or her career on investigating the unknown, especially if the phenomena suffer from a certain preconceived dogma of being nonsensical. As far as UAP is concerned, at least perceptions are slowly changing.

13.6 Are Wormholes, Hyperdrives, and White Holes Possible?

Space is vast. The distance to our closest star system, Alpha Centauri, is about 4 light-years. A typical spiral galaxy like our Milky Way is about 10^5 light-years across in diameter. The distance from our galaxy to the nearest major galaxy, the Andromeda Galaxy, is about 2.5 million light-years. Therefore even if extraterrestrial civilizations exist, conventional physics seem to rule out realistic physical communication, not to mention physical visits. How about *unconventional* physics? Science fiction stories have explored various options, such as wormholes and superluminal "hyperdrive" (or "warp drive"). None of these seem to be feasible as far as we know. A wormhole is like a black hole, with a mathematically defined surface (instead of a horizon, we have a wormhole "mouth"), which upon entering, you could reach the other side, which could be millions of light-years away, in a few minutes. The exact travel time depends on the geometry of the specific wormhole. One of the simplest examples of a wormhole is given by the metric

$$ds^2 = -dt^2 + dl^2 + (b^2 + l^2)\,(d\theta^2 + \sin^2\theta d\phi^2), \qquad (13.22)$$

where l is the radial coordinate, $-\infty < l < +\infty$, and b is the "throat radius". This is known as the Morris–Thorne wormhole, named after Michael Morris and Kip Thorne.[14] It describes a wormhole that connects two universes (hence l, unlike the spherical coordinate r, extends to negative infinity: $l < 0$ is the other asymptotically flat Universe).

The simplest model of a "hyperdrive" is the "Alcubierre drive", named after Miguel Alcubierre (1964–present), which describes a "bubble" in spacetime that contracts space in front of it and expands space behind it, thus allowing the spaceship inside the bubble to move across vast distances much faster than the speed of light. The metric of an Alcubierre bubble that moves in the positive z-direction can be written as

$$ds^2 = -dt^2 + dx^2 + dy^2 + [dz - v(t)f(x, y, z - z_0(t))dt]^2, \qquad (13.23)$$

where $v(t) = dz_0(t)/dt$ is the velocity of the bubble. f is a function that needs to vanish outside the bubble but equals unity inside. One specific example is

$$f(r) = \frac{\tanh[\sigma(r + R)] - \tanh[\sigma(r - R)]}{2\tanh(\sigma R)}, \qquad (13.24)$$

where $f(x, y, z - z_0(t)) = f(r(t))$, with $r(t) = \sqrt{x^2 + y^2 + (z - z_0(t))^2}$, and $R > 0$, $\sigma > 0$ being two arbitrary parameters. Physically, R represents the radius of the warp bubble, while σ is essentially inversely proportional to the thickness of the bubble wall (Alcubierre & Lobo 2017).

[14] As mentioned in Section 11.2, Kip Thorne was awarded the 2017 Nobel Prize in Physics for his contributions to the LIGO detector and the observation of gravitational waves, together with Rainer Weiss and Barry Barish. Kip Thorne was the scientific consultant and an executive producer for the movie *Interstellar* (2014). His work on wormholes was also incorporated by Carl Sagan in the 1997 movie *Contact*, which Sagan was an adviser to.

This form of superluminal travel does not violate special relativity because the expansion or contraction of space itself is not constrained by the speed of light; after all, during the inflationary era, the Universe expands exponentially fast. However, the "Alcubierre drive" is not really a "drive" in the sense of a conceptual engine design. It is just a mathematical solution of the Einstein field equations. Both wormholes and hyperdrives require some amount of "exotic matter" (matter that violates energy conditions) to be stable. It is not that exotic matter cannot exist, as we know that quantum processes can violate energy conditions, even the weakest classical energy condition of them all—the null energy condition. However, we do not have access to the copious amount of such matter to make a wrap bubble feasible. Even quantum matters cannot violate the energy condition *too much*; they still need to satisfy "quantum inequalities" that put constraints on how much negative energy density can be produced during a certain duration. They are like the uncertainty principle; you can't have too much negative energy for too long. It has been estimated that the total negative energy required for an Alcubierre drive exceeds the total mass of the visible Universe by a hundred billion times (Pfenning & Ford 2025).

Worse still, faster-than-light travel, whether using hyperdrives or wormholes or some other method, can always be devised into a time machine,[15] giving rise to a whole mess of paradoxes. For these reasons, theoretical physicists do not think that hyperdrives or macroscopic wormholes that can transport spaceships can ever be feasible. There are other problems with the "Alcubierre drive". For example, even if it can be built, it may not be stable due to the flux of particles that would be created inside the field (much like Hawking radiation). In addition, during the deceleration phase to stop the drive, a significant amount of radiation from the particles that piled up on the front of the bubble during its superluminal travel may be emitted forward and destroy whatever is in front of the spacecraft, thus making it a dangerous space weapon (McMonigal et al. 2012)! Perhaps more complicated designs can avoid such issues? In any case, future gravitational wave detectors can possibly detect the signatures of warp bubbles if they exist, but more research is needed to determine what such signals would look like (Clough et al. 2024).

---

[15] The technical term is a "closed timelike curve" (CTC). One way to—in theory—construct a CTC, is to have an infinitely long and sufficiently massive cylinder (the "Tipler–van Stockum cylinder") that rotates around on its axis of symmetry, in which case there is a CTC sufficiently far away from the cylinder. It should be emphasized that contrary to many explanations, it is *not* the frame-dragging caused by the rotation that directly causes the CTC to be formed. After all, if that's the case we should expect CTCs to form near the rotating mass and not far away; and also the CTC is *counter-rotating* with respect to the cylinder, *not* along with it! The same counter-rotating phenomenon has been observed in many spacetimes with CTCs (Andréka et al. 2008), but there is no direct relation between counter-rotating and the formation of CTCs (Duan et al. 2022). We should not be too alarmed about the lack of obvious "cause" from a "source", since such a situation is not new in general relativity. Recall the interpretation of the mass of a black hole, for example. Take the Schwarzschild black hole for simplicity, which is a *vacuum* solution to Einstein field equations, its mass is then just a property of the nontrivial curvature of the spacetime, and should not be regarded as being "sourced" by the singularity, which is after all, spacelike. Nontrivial spacetime geometries can give rise to surprising behaviors that while mathematically clear, sometimes simply do not admit a good intuitive, Newtonian-like, explanation.

What about a white hole? Despite its name, which suggests that a white hole would appear *white*, this is not the case. The maximally extended Schwarzschild black hole solution (the Kruskal–Szekeres manifold) has both a white hole and a black hole. Both have the *same* metric. In fact, classically a white hole has the same metric tensor as a black hole, and would appear the same to an exterior observer. The only difference is that instead of things falling into it, a white hole only allows things to come out from its horizon. Thus, if there is no matter around, both are indistinguishable. In the Kruskal–Szekeres spacetime, there is another asymptotically flat universe inside the black hole, which is connected to ours by the "Einstein–Rosen bridge", which is technically a non-traversable wormhole, as nothing can pass through it unless one travels at a speed greater than light. This is quite different from the wormholes discussed above, which are traversable. The reason why most physicists are skeptical of the existence of a white hole is that there is no known physical process that can form one. We know that gravitational collapse of a massive star can give rise to a black hole. However, a white hole is a time-reversed of a black hole, so it seems quite impossible to form one. One possible exception is at the microscopic scale, perhaps at the Planck scale, quantum fluctuation can make tiny black and white holes. It has also been suggested that a black hole remnant may quantum tunnel into a white hole when it reaches the Planck scale. One way to appreciate this, is to recall that the arrow of time only arises in macroscopic physics, so a small black hole should not distinguish itself from a small white hole, and thus should be in a quantum superposition of both. According to Loop Quantum Gravity calculation, the tunneling probability is $\exp(-S)$, where S denotes the Bekenstein–Hawking entropy of the black hole. Thus, the smaller the black hole is, the more likely it is to tunnel into a white hole (Rovelli & Vidotto 2025).

13.7 Are We Living In a Simulation?

In *The Grand Design* (Mlodinow & Hawking 2010), Stephen Hawking wrote:

> "A few years ago, the city council of Monza, Italy, barred pet owners from keeping goldfish in curved bowls... saying that it is cruel to keep a fish in a bowl with curved sides because, gazing out, the fish would have a distorted view of reality. But how do we know we have the true, undistorted picture of reality?'

The possibility that the world is not quite as it seems; that it is "just a dream", or an "illusion" of some kind have been with us for a long time, for example the "Butterfly Dream" of Zhuangzi (庄子) from ancient China, and the *Maya* of Indian and Buddhist philosophy. Perhaps all of the physical world is but the creative manifestation of our own consciousness. Maybe something external capable of manipulating our consciousness is producing the illusion, tricking us about our "reality". Descartes would argue that since God is not malicious, he would not play such a cruel joke on us. Still, if you think hard about this issue, you would quickly realize that we do not have a direct contact with the external world. We can perceive the world only through our sensory organs, which convert external stimuli to electrical signals that are then processed and interpreted by the brain. We also know very well how our brain can be easily tricked by various illusions. Therefore, it is not hard to imagine that a sufficiently advanced technology can directly stimulate our brain to have experiences as real as the actual one. In that case, is there even a difference about what is "real"?

Philosophers have proposed numerous thought experiments of this sort. For example, imagine that you are given a choice to enter a machine, which will stimulate your brain into believing that you are in the real world, living a life filled only with happiness and bliss, while also erasing your memory about having entered the machine. In other words, you will simply think that your life has started to turn for the better and remain good thereafter. Would you choose to enter such a machine? If so, why not? Most people would instinctively choose not to do so, because the experience is "not real". But what *is* reality, if not just what the brain perceives to be? A worse scenario is the "brain in a vat". What if you are just a brain in a laboratory, hooked up to a machine that produces all the sensations that you have always believed to be your life? Descartes' God is not malicious, but what about some advanced extraterrestrial mad scientist?

A more recent argument at the age of computer science, formally proposed by philosopher Nick Bostrom (1973–present), is that maybe we are just inside someone's simulation. The simulation argument begins with the claim that with sufficient computing power, we will eventually be able to simulate life forms in the digital world much like ours. If we can do it, maybe someone already did, and *we* are their simulation. In fact, since such a civilization capable of creating such a simulation would likely run many copies and variations thereof, it seems that the probability of us being a simulation is far larger than being an actual organic life form. Thus, unless for some reason such a simulation is prohibited by some laws of physics, perhaps pertaining to the limit of computation, or that no civilization makes it to that stage of technological advancement, it is overwhelmingly likely that we are living in someone's simulation. In Bostrom's words, "It is then possible to argue that, if this were the case, we would be rational to think that we are likely among the simulated minds rather than among the original biological ones." Some arguments were suggested that since the simulated world would be constrained by computational resources, there must be resolutions to the "pixels", so we would notice some aspects of discreteness in fundamental physics. Indeed, many quantum gravity theories predict that spacetime is discrete. Also, quantum mechanics means that we have a "resolution" bounded by the Planck constant, as in the case of the uncertainty principle. Of course, these are only consistent *with* the simulation hypothesis, not evidence *for* it.

We can in fact continue with the simulation argument onto the next level: some of the simulated realities would someday reach sufficiently advanced technology to run their own simulations, and so on, but this cannot continue on *ad infinitum*. Each level of the simulated realities would have less computing power (since they all inherited it from the base level reality), so at some point there would not be enough resources to do this. Now, there are always many more level N worlds than level $N-1$, so probabilistic argument would lead to the conclusion that it is overwhelmingly the case that we are living in a world that *cannot* simulate a reality. But didn't we start the whole argument by assuming that this is possible? Maybe this contradiction can be elevated somewhat if simulations are not necessarily classical Turing machines, but rather some advanced quantum computers of arbitrary power.

Having said that, we cannot prove that we are *not* inside a simulation, anymore than we can prove that we are in one (though perhaps the burden of proof should belong to the latter camp). Therefore ultimately I think that there is no point asking such a question, at least as far as physics is concerned. Even if we are in a simulated reality, it doesn't mean that it is useless to study the laws of physics in this simulated reality. To the contrary, look at what we have achieved by doing science. Maybe there is a "real world" outside ours, but we have no access to it, and as such *does not matter*. (In fact, even without the simulation argument, the philosopher Immanuel Kant argued in his 1781 book *Critique of Pure Reason* that we cannot know the world as it "really is"—the "noumenon"; as opposed to the "phenomenon"—because our minds actively shape and filter our perception of reality.) Sure, maybe we can somehow figure out that we are in a simulation and even manage to deduce some laws of the "real world"; like Hawking's goldfish, if smart enough, we can deduce how the world outside the tank looks like by accounting for light refraction. But then maybe *that* "real world" is also a simulation. What then? As Nietzsche wrote in *Beyond Good and Evil*,

> "It is nothing more than a moral prejudice that truth is worth more than semblance; it is, in fact, the worst proved supposition in the world."

In 2019, philosopher Preston Greene even suggested that it may be best not to find out if we are living in a simulation, for otherwise this might end the simulation (Greene 2019).

13.8 Are Physical Constants Really Constant?

The ultimate hope is that if there is a unified theory for fundamental physics, we can compute the values of physical constants like G, c, $\hbar$, and k_B directly, which also agree with the observed values. This remains a dream. It is definitely possible that these values are just "given", and that the Universe could have started with other values of these constants. If the values are drastically different, the Universe might collapse quickly, or stars could not form, or life could not arise; these we have already discussed. In this section we are interested in another possibility: is it possible that there are no true constants? Could the values of say, the speed of light in vacuum, be different in the distant past?

Indeed one can consider such a possibility, and many did, including Robert Dicke, as far back as 1957. There are newer versions of such "variable speed of light" (VSL) theories in recent years (Magueijo 2003). Indeed it has been argued that the horizon problem in cosmology can be solved with the VSL model in place of inflation. Similarly, varying-G theories have also been considered, as early as in 1937 by Paul Dirac. This would of course have implications for cosmology (Uzan 2011).

Hermann Weyl had an interesting idea about G being possibly "statistical". In his book *Philosophy of Mathematics and Natural Science* (Weyl 2009), Weyl essentially noted that the ratio of the magnitude of gravitational force to that of the electromagnetic force between two electrons is about $O(10^{-42})$; see Section 11.3. (In his book the number given is 10^{-41}, denoted as ε). Based on an analysis of a closed Universe whose radius is given by the scale of the cosmological constant, and assuming that all particles are just electrons for simplicity, he estimated that there are about $N = O(10^{81})$ particles in the Universe. Although we no longer think that our Universe is spatially closed, the number of particles often quoted as an estimate is still about 10^{80}, known as the "Eddington number", so Weyl's result is not that far off the current estimate. He then noticed that $\varepsilon \approx 1/\sqrt{N}$, and so he wondered whether a better theory of gravity would have G that depends on the number of particles in the Universe. He wrote:

> "The idea is not as strange as it may first sound. Long ago E. Mach tried to interpret the inertial mass of a body as an inductive effect of the other masses of the universe. Einstein's theory of gravitation does not satisfy Mach's postulate, though historically the latter played a role in its conception. A theory that meets Mach's challenge remains a desideratum (would it be a statistical theory of gravitation, as the square root in the law $\varepsilon \approx 1/\sqrt{N}$ seems to indicate?)"

Weyl's idea was based on the fact that the occurrence of $\sqrt{N}$ is common in probability and statistics, for example, the expected distance traveled after N steps is roughly $\sqrt{N}$ in a random walk. One obvious objection is: this seems to imply that G would be infinite if $N = 0$. But if there is no particle, there is nothing to experience G, so its value is not relevant. (Recall the relational view of space and time in Section 11.1.)

In his book, Weyl also mentioned an example in which a "constant" turned out not to be. Namely, Maxwell had assumed that the electric polarization P is proportional to the field strength E. This is written as $P = \varepsilon_0 \chi_e E$, where ε_0 is the electric permittivity of free space and χ_e is the "electric susceptibility". This is only correct for static and slowly varying fields, but when there are rapid oscillations, there would be a new phenomenon of dispersion. In this case χ_e, or equivalently the relative permittivity (which in older texts, including Weyl's book, is called the "dielectric constant") $\varepsilon_r = \chi_e - 1$, also becomes frequency dependent.

Another possibility is that G may depend on energy scale, as in the approach of "asymptotically safe gravity". Similarly, it has been suggested that even $\hbar$ could go to zero as temperature becomes sufficiently high. If so, the very early Universe at the Planck era could become classical (Hossenfelder (2013), contrary to the usual expectation that one requires quantum gravity to describe that epoch. Another famous quantity is the "fine structure constant" $\alpha \equiv e^2/\hbar c$, where e is the charge of the electron, whose possible variations have been studied extensively. Of course, a variation in α could be due to either the variation in c or e or $\hbar$ (though due to quantum mechanics being so well tested, usually one avoids changing $\hbar$ except possibly at very high energy scale). Incidentally, in SI units, the fine structure constant is

$$\alpha = \frac{1}{4\pi\varepsilon_0}\frac{e^2}{\hbar c} \approx \frac{1}{137}. \tag{13.25}$$

If this were to be an exact value, it would be a surprise and may contain a deeper secret, but alas the value is only approximate!

Of course, just like modifying our theories could lead to problems (as we have seen in Chapter 12), the same is true with changing the values of physical constants. There are usually tight constraints coming from various observations and experiments. In other words, we don't really know if physical constants can vary, the possibility is certainly not completely ruled out and is still very much an active field of research today.

References

Abramowicz, M., Bejger, M., Gourgoulhon, É., & Straub, O. 2020, NatSR, 10, 7054

Alcubierre, M., & Lobo, F. S. N. ed. F. Lobo 2017, Wormholes, Warp Drives and Energy Conditions, Fundamental Theories of Physics Vol. 189 (Cham: Springer)

Andréka, H., Németi, I., & Wüthrich, C. 2008, GReGr, 40, 1809

Borde, A., Guth, A. H., & Vilenkin, A. 2003, PhRvL, 90, 151301

Bouhmadi-Lopez, M., Errahmani, A., Martin-Moruno, P., Ouali, T., & Tavakoli, Y. 2015, IJMPD, 24, 1550078

Brown, A. R., & Dahlen, A. 2012, PhRvD, 85, 104026

Carroll, S. M., & Tam, H. 2010, [arXiv:1007.1417 [hep-th]] 2010arXiv1007.1417C/abstract

Cassé, M., & Silk, J. 2022, FoPh, 52, 1

Chen, Y., Ni, J., & Ong, Y. C. 2022, EPJP, 137, 1109

Clough, K., Dietrich, T., & Khan, S. 2024, OJAp, 7, 63

Coles, P., & Ellis, G. F. R. 2010, Is the Universe Open or Closed? The Density of Matter in the Universe (Cambridge: Cambridge Univ. Press)

Duan, Y., Liu, F., Wang, Y., & Ong, Y. C. 2022, Univ, 8, 28

Frampton, P. H., Ludwick, K. J., & Scherrer, R. J. 2011, PhRvD, 84, 063003

Geshnizjani, G., Ling, E., & Quintin, J. 2023, JHEP, 2023, 182

Greene, P. 2019, The New York Times, https://www.nytimes.com/2019/08/10/opinion/sunday/are-we-living-in-a-computer-simulation-lets-not-find-out.html

Helbig, P. 2021, EPJH, 46, 10

Helbig, P. 2012, MNRAS, 421, 561

Holman, M. 2018, FoPh, 48, 1617

Hossenfelder, S. 2013, PhLB, 725, 473

Hynek, J. A. 1953, JOSAA, 43, 311

Johnson, M. C., Wainwright, C. L., Aguirre, A., & Peiris, H. V. 2016, JCAP, 07, 020

Khoury, J., Ovrut, B. A., Steinhardt, P. J., & Turok, N. 2001, PhRvD, 64, 123522

Lyth, D. H., & Liddle, A. R. 2012, The Primordial Density Perturbation: Cosmology, Inflation and the Origin of Structure (Cambridge: Cambridge Univ. Press)

Magueijo, J. 2003, RPPh, 66, 2025
McInnes, B. 2007, NuPhB, 782, 1
McInnes, B. 2027, [arXiv:0711.1656 [hep-th]] 2007arXiv0711.1656M/abstract
McMonigal, B., Lewis, G. F., & O'Byrne, P. 2012, PhRvD, 85, 064024
Mlodinow, L., & Hawking, S. 2010, The Grand Design (New York: Bantam)
Mukhanov, V. 2012, Physical Foundations of Cosmology (Cambridge: Cambridge Univ. Press)
Nojiri, S., Odintsov, S. D., & Tsujikawa, S. 2005, PhRvD, 71, 063004
Osborne, S., Senatore, L., & Smith, K. 2013, [arXiv:1305.1964 [astro-ph.CO]] 2013arXiv1305.1964O/abstract
Padmanabhan, T. 2010, Gravitation: Foundations and Frontiers (Cambridge: Cambridge Univ. Press)
Page, D. N. 1983, Natur, 304, 39
Page, D. N. 1984, IJTP, 23, 725
Paulson, N., "Paul Gauguin, Where do we come from? What are we? Where are we going?," in Smarthistory, August 9, 2015, accessed September 23, 2025, https://smarthistory.org/gauguin-where-do-we-come-from-what-are-we-where-are-we-going/
Pfenning, M. J., & Ford, L. H. 1998, [arXiv:gr-qc/9805037]] 1998PhDT........38P/abstract
Rovelli, C., & Vidotto, F. 2024, [arXiv:2407.09584 [gr-qc]] 2024arXiv240709584R/abstract
Schelling, T. C. 1981, The Strategy of Conflict (Harvard: Harvard Univ. Press)
Sellers, L., Bobrick, A., Martire, G., Andrews, M., & Paulini, M. 2022, [arXiv:2212.02065 [astro-ph.IM]] 2022arXiv221202065S/abstract
Singal, A. K. 2024, EPJC, 84, 385
Taylor, W., & Wang, Y.-N. 2015, JHEP, 2015, 1
Uzan, J.-P. 2011, LRR, 14, 2
Wainwright, C. L., Johnson, M. C., Aguirre, A., & Peiris, H. V. 2014, JCAP, 10, 024
Webb, S. 2015, If the Universe Is Teeming with Aliens … WHERE IS EVERYBODY? Seventy-Five Solutions to the Fermi Paradox and the Problem of Extraterrestrial Life (Cham: Springer)
Weyl, H. 2009, Philosophy of Mathematics and Natural Science (Princeton, NJ: Princeton Univ. Press)

Chapter 14

Afterthoughts: Humanity and the Cosmos

"When the sound and wholesome nature of man acts as an entirety, when he feels himself in the world as in a grand, beautiful, worthy and worthwhile whole, when this harmonious comfort affords him a pure, untrammeled delight: then the universe, if it could be sensible of itself, would shout for joy at having attained its goal and wonder at the pinnacle of its own essence and evolution. For what end is served by all the expenditure of suns and planets and moons, of stars and Milky Ways, of comets and nebula, of worlds evolving and passing away, if at last a happy man does not involuntarily rejoice in his existence?" —Johann Wolfgang von Goethe.[1]

Let us end this book with some musings on how doing science, in particular cosmology and fundamental physics, have also impacted our culture, art, and other aspects of human civilizations. Science has progressed a lot in just a couple of centuries. Yet we must keep in mind that science is not just collections of hard facts, mindless experimentations, and senseless generalizations of existing theories. It is about our curiosity, and about understanding how the Universe works. We should not forget the human elements of doing science, of philosophizing life, and of appreciating art. Live a good life.

If neutrinos are Majorana fermions (see Section 8.4.3), this will permit the so-called "neutrinoless double beta decay", a type of nuclear decay that cannot otherwise occurs. To look for rare events like this requires that the detectors be shielded as much as possible from background signals. This can be achieved by carrying out the experiment deep underground (just like other detectors that search for dark matter, proton decay, etc.), so that the earth and rock above can shield the

[1] I could not find the exact source, but it is a nice paragraph!

doi:10.1088/2514-3433/ade963ch14

detectors against cosmic rays. The Cryogenic Underground Observatory for Rare Events (CUORE) is one such facility located underground at the Laboratori Nazionali del Gran Sasso in Assergi, Italy. Its upgraded version, CUORE Upgrade with Particle Identification program, is abbreviated as CUPID. The mountain granite provides an excellent shielding against cosmic rays. However, the granite itself is radioactive, and so there is a need to shield the detector against this. Lead can be a good shield if it does not contain the radioactive lead isotope, Pb-210. This cannot be avoided, but one could wait. The half-life of Pb-210 is 22.3 years, so if one uses *old* lead, it would be a better shield for the detector. The CUORE team ended up using the Roman lead ingots discovered in 2000, in the "Bou Ferrer shipwreck", which was traced back to 50 AD. There was a considerable debate about using old lead ingots for particle physics research, because it means destroying valuable archaeological findings (though the lead bars chosen for this purpose were the worst in shape, and "only" 10% of the total amount found, and all writings on them were carefully cataloged before the ingots were melted down to make new shields). In this case, a deal was reached with the archaeologists exchanging the ingots with monetary support that would fund the operation of bringing up the ship's artifacts. The physicists also helped to analyze the compositions of the lead, and thus assisted in determining when and where the material was most probably mined—the Sierra de Cartagena Mines in southern Spain (which was under the territory of the Roman empire). But is understanding the Universe more important than figuring out history? Would such a demand for ancient artifacts encourage illegal operations? Dilemmas like these are unavoidable, and probably require more dialogues between the two fields. Similar debates include whether we should build telescopes on mountains that are considered sacred sites, as in the case of Mauna Kea in Hawaii. These should not be seen as a conflict between culture and science, but that people have different needs and as such have to learn how to best accommodate each other. Not to mention that, archaeology and spiritual beliefs are both different attempts to figure out the ultimate questions concerning humanity's past and purpose. For the ingots that were used by physicists, I find it fitting that after so many centuries, they were finally being used for a project that the ancient Romans couldn't even have imagined. Ingots were meant to be used after all, and they have finally fulfilled their purpose. In some sense, it is the continuation of the human civilization, the modern one owes itself in more than one way to its ancient counterparts.

The human species has come a long way in our understanding of the Universe, yet in terms of geological timescale, a couple of hundred years of scientific advancements is but a blink of an eye. Yet, during this period, we have moved from believing that Earth is the center of the Universe to the appreciation that the cosmos is unimaginably large and filled with exoplanets orbiting other stars. This change of worldview was initiated by scholars like Copernicus, Descartes, Tycho, Kepler, Galileo, and Newton (see Chapter 2), as well as Digges and Bruno. Though Digges' illustration of the solar system already depicted the idea of an infinite universe (see Chapter 3), drawings that show other stars surrounded by their own planets are rare. One such example can be found in *Atlas Coelestis in quo Mundus Spectabilis* of 1742,

Figure 14.1. *Systema Solare et Planetarium* in the map collection *Atlas Novus Terrarum Orbis Imperia Regna et Status Exactis Fabulis Geographice Demonstrans* by Johann Homann around 1716, though most maps are undated. The volume is 12 cm thick. In this illustration, we can see that the upper right corner depicts exoplanets orbiting other stars. The upper left corner is an early attempt to show the planets to scale with the Sun. The main figure in the center is the Copernican model of the Universe. The Ptolemaic model and the hybrid model of Tycho Brahe can be seen in the lower right. A lunar eclipse was also shown in the lower right. Meanwhile, the lower left shows a solar eclipse. *Credit:* David Rumsey Historical Map Collection, David Rumsey Map Center, Stanford Libraries.

produced by the astronomer and cartographer Johann Gabriel Doppelmayr (1677–1750). It was based on an earlier work by Johann Homann (1664–1724), shown in Figure 14.1. The fact that exoplanets exist at all was only confirmed in 1992, with the discovery of one planet that orbits around a pulsar. As of April 30th, 2025, we have discovered 5885 exoplanets.[2] Thus far we have not found any planet that has clear, unambiguous signature of extraterrestrial life, but the search continues.

Comets were perceived by cultures all around the ancient world to be a bad omen, harbinger of death via diseases, famine, or war. The comic in Figure 14.2 depicts this fear. Today, we understood that comets are just giant snowballs made of ice, dust, and frozen gas, which develop their signature tails (often two tails: a dust one and an ion one) when they are sufficiently heated by the Sun. Of course, comets and asteroids can still pose danger to us if they collide with the Earth. A reminder of this

[2] You can check the latest number at https://science.nasa.gov/exoplanets/.

Figure 14.2. "- Adelaide, Adelaide.. I think I can see the comet coming!!... - Oh my God... this is the end of the world.... how annoying. They promised it wouldn't come until June 13!," plate 31 from Croquis Parisiens. *Credit:* Gift of Charles E. Worcester. *Source*: https://www.artic.edu/artworks/3400/adelaide-adelaide-i-think-i-can-see-the-comet-coming-oh-my-god-this-is-the-end-of-the-world-how-annoying-they-promised-it-wouldn-t-come-until-june-13-plate-31-from-croquis-parisiens

came in February 2013, with an asteroid exploded over Chelyabinsk, Russia, an event that was captured by many video cameras. The shock wave shattered windows and shook buildings, leading to the injury of about 1500 people. Today, we have the robotic survey ATLAS (the Asteroid Terrestrial-impact Last Alert System) funded by NASA, which scans the sky looking for potential threats; and planetary defense capability via altering an asteroid course in advanced by colliding it with a high-velocity projectile, as demonstrated by the Double Asteroid Redirection Test (DART) conducted by NASA in 2022. At the time of writing (April 2025), Asteroid 2024 YR4 is expected to get so close to us in December 2032 that there is a chance that it may hit the Moon, with a probability around 1.8%. The chance for it to hit Earth has decreased from the initial 3.1% estimate to a mere 0.001% (The Economic Times 2025). These numbers are likely to change further as modeling improves. It is natural to be worried about a possible apocalypse, but on the other hand, it is quite remarkable that in a few hundred years, we have advanced from being so ignorant about the true nature of comets to be able to make orbital prediction 7 years ahead of a newly discovered asteroid, *and* to be in within our capability to change its course if necessary. (Though it is also crucial to note that science can only make progress like this if there is enough funding from the governments. Private companies would only support research that can be applied in industries in the relatively immediate future, not fundamental science.)

There were other new advances concerning comets in the 21st century. On August 30, 2019, an amateur astronomer Gennady Borisov discovered a new comet, now called 2I/Borisov. Its orbit revealed that the comet is not gravitationally bound to our Sun, and therefore came from interstellar space. Subsequent analysis indicated that 2I/Borisov does not have as much water and diatomic carbon compared to other comets in our solar system. On the other hand, it contains more carbon monoxide and amines. One of the hypotheses proposed was that 2I/Borisov may come from a planetary system with a red dwarf star, which is the most common type of star in our Milky Way Galaxy (73% of all stars are red dwarfs; our Sun is a type-G star, which make up only about 6% of the stellar population). Contemplate that again: a visitor from another star system! This was the *second* time an interstellar object entered our solar system, the first being the asteroid called *Oumuamua* (formally designated 1I/2017 U1), discovered by Robert Weryk using the Pan-STARRS telescope in Hawaii in 2017. The name means "a messenger from afar arriving first" in Hawaiian.

In this book, we have seen how human beings have tried to make sense of the world since antiquity, from alchemy and esoteric ideas to modern science. Falling comets or asteroids often remind me of *Melencolia I* (1514), which depicted one such astronomical event. *Melencolia I*, as shown in Figure 14.3, is an engraving by the German Renaissance artist Albrecht Dürer (1471–1528). It is full of symbolism and various people have tried to interpret the work, including David Finkelstein (1929–2016), a physicist best known for his contribution in black hole physics (Finkelstein 2025). To me, *Melencolia I* illustrated some aspects in our attempt to understand the Universe. For example, there are many measuring devices depicted, including a scale for measuring weight, an hourglass for measuring time, and a compass in the hand of the angel for measuring length; quantitative measurements being crucial for science. The name "Melencolia", or melancholia, today means a state of deep sadness. In ancient beliefs, the four basic elements—fire, water, air, and earth, also gave rise to the idea of "humor" in human bodies. Each of these elements corresponds to a bodily fluid (in the same order): yellow bile, phlegm, blood and black bile (yellow bile was associated with the gallbladder, while black bile was associated with the spleen). Melancholia was thought to be the result of too much black bile. In alchemy, black bile is considered "dry and cold". To relieve someone of this state then, one should increase wetness and heat of the body. The adult angel is wearing a wreath of plants on her head, which was "wet". The magic square has the magic constant 34, i.e., the sum of the numbers in any horizontal, vertical and main diagonal is 34. This particular magic square also has the property that the sums in any of the four quadrants, as well as the sum of the middle four numbers, are all 34. The "1514" in the bottom row is the year of the engraving. Numerologically speaking, the number 34 makes the square a talisman of Jupiter, a planet whose power is also said to relieve melancholia. Such a worldview would eventually be replaced by modern chemistry and medicine. On the left, we see a huge polyhedron, which to me signifies the importance of geometry. Kepler was fascinated by

Figure 14.3. *Melencolia I* by Albrecht Dürer, 1514. Why the "I" in the title? One possible explanation is in reference to Heinrich Cornelius Agrippa's *De Occulta Philosophia*, in which he classified melancholic inspiration into three levels. The lowest level is imagination (associated with artistic genius), which is probably what Dürer was referring to by "I". The second level is reason, associated with the physicians and natural sciences, and the third (highest level) is the intuition of the theologians. The dog is also believed to be a symbol of melancholy. *Credit:* Clarence Buckingham Collection. *Source*: https://www.artic.edu/artworks/28260/melencolia-i

geometry, trying to fit the solar system in nested Platonic solids. Today, differential geometry is used in general relativity, and symmetries of group theory are employed in particle physics and gauge theories. So why is the angel in a melancholic state? Perhaps she is trying to figure out how the Universe works but is unable to? Melancholic state was often associated with artists, poets and scholars. Aristotle suggested that this is due to their "heightened sensitivity", which was what allowed them to perceive the world differently from other people. It is the view of the tragic genius—the gift that allows one to understand the world often also leads to suffering. For scientists, we may become frustrated if our research gets stuck, or when we temporarily run out of good ideas to pursue, or worse when we are too busy with other duties (like filling in useless bureaucratic forms) that take away our precious time that could be used for research. It is important to recognize that this is normal, and to try to find ways to relax. Life is more than what our works define. Years of pursuit of papers and grants, now almost a necessity to survive in academia, have the danger[3] that it may re-define what we *ourselves* may deem as a "successful" career, the initial dream to understand the cosmos gradually dimmed while pursuing mere shadows.[4] We must fight this and constantly re-evaluate what we really want to do in life, and why we chose to do science in the first place. Simone de Beauvoir (1908–1986), in her work *The Ethics of Ambiguity*, wrote:

> "The serious man gets rid of his freedom by claiming to subordinate it to values which would be unconditioned. He imagines that the accession to these values permanently confers value upon himself."

Therefore, don't be too serious (caught up in your work), leave some time for reflections, embrace your freedom by truly asking: is this what I want to do with my life? And if not, what can I do about it to improve the situation?

In any case, with all her heavenly powers, even Dürer's angel seems to be confused about how the Universe really functions. Nicholas of Cusa, whom we have met many times in this book, encouraged the use of mathematics to model physical phenomena, though like many of his contemporaries, he did not believe that we have the intellectual power to fully understand how the world actually works as designed by God. Nicholas of Cusa expressed this idea that "the more he knows that he doesn't know, the more learned he will be" in his book *De Docta Ignorantia*, or "On Learned Ignorance", published in 1440. This idea came to him like a divine epiphany when he was sailing between Constantinople and Venice in the winter of 1437. The notion that we may never know the ultimate workings of Nature looks pessimistic. Yet, precisely because we cannot hope to know the "truth", we are *free* to construct different theories and then test them to see if they fit observations. Modern science would share a similar view. There may be "the truth", but we no

[3] Another danger is that many scientists now fake results just to publish. This pollutes the literature and makes the quest for actual scientific understanding even more difficult.

[4] Keep in mind Tycho's lamentation that professors are "shadow chasers"; see Section 2.1.

longer speak of it as such. Mathematical modeling is in some sense, "world building", just like writers do in their stories. We use mathematics to construct a consistent worldview that can explain what we actually observe and measure in our own world. Whether the latter really works like that is a completely different question. As quoted in Aage Petersen's *The Philosophy of Niels Bohr* (Petersen 1963), Niels Bohr's view was that:

> "It is wrong to think that the task of physics is to find out how nature is. Physics concerns what we can say about Nature."

On the other hand, in Max Planck's book *Where is Science Going* first published in 1932 (Planck 1981), he advocated for a balanced approach (between theorizing freely and experimenting carefully):

> "[...] in the earlier stages of physical science, which are now called the classical physics, the opinion held sway that the actual measurement itself furnishes a direct glimpse into the real happenings of nature. But in this assumption, as we have already seen, there was a fundamental mistake which is the counterpart of the positivist error, namely, the paying of attention solely to the results given by experimental measurements and entirely ignoring the inner reality of natural processes. Yet while we recognize this as a mistake on the one side we must also realize on the other side that if we are to abandon the measuring method we have no way of coming into touch with the real happenings themselves. "

There are two lessons here: (1) we must keep in mind that our best theories and models are just that, they need not reflect how reality "actually is", and (2) this does not mean we have completely free reign of our imagination, as we are still constrained by observations of the real world. Even if our theory is only a simplified model (recall again the philosophy behind mathematical modeling explained in Section 1.5: always start simple), sometimes that's all we have. In this case, e.g., in quantum gravity, we can only push on and see where our simplified models lead us, and what it teaches us when it fails. On page 274 of Hermann Weyl's *Philosophy of Mathematics and Natural Science*, he wrote (Weyl 2009):

> "... do not take too literally such preliminary [...] schemes [...], useful as they are as a first guide in a seemingly incoherent mass of facts. A picture of reality drawn in a few sharp lines cannot be expected to be adequate to the variety of all its shades. Yet even so the draftsman must have the courage too draw the lines firm. Nicholaus Cusanus stressed in the *De docta ignorantia*: if the transcendental is accessible to us only through the media of image and symbols, let the symbols at least be as distinct and unambiguous as mathematics will permit."

As we have seen repeatedly in this book, natural philosophers and scientists have devised many clever ideas throughout history. Some of the ideas are used again when facing different problems. For the most part we are creatures stuck in our own time, since we learned from textbooks and from our teachers and professors how things work according to the present understanding. To go beyond the current way of thinking is easier said than done.[5] The idea of the luminiferous æther, for example, had been the mainstream for a long time, and even after the conception of special relativity, not everyone gave up æther.

Sometimes ideas come from our conscious creativity and ingenuity. Some other times, as experienced by Nicholas of Cusa, new ideas can come "in a flash", like a divine epiphany. This is more common than one might think (I have had quite a few such instances), and is usually the result of the subconscious mind that continues to work on a problem when we are doing something else altogether, like walking, cooking, or bathing. Of course, this can only happen after we have consciously worked on the problem for a substantial amount of time. The subconscious mind can then try to see how various aspects fit together and in doing so, prompts us in the right direction, when we are not drowning in our own conscious thought of the problem. The idea of trapped surfaces introduced in Section 10.4, for example, occurred to Roger Penrose when he was crossing a street in London.

Werner Heisenberg had an epiphany on matrix mechanics in 1925 when he was on the island of Helgoland, in between climbing and memorizing poems from Goethe's *West-östlicher Diwan*. He wrote (Pais 1991):

> "It was about three o' clock at night when the final result of the calculation [...] lay before me. [...] At first I was deeply shaken. [...] I was so excited that I could not think of sleep. So I left the house [...] and awaited the sunrise on the top of a rock."

I find it interesting that *West-östlicher Diwan* was in turn inspired by Goethe's readings of the Persian poet Hafez (1325–1390), who according to some legends, began a 40-day-and-night vigil without food, sitting in a circle when he was 60 years old. This is a spiritual practice called *Chilla-nashini* in the Persian traditions (some readers may recall that Jesus Christ also fasted for 40 days and nights in the wilderness after his baptism). The legend has it that, when he was offered a cup of wine on the 40th day, he attained some form of cosmic awareness. My point is that it is important to let our mind wander, instead of just keep working hard all the time. Creativity cannot be forced.

Sometimes ideas come in dreams. For example, the "Violin Sonata in G minor" by Giuseppe Tartini (1692–1770) is known as the *Devil's Trill Sonata* since the musician dreamed that the Devil played the tune (a better version that he could not replicate). Mathematical genius Srinivasa Ramanujan (1887–1920), who had no

[5] "The difficulty lies, not in the new ideas, but in escaping from the old ones, which ramify, for those brought up as most of have been, into every corner of our minds."—John Maynard Keynes, 'The General Theory of Employment, Interest and Money'.

formal training in pure mathematics, recalled dreams in which he was shown scrolls of mathematical equations, which he attributed to be of divine origin. Perhaps the most famous dream that led to a scientific discovery was the dream by the chemist Friedrich August Kekulé (1829–1896), in which he saw snakes biting their own tail, which led to his idea that benzene molecules are ring-shaped.

A snake that bites its own tail is an ancient symbol called the ouroboros. It often represents the cycle of life and death, of rebirth. The idea of ouroboros can be found in many cultures. For example, in Norse mythology, Jörmungandr is the "World Serpent" that was cast into the ocean by Odin, grew so large that it was also to eventually bite its own tail, having encircled the realm of Midgard (where humans live). The day Jörmungandr would release its tail is the day of Ragnarök, in which most gods (not to mention humans), along with the world, will be destroyed. Yet a new world will rise from the destruction. The British cosmologist and Astronomer Royal, Martin Rees (1942–present), introduced the "cosmic ouroborus" metaphor in his book *Just Six Numbers: The Deep Forces That Shape The Universe* (Rees 1999). Imagine the head of the serpent represents the largest scale of our cosmic particle horizon, which is at the order of 10^{28} cm, while the tail represents the tiniest Planck length, which is about 10^{-33} cm. Now subdivide the length of the serpent body into 12 parts (like a clock), each part increasing by a power of ten; humans occupy roughly the middle of it at 6 o'clock. Sir Martin Rees wrote:

> "Such an enormous range [of length scales] is actually a prerequisite for an 'interesting' universe. A universe that didn't involve large numbers could never evolve a complex hierarchy of structures: it would be dull, and certainly not habitable. And there must be long timespans as well. […] The tremendous timespans involved in evolution offer a new perspective on the question 'Why is our universe so big?' The emergence of human life here on earth has taken 4.5 billion years. Even before our Sun and its planets could form, earlier stars must have transmuted pristine hydrogen into carbon, oxygen and other atoms of the periodic table. This has taken about ten billion years. […] This is a startling conclusion. The very hugeness of our universe, which seems at first to signify how unimportant we are in the cosmic scheme, is actually entailed by our existence! This is not to say that there couldn't have been a smaller universe, only that we could not have existed in it[6]."

The largest scale and the smallest scale are not unrelated. The head of the serpent bites its own tail. As we have mentioned in Section 5.4, the tiniest quantum fluctuations in the very early Universe occurred when the Universe was around 10^{-35} seconds old. These fluctuations were "seeds" that led to the formation of large-scale structures like the filaments of galaxy superclusters. In the inflationary paradigm (Section 13.3), the small perturbations in the inflaton field were amplified

[6] On a similar note, physicist Emerson Pugh once said that "If the human brain were so simple that we could understand it, we would be so simple that we couldn't."

exponentially, leading to fluctuations across various length scales. As inflation came to an end, the Universe underwent "reheating" (the decay of the inflaton field into standard matter and radiation). By then, those fluctuations became classical density perturbation, which propagated through the baryon-photon fluid in the early Universe, creating sound waves in the plasma, the baryon acoustic oscillations. This oscillation was due to two competing forces: gravity that tried to attract the overdensed region, and radiation that exerted outward pressure and therefore work against gravity (recall that the radiative pressure inside a star stabilizes it against gravitational collapse). These "ripples" in the fluid of particles and radiation were frozen when the Universe cooled down and radiation decoupled from matter. The ripples triggered more active star formations compared to other regions. The expansion of the Universe then stretched the ripples to tremendous scales that can be seen when we observe the galaxies today. By studying galaxy distribution via, for example, ESA's Euclid (Figure 14.4), we can recover the information about the size of the ripples and in turn deduce their evolution throughout cosmic time, which would give us information about the matter-energy content of the Universe, including the amount of dark matter and dark energy (whatever their true nature might be; see Chapter 6).

In modern abstract art, there are paintings that consist of nothing but a single color (technically called "monochromatic painting"), which nevertheless are sometimes sold for millions of dollars. There are many factors that contribute to this, such

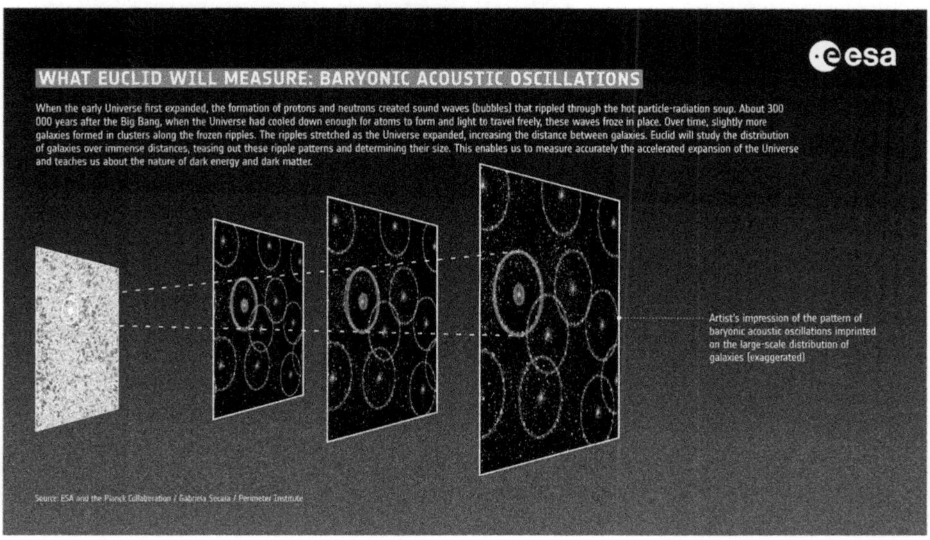

Figure 14.4. ESA's Euclid and the study of Baryon Acoustic Oscillations. *Credit:* ESA (acknowledgement: work performed by ATG under contract to ESA). CC BY-SA 3.0 hyperlink to https://creativecommons.org/licenses/by-sa/3.0/igo/

[7] Minimalism itself is a counter-movement to "abstract expressionism", deemed too intense and chaotic in its expression of emotions. What could be more orderly than a monochromatic painting?

as the reputation of the artist, the *idea* and the contexts behind the art, the modern yearn for simplicity and "minimalism"[7] etc. The *Blue Monochrome* of Yves Klein (1928–1962) in 1961, for example, shows a color now known as "International Klein blue"; while the 1913 work of Kasamir Malevich (1879–1935) shows a black square (and indeed entitled as such). Yet if you look closer at these paintings, they are not perfectly uniform. Even the controversial composition 4′33″ by John Cage (1912–1992), which is performed by *not playing any instrument* for precisely that duration of time, is not about silence itself. During its premier, the audience was reasonably angered for having just heard "nothing". However, Cage explained that they missed the point, that "there's no such thing as silence", that the piece was meant to make us aware of the background sounds all around us. Should we consider 4′33″ as music? Is an empty paper a piece of poetry? That depends on how we define those terms. Still, the point is that there is no such thing as emptiness, or nothing, or perfectly smooth canvases with perfectly uniform color. There is beauty in imperfection. The Universe would be eternally humdrum if it was perfectly smooth without any fluctuation whatsoever. Nothing would ever change. We *are* part of the imperfections of the cosmos (Figure 14.5). Perhaps this should remind us that in life too, we

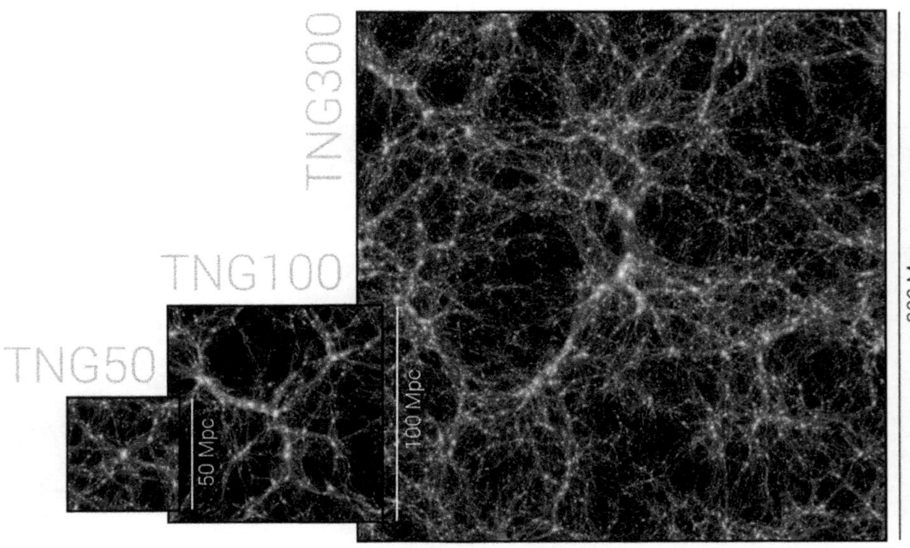

Figure 14.5. Comparison of the three TNG computer simulations of the large-scale structure of the Universe which depict the galaxy filaments: TNG50, TNG100, and TNG300, with the simulation box being 50, 100, and 300 megaparsecs across, respectively. For each projected dark matter density is shown. The three volumes are roughly though not exactly to scale. Filaments are located at the intersections of planar structures called "walls", which are formed from galactic superclusters. The relatively empty regions are called "voids". Galaxy filaments will dissolve in the far future if the accelerated expansion of the Universe continues. In the grand scheme of things, we are insignificant, "like a mote of dust in the morning sky" in Carl Sagan's words in his famous book *Cosmos*. Yet, as small as we are, we managed to discover all this, and will continue to find out even more. The torch of human curiosity will be passed on from one generation to the next. Perhaps one day, if we can get past our differences and work together instead of against each other, the human species will be among the stars. *Credit:* Reprinted with permission from the TNG collaboration. (c) 2023 The TNG Collaboration.

need not always insist on being perfect. Instead, there is virtue in accepting transience and imperfection, as in the traditional Japanese aesthetics of *wabi-sabi* (侘寂), which is deeply influenced by Buddhist's philosophy.

My hope is that readers of this book can learn not only about cosmology and theoretical physics in general, but also are able to now appreciate human's creativity in coming up with ideas, and our determination that we can figure out how the Universe works, even though the ultimate truth may be forever beyond our reach. In fact, it is precisely because of the unattainability of the ultimate understanding that our attempt to nevertheless try our best to glimpse even the slightest secrets thereof—in the words of Steven Weinberg that we quoted right at the beginning of Chapter 1—"lifts human life a little above the level of farce, and gives it some of the grace of tragedy." Immanuel Kant, in his *Critique of Pure Reason*, wrote that "the mind in the presence of the sublime, attempting to imagine what it cannot, has pain in the failure but pleasure in contemplating the immensity of the attempt". What is *sublime?* The paintings by Caspar David Friedrich (1774–1840) often depict such a concept, or rather, a feeling—the sublime[8] is an aesthetic experience of the feeling of something so immense and great that it instils a profound sense of wonder and awe, even anxiety and fear[9], in us; for we suddenly realize that we are so insignificant in comparison. For example, in Figure 14.6, a monk stands alone on the shore, dwarfed by the sea. And as Isaac Newton once said,

> "I do not know what I may appear to the world, but to myself I seem to have been only like a boy playing on the sea-shore, and diverting myself in now and then finding a smoother pebble or a prettier shell than ordinary, while the great ocean of truth lay all undiscovered before me."

I also hope that the readers have come to realize that different fields—science, arts, philosophy, etc.—can and do influence each other, and thus the unfortunately common tendency for some scientists to look down and belittle other fields that they somehow perceived to be "inferior" is uncalled for. Cross-disciplinary influences can sometimes be very unexpected. For example, politics have been influenced by astronomy indirectly. The starry heaven was of course used since antiquity by emperors across different civilizations and cultures to justify their rules as a divine

[8] The word, from the Latin *sublimis*, has the meaning of "to elevate". In chemistry, we still use the word "sublimation" to describe the transition of a substance from solid directly to gas. However, in the old days of alchemy and esoteric practices, sublimation has a deeper meaning—the human soul being transformed and elevated into a higher spiritual awareness. The main point, however, is that "sublime" is not just the feeling of "wow" when looking at something grand, it is also the feeling that propels us to achieve something greater, despite our relative insignificance.

[9] "Whatever is fitted in any sort to excite the ideas of pain and danger, that is to say, whatever is in any sort terrible, or is conversant about terrible objects, or operates in a manner analogous to terror, is a source of the sublime; that is, it is productive of the strongest emotion which the mind is capable of feeling. [...] When danger or pain press too nearly, they are incapable of giving any delight, and are simply terrible; but at certain distances, and with certain modifications, they may be, and they are, delightful, as we every day experience." —Edmund Burke, "A Philosophical Enquiry into the Origin of Our Ideas of the Sublime and Beautiful" (1757).

Figure 14.6. *Monk by the Sea*, by Caspar David Friedrich from either 1808 or 1810. He described the painting as follows. "It is a seascape. In the foreground, a desolate sandy beach, then, the choppy sea, and so is the air. A man, in black robes, walks thoughtfully on the beach; seagulls fly around him, screeching fearfully, as if to warn him not to venture out onto the turbulent sea. This was the description, now come the thoughts: Even if you pondered from morning to evening, from evening until the setting of midnight; you still would not conceive, would not fathom, the inscrutable beyond!" (quoted from: H. Zschoche, Caspar David Friedrich. The Letters 2005, p. 64). *Credit:* Staatliche Museen zu Berlin, Nationalgalerie / Andres Kilger. *Source*: https://recherche.smb.museum/detail/965511/m%C3%B6nch-am-meer

mandate (hence they employed court astronomers). However, with the discovery of Newtonian gravity, another vision emerged: if gravity is *universal*, and is felt equally between two bodies, then perhaps everyone in society should be subject to the same laws equally. Similarly, John Locke (1632–1704) proposed the "natural rights theory" in which he claimed that much like the orderly Universe, human rights are inherent in Nature. Then Auguste Comte (1798–1857) raised the possibility that since people should be subject to some natural laws, we should be able to study human behavior by applying mathematics. Though things are not as simple as Auguste thought, modern sociology does employ mathematics (Campion 2009). Modern game theory, which is employed in economics and decision theory, had a similar origin: the philosopher Karl Menger (1902–1985) was looking into how to treat ethics mathematically. He considered how to analyze the dynamics among a group of people with different positions. His book was then read by Oskar Morgenstern (1902–1977), an economist who later founded the field of game theory (Edmonds 2020). Even the idea of someone being psychologically "repressed", as if being pressed down by an external pressure, was argued to be influenced by the physicist and physician Hermann von Helmholtz (1821–1894) (Bernfeld 1944). The studies of the Universe also end up influencing what is down here on Earth in other aspects of our life, not just via scientific understanding and technological spin-off, but also via politics and culture. So let us be open-minded and embrace all the complexities of the world.

Creativity and imagination are as important in the sciences as in the arts. We must let our mind soar free to contemplate the Universe. In the face of the immensity of space and time, we should be humble but also ambitious, and believe that—in spite of all our differences between communities and nations—we as a species, can mature and achieve greater heights. Let me finish by quoting Hermann Weyl yet again, from his book *Philosophy of Mathematics and Natural Science* (Weyl 2009):

> "And yet science would perish without a supporting transcendental faith in truth and reality, and without the continuous interplay between its facts and constructions on the one hand and the imagery of ideas on the other."

References

Bernfeld, S. 1944, PQ, 13, 341

Campion, N. 2009, Proc. IAU, 5, 595

Edmonds, D. 2020, The Murder of Professor Schlick: The Rise and Fall of the Vienna Circle (Princeton, NJ: Princeton Univ. Press)

Finkelstein, D. R. 2007, [arXiv:physics/0602185 [physics.hist-ph]] 2006physics...2185R/abstract

Pais, A. 1991, Niels Bohr's Times, In Physics, Philosophy, and Polity (Oxford: Oxford Academic)

Petersen, A. 1963, BuAtS, 19, 8

Planck, M. 1981, Where Is Science Going? (Woodbridge, CT: Ox Bow Press)

Rees, M. 1999, Just Six Numbers: The Deep Forces that Shape the Universe (London: Weidenfeld & Nicolson)

The Economic Times, "Asteroid 2024 YR4 poses threat to Moon, NASA says chances of collision happens once 1,000 years. All details", March 4, 2025. https://economictimes.indiatimes.com/news/international/global-trends/asteroid-2024-yr4-poses-threat-to-moon-nasa-says-chances-of-collision-happens-once-1000-years-all-details/articleshow/118709431.cms

Weyl, H. 2009, Philosophy of Mathematics and Natural Science (Princeton, NJ: Princeton Univ. Press)

www.ingramcontent.com/pod-product-compliance
Ingram Content Group UK Ltd.
Pitfield, Milton Keynes, MK11 3LW, UK
UKHW051923100926
8556IPUK00021B/1347